ORGANIC CHEMISTRY:
THE NAME GAME
Modern Coined Terms and Their Origins

Related Pergamon Titles of Interest

BOOKS

Organic Chemistry Series

CARRUTHERS: Cycloaddition Reactions in Organic Synthesis
DAVIES: Organotransition Metal Chemistry: Applications to Organic Synthesis
DEROME: Modern NMR Techniques for Chemistry Research
DESLONGCHAMPS: Stereoelectronic Effects in Organic Chemistry
GIESE: Radicals in Organic Synthesis: Formation of Carbon–Carbon Bonds
HANESSIAN: Total Synthesis of Natural Products: The 'Chiron' Approach
PAULMIER: Selenium Reagents and Intermediates in Organic Synthesis
WILLIAMS: Synthesis of Optically Active α-Amino Acids
WONG & WHITESIDES: Enzymes in Synthetic Organic Chemistry

Major Works

ALLEN: Comprehensive Polymer Science
BARTON & OLLIS: Comprehensive Organic Chemistry
HANSCH: Comprehensive Medicinal Chemistry
KATRITZKY & REES: Comprehensive Heterocyclic Chemistry
MOO-YOUNG: Comprehensive Biotechnology
TROST: Comprehensive Organic Synthesis
WILKINSON *et al*: Comprehensive Organometallic Chemistry

Also of Interest

COETZEE: Recommended Methods for Purification of Solvents and Tests for Impurities
HERAS & VEGA: Medicinal Chemistry Advances
KATRITZKY: Handbook of Heterocyclic Chemistry
MIYAMOTO & KEARNEY: Pesticide Chemistry: Human Welfare and the Environment
PERRIN & ARMAREGO: Purification of Laboratory Chemicals, 3rd Edition
RIGAUDY & KLESNEY: Nomenclature of Organic Chemistry. 'The Blue Book'

JOURNALS

Tetrahedron (primary research journal for organic chemists)
Tetrahedron Letters (rapid publication preliminary communications journal for organic chemists)

Full details of all Pergamon publications/free specimen copy of any Pergamon journal available on request from your nearest Pergamon office.

ORGANIC CHEMISTRY: THE NAME GAME
Modern Coined Terms and Their Origins

ALEX NICKON

Vernon K. Krieble Professor of Chemistry
The Johns Hopkins University

and

ERNEST F. SILVERSMITH

Professor of Chemistry
Morgan State University

Illustrated by

LEANNE M. NICKON

PERGAMON PRESS

New York · Oxford · Beijing · Frankfurt ·
São Paulo · Sydney · Tokyo · Toronto

Pergamon Press Offices:

U.S.A.	Pergamon Press, Maxwell House, Fairview Park, Elmsford, New York 10523, U.S.A.
U.K.	Pergamon Press, Headington Hill Hall, Oxford OX3 0BW, England
PEOPLE'S REPUBLIC OF CHINA	Pergamon Press, Room 4037, Qianmen Hotel, Beijing, People's Republic of China
FEDERAL REPUBLIC OF GERMANY	Pergamon Press, Hammerweg 6, D-6242 Kronberg, Federal Republic of Germany
BRAZIL	Pergamon Editora, Rua Eça de Queiros, 346, CEP 04011, Paraiso, São Paulo, Brazil
AUSTRALIA	Pergamon Press (Aust.) Pty., P.O. Box 544, Potts Point, NSW 2011, Australia
JAPAN	Pergamon Press, 8th Floor, Matsuoka Central Building, 1-7-1 Nishishinjuku, Shinjuku-ku, Tokyo 160, Japan
CANADA	Pergamon Press Canada Ltd, Suite 271, 253 College St, Toronto, Ontario, Canada M5 T1RS

First printing 1987

Library of Congress Cataloging-in-Publication Data

Nickon, Alex.
Organic chemistry, the name game.
Includes bibliographical references and index.
1. Chemistry, Organic—Nomenclature.
2. Chemistry, Organic—Terminology.   I. Silversmith,
Ernest F.   II. Title.
QD291. N53   1987   547'.0014   86-30450

ISBN 0-08-034481-X Hardcover
ISBN 0-08-035157-3 Flexicover

Printed in Great Britain by Richard Clay Ltd, Bungay, Suffolk

CONTENTS

FOREWORD

Many of us who learn, teach and carry out research in organic chemistry struggle and juggle daily with endless numbers of facts, figures, structures, names and acronyms. To read a paper in the primary literature is often to test to the limit one's knowledge and understanding of the vocabulary of the subject. Indeed, some students claim to lose interest in the science of organic chemistry on the basis that they cannot cope with the necessary vocabulary. If true, this is most unfortunate, and perhaps a reflection of the manner in which they are taught, for the language of organic chemistry is rich, fascinating, and instructive; and even resorting to a standard dictionary is sufficient to reveal tantalizing glimpses of the history and derivation of some of the words we use every day. Thus, a minimum of investigation would uncover the origins of benzene, aniline, methyl, and ortho, meta and para; but no dictionary would list munchnones, bullvalene, magic methyl, or domino Diels–Alder reactions. Unravelling of the origins and significance of these names and terms, and of a large number of names and expressions for other organic chemicals, reactions and experimental phenomena, has involved the authors of this book in many years of patient and careful detective work, often with the help of those chemists who were the originators of the various terms. The result is a book to delight, amuse, intrigue, and inform every student and practitioner of organic chemistry, and one that is certainly the more valuable for the personal contributions of the many famous organic chemists who responded so enthusiastically to the authors' enquiries. As is evident even from a glance at the table of contents, the book is written in a light-hearted vein. It is also, however, a serious study of the development of many important aspects of organic chemistry during the last 50 years or so, in many ways a portrait of the evolution of twentieth century organic chemistry through the researches of the scientists who created it. This book is *not* a dictionary, and the authors are not lexicographers. They are distinguished practicing organic chemists, philologers who here elegantly demonstrate that, in science no less than in literature, language does not serve mankind only for communication any more than food serves only for nourishment. Here, through the language of practitioners, is a description of the more human side of organic chemical research that is not available in any other form of publication.

A. McKillop
University of East Anglia
Norwich, England

PREFACE

The seeds for this book were unwittingly sown at The Johns Hopkins University*
where one of us (A.N.) drew on anecdotes about modern coined names in his lectures in
basic organic chemistry. Michael Ross, an undergraduate in one class, became
fascinated and undertook, as a one-semester library project, to scan the literature for
first usage of the names and stories behind them. It soon became clear that we had to
contact the originators of the terms for inside information. Michael went on to
graduate school at the University of Illinois, and A.N. over the years continued writing
to chemists and gathering items from the literature. At first, the intent was to assemble
enough material to publish an educational article or two for the benefit of other
chemists and teachers. However, the responses to our inquiries were unexpectedly
enthusiastic and fruitful. Eventually, the accumulated information grew too long for an
article and reached book length. During the period 1980–1981 E.F.S. (Morgan State
University) spent a sabbatical year at Johns Hopkins and teamed up with A.N. on the
project.

Despite the light-hearted theme, this book has an earnest purpose: to serve as a
source and a permanent record of information that will not be easy to get after our
current generation of chemists is gone. Consequently, we strived to be accurate and to
document our material throughout. Please write us if you find any factual or perceptual
errors. We would be pleased to receive further information on these topics or on new
ones, perhaps for a future edition or an update article.

It was not our goal to assemble a glossary of technical terms and official definitions.

---

*You may be among those who have wondered about the distinctive "s" in Johns. It's quite simple really.
Johns Hopkins (1795–1873) was a Baltimore merchant descended from a prominent family in Maryland. His
great-grandmother was Margaret Johns, and he was given her last name as a first name. He amassed a
fortune and bequeathed it to establish the university and the hospital.

Outsiders often overlook the terminal "s," sometimes to the amusement of those affiliated with the
Hopkins institutions. On one occasion at a formal gathering in Pittsburgh, Milton S. Eisenhower, who was at
that time president of JHU, was introduced as being from "John Hopkins." Dr. Eisenhower began his speech
by saying how pleased he was to visit "Pitt-burgh."

Another instance involved a jest by noted American author Mark Twain. Speaking at Yale on June 26,
1889, at a banquet attended by Yale alumnus and first president of Johns Hopkins University, Daniel C.
Gilman, Twain quipped that the "public is sensitive to little things and wouldn't ever have full confidence in a
college that didn't know how to spell John" (Browning, R.P.; Frank, M.B.; Salamo, L. *Mark Twain's
Notebooks & Journals*, Anderson, F. (Ed.), University of California Press, Berkeley, CA, 1979, vol. III,
p. 456).

Various compendia of that sort already exist.* Instead, by an excursion through coined names and their origins, we glimpse a human side of chemistry that pervades its solemnity. It would please us if this volume provides educators with substance for enlivening lectures and enhancing interest in our science. You may find, as we did, that chemists (even Nobel laureates) are normal people. They often laugh at themselves, vie with one another, and find humor in their work. Anyone who synthesizes a novel structure, discovers a phenomenon, or develops a new technique experiences parental exhilaration at the birth. And the exhilaration is heightened by what Professor Philip Eaton called "the joy of christening" (*J. Am. Chem. Soc.* **1977**, *99*, 2751–2767, footnote 2). To all those chemistry "parents" who shared their joy with us, we gratefully dedicate this book.

Six appendixes (A–F) are included at the back. They contain items available elsewhere but handy to have in one volume for reference or reminder. The blank pages at the end of Appendix F ("Nobel Prizes in the Sciences") are not production slip-ups; they allow you to enter new laureates yearly and thereby keep the list up-to-date.

Leanne M. Nickon (daughter of A.N.) provided camera-ready original artwork for virtually all the structures, formulas, and drawings. We are indebted to her for these artistic skills. She used circles liberally to draw π clouds in benzenoid rings because it simplified the artwork. We are, however, mindful that this "aromatic sextet" notation, first suggested by James W. Armit and Robert Robinson (*J. Chem. Soc.* **1925**, *127*, 1604–1618), can be misleading for polycyclic molecules. And, finally, we acknowledge with thanks Barbara MacConnell, Janis J. Edwards, Renee Harrell, Jean Goodwin, and Audrey H. Ng, who typed the manuscript and saw it through all its revisions.

---

*Examples: *Glossary of Terms Used in Physical Organic Chemistry*, compiled and edited by V. Gold (*Pure and Applied Chemistry*, **1983**, *55*, 1281–1371); *The Vocabulary of Organic Chemistry*, written by M. Orchin, F. Kaplan, R.S. Macomber, R.M. Wilson, and H. Zimmer (John Wiley & Sons, New York, 1980); *Beilstein's Index: Trivial Names in Systematic Nomenclature of Organic Chemistry*, by F. Giese (Springer-Verlag, New York, 1986).

INTRODUCTION

Who hath not owned with rapture-smitten frame
The power of grace, the magic of a name?
　　　　　　　　　—Thomas Campbell

Undecacyclo[9.9.0.0^{2,9}.0^{3,7}.0^{4,20}.0^{5,18}.0^{6,16}.0^{8,15}.0^{10,14}.0^{12,19}.0^{13,17}]-eicosane!　This systematic name for a long-sought-after compound may be fine for a computer, but it is more than a bit cumbersome in a lecture or in casual conversation. Fortunately, the molecule is usually called "dodecahedrane," which is not only easier on the lips but is instantly more informative; its carbon atoms form a perfect pentagonal dodecahedron (**1**). You can build such a figure easily. Just cut its "net" (**2**) out of cardboard and fold it along all shared bonds. To get a perfect net by folding and cutting a sealed envelope, see Yamana, S. *J. Chem. Educ.* **1983**, *61*, 1058–1059.

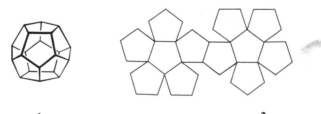

1　　　　　　　　　　　　　　　　　　　2

　　Dodecahedrane is but one example of a neat, descriptive euonym* coined for an organic molecule with a tongue-twisting title. For inspiration in nomenclature, chemists have turned not only to geometric shapes but also to such things as animals, head coverings, mechanical devices, musical instruments, dances, and persons (both famous and not-so-famous). Some names have fairly obvious origins. Many don't, however, and occasionally an interesting anecdote underlies the choice. We traced down many such a "story-behind-the-name" by writing to the chemists involved.

　　By far the largest category of descriptive terms is based on shape, as suggested by a molecular model or by a line formula. Often the relationship between the figure and the name is straightforward. But in other cases scientists used imagination. Our first eight chapters deal primarily with shape and focus on general themes from which chemists drew their inspirations. The remaining chapters follow no particular order; and the items within each are sometimes loosely threaded, but often are not. For the most part,

---

*A euonym is a name well suited to the person, place, or thing that it names.

we restricted our coverage to names coined by the present generation of organic chemists. And, except for occasional intrusions, we excluded natural products, generic names from commerce, and trade names. The field of natural products abounds with delightful terminology; but the names often reflect botanical or biological origins, and most of the "namers" are no longer with us.

Chapter 1

AN ANIMAL IS A CHEMIST'S BEST FRIEND

Andrew Gilbert and co-workers at England's University of Reading irradiated a mixture of 1, 2-dimethylenecyclohexane (**3**) and benzene, and they obtained the tricyclic triene **4**.[1] We understand this [4 + 4] cycloaddition, induced by light, in terms of the famous Woodward–Hoffmann rules of orbital symmetry.[2] When Dr. Gilbert and research colleague Robin Walsh[3] subsequently heated **4**, some of it simply reverted to **3** and benzene. (This reversal, if concerted, was "symmetry-forbidden"; orbital symmetry demands that a photochemical process should go into reverse photically but not thermally. Forbidden reactions do occur but with low probability.) The rest of the [4 + 4] adduct produced 1, 2, 3, 4-tetrahydronaphthalene (tetralin) and 1, 3-butadiene. To

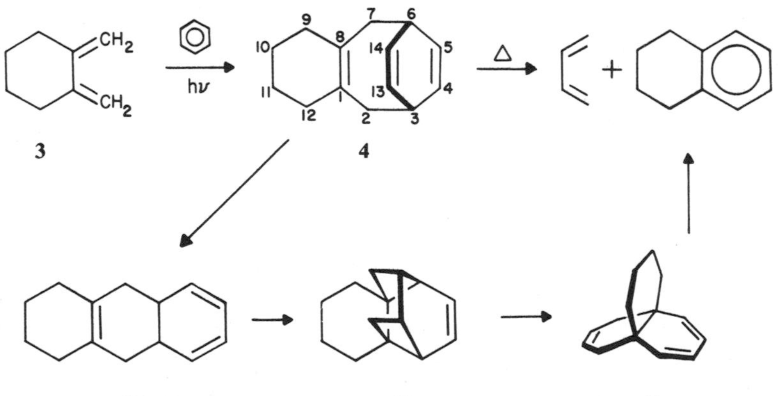

account for these last two products, the Reading researchers proposed a mechanism in step with the Woodward–Hoffmann canons. Thus, triene **4** gets to **5** by a 1, 3-sigmatropic shift of C7 (from C6 to C13); the statutes would require inversion at C7. The convolution of **5** → **6** proceeds via intramolecular [4 + 2] (Diels–Alder) cycloaddition. The remaining two steps (**6** → **7** → *products*) are retro Diels–Alder reactions.

Seeking a name for the unusual polycycle **6**, Gilbert and Walsh held a friendly competition in their research group.[4] A student, David A. Smith, mentally discarded the ring at the left, rotated the rest by 90°, and visualized the parent olefin **8**. A few more imaginative jumps transformed **8** into the face of a smiling cat, and so "felicene" for **8** was born. (*Felis* is Latin for "cat.") David Smith won the competition paws down!

8

Felicene's ears evolved from carbon bridges. But in Notre Dame, Indiana, they grew ears from lonesome electron pairs. In 1968, a team led by Ernest Eliel, then at the University of Notre Dame, discovered that the conformational equilibrium **9** ⇌ **10** preferred a chair with one axial N-methyl and one equatorial N-methyl, rather than with two equatorial N-methyls. They thought that repulsion between lone pairs on the two nitrogens might be a factor that disfavors **9**. When Professor Eliel discussed this "effect" at Dundee, Scotland, he drew the orbitals on a blackboard for a colleague. The next day the drawing (which resembled **11**) was still there, but some wag had written "brer rabbit" alongside it.[5] Our earnest chemist was amused by the notion of a "brer rabbit effect" and finally chose the modified expression "rabbit-ear" effect to describe destability brought on by parallel (i.e., *syn*-axial) orientation of unshared pairs on nonadjacent atoms.[6] However, Dr. Eliel's researchers later presented evidence that,

9   **10**

11   **12**

in fact, rabbit-ears contribute only in a minor way to the free-energy difference between **9** and **10**.[7] Some theoreticians think it's because the ears are clipped, i.e., a lone pair orbital on nitrogen is actually quasi-spherical and, therefore, nondirectional. In a theoretical study of the "anomeric effect,"* Saul Wolfe (Queen's University, Canada), Arvi Rauk (also then at Queen's), Luis Tel (Universidad de Valladolid, Spain), and Imre Csizmadia (University of Toronto) concluded that "the 'rabbit-ear' effect is actually an example of the well known Harvey phenomenon."[12] "Harvey" is the fictitious invisible rabbit in a stage play by Mary Chase.[13] And Dr. Wolfe learned about the play from a

---

*Raymond Lemieux adopted the phrase "anomeric effect" in 1959 when he was at the University of Ottawa.[8,9] Professor Lemieux, co-workers Paul Chü and Rudy Kullnig, and collaborating spectroscopists Harold Bernstein and William Schneider (National Research Council Laboratory at Ottawa) pioneered the use of ^{1}H NMR to study sugars and sugar derivatives. Among other things, they observed that axial anomers possess unexpected stability and attributed this "anomeric effect" at C1 to a stereoelectronic property of the acetal linkage.[10,11]

Queen's chemistry graduate student who was at that time moonlighting in a local summer production of Harvey.[14] The theorists pointed out that a spherical lone pair is likewise "invisible" from a stereochemical standpoint. (Trust theoreticians to help us view things more accurately but see them less clearly!)

Incidentally, chemical lobes did not begin at the University of Reading or at Notre Dame. In fact, symbol **12** drawn by Austrian physicist Joseph Loschmidt in 1861 to represent benzene unmistakably augured an "eary" future for chemistry.[15] According to science historians, Loschmidt used small circles for hydrogen, larger ones for carbon, two concentric circles for oxygen, and three concentric ones for nitrogen. In a booklet published in Vienna in 1861, and titled *Konstitutions Formeln der organischen Chemie in graphischer Darstellung, Chemische Studien, I*, Loschmidt, with keen insight, wrote out over 360 structures for organic compounds; these included **12** for benzene, **13** for acetylene, and **14** for benzyl benzoate. His cyclic formula for benzene resembles, at least superficially, the ring structure proposed by Friedrich August Kekulé four years (!)

13   14

later. Because of its small circulation, Loschmidt's brochure went virtually unnoticed; and August Kekulé (he used only his second given name) emerged as champion of the benzene ring.[16] By the way, even though the Kekulé name has an acute accent on the final e, the family was of Czech and not French origin. It seems that August Kekulé's father, Ludwig, was employed in Hesse, a grand duchy in southwest Germany at that time under the patronage of France. Ludwig added the accent, probably to keep the French from mispronouncing his name "Kekyl."[17]

Oyo Mitsunobu's research with diethyl azodicarboxylate (**15**) and triphenylphosphine (**16**), conducted initially at Tokyo Institute of Technology and then at Tokyo's Aoyama Gakuin University, has spawned some very useful chemistry.[18] These two reagents combine to form a salt (**17**) that effectively activates alcohols toward S_N2-type displacement by external or internal nucleophiles. The reaction, which may proceed via an alcohol–salt complex (**18**) or via pentavalent phosphorus species,[18c] has broad scope.[19] For example, the conversion, in 97% yield, of the steroidal 3β-ol **19** to the 3α-ol formate **20**,[20] developed in Ajay Bose's laboratory at Stevens Institute of Technology, is but one of its many applications.

Mitsunobu and his colleagues could hardly have guessed that their versatile reaction would acquire different nicknames. Chemists who took up work with diethyl azodicarboxylate soon adopted for this ester the acronym "DAD."[21] In 1972, William Hoffman, a student in Professor Bose's laboratory, somehow felt this term was disrespectful of fathers, so he expanded the alphabet to "DEADC." (Those days the Dead Sea scrolls were often in the news.)[21] Some years later Dr. Bose met Jan Sjövall at a World Health Organization conference in Stockholm and learned that Sjövall's co-

15   16   17

18

19   **20**

workers Josef Herz and Thomas Baillie at the Karolinska Institute had their own abbreviation for the Mitsunobu process: "DEADCAT." Its origin is not quite as suggested in the literature[22] but is *di*ethyl *azodi*carboxylate; *a*cid; *t*riphenylphosphine.[23-25] Amused by this acronym, Bose's bunch adopted and shared it with Frank DiNinno at Merck, Sharp and Dohme Laboratories, who included it in a paper in the *Journal of the American Chemical Society*.[26,27] Shortly thereafter, cartoonist Simon Bond hit the lay world with his picture-book, *101 Uses for a Dead Cat*.[28] Mr. Bond is not a chemist, so the events are quite unrelated. But his feline caricatures did not sit well with many cat lovers and may have contributed to chemists favoring the shorter acronym "DEAD"[29] or "DEADC"[30] in publications dealing with diethyl azodicarboxylate. Indeed, when Professor Bose submitted a communication to the *Journal of Organic Chemistry* he was informed that DEAD was the preferred shorthand.[21] Going one step further, he termed the reaction "DEAD-TPP" and so gave equal status to *t*riphenyl*p*hosphine, the other necessary reagent.[31] Dr. DiNinno would have opted for "DEAD-TRIPP," except that it sounded rather foreboding.[27] Incidentally, his colleague R.P. Volante at Merck, Sharp and Dohme found it experimentally advantageous to replace DEAD with *di*isopropyl *azodi*carboxylate in some applications of the Mitsunobu reaction.[32] This latter reagent has been dubbed "DIAD."[33] (But don't use it for diadic tautomerism, chapter 14.) Although the nickname DIAD seems simple enough, we wonder what onomastic evolution lies in store. Charles Darwin would have wondered too.

By the way, feline fans, although cats were dropped from DEAD methodology, they found secure perches elsewhere. For example, your hospital may have access to a CAT scanner (computerized axial tomography) for medical diagnoses. It's not cute, but you couldn't find anything nicer for a PET (positron emission tomography).[34]

Louis and Mary Fieser, renowned chemists, educators, and writers, loved Siamese cats and even graced their books with pictures of these regal animals. So you might think that when the Fiesers coined the term "cathylate,"[35] its first three letters meant another plug for the feline species. Actually, however, the word is but a contraction of carboethoxylate. Research in their laboratory had revealed that ethyl chlorocarbonate (dubbed "cathyl chloride")[35] selectively carboethoxylates equatorial secondary alcohols and not axial ones, so "cathylation" proved useful in chemical manipulation of polyhydroxysteroids.[36]

Nonetheless, cats do know their stereochemistry, thanks to Francis Carey of the University of Virginia and Martin Kuehne of the University of Vermont. These chemists coined "syncat" and "ancat" to specify relative chirality at sites in acyclic molecules.[37] For example, in an extended staggered (zig-zag) chain, two centers are "relative syncat" when they dispose their larger substituents on the same side of the planar chain; in "relative ancat" these groups project to opposite sides. Structure **21** shows a simple case, where x, y, and z would be replaced by the appropriate carbon

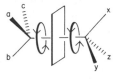

x,y — SYNCAT
y,z — ANCAT
x,z — ANCAT

21

numbers. Syncat and ancat are merely contractions of syn-catenoid and anti-catenoid (*catena* is Latin for "chain"). We are glad Professors Carey and Kuehne eschewed "anticat"; otherwise purrs might have epimerized to growls.

Configurational description based on substituent "size" serves well in many circumstances. But for rigorous identification of relative chirality these same researchers coined the descriptors "priority reflective" and "priority antireflective." Their respective acronyms are "pref" and "parf"; and priority refers to sequence-rule priority according to the Cahn–Ingold–Prelog convention.[38] In the Carey–Kuehne terminology, two contiguous chiral centers have a pref relationship if the order of decreasing priority of the three remaining substituents at one center is a mirror reflection of the order of decreasing priority of the groups at the other center (e.g., **22**). When two descending-order circuits lack reflection symmetry, the relative configuration is parf (e.g., **23**). Such designations of relative stereochemistry are independent of conformation because they compare priority-determined *order*, and not momentary *spatial positions*, of the substituents. This system obviates the well-known ambiguities associated with terms like threo and erythro and can accommodate structures with nonadjacent, or with multiple, chiral centers. In the latter case, you need only identify them with numbers and treat each pair of centers independently of all others. For example, of the four diastereomers of 3-(hydroxymethyl)-4-phenyl-2-pentanol, structure **24** is the 2, 3-*parf*-3, 4-*pref* isomer; and **25** is 2, 3-*pref*-3, 4-*pref*. And what about the trideuteriobutanol **26**? Parf–parf!

If you are not completely happy with syncat/ancat and *pref/parf*, try out a stereochemical notation developed by savants Dieter Seebach and Vladimir Prelog, at the ETH in Zurich. Whether you fancy reactants or products, the Seebach–Prelog notation for the steric course of asymmetric syntheses has much for you to "like" and "unlike," and employs the symbols *lk*, *l*; *ul*, *u*.[39a] Earlier, Ivar Ugi had recommended a system that used the descriptors *p* and *n*.[39b]

22   **23**

Cahn — Ingold — Prelog

priorities :   a > b > c
             x > y > z

By now you may think that the feline family does well by chemists. Quite right. But cats fare even better in physics, where they have appeared as authors on research papers. In 1975, Professor Jack H. Hetherington (Michigan State University) wrote a theoretical paper on his own and was about to send it to *Physical Review Letters*.[40] But a colleague warned that the manuscript would be returned because of an editor's rule that words like "we" and "our" should not be used in a publication with only one author. Dr. Hetherington did not relish revising and retyping the whole text, so, instead, he simply added a co-author: his Siamese cat Chester (sired by Willard). And for legitimacy, he tacked on two more initials, FD (from *Felix domesticus*), to create "F.D.C. Willard." The Hetherington–Willard article was duly published[41]; and Mrs. Hetherington went on sleeping with both authors.

Eventually, the cat had to be let out of the bag when a visitor came to campus to see Professor Hetherington, found him unavailable, and then asked to speak to Willard. The feline caper also unfolded internationally when J.H.H. mailed reprints to a few friends. The usual annotation "Compliments of the authors" included his signature and an imprint of pussy's inked paw![41,42]

Willard became bilingual and authored his second paper in 1980 in *La Recherche*.[43] This French article about antiferromagnetism in solid ^{3}He was intended for broad readership and was put together by a group of French and American physicists, including Professor Hetherington. But, because of some disagreement about its content, F.D.C. Willard ended up as the sole author. That way, if the paper had flaws, the cat could take the rap.[40] Occasionally one sees references to "F.D.C. Willard, private communication." And sometimes in acknowledgments he is thanked for "helpful discussions."[40] Although F.D.C.W.'s future in physics is uncertain, we like his style and hope he gets tenure.

In the field of chemistry, the American Chemical Society has virtually squelched any hope for mischief along "authorship" lines. Its "Ethical Guidelines to Publication of Chemical Research" establish criteria for co-authorship and stipulate that "no fictitious name should be listed as an author or co-author."[44] But we wonder whether co-authorship would be sanctioned for a cat that performs some of the laboratory work? An unlikely scenario you say? Perhaps unlikely in chemistry, but not in physics. In fact, Johns Hopkins physicist Robert W. Wood, an inventive genius and one of the founding fathers of spectroscopy, called upon feline help to clean out a spectroscopy tube he had built in a barn.[45] The wooden tunnel, 42 feet long and about 6 inches wide, projected outside the barn and accumulated cobwebs when not in use. Professor Wood popped the family cat in at one end of the shaft. Pussy squirmed through and thus swept the passage free of cobwebs. In 1912, Dr. Wood reported his cleaning method in *Philosophical Magazine* but did not name his feline assistant.[46]

The notion to conscript four-legged friends into laboratory service took hold. Around 1970, scientists at the Atomic Energy Commission's National Acceleration Laboratory, Batavia, Illinois, used a ferret, named Felicia, to clean long synchrotron tubes. The NAL engineers had originally planned to build an expensive pipe cleaner; instead, they bought Felicia for $35, fitted her with a harness, and let her drag cleaning cloths through the 300 ft tubular sections.[47] And telephone and electric companies were known to train, induce, or seduce small animals to scamper through long pipes dragging wires behind them.[48] But enough of this; let us return to chemistry and look to molecules for divertissement.

The coupled compound 1, 2-bis(9'-anthranyl)ethane (28) can be prepared from 9-(halogenomethyl)anthracenes (27) with Grignard reagents.[49] One or two side products

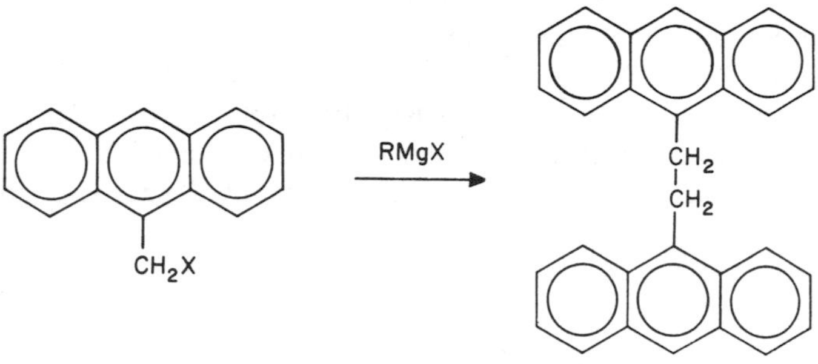

27   28

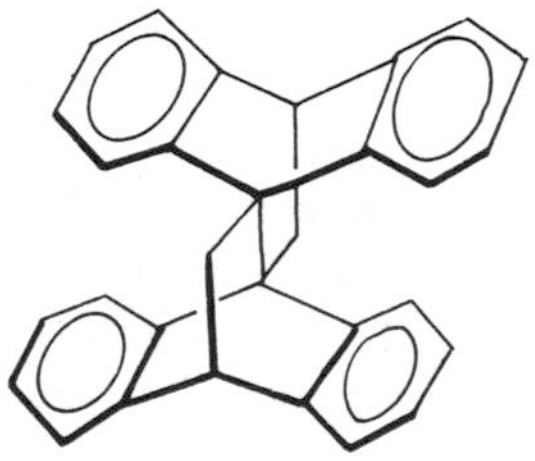

29 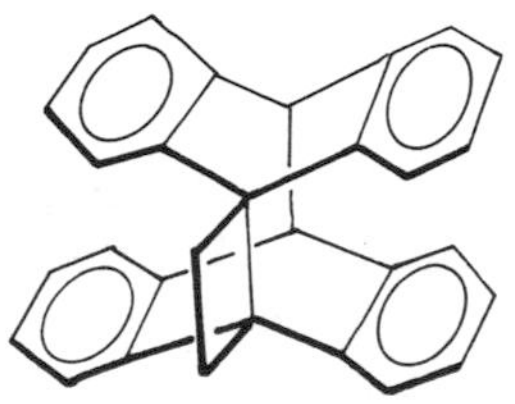   30

turn up, and early workers disagreed about their structures.[50-53] Henri Bouas–Laurent at the University of Bordeaux cleared up the controversy in 1975.[54] Evidently, when the reaction is conducted in the dark, the side product is 29. But in light, some of the main product, 28, isomerizes to 30; so both side products may be present in the mixture. Ultimately, X-ray crystallography proved the constitution of 29.[55] Interestingly, structure 29 had been proposed in a thesis at the University of Minnesota in 1962, but it was not published in a journal.[56]

Understandably, these formulas filliped the imagination of Dr. Bouas–Laurent. The phenylene rings reminded him of wings. In 29, each pair of wings is displaced from the other, as in a butterfly; so he called it "lepidopterene." (*Lepidoptera* is the order of insects consisting of butterflies and moths; it comes from Latin *lepis* ("scale") and Greek *pteron* ("wing").) On the other hand, in 30 the wings are superposed, just as in a biplane of World War I vintage. Hence, the name "biplanene" took to the air.[57]

Other syntheses of lepidopterene have been published.[58,59] Also, a lepidopterene-like molecule with clipped wings (31) was prepared at The Ohio State University by Leo

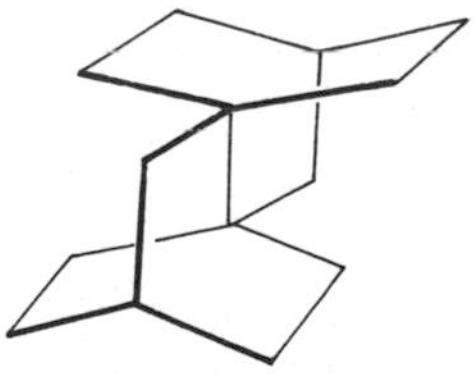

31

Paquette's group, who named it [2, 2, 2] "geminane" (see p. 179)[60]; it consists of two equal parts, and the Latin *geminus* means "twin."

While on the subject of butterflies, we note that the term "butterfly" also was applied, by Arnold Gordon and John Gallagher of The Catholic University of America, to the *syn* conformation of certain metacyclophanes such as **32**.[61] The oxygen bridge keeps the "butterfly" from becoming a "ladder" (**33**, the *anti* conformation[62]). Talk about metamorphosis! Inorganic complexes have enhanced the beauty and diversity of butterfly collections. For example, researchers in Milan, Italy, bagged an iron "butterfly" cluster.[63] And Malcolm Chisholm's coterie at Indiana University showed that the molybdenum atoms in solid $Mo_4Br_4(OiPr)_8$ place themselves as in **34**.[64] And you may come across an ephemeral "butterfly" if Robert Bach's bunch at Wayne State University can induce you to probe the mechanisms of alkene oxidation by peroxy acids.[65]

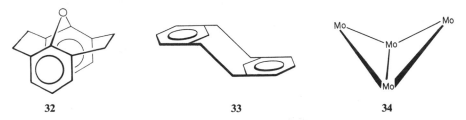

32	**33**	**34**

Even the lowly earthworm has inspired chemists. Jacques Simon and co-workers at Centre de Recherches sur les Macromoleculares, Strasbourg, found that **35** and **36** bind nicely with transition metal ions such as Co^{+3} to give complexes **37** and **38**, respectively.[66] Molecular models of these complexes reminded Dr. Simon of a washer-like ring, with a ball stuck in the hole and an earthworm dangling off the edge (see **39**). The "ball" is, of course, the metal ion. Because earthworms belong to the phylum of invertebrates known as Annelida, Dr. Simon adopted "annelides" for this class of compounds. As luck would have it, archaic French for "ring" is *anel* (derived in turn from the Latin word for ring, *anulus*; diminutive form, *annelus*).* Thus, the name

---

*The Latin *anulus* is sometimes spelled "annulus," and chemists use terms such as "annelation" or "annulation" for the synthetic process of building a ring onto a molecule. In an article titled "A Review of Annulation," UCLA's Michael Jung favored the "u" over the "e" spelling in accord with the dictionary preference.[67] However, Paul Caluwe at the State University of New York (Syracuse) took the opposite stance in his review titled "Heteroannelations with *o*-Aminoaldehydes."[68] Chemists seem to be divided on this "trivial" issue, and both spellings are common. Perhaps we should defer to the first use of the term in the organic chemical literature. A German article in 1919 by Polish researchers Wl. Baczyński and St. von Niementowski contains the phrase "durch lineare Annellierung der Ringe."[69] (Do you know of an earlier use?) However, Matthew Schlecht (Polytechnic Institute of New York) informed us that annellation (also spelled anellation and annelation) has appeared in contexts other than that of Baczyński and Niementowski's "ring-forming reactions."[70] For example, in 1932 Eric Clar described multicyclic aromatic systems as being related to simpler ones by linear or angular "anellation" (Anellierung), in other words, by formal ring addition.[71] And Jerry March's textbook *Advanced Organic Chemistry*[72] taught annellation as the "phenomenon, whereby some rings in fused systems give up part of their aromaticity to adjacent rings." So, avers Professor Schlecht, the term is thus rendered ambiguous, because it can mean "ring-forming reaction," a "ring-added relationship," or "aromatic ring character partitioning." In contrast, annulation has no meaning other than the formation of rings, and the spelling is more constant. Finally, for chronological record, Dr. Schlecht pointed out that the first principal chemical usage of the term annulation in English was by William S. Johnson and co-workers in a 1960 paper titled "The Nature of the Intermediary Ketols in the Robinson Annulation Reaction."[73] And annulation showed up initially in a 1970 communication by Elias Corey and Joel Shulman with the heading "A Method for the Introduction of Two Carbon Appendages at a Carbonyl Carbon. Application to Double Chain Branching and Spiro Annulation Operations."[74] The following year, Charles Scanio and Richmond Starrett titled a publication "A Remarkably Stereoselective Robinson Annulation Reaction."[75] Unless IUPAC or the like[76] deigns to proclaim a preference, annulation/annelation may remain a dichotomy in chemistry writing.

"annelides" actually embraces both the worm and the ring parts of these interesting molecules.[77]

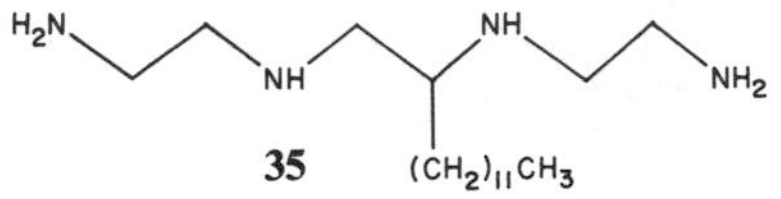

35      $(CH_2)_{11}CH_3$

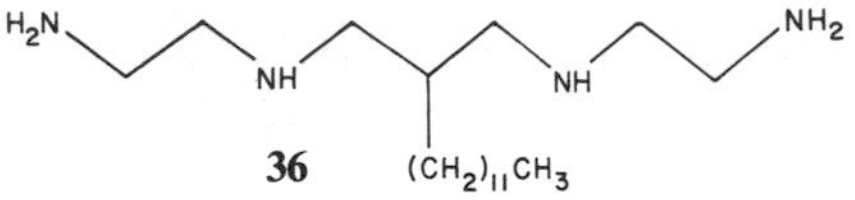

36      $(CH_2)_{11}CH_3$

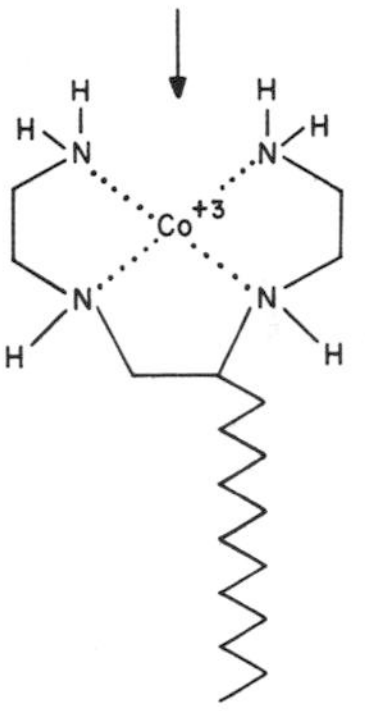

37

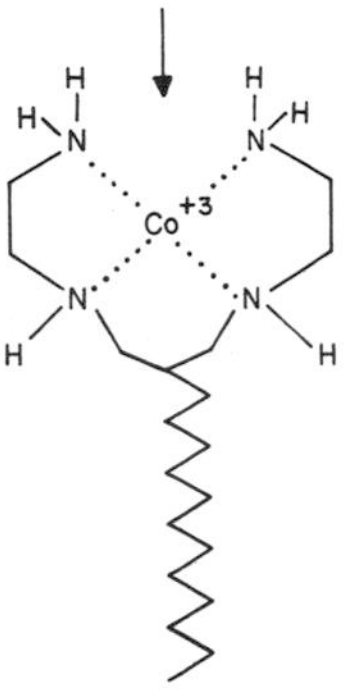

38

39

Because annelides sport polar heads and nonpolar tails, they are surfactants and tend to form "supermolecular structures" in which the polar tips of many molecules attract one another.[66] Such aggregates are also called micelles. Furthermore, the polar end with its encumbered ion is somewhat akin to a porphyrin–Fe complex, as in the heme portion of hemoglobin (40). And, not surprisingly, annelides 37 and 38 can bind molecular oxygen.[66]

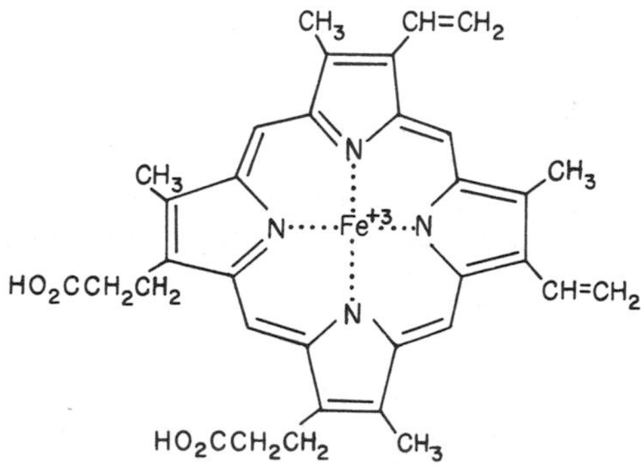

40

Some aquatic species became immortalized by chemical names. Molecular models of compounds 41 and 42, which Professor Leo Paquette and his co-workers synthesized in 1975,[78] reminded them of an open clam, or oyster, shell. Hence, they dubbed these structures "bivalvanes." Interestingly, 41 and 42 are DL and *meso* diastereomers. You can appreciate this structural difference better from the flattened formulas, 43 and 44.

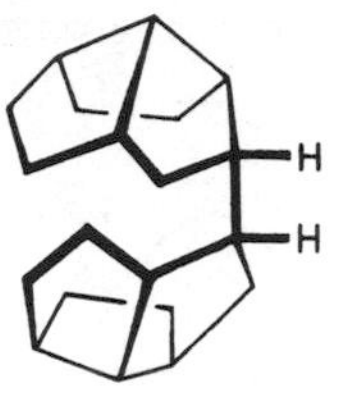

41 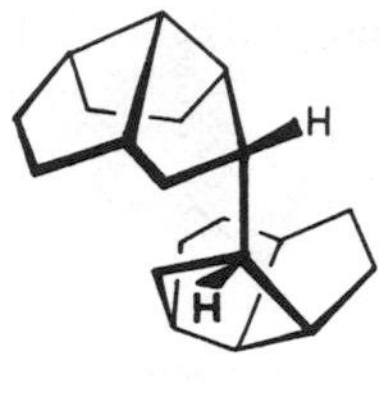   42

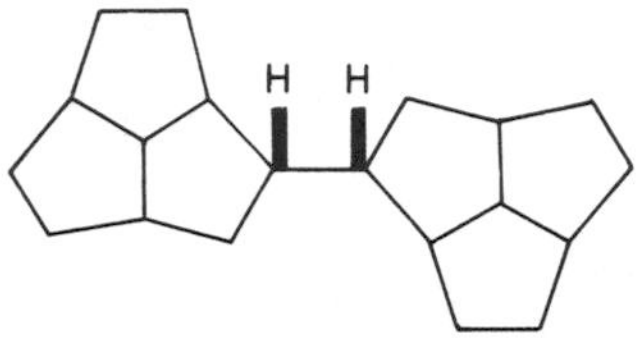

43 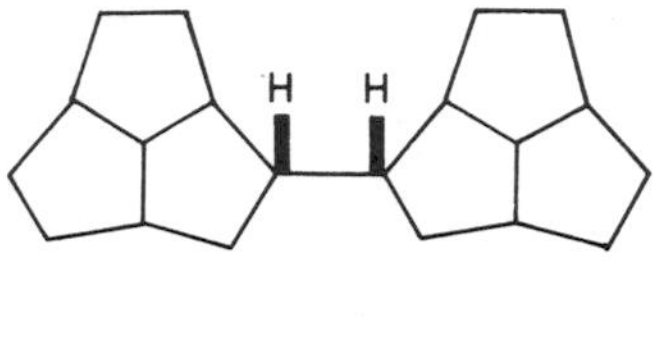   44

Dr. Paquette wanted to synthesize **41** because mental joining of the two halfshells by five more carbon–carbon bonds produces the fascinating dodecahedrane (**1**). His team first developed routes to the halfshells and then hooked them together by a pinacol-type reduction.[79] Removal of the hydroxyl groups left Dr. Paquette's beachcombers ready for their clambake.

Chemists Fritz Vögtle and Edwin Weber, then at the University of Würzburg, have constructed molecules such as **45**.[80] As Dr. Voḡtle is a compressed-air diver and an avid reader about marine life, it is not surprising that he called them "octopus molecules."[81] The word "octopus" comes from Greek *okto-* (eight) and *pous* (foot). Actually his

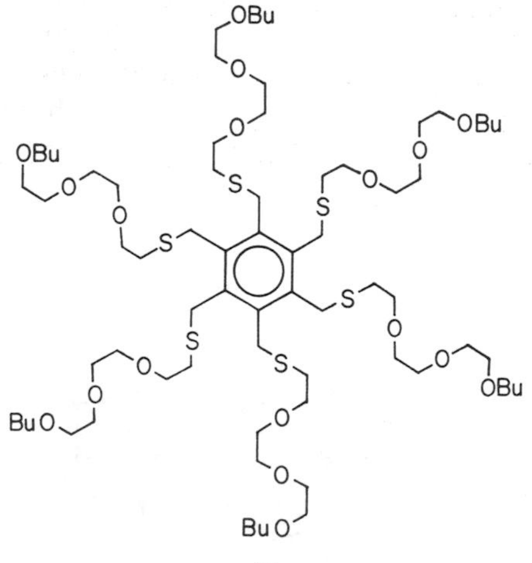

45

compounds own but six tentacles and might have been christened "hexapus." In any case, Dr. Vögtle's term is apt because the heteroatoms act like the suction cups with which a real octopus grasps food. For compound **45**, the meal consists of metallic ions, which the sulfur and oxygen donor atoms envelop with great tenacity. Its appetite for ions from main groups I and II of the periodic table[82] is more voracious than that of known crown ethers (chapter 2) and about the same as that of cryptates (chapter 17). A molecular octopus with shorter tentacles doesn't grasp ions as tightly. Hunger also wanes when you replace the oxygen suction cups by CH_2 groups.

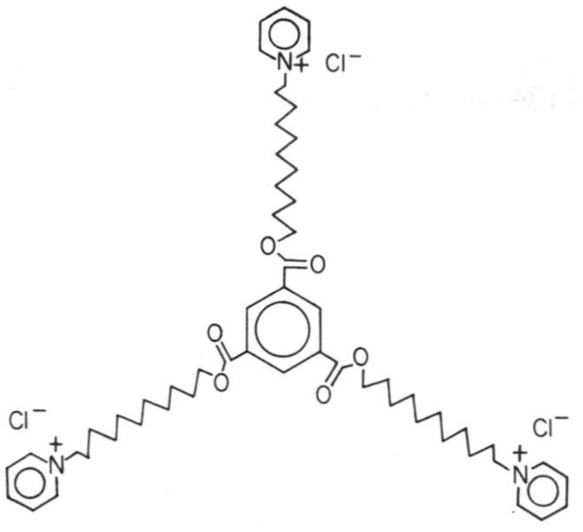

46 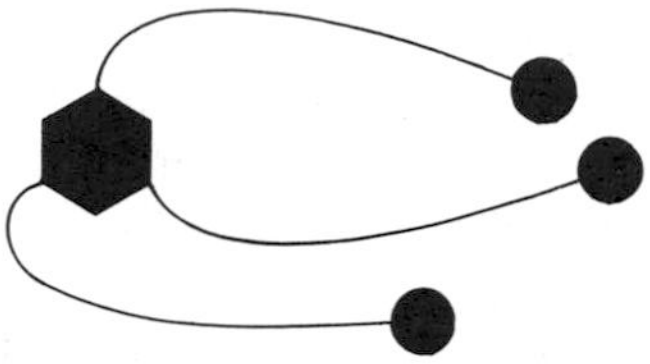   47

Since the appearance of the first octopus molecule, other species have invaded the chemical aquarium, including molecules with different types of bodies and tentacles.[83,84] The numerically more accurate term "hexapus" for molecules with six tentacles has appeared in print, thanks to Fred Menger of Emory University, who also contributed other hydronyms* (p. 121).[85] And Colin Suckling of the University of Strathclyde, Glasgow, thought that three tentacles may be enough to do a job. For example, he found that **46** forms complexes with small aromatic substrates, and can thereby influence their reactivity (e.g., **46** inhibits the chlorination of phenol.) Professor Suckling referred to **46** as a "tentacle molecule," in other words, a detergent-like molecule whose motion in restricted by attachment to a central framework (see **47**).[86] But you may call it triapus. Or even trigapus.[87]

If you like tentacles and need lots of them in a hurry, Professor Vögtle and colleagues at the University of Bonn advocate their principle of "cascade" synthesis.[88] In this approach, they manipulate a functionalized molecule to get a product armed with twice as many of those same functional groups. Then, repeat the manipulations and you again double the units. For example, bis(cyanoethylation) of RNH$_2$ gave dinitrile **48**, reduction of which afforded the bitentacled amine **49**. Carried through once more, the process converted this compound to **50** and then to **51**. Such heavily armed structures might be good ligands for grasping ions.

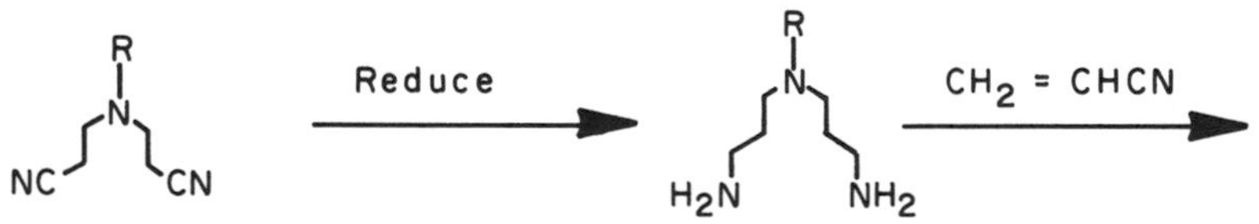

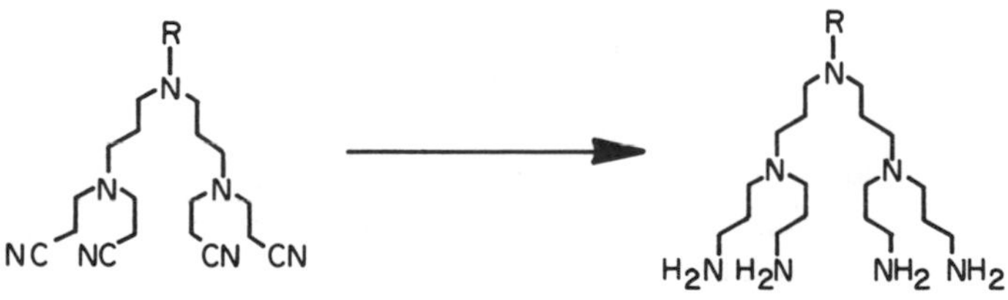

*A hydronym is a name of a body of water. We are stretching the term to include other aqueous items.

The reagent used for these cyanoethylations was $CH_2{=}CHCN$. If you ever need a large quantity of this chemical and want delivery without delay, it might be prudent to specify acrylonitrile rather than vinyl cyanide. A story recounted in *Chemical and Engineering News* tells us why.[89]

It seems that in the early days of the U.S. synthetic rubber industry a company ordered an 8000 gal tank car of vinyl cyanide for use as a copolymer with butadiene to produce an oil-resistant rubber. The Interstate Commerce Commission refused permission to ship the chemical by rail on the grounds that cyanides are poisonous. What would happen in the event of an accident? How would the public react to ICC's allowing a volatile cyanide to be transported by rail?

The impasse was broken by a minor chemist in the company, who suggested that an order be placed for a tank car of acrylonitrile. The company did so, and the chemical was delivered without objection from the ICC. The minor chemist later became an executive.[89]

The essence of cascade synthesis is progressive branching. Branches bring to mind trees, and that's just how George Newkome (then at Louisiana State University) and his colleagues got into new cascade approaches to unusual micellar structures. Their concept was based on an architectural model of trees (called the Leeuwenberg model), and they applied it to synthesize a highly branched prototype with an outer surface of polar groups.[90] Drawing **52** pictorializes this model as applied to micellar construction.

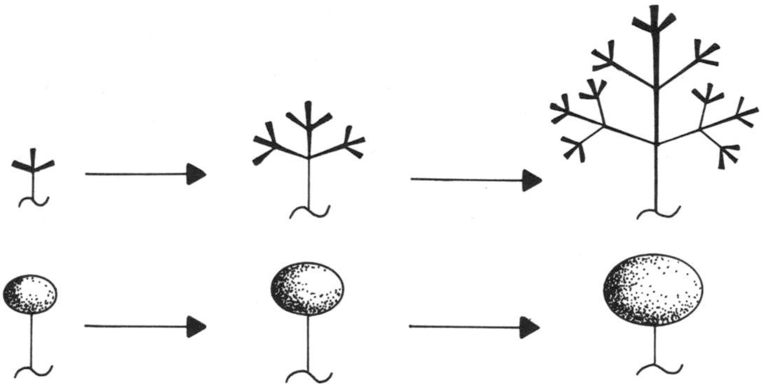

52

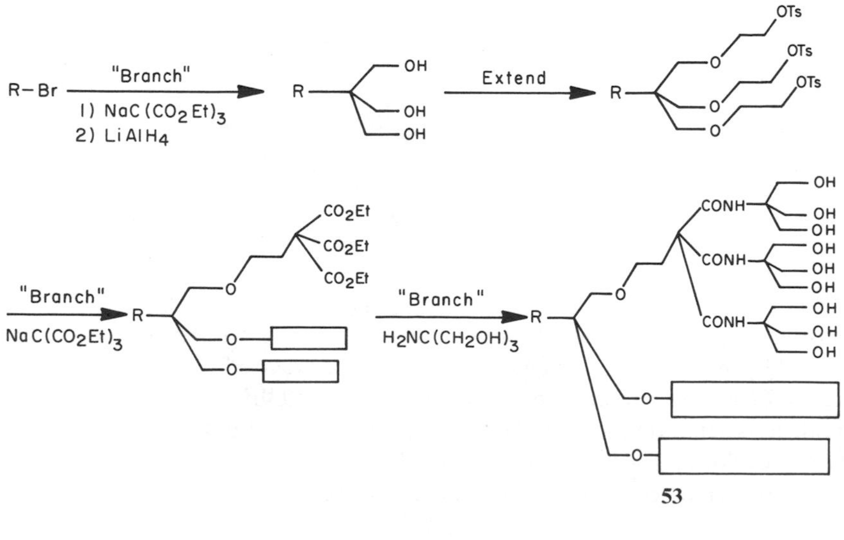

53

Specifically, Newkome and his rangers prepared **53** by a three-tiered construction sequence as shown. This multitendrilled monster turned out to be miscible with water despite its high molecular weight (> 1600). Because this unidirectional cascade is based on arboreal design (Latin *arbor* means "tree"), Professor Newkome and co-workers termed their products "arborols." Structure **53**, has 27 branches, so that makes it a [27]-arborol. (Tomalia *et al.* independently described a similar class of cascades termed "starburst-dendritic" polymers.[91] In Greek, *dendro* means "tree.") Professor Newkome likened a Leeuwenberg cascade in two directions to a tree with its accompanying root system,[90b] and he named these "sylvanols" (*silva* is Latin for "forest").[90]

The Louisiana State researchers also extended their chemistry to three directions by synthesizing **54**, in which a trio of cascades emanate from a benzene core. Light-scattering experiments indicated that aggregates of **54** in solution have micellar character.[92]

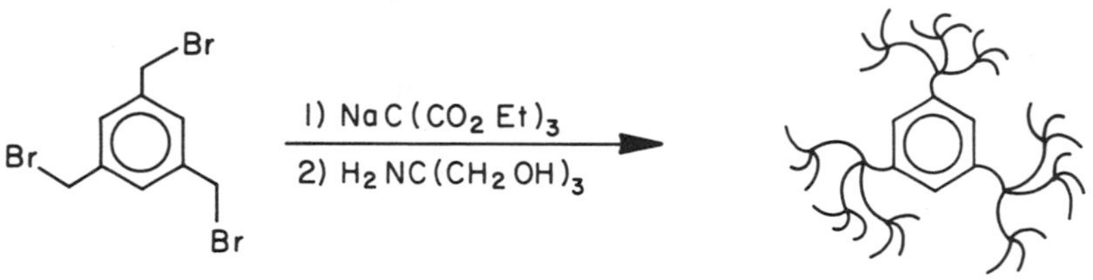

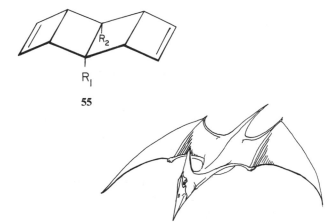

54

Animals that inspire names are not always found in trees, zoos, or oceans—at least not in our times. Consider compounds **55**, which their creator, Rowland Pettit of the University of Texas at Austin, dubbed "pterodactyladienes."[93,94] These structures reminded him of the extinct flying reptiles known as pterodactyli. This word is Latin, but stems from the Greek *pteron* (wing) and *daktylos* (finger). Such creatures had winged membranes extending from the side of the body to the end of the greatly enlarged fourth digit ("finger"). Professor Pettit and co-workers synthesized their

55

prehistoric creatures from cyclobutadiene, a molecule whose own existence was questioned for many years. When cyclobutadiene is released from its iron tricarbonyl complex (**56**) in the presence of ceric ions and a suitable alkyne, two successive cycloadditions generate the pterodactyladiene. Each cyclobutadiene ends up as a wing tip. If these cycloadditions are $[4 + 2]$, then Drs. Diels and Alder could be considered godfathers of Professor Pettit's prehistoric offspring.

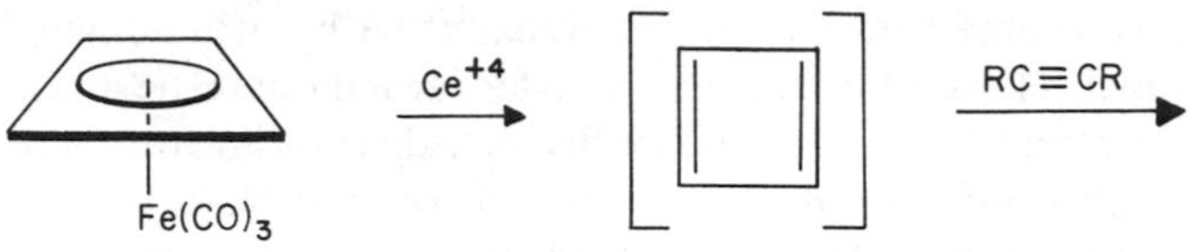

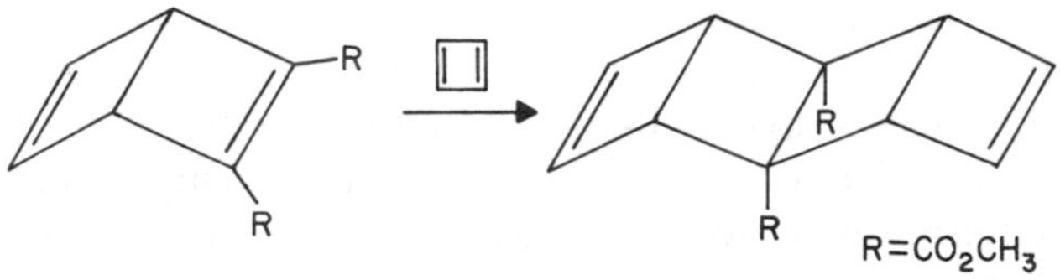

Hans–Dieter Martin and Mirko Hekman at the University of Würzburg increased our knowledge of these winged molecules. They prepared pterodactyladienes with several other R groups ("heads" and "tails")[95,96] and also hydrogenated the double bonds to produce the parent pterodactylanes.[97] In their intriguing synthetic variation,

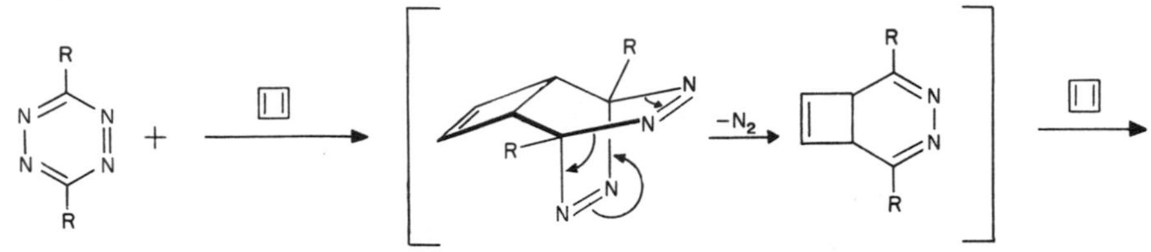

cyclobutadiene takes the opposite role in a Diels–Alder merger with **57**. Adduct **58** plays the organic chemist's favorite game, electron reorganization, to **59**. Another [4 + 2] cycloaddition gives **60**, a pterodactylus with a pot belly of two nitrogens. Then, a dash of UV light, more electron scrambling, and (voilà) the belly is gone. Weight watchers take note! Professor Martin continued his chemical ornithology at the University of Düsseldorf, where, in 1981, he and his colleagues worked out still another posthistoric route to this prehistoric bird.[98]

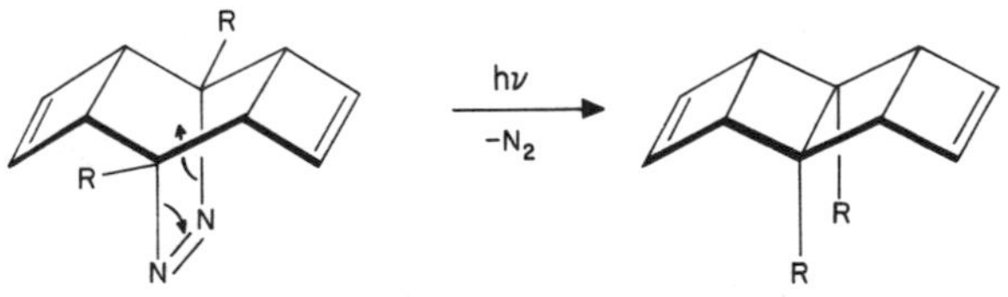

You might think that the titles "equinene" for **61** and "calfene" for **62** belong in our discussion of animal shapes. However, equinene was named after a two-legged "horse" rather than a four-legged one. And calfene (so labeled by Maynard Sherwin, a graduate student of Howard Zimmerman, in whose Wisconsin laboratory it was synthesized[99]) got its tongue-in-cheek name from the fact that it is an incompletely grown bullvalene (**63**) (i.e., it lacks one ethyleno bridge). More conservatively, the chemical literature usually refers to **62** as "semibullvalene." Interestingly, the story of bullvalene itself involves a two-legged creature; we discuss it along with equinene in chapter 10.

How about naming a molecule after a part of an animal? Professor Paquette has done just that with **64**, which he dubbed "snoutene."[100] This onomastic oddity was inspired by the realization that **64** can be made to look like Pinocchio (the young lad in the classic children's story whose nose grew each time he told a lie.) The flattened "nose" also reminded Paquette of a crocodile snout; and when you look at **64** from different directions it even resembles a maybasket (**65a**) and a large scoop (**65b**). Hence, the Ohio State University researchers also considered "maybasketene" and "large scoupene"; but "snoutene" finally won by a nose.[101] A key step in the synthesis of snoutene was rearrangement of **66** to **67**, a remarkable silver-ion-catalyzed skeletal change reported by Paquette[102] and by Dauben.[103]

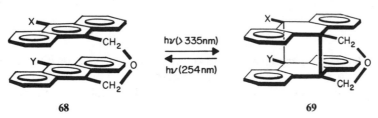

The photic interconversion of molecules **68** and **69** struck Henri Bouas–Laurent and co-workers at the University of Bordeaux as being similar to the closing and opening of a jaw.[104] Compounds **68** are yellow and **69** are colorless, so irradiation brings about a color change (photochromic). This last term comes from the Greek *phos* (light) and *chroma* (color). Therefore, Professor Bouas–Laurent called their system "jaw photo-chromic."[104] The jaw closure is a photochemically allowed [4 + 4] cycloaddition, suprafacial in both components; jaw opening is just the reverse. Furthermore, the closure occurs at wavelenghts > 335 nm (i.e., with light absorbed by the yellow reactant but not by the colorless product). To force the jaw to open, inject light that *is* absorbed by **69**; and 254 nm "fills the bill" perfectly. We're dealing with a very intelligent jaw.

The name "cristane," bequeathed to tricyclo[5.3.0.0^{3,9}]decane (**70**),[105] emerges from an animal part but not because of the molecule's shape. Langley Spurlock of Brown University, originator of cristane, put the matter most delicately.[106] The word derives from "crissum," the area around the cloacal opening under a bird's tail. It happened that on the evening the compound was chemically identified someone left a laboratory window open. During the night one of the pigeons frequenting the window ledge got into the room and apparently spent much time trying to find its way out. Next morning the laboratory benches and floor were covered with unwelcome evidences of the bird's anxiety while it had attempted to escape. Spurlock's clean-up crew, therefore, named their new tricyclic molecule in honor of the bird's anatomical region that provided the "surprisingly abundant gift." We salute the Spurlock team for coining a name for posterity.

70 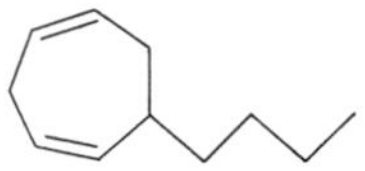   71

We may stretch things a bit to refer to sperm as a part of an animal, but not if we think of the very beginning of the animal's existence. In any case, when Subramania Ranganathan of the Indian Institute of Technology of Kanpur noted structure **71** (a constituent of an essential oil of algae of the genus *Dictyopteris* and hence also known as dictyopterine C′ in the literature[107]), he jokingly dubbed it "spermane."[108] This molecule possesses a "head" and hydrocarbon "tail," just like the annelides we described earlier. Obviously, the title given a compound is based on what is in the mind of its beholder.

The name "bimanes," derived from the Latin *bi* (two) and *manus* (hand), was assigned to molecules with structure **72** by their inventor, Edward Kosower, because they resembled two hands.[109] These compounds arise from **73** by the action of base. After

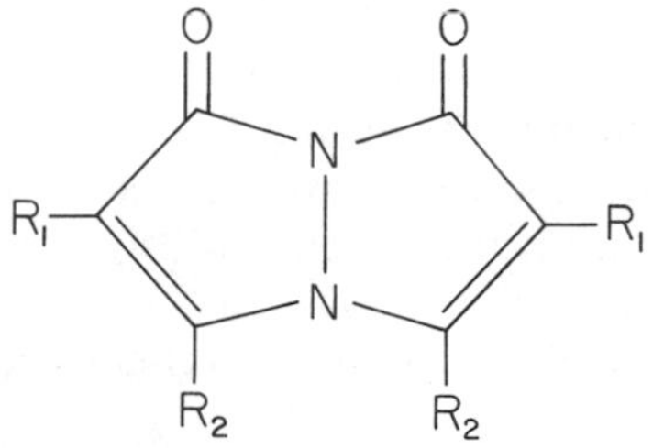

72 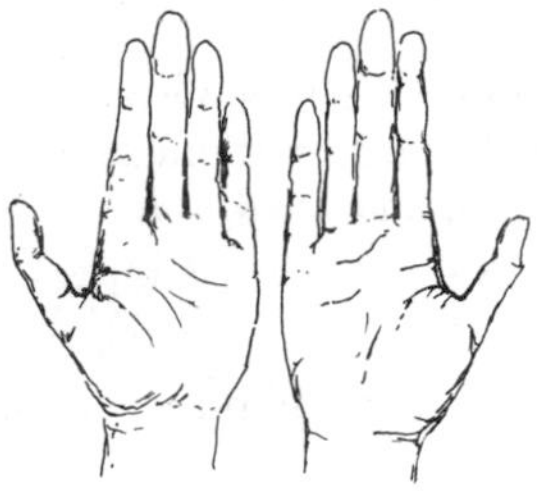 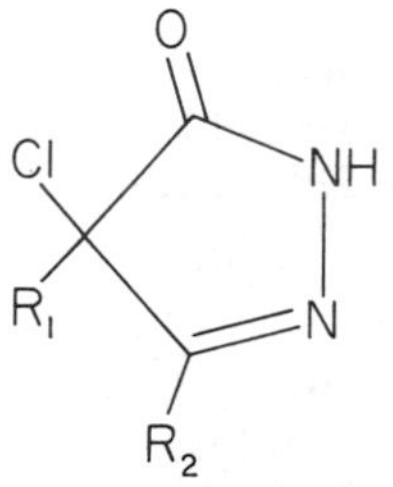   73

looking at the clever mechanism Professor Kosower proposed for this reaction, you will undoubtedly want to give him a hand, or perhaps even two (one for each of his two academic homes: Tel Aviv University and the State University of New York at Stony Brook).

REFERENCES AND NOTES

1. Berridge, J.C.; Bryce–Smith, D.; Gilbert, A. *Tetrahedron Lett.* **1975**, 2325–2326.
2. See, for example, (a) Morrison, R.T.; Boyd, R.N. *Organic Chemistry*, 4th ed. (Allyn and Bacon, Boston, MA, 1983), chapter 33, (b) Woodward, R.B.; Hoffmann, R. *The Conservation of Orbital Symmetry* (Verlag Chemie GmbH, Weinheim/Bergstr., 1970).
3. Gilbert, A.; Walsh, R. *J. Am. Chem. Soc.* **1976**, *98*, 1606–1607.
4. Gilbert, A., private communication, June, 1977; Walsh, R., private communication, November, 1977.
5. Eliel, E., private communication, March, 1984.
6. Hutchins, R.O.; Kopp, L.D.; Eiel, E.L. *J. Am. Chem. Soc.* **1968**, *90*, 7174–7175.
7. Eliel, E.L.; Kopp, L.D.; Dennis, J.E.; Evans, S.A. Jr. *Tetrahedron Lett.* **1971**, 3409–3412.
8. Lemieux, R.U. *Abstracts of Papers, 135th National Meeting of the American Chemical Society*, Boston MA (American Chemical Society, Washington, DC, **1959**), p. 5E.
9. Lemieux, R.U., private communications, March and June, 1983.
10. (a) Lemieux, R.U.; Kullnig, R.K.; Bernstein, H.J.; Schneider, W.G. *J. Am. Chem. Soc.* **1957**, *79*, 1005–1006. (b) Lemieux, R.U.; Chü, N.J. *Abstracts of Papers, 133rd National Meeting of the American Chemical Society*, San Fransisco (American Chemical Society, Washington, DC, **1958**), p. 31N. (c) Lemieux, R.U.; Kullnig, R.K.; Bernstein, H.J.; Schneider, W.G. *J. Am. Chem. Soc.* **1958**, *80*, 6098–6105.
11. Kirby, A.J., *The Anomeric Effect and Related Stereoelectronic Effects at Oxygen* (Springer-Verlag, New York, 1983).
12. Wolfe, S.; Rauk, A.; Tel, L.M.; Csizmadia, I.G. *J. Chem. Soc. B* **1971**, 136–145.
13. Chase, M. *White Rabbit*, (copyright 1943), and *Harvey*, (copyright 1944). (Dramatists Play Services, Inc., New York, NY).
14. Wolfe, S., private communication, August, 1979.
15. (a) Crosland, M.P. *Historical Studies in the Language of Chemistry* (Dover Publications, New York, 1978), chapter 3, (b) Mason, H.S. *Isis* **1943**, *34*, 346–354, (c) Partington, J.R. *A History of Chemistry* (Macmillan & Co., New York, 1964), vol. 4, pp. 532–565.
16. In 1886, a delightful spoof pamphlet was published in Germany to honor Kekulé, who was at that time president of the German Chemical Society. Among other things, the booklet depicted a benzene ring as six monkeys holding hands and tails. For an account see Wilcox, D.H. Jr.; Greenbaum, F.G. *J. Chem. Educ.* **1965**, *42*, 266–267; and also Wotiz, J.H.; Rudofsky, S. *Chem. Brit.* **1984**, *20*, 720–723.
17. (a) Reese, K.M. *Chem. Eng. News* **1984**, March 19, p. 120; April 2, p. 60, (b) Kincl, F.A. *Chem. Eng. News* **1986**, February 10, p. 47.
18. (a) Mitsunobu, O.; Yamada, M. *Bull. Chem. Soc. Jpn.* **1967**, *40*, 2380–2382, (b) Mitsunobu, O.; Wada, M.; Sano, T. *J. Am. Chem. Soc.* **1972**, *94*, 679–680, (c) Grochowski, E.; Hilton, B.D.; Kupper, R.J.; Michejda, C.J. *J. Am. Chem. Soc.* **1982**, *104*, 6876–6877.
19. Mitsunobu, O. *Synthesis* **1981**, 1–28.
20. Bose, A.K.; Lal, B.; Hoffman, W.A., III; Manhas, M.S. *Tetrahedron Lett.* **1973**, 1619–1622.
21. Bose, A.K., private communication, September, 1982.
22. Lal, B.; Pramanik, B.N.; Manhas, M.S.; Bose, A.K. *Tetrahedron Lett.* **1977**, 1977–1980.
23. Sjövall, J., private communication, October, 1982.
24. Baillie, T.A., private communication, December, 1982.
25. Herz, J.E., private communication, January, 1983.
26. DiNinno, F. *J. Am. Chem. Soc.* **1978**, *100*, 3251–3252.
27. DiNinno, F., private communication, July, 1979.
28. Bond, S. *101 Uses for a Dead Cat* (Clarkson N. Potter, Pub., New York, NY, 1981). A sequel, *101 More Uses for a Dead Cat*, appeared in 1982.
29. Bass, R.J.; Banks, B.J.; Leeming, M.R.G.; Snarey, M. *J. Chem. Soc. Perkin Trans 1* **1981**, 124–131.
30. Hoffman, W.A., III *J. Org. Chem.* **1982**, *47*, 5209–5210.
31. Bose, A.K.; Sahu, D.P.; Manhas, M.S. *J. Org. Chem.* **1981**, *46*, 1229–1230.
32. Volante, R.P. *Tetrahedron Lett.* **1981**, *22*, 3119–3122.
33. Aldrich Chemical Co. advertisement in *J. Org. Chem.* **1982**, *47*, March 12 issue.
34. (a) *Chem. Eng. News*, **1979**, May 21, p. 16, (b) *The ACS Style Guide: A Manual for Authors and Editors*, Dodd, J.S. (Ed.), (American Chemical Society, Washington, DC, 1986), p. 63.
35. Fieser, L.F.; Fieser, M. *Reagents for Organic Synthesis* (John Wiley & Sons, New York, 1967), pp. 364–367.

36. Fieser, L.F.; Rajagopalan, S. *J. Am. Chem. Soc.* **1950**, *72*, 5530–5536.
37. Carey, F.A.; Kuehne, M.E. *J. Org. Chem.* **1982**, *47*, 3811–3815.
38. Cahn, R.S.; Ingold, C.; Prelog, V. *Angew. Chem. Int. Ed. Engl.* **1966**, *5*, 385–415.
39. (a) Seebach, D.; Prelog, V. *Angew. Chem. Int. Ed. Engl.* **1982**, *21*, 654–660, (b) Ugi, I. Z. *Naturforsch.* **1965**, *20B*, 405–409.
40. Hetherington, J.H., private communication, September, 1983.
41. Hetherington, J.H.; Willard, F.D.C. *Phys. Rev. Lett.* **1975**, *35*, 1442–1444.
42. The Willard episode has also been recounted by Weber, R.F., *More Random Walks in Science* (The Institute of Physics, London, 1982), pp. 110–111.
43. Willard, F.D.C. *La Recherche*, **1980**, No. 114, 972–973.
44. (a) *Acc. Chem. Res.* **1985**, *18*, 356–357, (b) *J. Org. Chem.* **1986**, *51*, No. 7, pp. 7A–8A.
45. (a) Seabrook, W. *Doctor Wood, Modern Wizard of the Laboratory* (Harcourt, Brace & Company, New York, 1941), chapter 10, (b) Reese, K.M. *Chem. Eng. News*, **1971**, November 29, p. 52.
46. Wood, R.W. *Philos. Mag.* **1912**, *24*, 673–693.
47. Reese, K.M. *Chem. Eng. News*, **1971**, November 8, p. 44.
48. Reese, K.M. *Chem. Eng. News*, **1972**, January 24, p. 32; February 14, p. 36.
49. Stewart, F.H.C. *Austral. J. Chem.* **1961**, *14*, 177–181; *ibid*, **1968**, *21*, 1107–1108.
50. Roitt, I.M.; Waters, W.A. *J. Chem. Soc.* **1952**, 2695–2705.
51. Beckwith, A.L.J.; Waters, W.A. *J. Chem. Soc.* **1956**, 1108–1115.
52. Cristol, S.J.; Perry, J.S., Jr. *J. Am. Chem. Soc.* **1967**, *89*, 3098–3100.
53. Badger, G.M.; Pearce, R.S. *J. Chem. Soc.* **1950**, 2314–2318.
54. Felix, G.; Lapouyade, R.; Castellan, A.; Bouas–Laurent, H. *Tetrahedron Lett.* **1975**, 409–412.
55. Gaultier, P.J.; Hauw, C. *Acta. Cryst.* **1976**, *B32*, 1220–1223.
56. Henderson, W.W., Ph.D. Thesis, University of Minnesota, 1962.
57. Bouas–Laurent, H., private communication, November,1976.
58. Becker, H.–D.; Sandros, K.; Arvidsson, A. *J. Org. Chem.* **1979**, *44*, 1336–1338.
59. Couture, A.; Lablache–Combier, A.; Lapouyade, R.; Felix, G. *J. Chem. Res. (S)* **1979**, 2887–2897.
60. Park, H.; King, P.T.; Paquette, L.A. *J. Am. Chem. Soc.* **1979**, *101*, 4773–4774.
61. Gordon, A.J.; Gallagher, J.P. *Tetrahedron Lett.* **1970**, 2541–2544.
62. Hess, B.A., Jr.; Bailey, A.S.; Bartusek, B.; Boekelheide, V. *J. Am. Chem. Soc.* **1969**, *91*, 1665–1672.
63. Manassero, M.; Sansoni, M.; Longoni, G. *J. Chem. Soc., Chem. Commun.* **1976**, 919–920.
64. Chisholm, M.H.; Errington, R.J.; Folting, K.; Huffman, J.C. *J. Am. Chem. Soc.* **1982**, *104*, 2025–2027.
65. Bach, R.D.; Wolber, G.J. *J. Am. Chem. Soc.* **1984**, *106*, 1400–1415.
66. Simon, J.; Le Moigne, J.; Markovits, D.; Dayantis, J. *J. Am. Chem. Soc.* **1980**, *102*, 7247–7252.
67. Jung, M.E. *Tetrahedron*, **1976**, *32*, 1–29.
68. Caluwe, P. *Tetrahedron*, **1980**, *36*, 2359–2407.
69. Baczyński, Wl.; Niementowski, St. v. *Chem. Ber.* **1919**, *52*, 461–484. Dr. P. Caluwe provided helpful insight and brought this paper to our attention in private communications, April and October, 1979.
70. (a) Schlecht, M.F., private communication, April, 1985, (b) *Chem. Eng. News* **1985**, April 15, p. 4.
71. Clar, E. *Chem. Ber.* **1932**, *65*, 503–519.
72. March, J. *Advanced Organic Chemistry*, 3rd ed. (John Wiley & Sons, New York, 1985), p. 41.
73. Johnson, W.S.; Korst, J.J.; Clement, R.A.; Dutta, J. *J. Am. Chem. Soc.* **1960**,*82*, 614–622.
74. Corey, E.J.; Shulman, J.I. *J. Am. Chem. Soc.* **1970**, *92*, 5522–5523.
75. Scanio, C.J.V.; Starrett, R.M. *J. Am. Chem. Soc.* **1971**, *93*, 1539–1540.
76. Possibly an American Chemical Society subcommittee to deal with pronunciation; see individual letters by J.F. Gall and J.H. Stocker in *Chem. Eng. News*, **1985**, May, 13, p. 36.
77. Simon, J., private communication, February, 1981.
78. Paquette, L.A.; Itoh, I.; Farnham, W.B. *J. Am. Chem. Soc.* **1975**, *97*, 7280–7285.
79. Paquette, L.A.; Meehan, G.V.; Marshall, S.J. *J. Am. Chem. Soc.* **1969**, *91*, 6779–6784.
80. Vögtle, F.; Weber, E. *Angew. Chem. Int. Ed. Engl.* **1974**, *13*, 814–816.
81. Vögtle, F., private communication, October, 1975.

82. A revised format for the periodic table of the elements has been approved by the American Chemical Society on Nomenclature; see Loenig, K.L. *J. Chem. Educ.* **1984**, *61*, 136.
83. Hyatt, J.A. *J. Org. Chem.* **1978**, *43*, 1808–1811.
84. Weber, E.; Müller, W.M.; Vögtle, F. *Tetrahedron Lett.* **1979**, 2335–2338.
85. Menger, F. M.; Takeshita, M.; Chow, J.F. *J. Am. Chem. Soc.* **1981**, *103*, 5938–5939.
86. Suckling, C.J. *J. Chem. Soc., Chem. Commun.* **1982**, *40*, 785–791.
87. Menger, F.M.; Angel de Greiff, D.A.; Jaeger, D.A. *J. Chem. Soc., Chem. Commun.* **1984**, 543–544.
88. Buhleier, E.; Wehner, W.; Vögtle, F. *Synthesis* **1978**, 155–158.
89. Reese, K.M. *Chem. Eng. News* **1978**, March 6, p. 50.
90. (a) Newkome, G.R.; Yao, Z.-Q.; Baker, G.R.; Gupta, V.K. *J. Org. Chem.* **1985**, *50*, 2003–2004, (b) Newkome, G.R.; Baker, G.R.; Saunders, M.J.; Russo, P.S.; Gupta, V.K.; Yao, Z.-Q.; Miller, J.E.; Bouillion, K. *J. Chem. Soc., Chem. Commun.* **1986**, 752–753.
91. Tomalia, D.A.; Baker, H.; Dewald, J.; Hall, M.; Kallos, G.; Martin, S.; Roeck, J.; Ryder, J.; Smith, P. *Polym. J.* **1985**, *17*, 117.
92. Newkome, G.R.; Yao, Z.-Q.; Baker, G.R.; Gupta, V.K.; Russo, P.S.; Saunders, M.J. *J. Am. Chem. Soc.* **1986**, *108*, 849–850.
93. *Chem. Eng. News*, **1965**, August 23, pp. 38–39.
94. Pettit, R., private communication, May, 1979.
95. Martin, H.-D.; Hekman, M. *Chimia* **1974**, *28*, 12–15.
96. Martin, H.-D.; Hekman, M. *Tetrahedron Lett.* **1978**, 1183–1186.
97. Martin, H.-D.; Hekman, M. *Angew. Chem. Int. Ed. Engl.* **1976**, *15*, 431–432.
98. Martin, H.-D.; Mayer, B.; Pütter, M.; Höchstetter, H. *Angew. Chem. Int. Ed. Engl.* **1981**, *20*, 677–678.
99. Zimmerman, H.E.; Grunewald, G.L. *J. Am. Chem. Soc.* **1966**, *88*, 183–184.
100. Paquette, L.A.; Stowell, J.C. *J. Am. Chem. Soc.* **1971**, *93*, 2459–2463.
101. Paquette, L.A., private communication, September, 1975.
102. Paquette, L.A.; Stowell, J.C. *J. Am. Chem. Soc.* **1970**, *92*, 2584–2586.
103. Dauben, W.G.; Buzzlini, M.G.; Shallhorn, C.H.; Whalen, D.L.; Palmer, K.J. *Tetrahedron Lett.* **1970**, 787–790.
104. Castellan, A.; Lacoste, J.M.; Bouas–Laurent, H. *J. Chem. Soc. Perkin Trans. 2*, **1979**, 411–419.
105. Henkel, J.G.; Spurlock, L.A. *J. Am. Chem. Soc.* **1973**, *95*, 8339–8351.
106. Spurlock, L.A., private communication, July, 1978.
107. Billups, W.E.; Chow, W.Y.; Cross, J.H. *J. Chem. Soc., Chem. Commun.* **1974**, 252.
108. Ranganathan, S., private communication, March, 1976.
109. Kosower, E.M.; Pazhenchevsky, B. *J. Am. Chem. Soc.* **1980**, *102*, 4983–4993.

Chapter 2

HATS OFF TO ORGANIC!

Organic chemists need to keep a great many facts and concepts in their heads. Not surprisingly, therefore, they sometimes name compounds after head coverings, protecting, as it were, this storehouse of knowledge.

To begin with, consider hexacyclo[4.4.0.0^{2,10}.0^{3,5}.0^{4,8}.0^{7,9}]decane (**2**), obtained by Armin de Meijere's research group in 1971 at the University of Göttingen upon irradiation of snoutene (**1**).[1] Professor de Meijere and co-workers thought **2** resembled a crown and dubbed it "diademane," from the Greek word for crown, *diadema*.[2] Formally, the **1** → **2** change is akin to an internal [2 + 2] cycloaddition except that a cyclopropane σ bond takes the place of an alkene π bond. Such photorearrangements were first described in 1965 by Horst Prinzbach *et al.*, who also extended them to oxirane and aziridine analogs.[3]

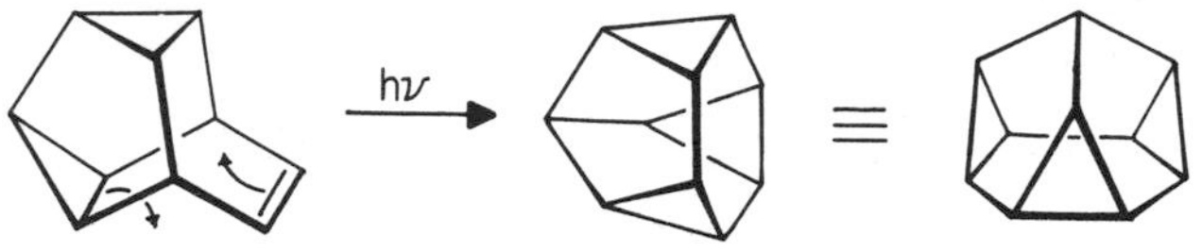

A few months before Dr. de Meijere's paper appeared, Professor Prinzbach's team at the University of Freiburg was attempting the **1** → **2** photoisomerization and had called **2** "mitrane,"[4] from "mitra," a kind of hat sometimes worn by Catholic bishops.[5] When Dr. Prinzbach mentioned this etymology in a colloquium at the California Institute of Technology in 1971, he joked that perhaps the clergical connotation would help him get research funds from the Archbishop of Freiburg. Professor George Hammond piped up from the audience that, as he knew the Catholic church in the United States, an archbishop would not likely come through with any money but only with prayers for higher yields.[5] Because Dr. Prinzbach did not then have **2** in hand, whereas Dr. de Meijere had succeeded in isolating it, these two chemists agreed to let "diademane" stand as the trivial name.[2,5] (In a way, this regal term even does honor to Dr. Prinzbach, because the first five letters of his surname are German for "prince.")

Fashion designers may balk at the thought of reshaping one type of hat to another, but not organic milliners. For example, when heated near its melting point (96–97°C), diademane (**2**) converts to triquinacene, **3** (via an allowed [2s + 2s + 2s] cycloreversion).[6] And then irradiation of triquinacene (see p. 201 for this term) produced another hat-like molecule **4**, which Bosse and de Meijere dubbed "barettane."[6] This appellation was suggested by Dr. Wolfgang Walter of the University of Hamburg after a lecture by Professor de Meijere at Stuttgart in March, 1974.[2] It derives from the Spanish word *baretta*, for a cap originally worn by Basque people in northern Spain and southern France. In their paper, Bosse and de Meijere showed no mechanism for the **3→4** conversion.[6] No wonder! But formally we can get there by using two π electrons from C2—C3 to form a bond between C3 and C5, a pair from C5—C6 to join C6 and C9, and two from C8—C9 to link C1 and C8. Finally, move the C1—C10 electrons between C2 and C10. After all this sartorial grooming the baretta had better fit!

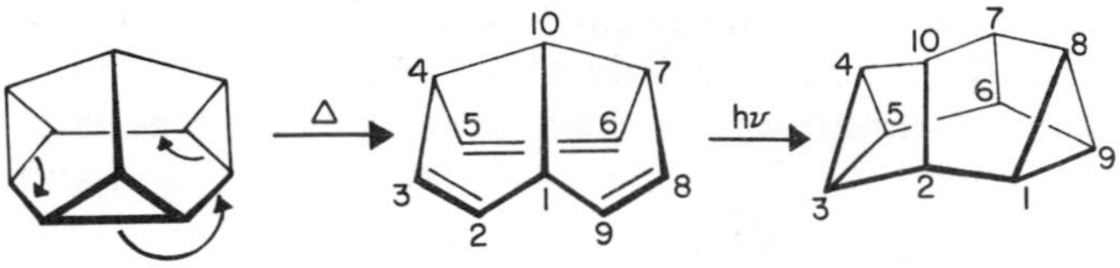

Perhaps the best-known head coverings are the "crown" ethers, reported by Charles Pedersen of the Du Pont Company.[7] These large-ring molecules (e.g., **5**) contain heteroatoms (e.g., oxygen) with unshared electrons. A molecular model of **5** looks very much like a crown. Because its systematic name (1, 4, 7, 10, 13, 16-hexaoxacyclooctadecane) is a mouthful, Dr. Pedersen introduced shorter terminology based on royalty. Thus, **5** is 18-crown-6 (18 atoms in the ring, 6 of which are hetero elements). Some writers prefer to bracket the first number; and a few abbreviate the name even further to 18C6.

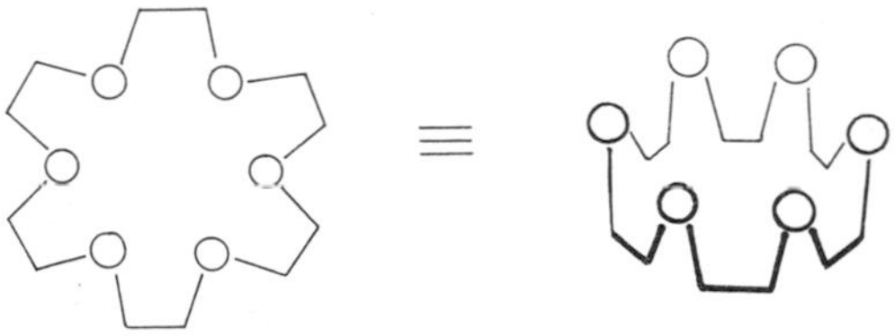

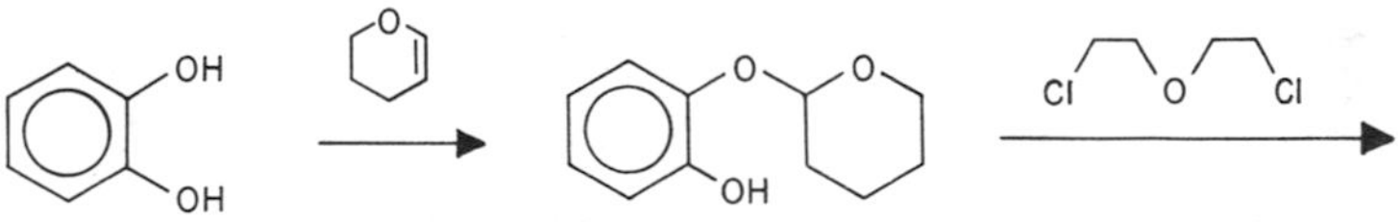

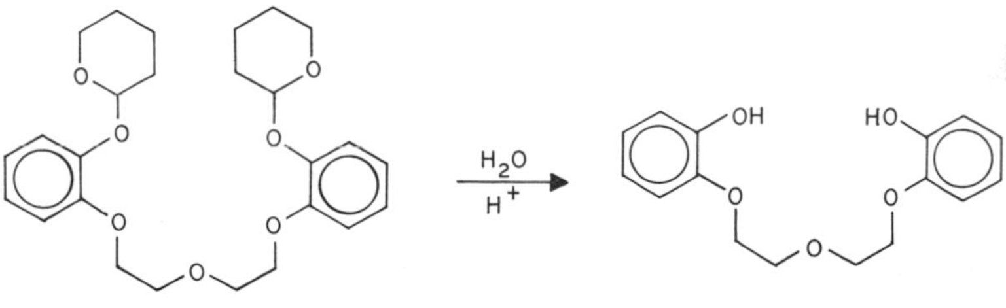

The discovery of crown ethers illustrates serendipity.[8] Dr Pedersen intended to prepare the diol **9** by protection of one hydroxyl group in catechol (**6 → 7**), followed by treatment with bis(2-chloroethyl) ether and hydrolysis of the protective group (**7 → 8 → 9**). The first reaction, however, did not go to completion; so **7** in the next step was contaminated with starting catechol, which did its thing with the chloride reagent to give dibenzo-18-crown-6 (**10**). Instead of simply discarding this unwanted by-product, the Du Pont researcher explored it and thereby gave birth to an exciting and important area of chemistry. We do not advocate use of impure reagents as a prescription for scientific breakthroughs, but no one can knock Dr. Pedersen's success in this case. His discovery may have been accidental but, as Louis Pasteur said, "In the fields of observation, chance favors only the mind that is prepared."[9]

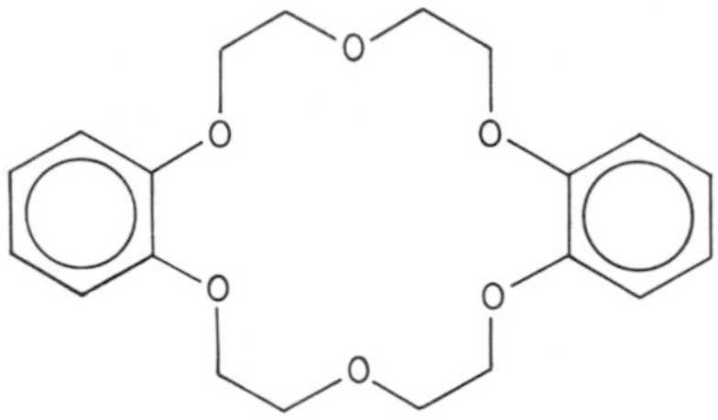

10

A fascinating and beneficial aspect of these crowns is that they seem to violate the adage, "like dissolves like." Teachers stress that polar likes polar, nonpolar likes nonpolar, but never the twain shall mix. Students lose points on examinations if they suggest that KOH, $KMnO_4$, NaI, CsOH, and other "ionic" substances dissolve in benzene. And yet, in the presence of crown ethers such ionic solutes *do* dissolve in benzene (and in other solvents of low polarity).[7] Actually, "like dissolves like" is not really violated; the crown oxygens bind the metal cation and thus surround it with (relatively unpolar) hydrocarbon units.[10] The cation is like a royal head that has donned a chemical crown. This ability to solubilize ionic compounds makes crown ethers enormously valuable in synthesis. For example, esters of 2, 4, 6-trimethylbenzoic acid are sterically hindered and ordinarily cannot be hydrolyzed by KOH in hydroxylic media; but KOH and dicyclohexano-18-crown-6 (saturated **10**) in benzene do the trick.[7] The unsolvated HO^- ion, freed of its K^+ counterion by the crown system, is much more reactive than when the anion is solvated by a polar milieu.

As expected, an outburst of papers followed these exciting developments. Chemists have altered the size of the crown, the kind and number of heteroatoms, the metal ion, and the groups attached to the ring.[11–13] Edwin Weber of the University of Bonn came up with "multiloop" versions and also with different-sized crowns in the same molecule (e.g., **11**); these kindred reagents can complex more than one type of cation at the same

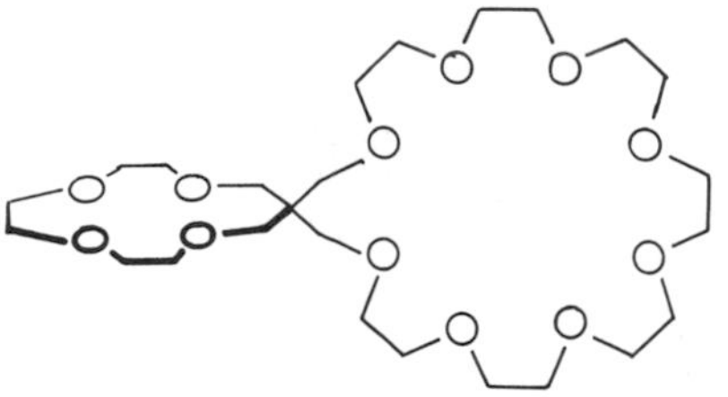

11

time.[14,15] We even have a "rope-skipping" monarch **12**, fashioned by Lynn Sousa and his band at Michigan State University.[16] In the conformation **12a**, hydrogens A (which point toward the naphthalene peri position) are not the same as hydrogens B (which point the other way). Yet ^{1}H NMR revealed that H_A and H_B are equivalent at room temperature. Naphthalene uses the large ring as a jump rope!

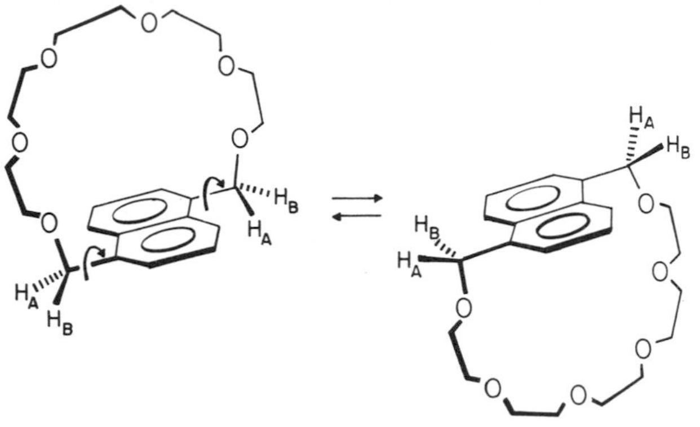

12a   12b

Dr. Sousa subsequently moved to Ball State University, Indiana. And, according to the title of a paper he delivered at a national chemistry meeting, his team at BSU conducted "A Quest for Flashy Crowns...."[17] That research entailed design, synthesis, and study of crown ethers equipped with fluorescent chromophores that would light up more intensely when triggered by bound cations like Na^+, K^+, and Ca^{++}. A nice notion, because such flashy crowns when added to biological fluids might signal the presence of specific cations or changes in their concentrations. At one seminar he presented, Professor Sousa's brilliant title evidently drew extra attendance; even an archeologist and a map librarian showed up for the talk. But these nonchemists did not stay long—probably because they failed to see the light.[18]

George Gokel's posse, then at the University of Maryland, attached to crowns a flexible chain containing one or more oxygens (e.g., **13**) and produced "lariat ethers."[19] Cowhands use lariats to rope animals, but Dr. Gokel's nooses adeptly snare sodium and potassium ions. And they sometimes are better at it than the simple crowns, especially if the ring "ropes" the cation (**14→15**) and the line pivots (**15→16**) to help "tie" it more securely.[19-21] When a chain also carried a free phenolic O to confine a K ion inside the crown, Professor Gokel dubbed it the "OK Corral."[22] We suppose a lariat that complexes Ag^+ ion should be called "Hi-ho Silver!"

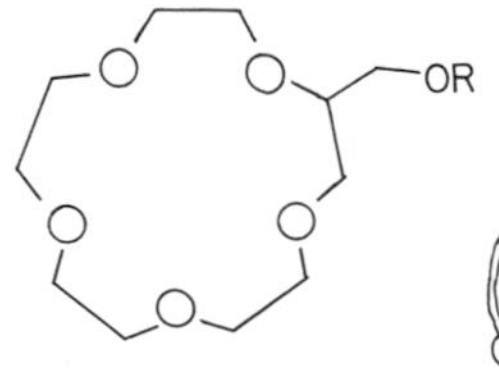

13

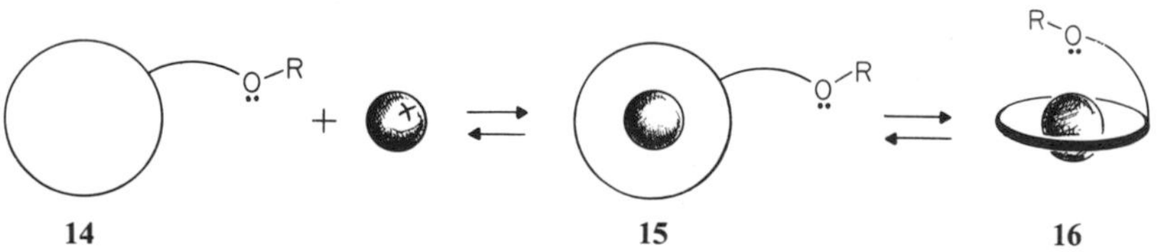

14   15   16

A molecular rope with a terminal arenediazonium unit (**17**) can swish about and imbed the cationic nitrogens into its own crown cavity if there is room (**18**). Beadle and Gokel labeled these species "ostrich complexes" because of the popular belief that this bird pokes its head into a hole when endangered.[23]

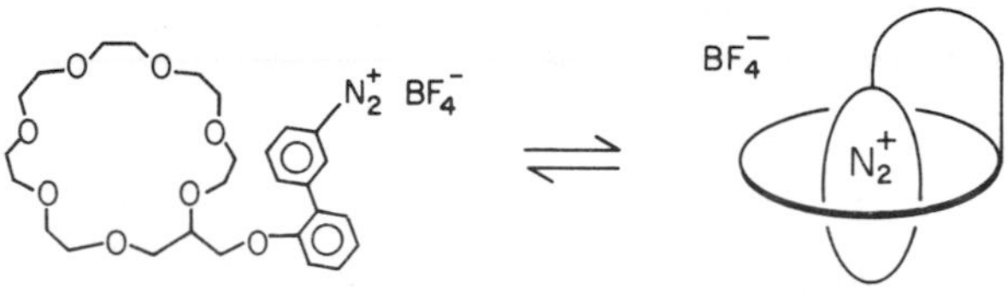

17   18

Japanese chemists also know how to twirl ropes.[24] For example, outfitters Seiji Shinkai, Osamu Manabe, and co-workers at Nagasaki University fashioned a lariat ether having an E-azobenzene unit in the line and an NH_3^+ at the very end (**19**). Then, a photic E → Z switch at the azo link permitted the cationic head to reach back and savor its own oxygens (**20**). The Nagasaki ranchhands termed this process "tail biting."[25] You can read about all kinds of photofunctional crowns in their review article.[26]

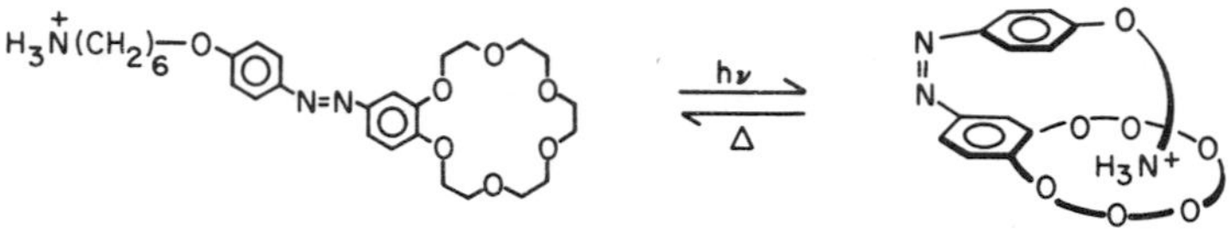

E   Z
19   20

A lariat's line can be thought of as an arm; and two of them could offer a statistical advantage for clasping a cation.[15,27] Independently, Hiroshi Tsukube at Okayama University and George Gokel (then at the University of Maryland) investigated diaza crown ethers equipped with a pair of N-pivot limbs (**21**). The Japanese researcher called them "double armed crown ethers,"[28] whereas Gatto and Gokel looked at Latin to brand these ligands "bibracchial lariat ethers" (*bracchium* means "arm"). Their recommended abbreviation is "BiBLEs."[29]

This acronym signals us not to overlook the scriptures for other insights. Friedrich Helfferich pointed out that according to the Holy Bible, Moses may have been the first to learn one of the practical benefits of ion complexation.[30] Exodus 15: 23–25 reads:

> ... they could not drink of the waters of Marah, for they were bitter.... And he cried unto the Lord; and the Lord showed him a tree, which he had cast into the waters, the waters were made sweet....

Thus Moses was shown to make brackish water potable, seemingly by ion exchange. About 1000 years later, Aristotle alluded to a similar phenomenon when he stated that sea water loses part of its salt content when percolating through certain sands.[30]

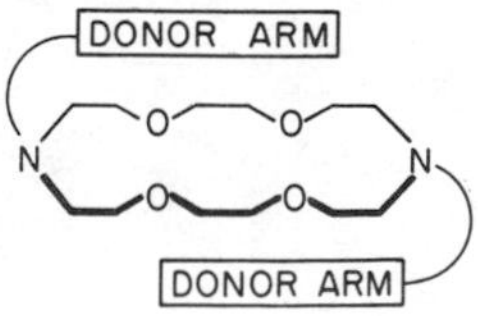

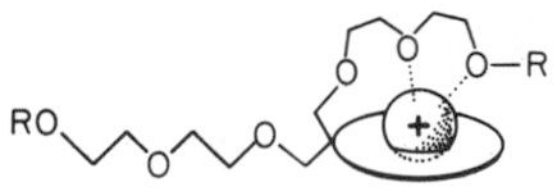

21   22

At Himeji Institue of Technology, a team guided by Yoshihisa Inoue and Tadao Hakushi probed metal-ion complexation by crowns fitted with geminal side arms. Their findings suggested that a cation too large to settle deep into an annular cavity might at least use the crown unit as a "tray." And donor atoms at elbow's length keep the ion from sliding off too readily (**22**).[31]

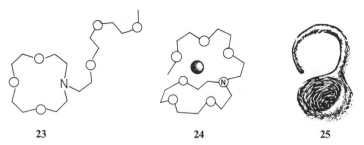

| 23 | 24 | 25 |

Thanks to caterers George Gokel (University of Miami, Florida) and Richard Gandour (Louisiana State University), you can choose ladles instead of trays to serve your guests metal ions. These entrepreneurs demonstrated that the seven hetero-atoms in the lariat ligand **23** engulf Na^+ or K^+ ions nicely. X-ray study of the potassium iodide complex revealed that the cation sits on the crown ring, and the "handle" swoops unsymmetrically across the top (**24**).[32] They dubbed this a "calabash" complex because of its similarity in shape to a Ugandan calabash ladle (**25**).[33] Tropical natives make such utensils from the dried hard shells of calabash gourds, which grow on climbing vines.

The crown concept is so well entrenched[34] that chemists from (appropriately enough) England titled a paper "Coronation of Ligating Acetonitrile by 18-Crown-6."[35] Presumably, a publication on the thermodynamics of crown ethers could be "A Study of the Crown Joules."[36]

And while we are in royal territory, let us see how a monarch might go about slicing a symmetrical (i.e., achiral) apple. First, make two perpendicular vertical half-cuts: one from the top to the equator, and the other from the bottom to the equator. Then make two nonadjacent quarter slices horizontally along the equator to connect the two vertical incisions (**26**). Voilà! The fruit separates into identical halves (**28** and **29**). French families know this parlor trick as "la coupe du roi" ("the king's cut").[37] Before sinking your teeth into the pieces, look at them closely. In 1937, chemist Alain Horeau (then working in the laboratory of Marcel Delépine) observed that two such halves are individually chiral, yet superposable! In modern lingo they are "homochiral."

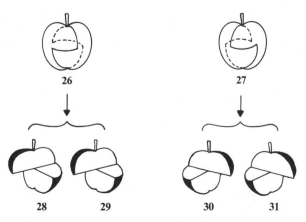

| 26 | | 27 |
| 28 | 29 | 30 | 31 |

Now, perform similar surgery on a second identical apple; only this time reverse the directions of the horizontal incisions (27). You get a second pair of homochiral halves (30 and 31). But you cannot mesh a member from the first pair with one from the second pair to reconstitute a complete apple. Such segments are "heterochiral" (e.g., 29 and 30). If *la coupe du roi* appeals to you, look up an incisive paper by Frank Anet, Steve Miura, Jay Siegel, and Kurt Mislow.[37] They show how to divide geometric objects into "isometric" segments and steer you to words like "anisometric," "pairwise and nonpairwise heterochiral," "parallelogons," and "planigons." And Professor Mislow, with graduate student Jay Siegel, has bestowed upon venerable asymmetric carbons the attributes "stereogenicity" and "chirotopicity."[38] It's not just regal fun and games. The UCLA and Princeton scholars suggested possible implications in chemistry and biology.[37,38]

REFERENCES AND NOTES

1. de Meijere, A.; Kaufmann, D.; Schallner, O. *Angew. Chem. Int. Ed. Engl.* **1971**, *10*, 417–418.
2. de Meijere, A., private communication, June, 1977.
3. (a) Prinzbach, H.; Eberbach, W.; von Veh, G. *Angew. Chem.* **1965**, *77*, 454–455, (b) Prinzbach, H. *Chimia* **1967**, *21*, 194–199, (c) Prinzbach, H.; Hunkler, D. *Angew. Chem.* **1967**, *79*, 232–233, (d) Prinzbach, H.; Klaus, M. *Angew. Chem.* **1969**, *81*, 289–291, (e) Prinzbach, H.; Klaus, M.; Mayer, W. *Angew. Chem.* **1969**, *81*, 902–903.
4. Prinzbach, H.; Stusche, D. *Helv. Chim. Acta* **1971**, *54*, 755–759.
5. Prinzbach, J., private communications, June, 1977 and March, 1984.
6. Bosse, D.; de Meijere, A. *Angew. Chem. Int. Ed. Engl.* **1974**, *13*, 663–664.
7. Pedersen, C.J. *J. Am. Chem. Soc.* **1967**, *89*, 7017–7036.
8. Pedersen, C.J. *Aldrichimica Acta* **1971**, *4*, (No. 1), 1–4. (A publication of the Aldrich Chemical Company, Milwaukee, WI)
9. Bartlett, J. *Familiar Quotations*, 14th ed. (Little, Brown and Co., Boston, 1968), p. 718.
10. Host Guest Complex Chemistry III. In *Topics in Current Chemistry*, Vögtle, F., Weber, E. (Eds.) (Springer-Verlag, Berlin, 1984), vol. 121.
11. Gokel, G.W.; Durst, H.D. *Aldrichimica Acta* **1976**, *9*, 3–12. (A publication of the Aldrich Chemical Company, Milwaukee, WI)
12. Pedersen, C.J.; Frensdorff, H.K. *Angew. Chem. Int. Ed. Engl.* **1972**, *11*, 16–25.
13. Vögtle, F.; Neumann, P. *Chem. Zeitung* **1973**, *97*, 600–610.
14. Weber, E. *Angew. Chem. Int. Ed. Engl.* **1979**, *18*, 219–220.
15. Weber, E. *J. Org. Chem.* **1982**, *47*, 3478–3486.
16. Brown, H.S.; Muenchausen, C.P.; Sousa, L.R. *J. Org. Chem.* **1980**, *45*, 1682–1686.
17. Sousa, L.R.; Son, B.; Beeson, B.E.; Barnell, S.A. *Abstracts of Papers, 190th National Meeting of the American Chemical Society, Chicago, IL* (American Chemical Society, Washington, DC, 1985), ORGN 280.
18. Sousa, L.R., private communication, August, 1985.
19. Gokel, G.W.; Dishong, D.M.; Diamond, C.J. *J. Chem. Soc., Chem. Commun.* **1980**, 1053–1054.
20. Dishong, D.M.; Diamond, C.J.; Gokel, G.W. *Tetrahedron Lett.* **1981**, *22*, 1663–1666.
21. Dishong, D.M.; Diamond, C.J.; Cinoman, M.I.; Gokel, G.W. *J. Am. Chem. Soc.* **1983**, *105*, 586–593.
22. Gokel, G.W., private communication, January, 1984.
23. Beadle, J.R.; Gokel, G.W. *Tetrahedron Lett.* **1984**, *25*, 1681–1684.
24. Nakatsuji, Y.; Kobayashi, H.; Okahara, M. *J. Chem. Soc., Chem. Commun.* **1983**, 800–801.
25. Shinkai, S.; Ishihara, M.; Ueda, K.; Manabe, O. *J. Chem. Soc., Chem. Commun.* **1984**, 727–729.
26. Shinkai, S.; Manabe, O. *Top. Curr. Chem.* **1984**, *121*, 67–104.
27. Weber, E. *Liebigs Ann. Chem.* **1983**, 770–801.
28. (a) Tsukube, H. *J. Chem. Soc., Chem. Commun.* **1984**, 315–316, (b) Tsukube, H.; Takagi, K.; Higashiyama, T.; Iwachido, T.; Hayama, N. *Tetrahedron Lett.* **1985**, *26*, 881–882.
29. Gatto, V.J.; Gokel, G.W. *J. Am. Chem. Soc.* **1984**, *106*, 8240–8244, (b) Gatto, V.J.; Arnold,

K.A.; Viscariello, A.M.; Miller, S.R.; Gokel, G.W. *Tetrahedron Lett.* **1986**, *27*, 327–330, (c) Professor Gokel (private communication, June, 1986) intended BiBLe nomenclature to be semisystematic. Three-armed compounds are TriBLEs; four-armed are TetraBLEs; six-armed are HexaBLEs, and so on.

30. Helfferich, F. *Ion Exchange* (McGraw-Hill, New York, 1962), pp. 1–4.
31. Ouchi, M.; Inoue, Y.; Wada, K.; Hakushi, T. *Chem. Lett.* **1984**, 1137–1140.
32. White, B.D.; Arnold, K.A.; Fronczek, F.R.; Gandour, R.D.; Gokel, G.W. *Tetrahedron Lett.* **1985**, *26*, 4035–4038.
33. Fournier, R. *Illustrated Dictionary of Pottery Form* (Van Nostrand Reinhold, New York, 1981), p. 41.
34. de Jong, F.; Reinhoudt, D.N. Stability and Reactivity of Crown Ether Complexes, in *Advances in Physical Organic Chemistry*, Gold, V.; Bethell, D. (Eds.) (Academic Press, New York, 1980), vol. 17, pp. 279–434.
35. Colquhoun, H.M.; Stoddart, J.F.; Williams, D.J. *J. Am. Chem. Soc.* **1982**, *104*, 1426–1428.
36. *Chem. Eng. News* **1978**, Oct. 9, p. 56.
37. Anet, F.A.L.; Miura, S.S.; Siegel, J.; Mislow, K. *J. Am. Chem. Soc.* **1983**, *105*, 1419–1426.
38. Mislow, K.; Siegel, J. *J. Am. Chem. Soc.* **1984**, *106*, 3319–3328.

Chapter 3

TOOLS OF OUR TRADE

Isn't it strange that princes and kings
And clowns that caper in sawdust rings
And common folks like you and me
    Are builders of eternity?
To each is given a bag of tools,—
A shapeless mass and a book of rules;
And each must make, ere life is flown,
A stumbling-block or a stepping-stone.

—Roy L. Sharpe[1]

This highly technological age surrounds us with mechanical devices, tools, and sundry gadgets. Small wonder that molecules have been named on account of their resemblance to such items.

Even the most complex mechanical marvels are based on one or more of six simple machines: the wedge, wheel and axle, pulley, inclined plane, screw, and lever.[2] And this fact did not escape Philip Eaton and his craftsmen at the University of Chicago when they synthesized **2** by rearranging cubane (**1**) with silver perchlorate in benzene.[3] A

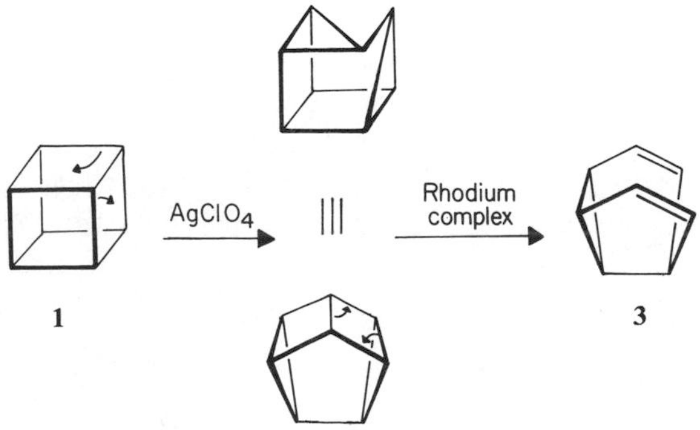

molecular model of **2** resembles a wedge, so the compound was dubbed "cuneane." (*Cuneus* is Latin for "wedge.") Treatment of cuneane with a suitable rhodium complex ruptures two cyclobutane bonds at the top (overall like a reverse [2 + 2] cycloaddition) and gives birth to calfene (**3**). Not a bad move, considering the price of veal.

Another simple machine, the wheel and axle, inspired Gottfried Schill and co-workers of the University of Freiburg to coin "rotaxane" for compounds such as **4**.[4] They had in mind a wheel-like large ring (Latin *rota* = "wheel") free to *rotate* about a long hydrocarbon chain ("*axle*").[5] The wheel can't slip off because the bulky

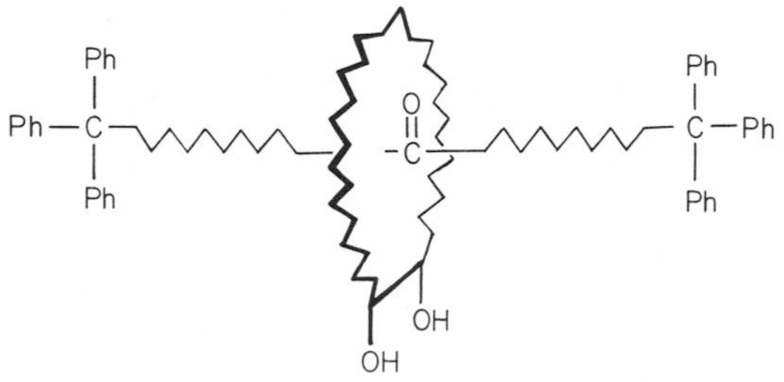

4

triphenylmethyl groups at the ends act as cotter pins. Professor Schill's synthesis involved common reactions but is ingenious because the wheel remained fixed to the axle until released in the very last step. They first built the hoop (**5 → 6**), and then glued an axle temporarily to its rim (**6 → 7**) through a ketal link. In a few molecules the axle

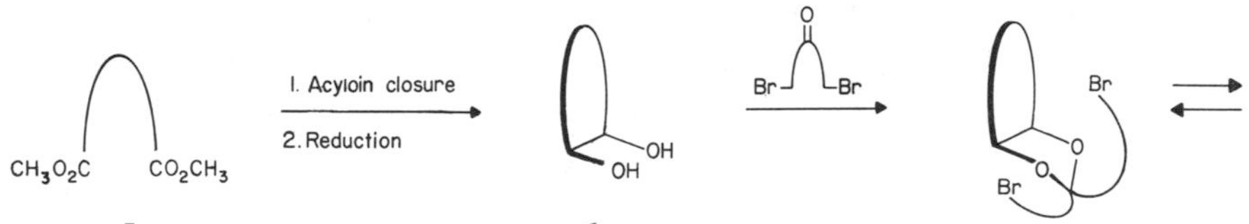

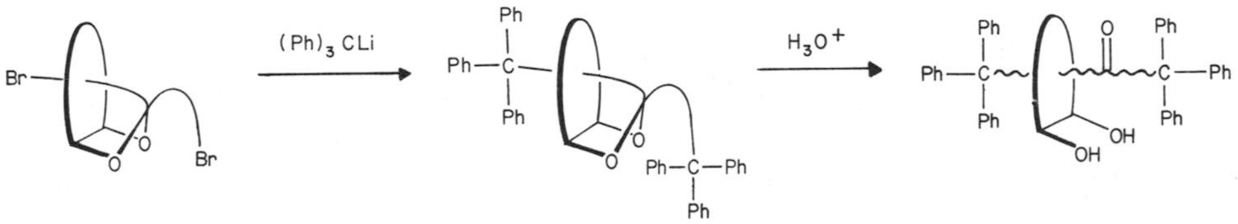

chain manages to thread its way through the wheel (**7 → 8**). The cotter pins were put in place (**8 → 9**) and the ketal "glue" was dissolved to separate axle and wheel (**9 → 4**). Molecules with the axle outside the circle (**7**) underwent the same chemistry, but no rotaxane resulted because the components part company when the glue is removed.

At about the time of Dr. Schill's initial work,[6] Ian Harrison and Shuyen Harrison at Syntex Research, Palo Alto, California reported the synthesis of a very similar molecule (**10**).[7] They chose the class name "hooplane" after "hoopla," a British term for the

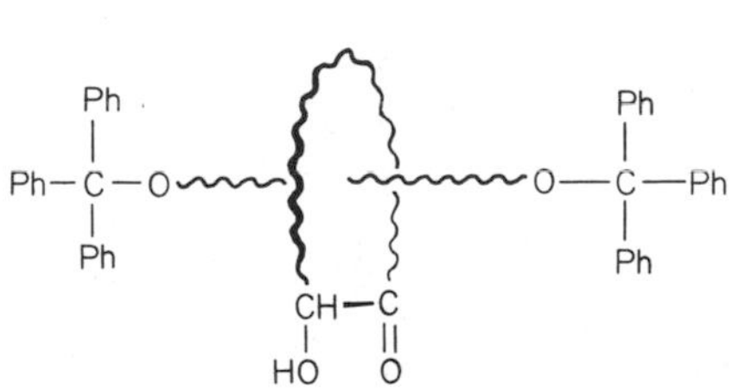

10

pastime known in the United States as "ring toss."[8] In this molecular game, the ring sticks on the stake until a chemist comes along and plucks off a triphenylmethyl knob.

Several types of structures have reminded chemists of propellers. In 1966, Jordan Bloomfield, then at the University of Oklahoma, felt that compound **11**, which he synthesized,[9] should be immortalized by the name "propellerane," even though the

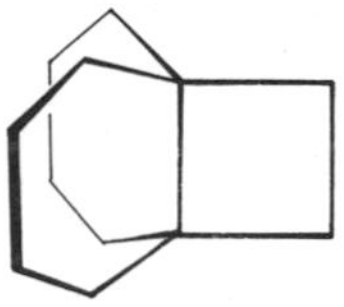

11

three blades are not equivalent. He stated so in a paper submitted for publication, but an editor nixed the notion.[10] So Dr. Bloomfield was forced to settle for the following sentence in a footnote: "We have coined the trivial name 'Propelleranes' to describe these molecules in our laboratory."

Soon thereafter, however, David Ginsburg in Israel was invited by the editors of *Tetrahedron* to contribute to a special issue (*Festschrift*) that honored Sir Robert Robinson on his 80th birthday.* Professor Ginsburg chose to describe work from his own laboratory at the Israel Institute of Technology in Haifa on a variety of molecules similar to **11** (e.g., **12–14**).[12a] Because these invited papers are not always refereed and

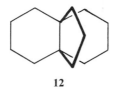

12

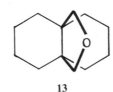

13

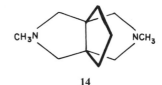

14

because Dr. Ginsburg, himself, was an honorary editor of this *Festschrift*, no one objected when he titled his article "Propellanes."[12b] (Note: The word has two letters fewer than the name Dr. Bloomfield chose.)

Propellanes are intriguing and later showed up in nature. Leon Zalkow's research team at the Georgia Institute of Technology isolated **15** from the toxic plant *Isocome wrightii* (rayless goldenrod) and named the compound "modhephene" (pronounced mod-heff-een), after the Hebrew word for propeller, *modheph*.[13] Dr. Zalkow's crew wanted to relate **15** to the propellane class and felt that modhephene sounded better than other foreign language translations of the word propeller.[14] Synthetic modheph-

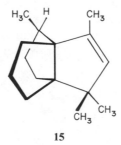

15

16

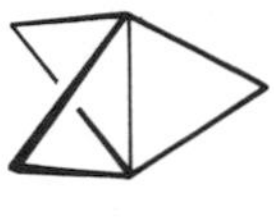

17

*Sir Robert Robinson, a distinguished British organic chemist, received the Nobel Prize in chemistry in 1947. He may have originated the "curved arrow" symbolism for electron displacement and used it in a landmark paper in 1922.[11]

ene first rolled off the assembly line in early 1980,[15] and production has spun right along.[16]

Yale's Kenneth Wiberg and Frederick Walker collaborated with Joseph Michl (then at the University of Utah) and shrunk the propeller blades with their synthesis of [2, 2, 1]propellane (16).[17] Backed by the National Science Foundation, the Yale mechanics have even fashioned the ultimate propellane, tricyclo[1.1.1.0]-pentane (17).[18] This highly strained oddity doesn't seem to know how to rid itself of energy, so it persists. And there was even talk that the central C–C bond is "phantom" and really may not be there.[19,20] If so, NSF should require Yale to account for the missing link. After all, tax dollars paid for it.

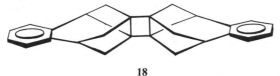

18

Compounds with two propellane units and a common ring (such as 18) are also available.[21] James Vincent and Joel Liebman of the University of Maryland–Baltimore County[22] call those with a common cyclobutane ring (e.g., 18) "buttaflanes," as the central portion is reminiscent of a butterfly (cf. lepidopterene, chapter 1). Leave it to chemists to make butterflies and propellers so beautifully compatible. Two more buttaflanes grew out of pupae in 1982, thanks to Yoshinobu Odaira and cohorts at Osaka University.[23]

Kurt Mislow of Princeton University aptly used the term "molecular propellers" for structures such as triphenylmethane (19), in which the three benzenoid units twist out of plane because of steric interference among the ortho hydrogens.[24] As a result, these rings resemble the blades of a screw propeller.

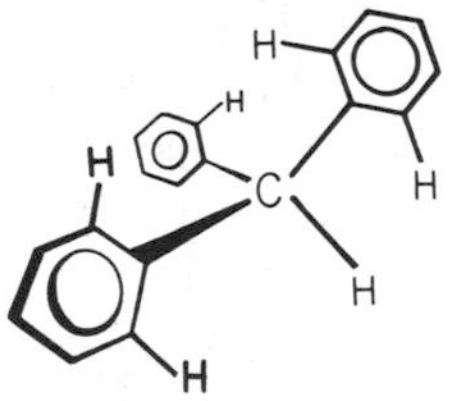

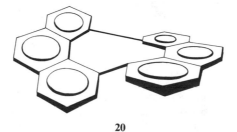

19

Still another propeller-like molecule is exemplified by 20, synthesized by Bengt Thulin and Olof Wennerström of Göteborg University in Sweden.[25] Nonbonded

20

repulsions force each phenanthrene moiety to adopt a helix-like twist and to produce what is referred to as a "helicene part" of the molecule. Because the whole structure has the symmetry of a two-bladed propeller, Drs. Thulin and Wennerström at first came up with "propellehelicene," but they bumped off *heli* to leave their smoother-sounding designation, "propellicene."[26] Was this an act of helicocide? Not according to Nakanishi and Kubo, who report that helicocides are chemicals that kill snails.[27] But what about the word helicene?

"Helicene" terminology sprang up in 1955 when Melvin Newman at Ohio State University and his students Wilson Lutz and Daniel Lednicer synthesized the twisted polycycle **21** and resolved it into optically active form ($\alpha_D = 3640°$).[28] This classic achievement marked the first resolution of an aromatic hydrocarbon whose chirality arose from intramolecular overcrowding. In their preliminary communication they wanted to call **21** "hexahelicene." But the editor resisted, so it became phenanthro[3,4-c]phenanthrene.[29] However, the whole matter aroused considerable discussion in American and international nomenclature committees, which began to ponder possible official adoption of "helicene" to designate such helix-like molecules. By the time the Ohio chemists submitted their full publication, this euphonious short name had gained enough inroads to allow them to title the paper "The Synthesis and Resolution of Hexahelicene."[30]

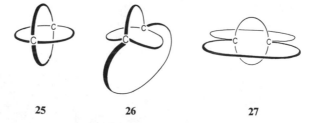

21   22   23   24

By the way, "Newman projections" for conformational isomers of open-chain and cyclic structures are also brain-children of the same Ohio professor. Typical drawings for staggered ethane (**22**), ethylene oxide (**23**), and chair cyclohexane (**24**) are shown. Doctoral student Peter Hay, in his Ph.D. thesis at Ohio State University in 1952, suggested "conformer" as a shorthand for conformational isomer.[31] And "conformation" in today's stereochemical sense slipped quietly into chemistry in 1929 through the courtesy of Walter Haworth at the University of Birmingham. He used that expression to describe shapes of sugar molecules.[32] In similar vein, Switzerland's Vladimir Prelog and Leopold Ruzicka spoke of the "constellations" of medium-sized rings, but that term did not gain popularity on our planet.[33] Now, back to propellers.

Let us mentally replace the central bond of a propellane with a fourth blade to get the general structure **25**. Jordan Bloomfield at the Monsanto Company in St. Louis did just that and came up with the term "paddlane." The Monsanto researcher may or may not

25   26   27

have had boats on the brain at the time.[10] But, Professor David Ginsburg recalled that Dr. Bloomfield churned up that class name while the two were dining in St. Louis overlooking the Mississippi River.[34] Paddlanes can take on different appearances (e.g., **26**, **27**) according to the length of each rim.[35]*

*In 1981, Gund and Gund categorized polycyclic systems on the basis of how rings can share a common atom. Their scheme classifies paddlanes as "pontetricyclic" (from the French *ponté*, for "bridged").[36]

David Ginsburg's go-getters, several thousand miles away, also had paddles in mind and attempted to prepare paddlane **29** from **28**.[37] Even with high-dilution techniques, **28** and sebacoyl chloride gave only double-paddlane **30**, which at least ensured faster travel.

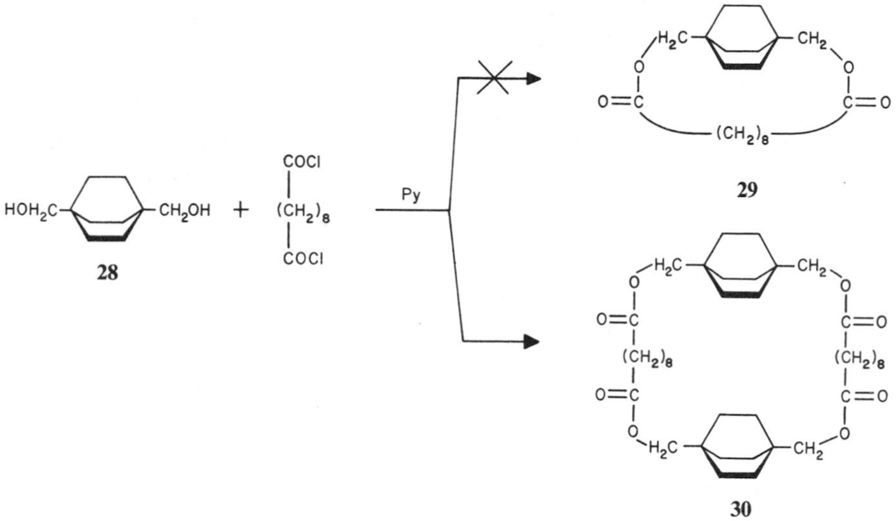

Chemists in Germany and the Soviet Union (headed, respectively, by H. Meier and by T. Skorochodowa) joined forces to fashion the novel structure **32**.[38] It is not a paddlane, but the scoops look big enough to propel a Mississippi steamer. The binational team got this paddle wheel by simple irradiation of (E, E, E)-1, 3, 5-tristyrylbenzene (**31**). A threefold [2 + 2] cycloaddition did the job; and the outcome was proved by X-ray analysis. Keep in mind the shape of **32** as you read further and you will note an interesting similarity to manxane (p. 122). And later still (chapter 16) you will appreciate that, paddles notwithstanding, **32** is in fact, a cyclophane.

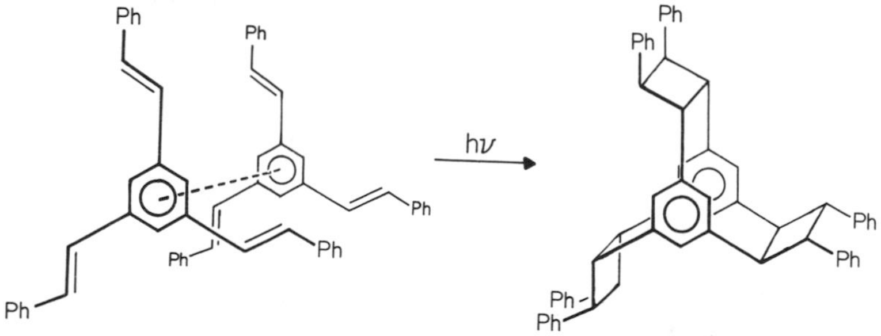

In 1969, Jean–Marie Conia and co-worker Jean–Louis Ripoll reported the synthesis of hydrocarbon **33**.[39] The cyclopropyl rings lie perpendicular to the five-membered frame and therefore reminded Dr. Conia of a paddle wheel, quite different from the type related to paddlanes.

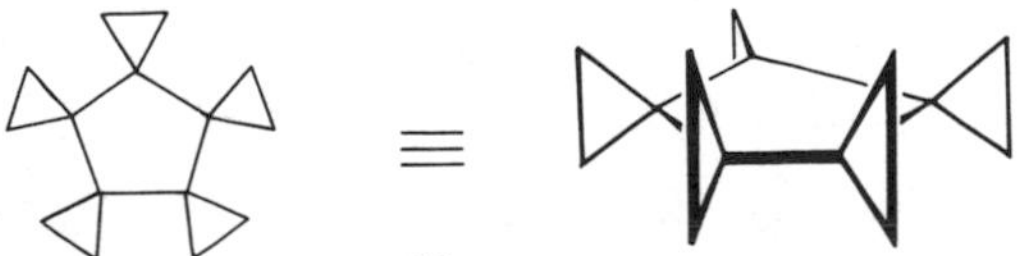

The French term for paddle wheel is *roue à aube*, but Dr. Conia felt that a name based on this derivation would lack international appeal. He therefore settled for "rotane," from the Latin *rota* (wheel).[40] Carbocycle **33**, with five paddles, is [5]rotane.

Other rotanes soon followed, including [4]rotane,[41,42] [3]rotane,[43,44] and [6]rotane.[45,46] In [3]rotane and [4]rotane, the inner ring is flat and the endocyclic bonds are shorter than usual.[47] In one synthesis, [4]rotane (**35**) was forged by thermal dimerization of biscyclopropylidene (**34**), which itself can be thought of as [2]rotane, the start of the homologous series.[42] In Dr. Conia's laboratory, bow-tie hydrocarbon **34** is simply *la cravate*.[40] University of Göttingen chemist Lutz Fitjer prepared [6]rotane from the [5] analog by a sequence that could, in principle, be repeated again

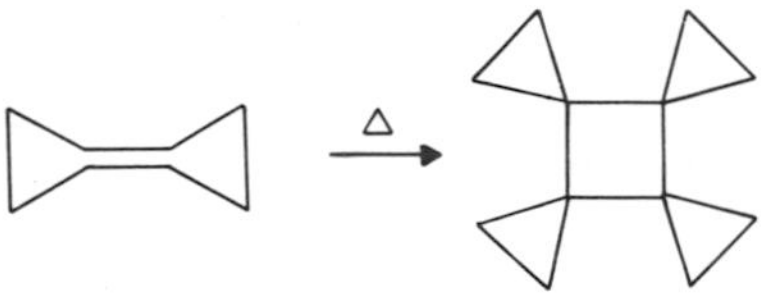

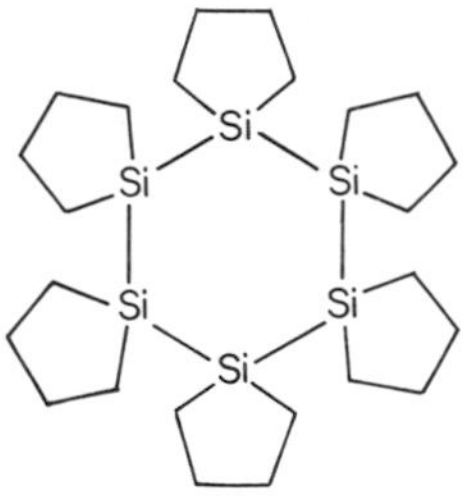

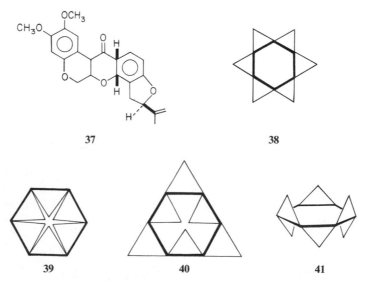

34   **35**   **36**

and again to get wheels of any size.[45] Robert West, at the University of Wisconsin, got the notion that rotanes would roll smoothly if lubricated with silicon. So he and his artisans assembled "organosilicon rotanes" such as **36**.[48,49] We truly have molecules being geared for action. By the way, rotanes are in no way related to a class of natural products known as "rotenoids." A typical member is "rotenone" (**37**), which found use as a pesticide.[50]

Professor Conia launched rotanes at the University of Caen and continued this fascinating research at the Laboratoire des Carbocycles, University of Paris–South at Orsay, where researchers proudly called themselves the "carbocyclistes." And special- ists in C_3 and C_4 rings were, of course, "les petits cyclistes,"[40] whose preoccupations

37   **38**

39   **40**   **41**

included the six-pronged beauty **38**. At conferences, Dr. Conia delighted listeners by addressing this heptacycle "davidane," as it looks like Israel's Star of David (see also

p. 295). You have probably noticed that cyclopropyl units lend themselves nicely to geometric imagery. For example, in two-dimensional drawings, **38** could just as well aim its tips inward (as in **39**) or could alternate them in and out (as in **40**). However, *petits cyclistes* have perspective and realize that adjacent points would actually project to opposite sides of the frame (see **41**), just like pedals on a bicycle.

Cyclists at Göttingen also know their wares. Driven by Lutz Fitjer, these experts would recognize "davidane" as but one member of a structural brotherhood in which an *m*-sided central ring shares each border with an *n*-sided external ring. If all fusions are *cis*, *m* is necessarily an even number. In 1979 Fitjer and cohort Detlef Wehle titled such centrosymmetric species [*m, n*]coronanes, from Latin *corona* meaning "crown," or "wreath."[51] Thus **42** would be [6,4]coronane, and **43** would be [6,5]coronane. The

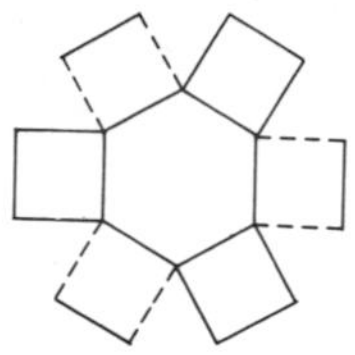

42 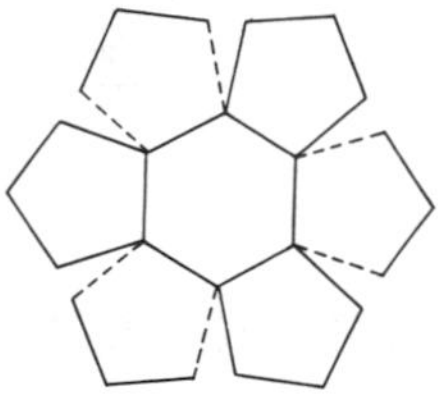   **43**

German researchers developed routes to coronanes involving sequential cationic ring expansions, cascade fashion, from rotane-like precursors.[52] In 1983, Marshall, Peterson, and Lebioda proposed that structures like **42** and **43** as well as analogs with nonidentical external rings be called "perannulanes."[53] Initially these authors did not cite any papers by Fitjer *et al.* and so presumably were unaware that baptism of this class of compounds had begun several years earlier.

If you like paddlanes and rotanes, then molecules machined at Texas A & M University and at Hebrew University School of Pharmacy may titillate you. For example, the Texas crew isolated alkali metal salts of pentaphenylcyclopentadiene and depicted them as pentapaddled structures (**44**).[54] And the Israeli team attached six

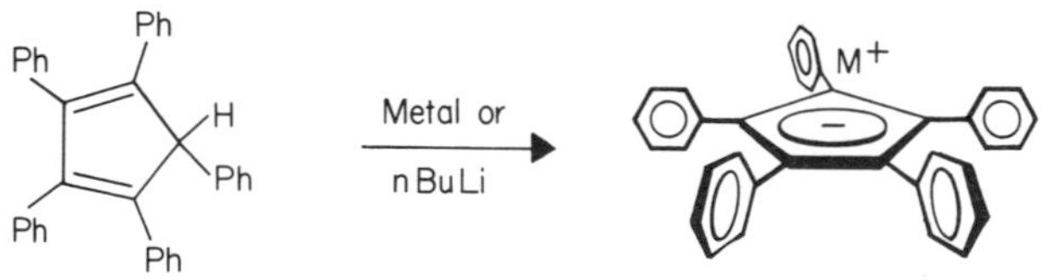

44

small paddles to benzene to create hexacyclopropylbenzene (**45**).[55] This esthetic creature can orient its flippers different ways by bond rotations; and only two conformers are shown. Such delightful structures seem to beg for picturesque names.

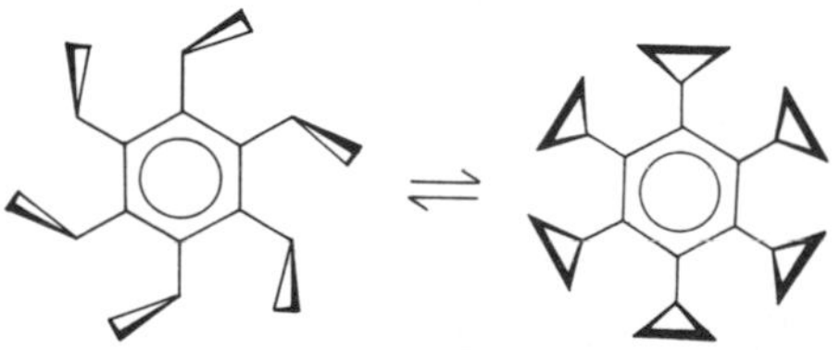

45

Perhaps the most dramatic event in modern times involving mastery of mechanics was the first landing on the moon on July 20, 1969, by American astronauts Neil

Armstrong, Edwin "Buzz" Aldrin, and Michael Collins. Of the millions who followed the progress of their Apollo 11* lunar mission, many will remember Armstrong's words, "The Eagle has landed," and "That's one small step for man, one giant leap for mankind." An unforgettable event! But what has it to do with chemical nomenclature?

Just this. As that "Eagle" landed, a Johns Hopkins research team headed by Alex Nickon was drafting a paper[56] for the *Journal of the American Chemical Society* on the structure of a sesquiterpenoid alcohol known since 1922 by the "temporary" tag, "α-caryophyllene alcohol." Japanese chemists had assigned this label to distinguish this alcohol as one of three (α, β, and γ) produced when commercial caryophyllene was

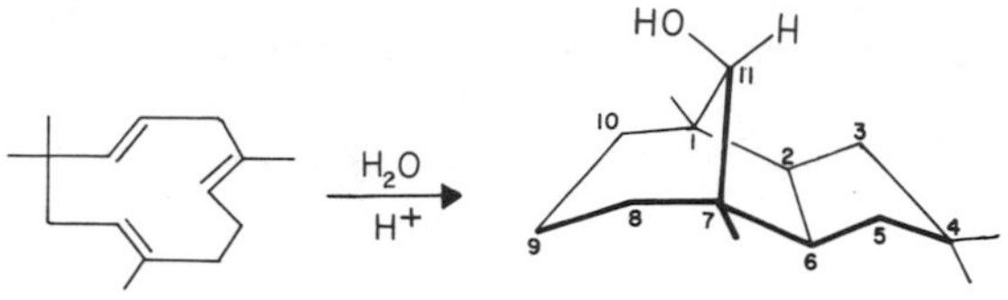

46   47

treated with aqueous acid.[57] The Hopkins group determined the "α" alcohol structure to be **47** and showed that it does not arise from caryophyllene but instead from the related sesquiterpene humulene (**46**), which often contaminates commercial caryophyllene.[56] Clearly, the old name α-caryophyllene alcohol was misleading and had to go. The molecule has appealing symmetry; and a flat drawing of the ring system (**48**) strikingly resembles a rocket, with side fins and exhaust tail. So Nickon dubbed the parent alkane "apollane" in timing with the Apollo 11 moon landing.[56] By happy coincidence, proper numbering of **47** locates the —OH at C11 and thus cemented further the onomastic link with Apollo 11. In fact, Neil Armstrong's personal memorabilia include a reprint of that chemistry publication.[58] Now, back to earth.

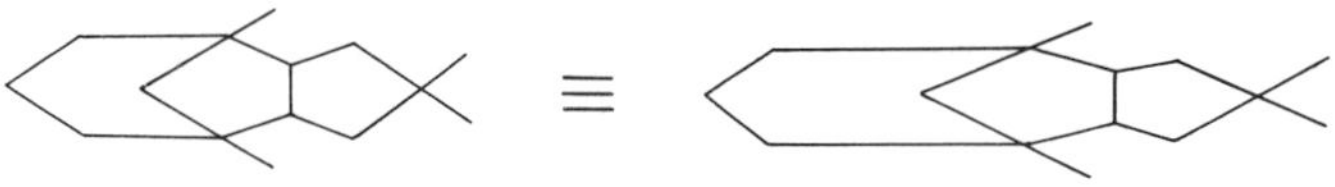

48

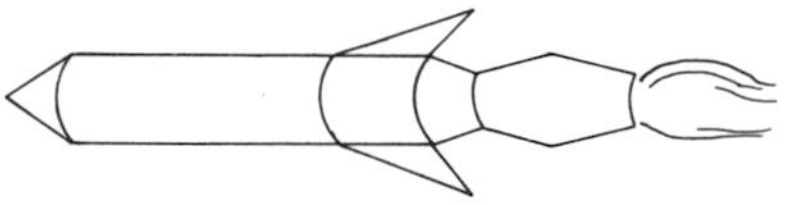

Evan Allred at the University of Utah found that irradiation of **49** ("basketene," see chapter 5) or **50** produces **51** (among other products).[59] Professor Allred and his

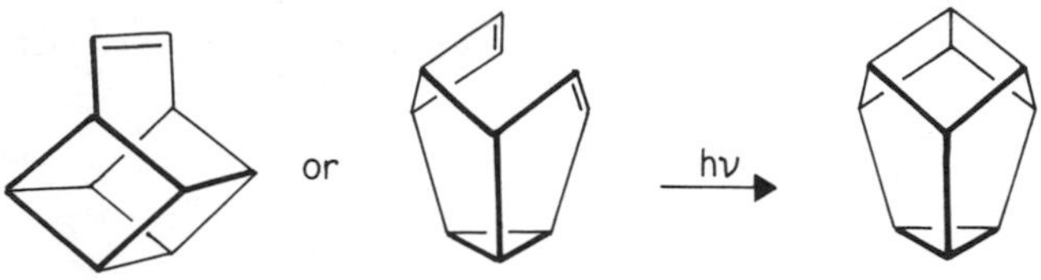

49   50   51

research squad submitted the findings for publication and referred to **51** as "lampane" because it resembles the shades often found on outdoor ornamental lamps (**52**) in the

---

*In Greek mythology, Apollo was (among other things) the god of sunlight, music, poetry, and prophesy.

United States and because a photo lamp is needed to get the compound.[60] Alas, the referees could not see the resemblance; so "lampane" was extinguished from the article and did not appear in print. Later papers by Dr. Allred didn't use the term either.[61,62]

Amusingly enough, the name "homolampane" did make the press in a published abstract of research described by Alan Marchand, Teh–Chang Chou, and Michael Barfield at the December, 1975 meeting of the American Chemical Society in Mexico City.[63] Dr. Marchand, then at the University of Oklahoma, had heard of "lampane" from Allred's colleague, David Grant; and in a weak moment Marchand titled the abstract "Homolampane—A New $C_{11}H_{12}$ Hydrocarbon."[64] The compound is **53**, a lamp with a methylene handle; so "ansalampane" would have fit as well. (See chapter 5 for *ansa*.) In organic chemistry, the prefix *homo-* frequently indicates the presence of one more carbon atom (cf. *homo*log, etc.). Mindful of lampane's ill fate at the hands of referees and editors, Marchand's team judiciously avoided "homolampane" in their

52   **53**

subsequent full publication and instead named **53**: hexacyclo[$5.4.0^{2,6}.0^{4,11}.0^{5,9}.0^{8,10}$] undecane.[65a] Less illuminating, don't you think? In any case, the term "isolampsic" (Greek *iso*, "equal"; *lampein*, "to light") is legitimate. It was coined by Henri Bouas–Laurent's team (University of Bordeaux) along with Edwin Chandross (Bell Laboratories).[65b] Isolampsic is to emission spectra what isosbestic is to absorption spectra, namely, a crossing point for equal intensities.

In 1968, Harold Kwart and his students at the University of Delaware generated the dimesitylmethyl carbocation (**54**) by placing dimesitylmethanol into a strongly acidic medium.[66] A space-filling model of **54** shows that steric interference of methyl groups prevents the rings from being coplanar. A wobbly perpendicular conformation (**55**) should result in three different ortho methyl environments. Actually, however, all four ortho methyls experience the same average surroundings because they appear as one singlet in the ^1H NMR, even at $-60°C$. Evidently, both rings can rotate so that each takes its turn being in the plane and perpendicular to the plane of the paper. A molecular model suggests this motion could avoid severe congestion only if both rings spin in unison. So, with mesityl groups, it may take two to tango! This situation reminded Dr. Kwart of meshed cogwheels that turn together; he thus dubbed the

54   **55**

phenomenon the "cogwheel effect."[66,67] If you look inside a spring-wound watch, you'll see cogwheels in action.

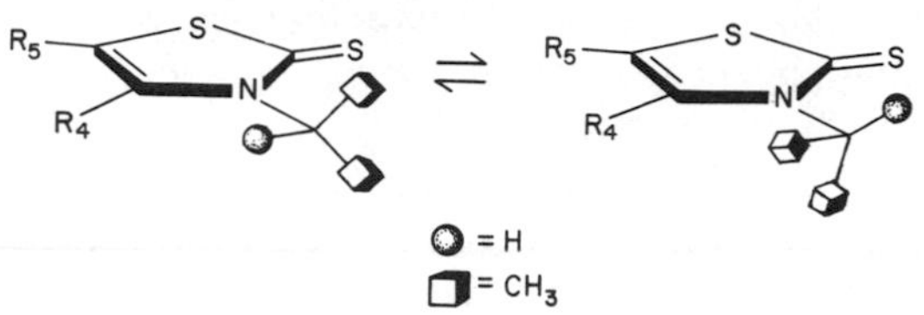

$\bigcirc = H$

$\square = CH_3$

56   **57**   **58**

In 1971, Christian Roussel, Michel Chanon, and colleagues at the University of Aix–Marseille III came upon a similar phenomenon when they examined ^{1}H NMR spectra of several 3-isopropylthiazoline-2-thione derivatives (**56**).[68] At room temperature these heterocycles showed one doublet for the methyl signal of the isopropyl group. But at $-10°C$ the NMR "camera" saw the same methyls as two doublets, which Drs. Roussel and Chanon attributed to slow interconversion of conformers **56** and **57**. The nature of R_4 and even of R_5 influenced the relative areas of these two doublets and, therefore, the proportions of the two forms.[69] Evidently, rotational motions of R_4, R_5, and the isopropyl are interlocked—somewhat like gears (**58**). The French chemists labeled this type of conformational transmission the "gear effect."[68,69]

Molecular gears can slip, but sometimes slippage is slow at room temperature. Research teams managed by Hiizu Iwamura in Japan[70] and by Kurt Mislow in the United States[71] independently found they could isolate conformational isomers of bis(triptycyl) molecules at ambient temperatures. For example, the Japanese contingent separated the *meso* (**59**) and DL (**60**) "phase" diastereomers of bis(4-chloro-1-triptycyl) ether, and the corps in America did likewise for the DL (**61**) and *meso* (**62**) forms of bis(2,3-dimethyl-9-triptycyl)methane. So the two triptycyl moieties in such

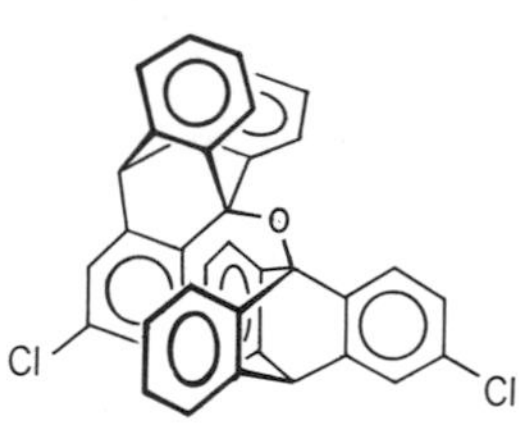

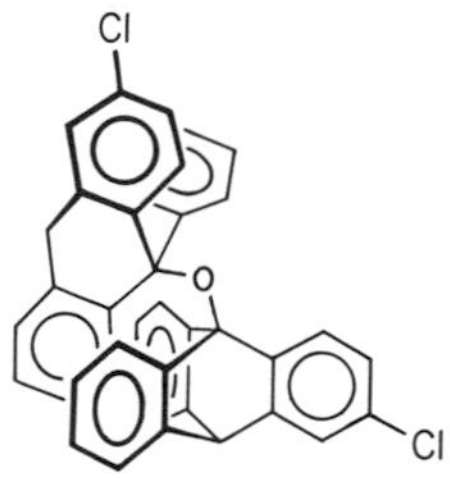

59   **60**

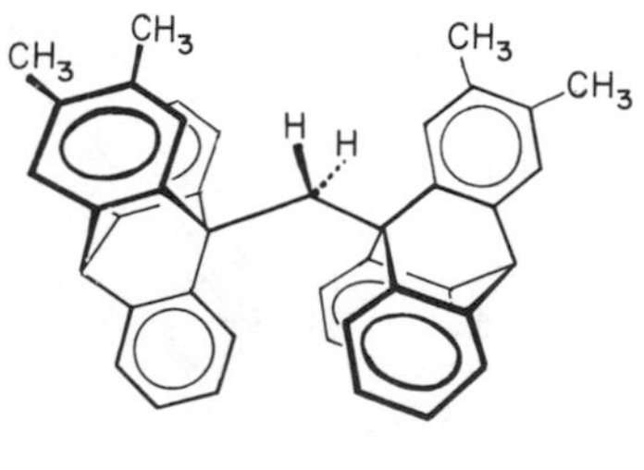

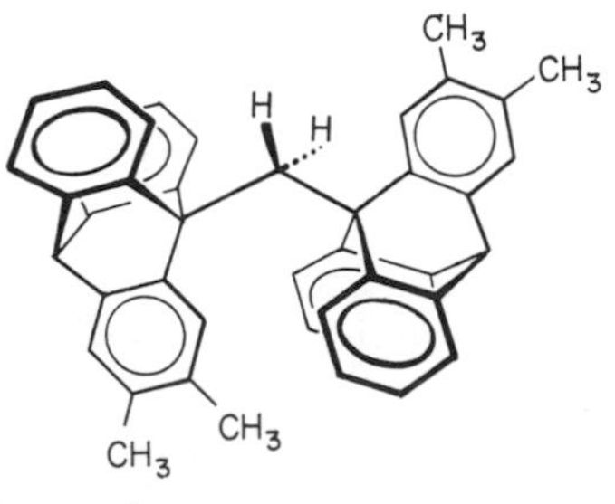

61   **62**

structures stay meshed as they twirl. This correlated motion was "advertised" in the Proceedings of the National Academy of Sciences, U.S.A.,* as a "bevel gear system."[71b] We've pictured a mechanical counterpart in **63**. Zvi Rappoport's chemical machinists at The Hebrew University in Jerusalem used Mislow's concept of "correlated rotation"[75] to describe a similar motion they observed in trimesitylethenol (**64**).[76]

63   **64**

But, correlated movement is not always the best way for molecules to get around. Consider hexaisopropylbenzene, a fairly rigid, statically geared molecule investigated by Mislow *et al.*[77] Theoretical calculations indicate that if all six isopropyls were to rotate in concert, the energy barrier would be *ca.* 35 kcal/mol. However, ^{1}H NMR coalescence studies on d-labeled material led to a homomerization barrier of about 22 kcal/mol, so the isopropyl units probably rotate one at a time. Seemingly, the grinding of gears is not so bad for molecular machinery.

As long as we are in a mechanical mood, let us glance at a conformational wiggle (**65 ⇆ 66**) that polymers undergo in solution. This change requires rotation around bonds x and y and resembles the turning of an automobile crankshaft; hence, polymer chemists call it "crankshaft motion."[78]

65   **66**

Who would have thought that a common thing like an automobile windshield wiper would become the focus of intense debate in organic chemistry? The controversy centered around carbocations produced in certain solvolyses. But first, a little background. "Classical" structures (for example, of carbocations) are those drawn with only venerable single, double, or triple bonds, which chemists denote by solid lines. A conventional single bond implies two nuclei held together by two paired electrons.

---

*Because the publication costs were defrayed in part by payment of page charges, postal regulations required that articles in that journal be labeled "advertisement."[71b] Incidentally, some scientists have announced their research as "advertisement" for other reasons. Science historian William Garvey cited the case of a respected physicist whose manuscript to *Journal of the Optical Society of America* was rejected on the grounds the work was too applied.[72] The editor recommended submission to *Journal of Applied Optics*, but the author did not accept this suggestion. Instead, he purchased space in the advertising section of *J. Opt. Soc. Am.* and published his paper as "an advertisement." Some eyebrows were raised, but the journal upheld the author's right to report results in that manner so long as he clearly marked the material "advertisement." For years, that author and others from the same institution continued to submit brief communications this way to *J. Opt. Soc. Am.*[73] These papers became cited in the literature,[74] and their roman numeral pagination was the only clue to distinguish them from conventional articles. If you are thinking of publishing in that manner, be sure what you report is accurate. Otherwise you could run afoul of the law in the matter of "truth in advertising."

"Nonclassical"* structures contain one or more "unconventional" single bonds and chemists usually depict them with dotted lines or other symbols.

A nonclassical representation for an organic cation (specifically the camphyl cation) first emanated from University College, London, in a 1939 paper by Christopher L. Wilson and fellow workers Thomas P. Nevell and Eduardo de Salas.[83] Whether Wilson's group originated the idea alone or through discussions with faculty colleague Christopher K. Ingold is not clear, but published recollections favor the former view.[84] The controversy we are about to describe split chemists into two "schools"; Herbert Brown[†] was headmaster of the "Classical School," and Saul Winstein was his counterpart in the "Nonclassical School."

The imbroglio began around 1949 when Dr. Winstein and his pupils found that exo-2-norbornyl brosylate (**67**, Bs = p-$BrC_6H_4SO_2$—) solvolyzes in acetic acid 350 times faster than does its endo isomer, **68**.[85] Furthermore, the product was, virtually completely, exo-2-norbornyl acetate (**69**).[86] When optically active **67** was used, the

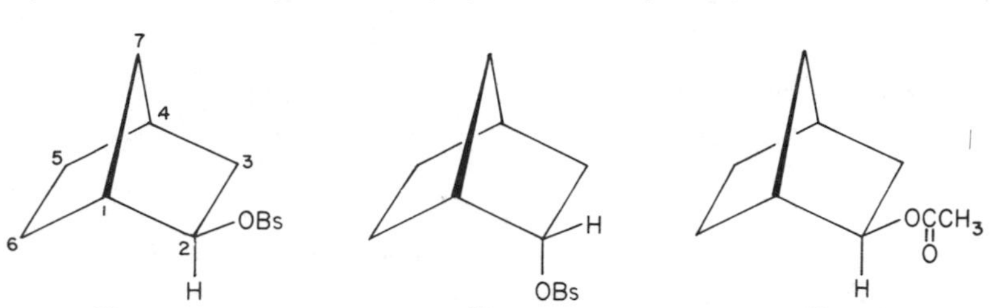

67   **68**   **69**

product was racemic.[86] (Later, more sophisticated analyses placed a lower limit of 99.98% on the amount of exo solvolysis product in aqueous acetone and an upper limit of 0.05% on its optical activity.[87])

Professor Winstein held that the greater reactivity of **67** relative to **68** could be understood if the C1—C6 bonding electrons *assist* ionization to produce a nonclassical cation (**70**) with delocalized sigma electrons and with positive charge spread largely between carbons 1 and 2.[88] This phenomenon was dubbed "anchimeric assistance" by Dr. Winstein's group[89] and "synartetic acceleration" by Sir Christopher K. Ingold

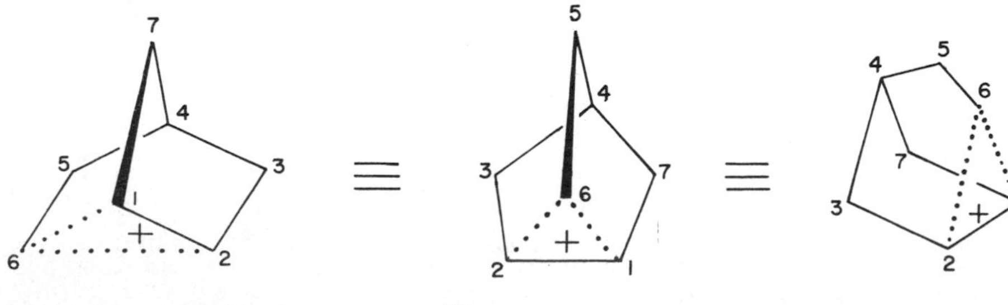

70

(chapter 18).[90] Ion **70** is symmetric, so attack by acetic acid at C1 and C2 is equally probable and accounts for racemic acetate. Dr. Winstein has summarized other

---

*In general, chemists think of an ion as "nonclassical" if its ground state has sigma electrons appreciably delocalized. John D. Roberts is credited with inventing the term in this context.[79] George Olah recommended "carbocation" as a generic title for all positive ions of carbon compounds; subclasses include "carbenium" for trivalent carbocations (such as CH_3^+) and "carbonium" for carbocations with higher coordination (such as CH_5^+).[80] These usages are sanctioned by IUPAC.[81] Jean Mathieu and André Allais coined the word "carbocation" in 1957.[82]

†Nobel Prize in chemistry, 1979, shared by Herbert C. Brown and George Wittig.

arguments for the mesomeric structure of the norbornyl cation, **70**;[88] and subsequent low-temperature NMR studies in superacid media by others seemingly backed him.[91]

Professor Herbert Brown, however, believed that these facts (and many others from his own laboratory) did not augur an unusual cation during solvolysis and could be explained just as well by rapid interconversion of classical ions **71** and **72**; his views, with supporting evidence, are summarized in several papers.[92] Look at the right-hand drawings for these ions and imagine the heavy bond wiping back and forth between C1 and C2. Do you get the picture? Someone attending a lecture by Dr. Brown at a mechanisms conference in 1962 did,[93] and the immortal words, "windshield wiper effect," "popped out of the audience floating on a burst of laughter."[94a] Several attendees and other correspondents could not recall who the popper was, so credit for this apt designation can't be given until that "someone" comes forward.[94]

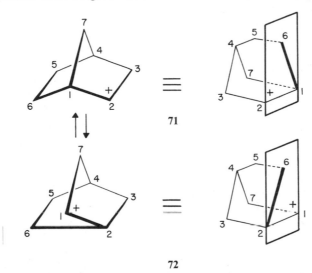

This term, so ingrained in mechanistic organic chemistry, was applied in 1978 to describe quite a different phenomenon. French chemists led by Jacques-Emile Dubois found that the methyl hydrogens in symmetrical tetra-t-butylacetone give two NMR signals at room temperature but only one at 41°C.[95] They attributed this behavior to the equilibrium **73** ⇌ **74**. At room temperature, the molecule is thought to linger in conformation **73**, with methyls a and b different. As the sample is warmed, interconversion with **74** becomes more rapid, and all methyls now see the same averaged traffic. The authors thought that **73** and **74** interconvert most simply by a wiper motion rather than by full rotations.[95] (Actually, there are *two* identical wipers in this case, one for the rear window and one for the front window—both standard equipment on the French model.)

Prostaglandins are compounds synthesized in the prostate gland and were first isolated from semen.[96] They occur elsewhere in the body, but the original name propsed by Ulf S. von Euler at Stockholm's Karolinska Institute has stuck. They arise *in vivo* from C_{20} polyunsaturated fatty acids such as arachidonic acid (**75**; 5Z, 8Z, 11Z, 14Z-eicosatetraenoic acid) and have been implicated in numerous biological processes. One of the first prostaglandins isolated in crystalline form—in 1958 by Sweden's Sune Bergström* and his collaborators—was PGE_1 (**76**). In this jargon PG stands for "prostaglandin," E means "keto-alcohol," and the subscript codes the carbon–carbon double bonds in the two chains.[†] If you draw prostaglandin PGE_1 as in **76**, it looks like a hairpin. This resemblance led Niels Andersen, Israel Rabinowitz, John Smythies, and Peter Ramwell to speak of "hairpin conformation" in research discussions at ALZA

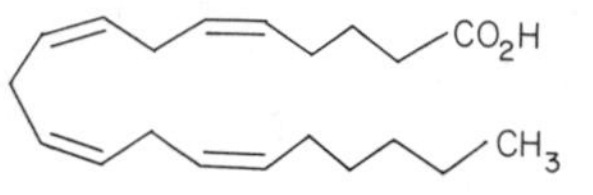

75 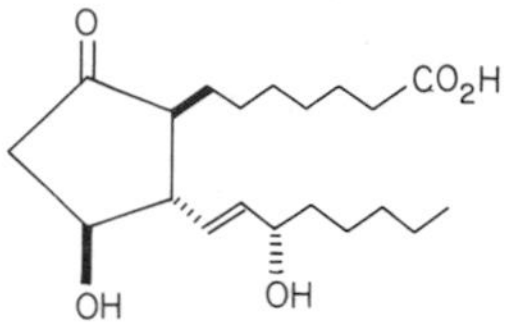   76

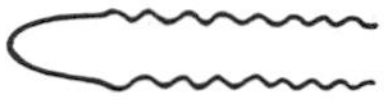

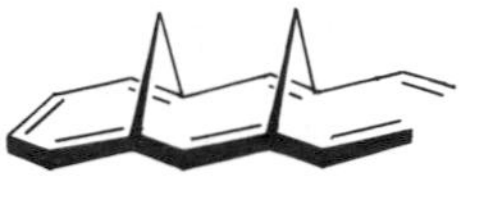

77 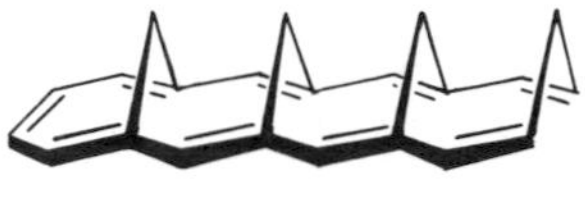   78

Corporation, Palo Alto, California, in 1970.[99] Later, X-ray[100] and circular dichroism[101] studies showed that the two chains actually diverge more than **76** depicts and also that the unsaturated one meanders away from the plane of the ring. So, the molecular hairpin is somewhat splayed and twisted[101]—not an unusual fate for the real gadget as well.

Edgar Heilbronner (University of Basel) felt that "hairpin" is a good way to describe the shape of conjugated polyenes that make a U-turn at a *cis*-olefinic link. He passed the word to researchers at the University of Cologne and at the University of Utah, who had collaboratively synthesized and investigated constrained analogs of the types **77**

---

*Nobel Prize in medicine, 1982, Shared by Sune Bergström, Bengt I. Samuelsson, and John R. Vane.

[†]Those who wince at this terminology may shudder to hear researchers in this field discussing "slow-reacting substances" (often shortened to SRS in print).[97] Such words conjure up images of lethargic molecules. But, in fact, SRS denotes a family of eicosanoids produced *in vivo* from arachidonic acid as part of the body's immunochemical response. And the odd designation "slow-reacting substances" derives from their characteristic ability to cause slow and prolonged contraction of a test strip of tissue kept under tension in a

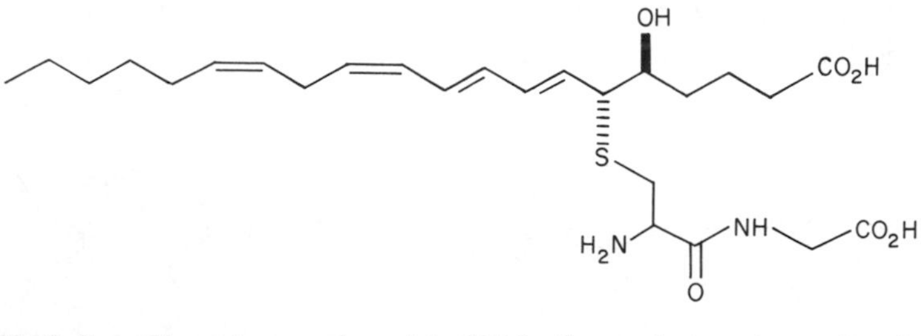

physiological bath. Several important members of the SRS family contain three, four, or five alkene units; yet all these members are referred to as leucotrienes (sometimes spelled with a k). For example, leucotriene D_4 (abbreviated LTD_4) is shown here. Letters stand for structurally different leucotrienes; and the subscript reveals the number of olefinic bonds. The isolated double bond gets slighted in the term "leucotriene" because it is not part of the easily detected (λ *ca.* 280 mm), conjugated triene chromophore.[98]

and **78**. Accordingly, the latter two teams titled a joint paper in 1983 "Excited Singlet States of 'Hairpin' Polyenes."[102]

Tweezers have a lot in common with hairpins, but not necessarily at the molecular bench. University of Wisconsin's Howard Whitlock, Jr. and C.–W. Chen designed "molecular tweezers" that deftly pluck ions like 1, 3-dihydroxy-2-naphthoate or 2, 6-dihydroxybenzoate.[103] For pincers they chose caffeine units, stuck to a diacetylene as in **79**. The rigid diyne "spacer" prevents self-association of the heterocycles and pries them suitably apart (~ 7 Å—oops, 700 pm) to lure in a flat π-body and to sandwich it tightly as a hydrophobic complex (**80 → 81**).

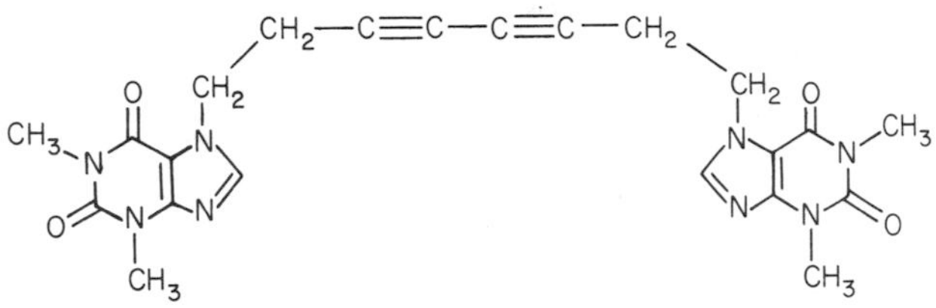

79

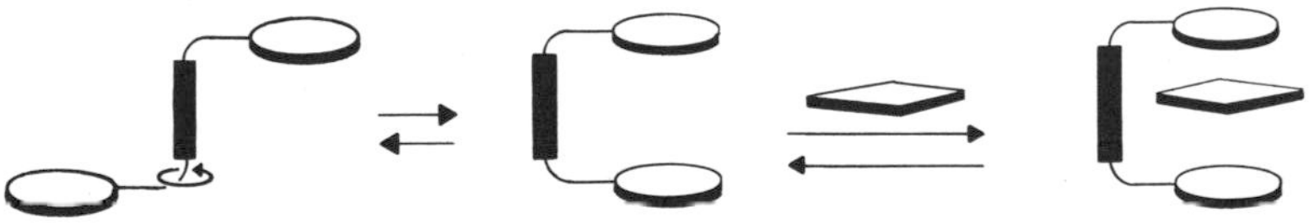

80   **81**

At Osaka University, Masahiro Irie and co-worker Masatoshi Kato learned how to control molecular tweezers with light switches. They began by demonstrating that the thioindigo derivative **82** has no noticeable talent for complexing metal ions. Then they exposed this E-alkene to 550 nm photons. This irradiation produced a Z-rich photostationary mixture that transported ions (e.g., K^+) from water to organic media (like 1, 2-dichloroethane). Evidently, in the Z isomer (**83**) the sulfurs and oxygens in one

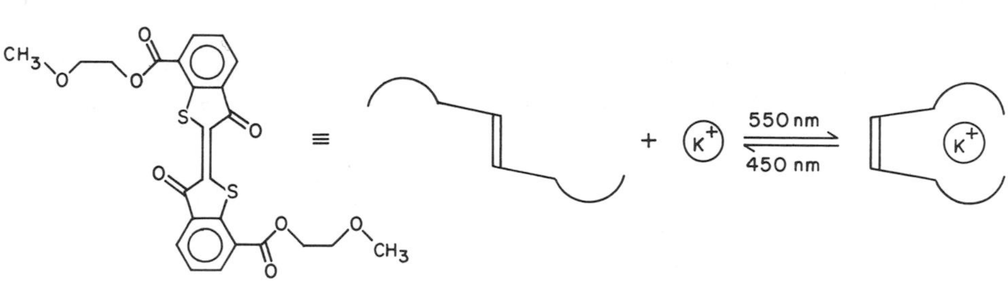

82   **83**

tweezer prong cooperate with those in the other to clasp an ion of suitable size. Switching to 450 nm light substantially reverted the Z isomer to the impotent E form. Researchers Irie and Kato proclaimed such systems "photoresponsive molecular tweezers."[104]

In 1964, Owen Mills and Gordon Robinson at England's University of Manchester used X-rays to reveal the structure of the complex $Co_2(CO)_4(C_2HtBu)_2(C_2H_2)$. Its six-

carbon chain spans the Co—Co bond much like a roadway passes over a railroad track
(see **84**), so they used "flyover" to describe this arrangement.[105] ("Flyover" is the British
equivalent of the American word "overpass.") As is often the case, the highway need not
span the railroad perpendicularly. In fact, for **84** the crossover angle is 28°.[105]

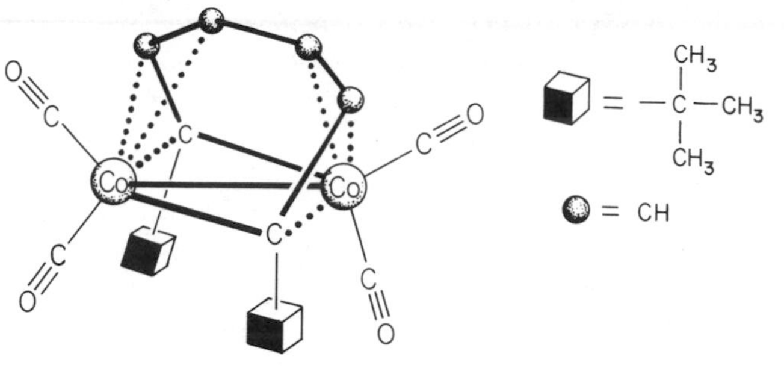

84

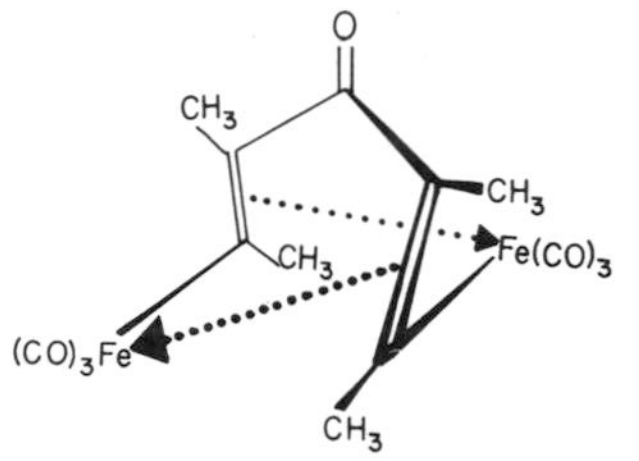

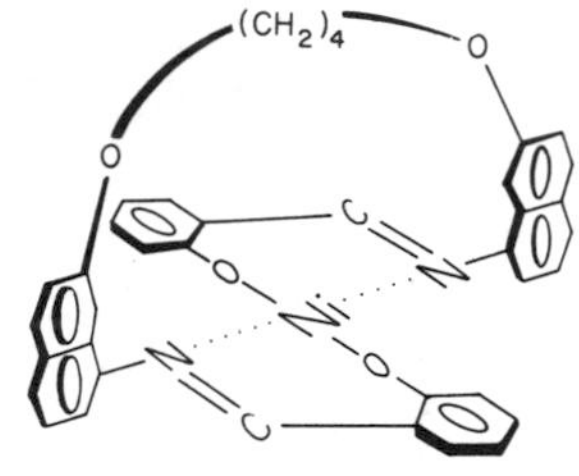

85   **86**

This apt term flew over the ocean and appealed to Roald Hoffmann* and David
Thorn at Cornell, who referred to molecules like **85**[106] as "flyover bridges."[107] In such
structures, a skeletally planar organic ligand domes a metal–metal bond.

The term also captured the attention of Australian National University chemist
Raymond Martin, who, in 1979, called the ligand in **86** a "fly-over ligand."[108] In 1981,
Charles Kraihanzel of Lehigh University, Ekk Sinn of the University of Virginia, and
Gary Gray of J.T. Baker Chemical Company used "fly-over chain" in reference to Dr.
Martin's complexes.[109]

In 1973, Melvin Goldstein (then at Cornell University) and Stanley Kline found that
low-temperature reaction of **87** with the powerful acid FSO_3H gives a most unusual

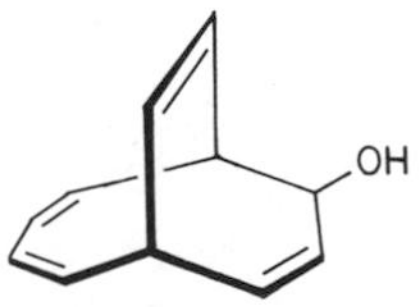

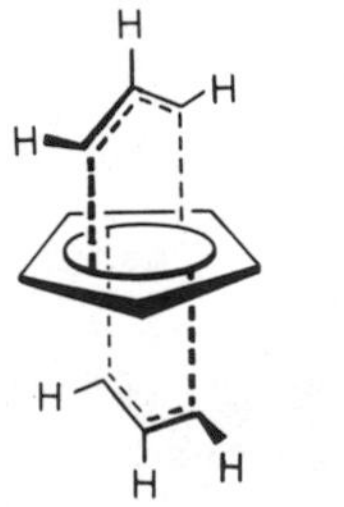

87   **88**

---

*Nobel Prize in chemistry, 1981, shared by Kenichi Fukui and Roald Hoffmann.

$C_{11}H_{11}^+$ rearranging ion.[110] Their ^1H and ^{13}C NMR data seemed to fit structure **88**—a completely organic "sandwich" (see chapter 8). It consists of two mutually perpendicular 1,3-dehydroallyl cation ligands, above and below (and orthogonal to) the plane of a cyclopentadienyl anion.[110] This 3:5:3 sandwich deserved christening, so Dr. Goldstein beseeched suggestions from colleagues and friends.[111] No one quite hit the mark, but daughter Deborah (then aged 12) came close one evening at the Goldstein dinner table. She piped up, "Why don't you name it after the Japanese haiku, a verse written in three lines with three syllables in the first, five in the second, and three in the third?" Papa Goldstein's mind flashed: haikuene! However, a tribunal of smiling Japanese co-workers in his research group shook their heads disapprovingly. A haiku has five, seven, and five! Downcast, Professor Goldstein likely had words with Deborah about her numerology. But, finally he decided that the peculiar topography of their proposed $C_{11}H_{11}$ cation bore a resemblance—however far-fetched—to an armillary sphere, used by astronomers since ancient times to depict the celestial sphere.[111] Hence, he came up with "armilenyl ion" for **88**.[110] The key parts of an armillary sphere are two rings (Latin *armilla* = "bracelet") that represent great circles such as tropics and meridians. In **88** the horizontal armilla corresponds to the cyclopentadienyl ring,

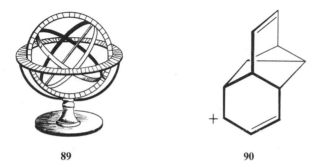

89   90

whereas the vertical one passes through the two dehydroallyl ligands. Actually, in an armillary sphere the two vertical halves form a continuous circle (see **89**); but in armilenium these sections are mutually orthogonal. Who said analogies have to be perfect? Alcohols other than **87** were also thought to furnish this "chemical universe" when treated with FSO_3H.[112,113]

But the universe was doomed. Dr. Goldstein and graduate student Eric Pressman later analyzed the cationic rearrangement of a deuterium-labeled precursor and raised doubts about structure **88** for the armilenium ion.[111,114] And, in the year of Orwell, 1984, Goldstein and pupil Joe Dinnocenzo dealt the organic sandwich its coup de grace after an incisive low-temperature ^{13}C NMR study, which favored assembly **90** for the armilenyl cation.[115] A 20-fold degenerate rearrangement of **90** leads to the same averaging process expected of the sandwich. But the morsel is never there.

REFERENCES AND NOTES

1. Printed in *Harvard Magazine*, **1983**, March/April, p. 104. Roy L. Sharpe (*ca.* 1880–1939) was a salesman for a wholesale drug firm, and he may have written only this one poem. Sharpe's own title was "A Bag of Tools." We thank his nephew, William L. Davenport, for information in correspondence, June, 1986.
2. *The World Book Encyclopedia* (Field Enterprises Educational Corp., Chicago, 1966), vol. 13, p. 11.
3. Cassar, L.; Eaton, P.E.; Halpern, J. *J. Am. Chem. Soc.* **1970**, *92*, 6366–6368.
4. Schill, G.; Beckmann, W.; Vetter, W. *Chem. Ber.* **1980**, *113*, 941–954.

5. Schill, G., private communication, August, 1977.
6. Schill, G. *Nachr. Chem. Techn.* **1967**, *15*, 149.
7. Harrison, I.T.; Harrison, S. *J. Am. Chem. Soc.* **1967**, *89*, 5723–5724.
8. Harrison, I.T., private communication, September, 1975.
9. Bloomfield, J.J.; Irelan, J.R.S. *Tetrahedron Lett.* **1966**, 2971–2973.
10. Bloomfield, J.J., private communication, March, 1984.
11. Kermack, W.O.; Robinson, R. *J. Chem. Soc.* **1922**, *121*, 427. For historical accounts, see Saltzman, M.D. *Chem. Brit.* **1986**, 543–548; *Natural Product Rep.* **1987**, *4*, 53–60.
12. (a) Altman, J.; Babad, E.; Itzchaki, J.; Ginsburg, D. *Tetrahedron* Suppl. 8, Part I, **1966**, 279–304, (b) Ginsburg, D., private communication, January, 1977.
13. Zalkow, L.H.; Harris, R.N. III; Van Derveer, D. *J. Chem. Soc., Chem. Commun.* **1978**, 420–421.
14. Zalkow, L.H., private communication, January, 1977.
15. Smith, A.B. III; Jerris, P.J. *J. Org. Chem.* **1982**, *47*, 1845–1855.
16. (a) Karpf, M.; Dreiding, A.S. *Tetrahedron Lett.* **1980**, 4569–4570, (b) Schostarez, H.; Paquette, L.A. *J. Am. Chem. Soc.* **1981**, *103*, 722–724, (c) Oppolzer, W.; Marazza, F. *Helv. Chim. Acta* **1981**, *64*, 1575–1578, (d) Oppolzer, W.; Bättig, K. *Helv. Chim. Acta* **1981**, *64*, 2489–2491, (e) Wender, P.A.; Dreyer, G.B. *J. Am. Chem. Soc.* **1982**, *104*, 5805–5807, (f) Wrobel, J.; Takahashi, K.; Konkan, V.; Lannoye, G.; Cook, J.M.; Bertz, S.H. *J. Org. Chem.* **1983**, *48*, 139–141, (g) Tobe, Y.; Yamashita, S.; Yamashita, T.; Kakiuchi, K.; Odaira, Y. *J. Chem. Soc., Chem. Commun.*, **1984**, 1259–1260, (h) Wilkening, D.; Mundy, B.P. *Tetrahedron Lett.* **1984**, *25*, 4619–4622, (i) Mehta, G.; Subrahmanyam, D. *J. Chem. Soc., Chem. Commun.* **1985**, 768–769.
17. Walker, F.H.; Wiberg, K.B.; Michl, J. *J. Am. Chem. Soc.* **1982**, *104*, 2056–2057.
18. Wiberg, K.B.; Walker, F.H. *J. Am. Chem. Soc.* **1982**, *104*, 5239–5240.
19. *Chem. Eng. News*, **1982**, Oct. 25, p. 25.
20. Mlinarić–Majerski, K.; Majerski, Z. *J. Am. Chem. Soc.* **1983**, *105*, 7389–7395, (b) Jackson, J.E.; Allen, L.C. *J. Am. Chem. Soc.* **1984**, *106*, 591–599, (c) Lee, I.; Yang, K.; Kim, H.S. *Tetrahedron* **1985**, *41*, 5007–5010.
21. Greenhouse, R.; Borden, W.T.; Hirotsu, K.; Clardy, J. *J. Am. Chem. Soc.* **1977**, *99*, 1664–1666.
22. Greenberg, A.; Liebman, J.F. *Strained Organic Molecules* (Academic Press, New York, 1978), p. 365.
23. Kimura, K.; Ohno, H.; Mirokawa, K.; Tobe, Y.; Odaira, Y. *J. Chem. Soc., Chem. Commun.* **1982**, 82–83.
24. Mislow, K.; Gust, D.; Finocchiaro, P.; Boettcher, R.J. *Top. Curr. Chem.* **1974**, *47*, 1–28.
25. Thulin, B.; Wennerström, O. *Tetrahedron Lett.* **1977**, 929–930.
26. Thulin, B. and Wennerström, O., private communication, August, 1977.
27. Nakanishi, K.; Kubo, I. *Isr. J. Chem.* **1977**, *16*, 28–31.
28. Newman, M.S.; Lutz, W.B.; Lednicer, D. *J. Am. Chem. Soc.* **1955**, *77*, 3420–3421.
29. Newman, M.S., private communication, May, 1983.
30. Newman, M.S.; Lednicer, D. *J. Am. Chem. Soc.* **1956**, *78*, 4765–4770.
31. (a) Newman, M.S. *Record Chem. Progr.* (*Kresge–Hooker Sci. Lib.*) **1952**, *13*, 111–116, (b) Newman, M.S. *J. Chem. Educ.* **1955**, *32*, 344–347.
32. Haworth, W.N. *The Constitution of Sugars* (Edward Arnold & Co., London, 1929), p. 90.
33. Kobelt, M.; Barman, P.; Prelog, V.; Ruzicka, L. *Helv. Chim. Acta* **1949**, *32*, 256–265.
34. Ginsburg, D., private communication, March, 1984. See also reference 37.
35. Eaton, P.E.; Leipzig, B.D. *J. Am. Chem. Soc.* **1983**, *105*, 1656–1658.
36. Gund, P.; Gund, T.M. *J. Am. Chem. Soc.* **1981**, *103*, 4458–4465.
37. Hahn, E.H.; Bohm, H.; Ginsburg, D. *Tetrahedron Lett.* **1973**, 507–510.
38. Juriew, J.; Skorochodowa, T.; Merkuschew, J.; Winter, W.; Meier, H. *Angew. Chem. Int. Ed. Engl.* **1981**, *20*, 269–270.
39. (a) Ripoll, J.L.; Conia, J.M. *Tetrahedron Lett.* **1969**, 979–980, (b) Ripoll, J.L.; Limasset, J.C.; Conia, J.M. *Tetrahedron*, **1971**, *27*, 2431–2452.
40. Conia, J.M., private communication, July, 1977.
41. (a) Conia, J.M.; Denis, J.M. *Tetrahedron Lett.* **1969**, 3545–3546, (b) Denis, J.M.; Le Perchec, P.; Conia, J.M. *Tetrahedron*, **1977**, *33*, 399–408.
42. Le Perchec, P.; Conia, J.M. *Tetrahedron Lett.* **1970**, 1587–1588.
43. Fitjer. L.; Conia, J.M. *Angew. Chem. Int. Ed. Engl.* **1973**, *12*, 761–762.
44. Fitjer, L. *Angew. Chem. Int. Ed. Engl.* **1976**, *15*, 762–763.

45. Fitjer, L. *Angew. Chem. Int. Ed. Engl.* **1976**, *15*, 763–764.
46. Proksch, E.; de Meijere, A. *Tetrahedron Lett.* **1976**, 4851–4854.
47. (a) Pascard C.; Prangé, T.; de Meijere, A.; Weber, W.; Barnier, J.P.; Conia, J.M. *J. Chem. Soc. Chem. Commun.* **1979**, 425–426, (b) Prangé T.; Pascard, G.; de Meijere, A.; Behrens, U.; Barnier, J.–P.; Conia, J.M. *Nouv. J. Chim.* **1980**, *4*, 321–327.
48. (a) West, R. *Abstracts of Papers, 184th National Meeting of the American Chemical Society, Kansas City, MO* (American Chemical Society, Washington, DC, 1982), INORG 78. A news release appeared in *Chem. Eng. News*, **1982**, Sept. 27, pp. 23–24. (b) Carlson, C.W.; West, R.; Zhang, X.–H. *Organometallics*, **1983**, *2*, 453–454. (c) Carlson, C.W.; Haller, K.J.; Zhang, X.–H.; West, R. *J. Am. Chem. Soc.* **1984**, *106*, 5521–5531.
49. West, R. private communication, October, 1982.
50. Crombie, L.; Holden, I.; Kilbee, G.W.; Whiting, D.A. *J. Chem. Soc. Perkin Trans 1* **1982**, 789–797.
51. Fitjer, L.; Wehle, D. *Angew. Chem.Int. Ed. Engl.* **1979**, *18*, 868–869.
52. (a) Fitjer, L.; Giersig, M.; Clegg, W.; Schormann, N.; Sheldrick, G.M. *Tetrahedron Lett.* **1983**, *24*, 5351–5354, (b) Wehle, D.; Fitjer, L. *Angew. Chem. Int. Ed. Engl.* **1987**, *26*, 130–132.
53. (a) Marshall, J.A.; Peterson, J.C.; Lebioda, L. *J. Am. Chem. Soc.* **1983**, *105*, 6515–6516, (b) *Ibid.*, **1984**, *106*, 6006–6015.
54. Zhang, R.; Tsutsui, M.; Bergbreiter, D.E. *J. Organometal. Chem.* **1982**, *229*, 109–112.
55. Usieli, V.; Victor, R.; Shalom, S. *Tetrahedron Lett.* **1976**, 2705–2706.
56. Nickon, A.; Iwadare, T.; McGuire, F.J.; Mahajan, J.R.; Narang, S.A.; Umezawa, B. *J. Am. Chem. Soc.* **1970**, *92*, 1688–1696.
57. Asahina, Y.; Tsukamoto, T. *J. Pharm. Soc. Jpn.* **1922**, *484*, 463–473; *ibid.*, **1929**, *491*, 1202.
58. Armstrong, N.A., private communication, July, 1979.
59. Allred, E.L.; Beck, B.R. *J. Am. Chem. Soc.* **1973**, *95*, 2393–2394.
60. Allred, E.L., private communication, January, 1976.
61. Allred, E.L.; Beck, B.R. *Tetrahedron Lett.* **1974**, 437–440.
62. Allred, E.L.; Beck, B.R.; Mumford, N.A. *J. Am. Chem. Soc.* **1977**, *99*, 2694–2700.
63. Marchand, A.P.; Chou, T.; Barfield, M. *Abstracts, 1st Chemical Congress of the North American Continent, Mexico City* (Port City Press, Baltimore, MD, *1975*), Organic Chemistry Division, paper 67.
64. Marchand, A.P., private communication, January, 1976.
65. (a) Marchand, A.P.; Chou, T.–C.; Ekstrand, J.D.; van der Helm, D. *J. Org. Chem.* **1976**, *41*, 1438–1444. (b) Bouas–Laurent, H.; Lapouyade, R.; Castellan, A.; Nourmamode, A.; Chandross, E.A. *Z. Phys. Chem. (Wiesbaden)* **1976**, 101 (39–44).
66. Kwart, H.; Alekman, S. *J. Am. Chem. Soc.* **1968**, *90*, 4482–4483.
67. Kwart, H., private communication, August, 1977.
68. (a) Roussel, C.; Chanon, M.; Metzger, J. *Tetrahedron Lett.* **1971**, 1861–1864. (b) Berg, U.; Liljefors, T.; Roussel, C.; Sandström, J. *Acc. Chem. Res.* **1985**, *18*, 80–86.
69. Roussel, C., private communications, September, 1977 and March, 1984.
70. (a) Kawada, Y.; Iwamura, H. *J. Am. Chem. Soc.* **1981**, *103*, 958–960; *ibid.*, **1983**, *105*, 1449–1459, (b) Iwamura, H. *J. Mol. Struct.* **1985**, *126*, 401–412.
71. (a) Cozzi, F.; Guenzi, A.; Johnson, C.A.; Mislow, K.; Hounshell, W.D.; Blount, J.F. *J. Am. Chem. Soc.* **1981**, *103*, 957–958, (b) Hounshell, W.D.; Johnson, C.A.; Guenzi, F.; Cozzi, F.; Mislow, K. *Proc. Natl. Acad. Sci. USA* **1980**, 77, 6961–6964, (c) Guenzi, A.; Johnson, C.A.; Cozzi, F.; Mislow, K. *J. Am. Chem. Soc.* **1983**, *105*, 1438–1448.
72. Garvey, W.D. *J. Res. Commun. Stud.* **1981**, *3*, 257–271.
73. For some advertisements during 1965 and references to earlier instances see (a) Mertz, L. *J. Op. Soc. Am.* **1965**, *55*, Jan., p. viii, (b) Mertz, L. *J. Op. Soc. Am.* **1965**, *55*, Feb., p. iv, (c) Young, N. *J. Op. Soc. Am.* **1965**, *55*, Mar., p. vi, (d) Whalen, R.L. *J. Op. Soc. Am.* **1965**, *55*, April, p. ix, (e) Mertz, L.; Curbelo, R. *J. Op. Soc. Am.* **1965**, *55*, May, p. iii, (f) Mertz, L. *J. Op. Soc. Am.* **1965**, *55*, June, p. v, (g) Persky, M.J. *J. Op. Soc. Am.* **1965**, *55*, Aug., p. xi, (h) Wyntjes, G. *J. Op. Soc. Am.* **1965**, *55*, Nov., p. vi, (i) Coleman, I. *J. Op. Soc. Am.* **1965**, *55*, Dec., p. viii.
74. Kozma, A.; Massey, N. *Appl. Opt.* **1969**, *8*, 393–397.
75. Mislow, K. *Acc. Chem. Res.* **1976**, *9*, 26–33.
76. Biali, S.E.; Rappoport, Z. *J. Am. Chem. Soc.* **1982**, *104*, 7350–7351.
77. Siegel, J.; Gutiérrez, A.; Schweizer, W.B.; Ermer, O.; Mislow, K. *J. Am. Chem. Soc.* **1986**, *108*, 1569–1575.
78. Pugh, D.; Pethrick, R.A. *Chem. Brit.* **1981**, *17*, 70–75.

79. Baum, R. *Chem. Eng. News*, **1983**, May 23, pp. 32–33.
80. Olah, G. *J. Am. Chem. Soc.* **1972**, *94*, 808–820.
81. Gold, V. *Pure Appl. Chem.* **1983**, *55*, 1281–1371.
82. (a) Mathieu, J., private communication, March, 1984, (b) Mathieu, J.; Allais, A. *Principes de Synthèse Organique* (Masson, and Co., Paris, 1957).
83. Nevell, T.P.; de Salas, E.; Wilson, C.L. *J. Chem. Soc.* **1939**, 1188–1199.
84. (a) Nevell, T.P. *Chem. Brit.* **1982**, *18*, 859, (b) Akeroyd, M. *Chem. Brit.* **1982**, *18*, 481–482, (c) Saltzman, M.D.; with the aid of Wilson, C.L. *J. Chem. Educ.* **1980**, *57*, 289–290, (d) Wilson, C.L. *Chem. Brit.* **1983**, *19*, 292, (e) Akeroyd, M. *Chem. Brit.* **1983**, *19*, 292, (f) Barber, N.B. *Chem. Brit.* **1983**, *19*, 644, (g) Interestingly, in 1920 Robert Robinson described Wagner-Meerwein rearrangements using delocalized structures somewhat akin to that proposed by Wilson in 1939. See Saltzman, M.D. *Natural Products Rep.* **1987**, *4*, 53–60 for a historical account and references.
85. Winstein, S.; Trifan, D.S. *J. Am. Chem. Soc.* **1949**, *71*, 2953.
86. Winstein, S.; Trifan, D.S. *J. Am. Chem. Soc.* **1952**, *74*, 1154–1160.
87. Winstein, S.; Clippinger, E.; Howe, R.; Vogelfanger, E. *J. Am. Chem. Soc.* **1965**, *87*, 376–377.
88. Winstein, S. *J. Am. Chem. Soc.* **1965**, *87*, 381–382.
89. Winstein, S.; Lindegren, C.R.; Marshall, H.; Ingraham, L.L. *J. Am. Chem. Soc.* **1953**, *75*, 147–155.
90. Brown, F.; Hughes, E.D.; Ingold, C.K.; Smith, J.F. *Nature (London)* **1951**, *168*, 65–67.
91. (a) Olah, G.A.; Prakash, G.K.S.; Arvanaghi, M.; Anet, F.A.L. *J. Chem. Soc.* **1982**, *104*, 7105–7108, (b) Yannoni, C.S.; Macho, V.; Myhre, P.C. *J. Am. Chem. Soc.* **1982**, *104*, 7380–7381, (c) For a news article about the NMR evidence in refs. (a) and (b), see *Chem. Eng. News* **1983**, April 4, pp. 21–23; for follow-up letters from G.A. Olah and from H.C. Brown see *Chem. Eng. News* **1983**, May 23, pp. 2, 38, 39, (d) Saunders, M.; Kates, M.R. *J. Am. Chem. Soc.* **1983**, *105*, 3571–3573.
92. Brown, H.C.; Chloupek, F.J., Rei, M. *J. Am. Chem. Soc.* **1964**, *86*, 1246–1247; *ibid.*, **1964**, *86*, 1247–1248; *ibid.*, **1964**, 1248–1250.
93. Brown, H.C. Presented at the Reaction Mechanisms Conference, Brookhaven National Laboratory, September 5, 1962 (unpublished).
94. (a) Berson, J.A., private communication, November, 1975, (b) Brown, H.C., private communications, August, 1977 and July, 1982, (c) Roberts J.D., private communications, June, 1977 and August, 1982, (d) Bartlett, P.D., private communication, April, 1983; Olah, G.A., private communication, May, 1983.
95. Dubois, J.–E.; Doucet, J.; Tiffon, B. *Tetrahedron Lett.* **1978**, 3839–3840.
96. (a) Bergström, S. Prostaglandins from Bedside Observation to a Family of Drugs, chapter III in *Proceedings of the Robert A. Welch Foundation Conferences on Chemical Research XXIV. The Synthesis, Structure, and Function of Biochemical Molecules* (The Robert A. Welch Foundation, Houston, TX, 1980), (b) Bergström, S. *Angew. Chem. Int. Ed. Engl.* **1983**, *22*, 858–866.
97. Corey, E.J. Slow Reacting Substances, chapter IV in reference 96a.
98. For reviews on leucotrienes see (a) Green, R.H.; Lambeth, P.F. *Tetrahedron*, **1983**, *39*, 1687–1721, (b) Samuelsson, B. *Angew. Chem. Int.Ed. Engl.* **1982**, *21*, 902–910, (c) Samuelsson, B. *Angew. Chem. Int. Ed. Engl.* **1983**, *22*, 805–815.
99. Andersen, N.H., private communication, December, 1975.
100. Rabinowitz, I.; Ramwell, P.; Davison, P. *Nature (London) New Biol.* **1971**, *233*, 88–90.
101. Leovey, E.M.K.; Andersen, N.H. *J. Am. Chem. Soc.* **1975**, *97*, 4148–4150.
102. Frölich, W.; Dewey, H.J.; Deger, H.; Dick, B.; Klingensmith, K.A.; Püttmann, W.; Vogel, E.; Hohlneicher, G.; Michl, J. *J. Am. Chem. Soc.* **1983**, *105*, 6211–6220.
103. Chen, C.–W.; Whitlock, H.W., Jr. *J. Am. Chem. Soc.* **1978**, *100*, 4921–4922.
104. Irie, M.; Kato, M. *J. Am. Chem. Soc.* **1985**, *107*, 1024–1028.
105. Mills, O.S.; Robinson, G. *Proc. Chem. Soc.* **1964**, 187.
106. Piron, J.; Piret, P.; Meunier-Piret, J.; Van Meerssche, M. *Bull. Soc. Chim. Belg.* **1969**, *78*, 121–130.
107. Thorn, D.L.; Hoffmann, R. *Inorg. Chem.* **1978**, *17*, 126–140.
108. Baker, A.T.; Martin, R.L.; Taylor, D. *J. Chem. Soc., Dalton Trans.* **1979**, 1503–1511.
109. Kraihanzel, C.S.; Sinn, E.; Gray, G.M. *J. Am. Chem. Soc.* **1981**, *103*, 960–962.
110. Goldstein, M.J.; Kline, S.A. *J. Am. Chem. Soc.* **1973**, *95*, 935–936.
111. Goldstein, M.J., private communication, April, 1981.

112. Goldstein, M.J.; Tomoda, S.; Warren, D.P. *Abstracts, 26th Congress of the International Union of Pure and Applied Chemistry, Tokyo, Japan, September, 1977*, p. 107.

113. Fujise, Y.; Yashima, H.; Sato, T.; Ito, S. *Tetrahedron Lett.* **1981**, *22*, 1407–1408.

114. (a) Goldstein, M.J.; Tomoda, S.; Pressman, E.J.; Dodd, J.A. *J. Am. Chem. Soc.* **1981**, *103*, 6530–6532, (b) Goldstein, M.J.; Pressman, E.J. *J. Am. Chem. Soc.* **1981**, *103*, 6533–6534.

115. (a) Goldstein, M.J.; Dinnocenzo, J.P. *Abstracts of Papers, 187th National Meeting of the American Chemical Society, St. Louis, MO* (American Chemical Society, Washington, DC, 1984), ORGN 198, (b) Goldstein, M.J.; Dinnocenzo, J.P. *J. Am. Chem. Soc.* **1984**, *106*, 2473–2475.

Chapter 4

EDIFICE COMPLEXES

Chemists spend a good deal of time inside buildings, so it is natural for them to relate the shapes of molecules to these ubiquitous structures.

The term "housane" was coined by Paul Schleyer (not by Georgia Institute of Technology Professor Herbert House!); it has been used for both of the domicile-like molecules, **1**[1] and **2**.[2] (Actually, for the latter framework, the publication gives

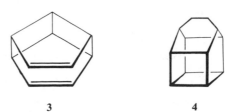

1   **2**

"housone" for the compound with a keto group at one of the lower carbons.) Structure **1** has also been called "pentaprismane";[3,4] and the flatter drawing emphasizes the *prism* containing two regular *penta*gons in parallel planes. This latter perspective resembles the famous Pentagon building, headquarters for United States military operations. Regardless of name, the molecule is loaded with strain;[4] attempts by Rowland Pettit's team to prepare it by intramolecular [2 + 2] cycloaddition of hypostrophene (**3**, p. 298) failed.[5] In 1981, however, Philip Eaton's crew at the University of Chicago erected **1** by a different method.[6] And how about the chic architectural style of dwelling **4**? To cut heating costs it features a large roof extension for solar panels. This compound was

3   **4**

constructed more recently[7] in Switzerland, and we can regard it as a contemporary housane.

Putting a steeple on housane converts it to "churchane," a term used reverently by Gerald Kent of Rider College, New Jersey to describe **5**.[8] However, others in the field think of **5** as "homopentaprismane";[9-13] perhaps these latter chemists have government grants and want to maintain strict separation of church and state. Several research groups simultaneously strived to synthesize **5** (or its precursor, homohypostrophene, **6**), and the race ended successfully in a three-way dead heat.[11-13] Irradiation of **6** *did* produce churchane,[12,13] whereas the analogous approach to housane failed.[4,5] Score one for the Reverend Billy Graham, who might have quoted Psalms 127:1, to wit: "Except the Lord build the house, they labor in vain that build it."

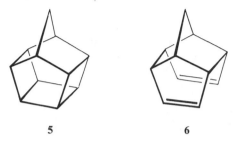

5   6

Philip Eaton's choice of "peristylane" for **7** cleverly connotes its structure and the use he envisaged for this compound. The name comes from *peristelon*, Greek for "a group of columns arranged about an open space and designed to support a roof."[14] The "roof" peristylane was meant to hold up is a five-membered ring which, when mentally placed atop **7**, would provide dodecahedrane (p. xi). Dr. Eaton's team first built a

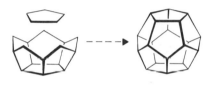

7

diketoacetoxyperistylane[14] and subsequently peristylane itself.[15] Both syntheses are true *tours de force* and required many steps.

England's Peter Garratt, of University College, London, extended the peristylane nomenclature.[16] He referred to **7** as [5]peristylane because its base-ring has 5 atoms connected to alternate carbons of a 10-membered zigzag ring above. An [n]peristylane has an n-membered base joined by n bonds to alternate atoms of a 2n-membered cycle. See triaxane (p. 295) for another example.

A birdcage may not strike you as a building, unless you happen to be a bird. So, for bird-loving chemists we include the "birdcage hydrocarbon" (**8**) as a home for winged creatures.[17] (Our feathered friend obligingly pops inside if you move drawing **9** up to your nose.)* Saul Winstein noticed the resemblance between **8** and a birdcage.[17] His researchers at the University of California, Los Angeles,[19] and a team from the Shell Development Company[20] simultaneously reported its synthesis. The UCLA workers had exploited polycyclic molecules such as **10** to study nonclassical cations and the windshield wiper effect (chapter 3). They obtained their birdcage by heating a solution of **10**. The right-hand portion of **10** contains a potential windshield wiper, and because wipers often remove aviary droppings, the onomastic tie-in is practical as well as chemical. In a suggested mechanism, **10** produces the nonclassical ion **11** (à la

---

*But if it doesn't, have your inorganic colleagues entice it with their fascinating array of polyhedral cages, termed *closo-* (closed); *nido-* (from Latin, meaning "nest"); *hypho-* (from Greek, meaning "web"); and *arachno-* (from Greek, meaning "net").[18] One way or another, chemists can offer your pet a good home.

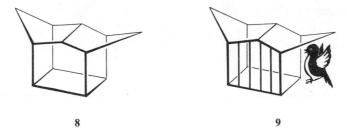

8   9

Winstein), for which **12** is an "extreme classical form."[19] Addition of the positive carbon to the double bond, followed by 1, 4-proton loss, gives the (upside-down) birdcage (**8**).

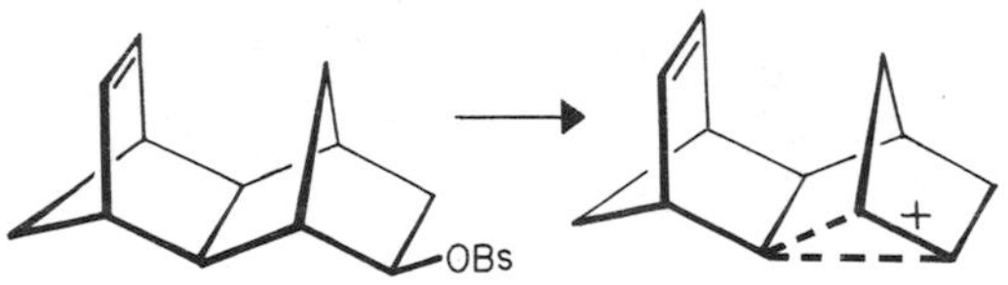

10   11

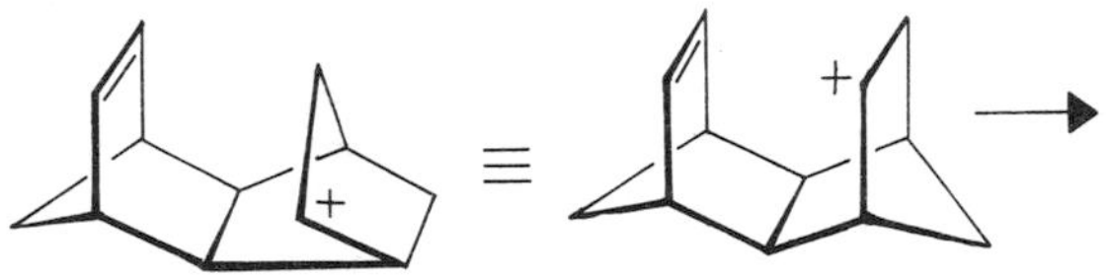

12

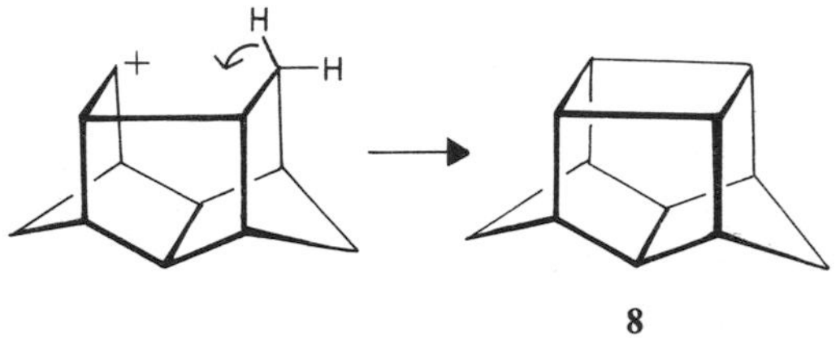

8

The Shell team obtained hexachlorobirdcage **14** from isodrin (**13**) by treatment with bromine or hydrogen bromide.[20] To synthesize the unadorned hydrocarbon, they first removed the chlorines in isodrin (**13** → **15** → **16**), then added iodine in a 1, 4-manner (**16** → **17**), and stripped the iodines with zinc. Now even undergraduates can build the birdcage hydrocarbon by a simplified route recommended by Dao Cong Dong.[21] It involves photochemical ring closure of isodrin (**13**) to **14** and then dechlorination with zinc. Cookson and Crundwell were the first to report this photic conversion, and they even called **14** "photodrin."[22] Isodrin is an insecticide,[23] and we presume that cages built from it are termite-proof.

But how do such insecticides function at the molecular level? Nicholas Turro of Columbia University provided keen insight at a conference at Bürgenstock, Switzerland, in May 1978. His lecture stressed the impact photochemistry can have on humanity. By way of illustration, he described how knowledge of photochemical action can lead to a humane—yet ecologically safe—insecticide. Specifically, in isodrin (**18**) an insect enters the cavity and spots a secluded resting place on bond a—b. Landing there perturbs the olefin links, which promptly undergo a [2 + 2] sunlight-induced closure that traps the intruder (**19**). But, how is this humane? Will the creature not

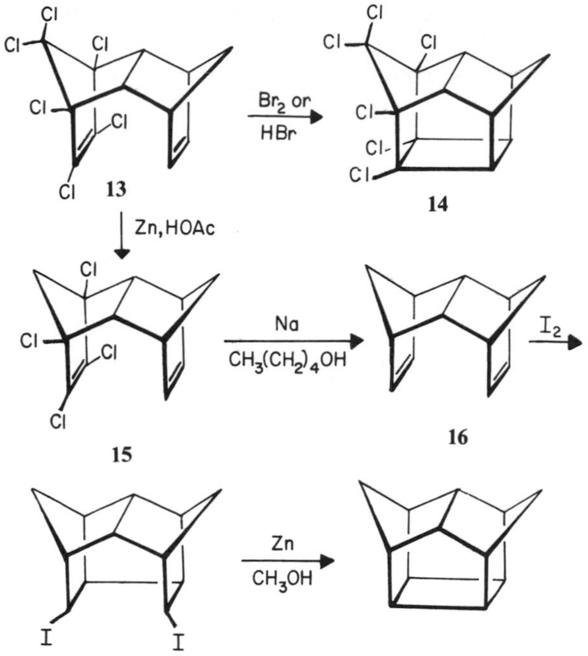

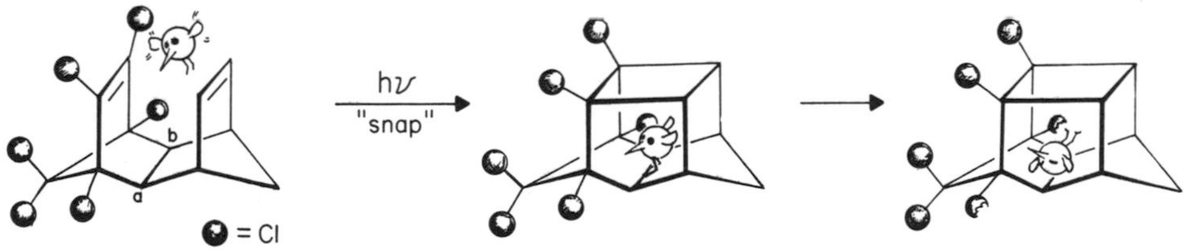

starve slowly? Our artful photochemist revealed a compassionate solution. The boxed bug feeds leisurely on the nearby chlorines until it falls into a peaceful slumber and expires (**20**). Professor Turro supports this dietary phase by noting that the chlorine content of an insecticide diminishes after use. By the way, he found fluorines less effective, although they did diminish tooth decay among insects. A translation of Dr. Turro's delightful joke lecture appeared in 1979 in a German news magazine (April issue!).[24,25] (See chapter 5 for more about spoof articles.)

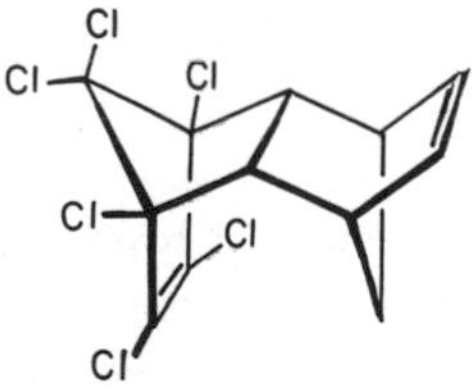

Incidentally, isodrin is a stereoisomer of another insecticide, aldrin (**21**),[23] named after Kurt Alder of Diels–Alder fame. Since its discovery, the versatile Diels–Alder reaction has been a workhorse for entry to bridged molecules. In fact, aldrin[26] and isodrin[21] are synthesized by [4 + 2] cycloadditions of suitable dienes and dienophiles. And chemists have used sequential inter- and intramolecular Diels–Alder methodology to assemble complex ring systems in one swoop.[27]

As a case in point, consider the reaction between 9,10-dihydrofulvalene (**22**) and dimethyl acetylenedicarboxylate, studied by Hedaya and co-workers at Union Carbide Corporation[28] and then by Paquette's laboratory at Ohio State University.[29] Under appropriate temperature control, both teams obtained a mixture of pentacycles **25** and **26**. Isomer **25** arises by an external Diels–Alder step (**22 → 23**) followed by internal [4 + 2] cycloaddition (**23 → 25**), in which the nonconjugated olefin born in the first event serves as dienophile in the second. Professor Paquette and student Mathew Wyvratt dubbed such consecutive cycloadditions "domino Diels–Alder" reactions.[29,30]

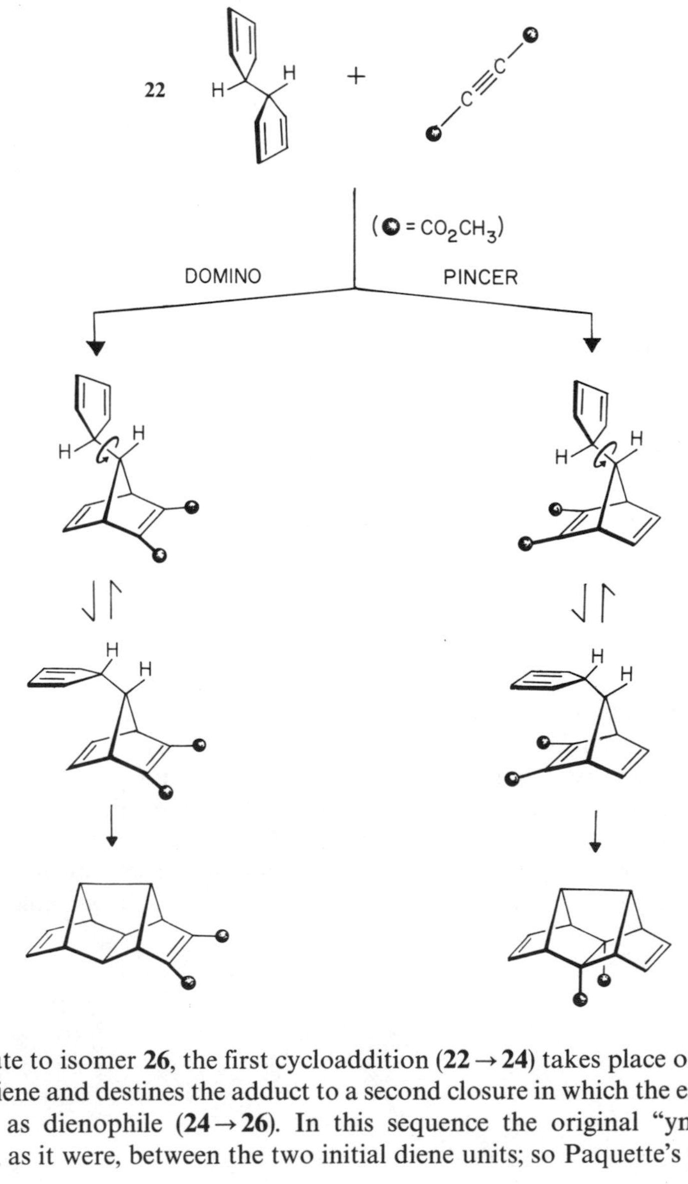

In the route to isomer **26**, the first cycloaddition (**22 → 24**) takes place on a different face of the diene and destines the adduct to a second closure in which the ester-bearing alkene acts as dienophile (**24 → 26**). In this sequence the original "yne" ends up compressed, as it were, between the two initial diene units; so Paquette's people term these cycloadditions "pincer Diels–Alder" reactions.[30] Whichever game the molecules play, it's clear that Diels and Alder control the dice.

One of the first things we learn in organic chemistry is that a saturated carbon has tetrahedral bonds. A square-planar arrangement is quickly ruled out by nonexistence of isomers of dichloromethane. And theoretical calculations support our expectation that a flattened carbon would be highly strained.[31–34] Yet, the possibility of preparing compounds in which saturated carbon is compelled to be planar has long intrigued chemists. Let's peer into this matter with the molecule 27, which Vlasios Georgian at Tufts University christened "fenestrane," from the Latin *fenestra* ("window").[35] Professor Georgian is of Greek descent, so he first considered "parathirane," because the Greek word for window is *parathiro*. But he felt that a four-syllable name would not

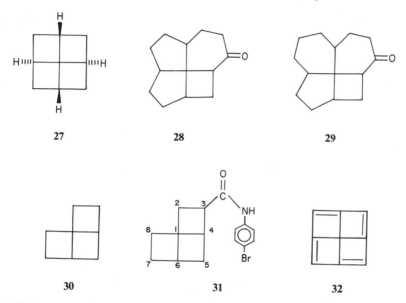

27   28   29

30   31   32

sell; and "fenestrane" came to him like a flash.[36] Yale's Kenneth Wiberg referred to 27 as "windowpane."[33] The four cyclobutane rings might force the central carbon to approach a square-planar geometry.[33–35]

Georgian's team synthesized the fenestrane-like analogs 28 and 29;[35] and Steven Wolff and William Agosta at the Rockefeller University[37] prepared several derivatives of what reviewers Joel Liebman and Arthur Greenberg called a "broken window," 30.[38] X-ray study of the *p*-bromoanilide 31 revealed that the carbons in this broken window are not flush.[37] In 1981, Paul Schleyer's glaziers at the University of Erlangen–Nürnberg predicted, by molecular orbital calculations, that framing the window with an annulene perimeter, as in 32, will not level the atoms either.[39] Theoretical scrutiny by others gave clues on how fenestrane might distort its bonds.[40] Therefore, the flatness implied by words like fenestrane and windowpane, and by other drawings like 28–31, must be taken with many grains of NaCl.

Chemists looking into fenestranes in the United States and abroad have used numbers to specify ring sizes.[41] For example, analog 33, constructed at Rockefeller University, was termed a [4.4.4.5]fenestrane derivative.[41a] Professor Georgian, however, had intended that the name "fenestrane" be reserved specifically for prototype 27 to recognize its unique standing in the community.[35] For other skeletons with a central carbon common to four rings (i.e., general structure 34), he later preferred the term "rosettane" (from rosette: a cluster of petals emanating from a central point).[42]

Accordingly, the Tufts floral expert regarded **28** as a [4.5.5.6]rosettane; and **33** as a [4.4.4.5]rosettane.

Mother nature assembled a [5.5.5.7]rosettane (**35**) eons ago, but chemists (in New Zealand) proved it only recently. They address this diterpenoid as "laurenene," presumably after D.R. Lauren, a Ph.D. candidate who worked on its structure.[43]

33   34   35

REFERENCES AND NOTES

1. Engler, E.M.; Andose, J.D.; Schleyer, P.v.R. *J. Am. Chem. Soc.* **1973**, *95*, 8005–8025.
2. *Chem. and Chem. Ind.* (*Tokyo*) **1975**, *28*, 735. We are indebted to Professor E. Ōsawa for drawing our attention to this article.
3. Schultz, H.P. *J. Org. Chem.* **1965**, *30*, 1361–1364.
4. Ōsawa, E.; Aigami, K.; Inamoto, Y. *J. Org. Chem.* **1977**, *42*, 2621–2626.
5. Schmidt, W.; Wilkins, B.T. *Tetrahedron* **1972**, *28*, 5649–5654; see footnote 9.
6. (a) Eaton, P.E.; Or, Y.S.; Branca, S.J. *J. Am. Chem. Soc.* **1981**, *103*, 2134–2136, (b) Eaton, P.E.; Or, Y.S.; Branca, S.J.; Shankar, B.K.R. *Tetrahedron* **1986**, *42*, 1621–1631.
7. Sedelmeier, G.; Prinzbach, H.; Martin, H.D. *Chimia* **1979**, *33*, 329–332.
8. Kent, G.J., private communication to Ōsawa, E. We thank Dr. Ōsawa for informing us in November, 1975 of this correspondence.
9. Underwood, G.R.; Ramamoorthy, B. *J. Chem. Soc., Chem. Commun.* **1970**, 12–13.
10. Ward, S.J.; Pettit, R. *J. Am. Chem. Soc.* **1971**, *93*, 262–264.
11. Smith, E.C.; Barborak, J.C. *J. Org. Chem.* **1976**, *41*, 1433–1437.
12. Marchand, A.P.; Chou, T.–C.; Ekstrand, J.D.; van der Helm, D. *J. Org. Chem.* **1976**, *41*, 1438–1444.
13. Eaton, P.E.; Cassar, L.; Hudson, R.A.; Hwang, D.R. *J. Org. Chem.* **1976**, *41*, 1445–1448.
14. Eaton, P.E.; Mueller, R.H. *J. Am. Chem. Soc.* **1972**, *94*, 1014–1016.
15. Eaton, P.E.; Mueller, R.H.; Carlson, G.R.; Cullison, D.A.; Cooper, G.F.; Chou, T.–C.; Krebs, E.P. *J. Am. Chem. Soc.* **1977**, *99*, 2751–2767.
16. Garratt, P.J.; White, J.F. *J. Org. Chem.* **1977**, *42*, 1733–1736.
17. Winstein, S. *Experientia, Supplement II* **1955**, 137–155. Professor Winstein discussed "birdcage" at a symposium on "Dynamic Stereochemistry" held at Manchester, England, March 31, 1954. (He hyphenated the term.) For a news account of that meeting see *Chem. Ind.* (*London*) **1954**, 562–563.
18. (a) Adams, R.M.; Loening, K.L. *Inorg. Chem.* **1968**, *7*, 1945–1964, (b) Greenwood, N.N., in *Comprehensive Inorganic Chemistry*, Trotman–Dickenson, A.F. (Exec. Ed.), (Pergamon Press, Oxford **1973**), vol. 1, pp. 734–735.
19. de Vries, L.; Winstein, S. *J. Am. Chem. Soc.* **1960**, *82*, 5363–5376.
20. Soloway, S.B.; Damiana, A.M.; Sims, J.W.; Bluestone, H.; Lidov, R.E. *J. Am. Chem. Soc.* **1960**, *82*, 5377–5385.
21. Dong, D.C. *J. Chem. Educ.* **1982**, *59*, 704–705.
22. Cookson, R.C.; Crundwell, E. *Chem. Ind.* (*London*) **1959**, 1004. For the nonchlorinated counterpart, see Jones, G., II.; Becker, W.G.; Chiang, S.–H. *J. Am. Chem. Soc.*, **1983**, *105*, 1269–1276.
23. *The Merck Index*, 9th ed. (Merck and Co., Rahway, NJ, 1976), pp. 32, 677.
24. Turro, N.J. *Nachr. Chem. Tech. Lab.* **1979**, *27*, 185–186.
25. Turro, N.J., private communication, June, 1979.

26. Korte, F.; Rechmeier, G. *Liebigs Ann. Chem.* **1962**, *656*, 131–135.
27. For an early example see Cram, D.J.; Knox, G.R. *J. Am. Chem. Soc.* **1961**, *83*, 2204–2205.
28. McNeil, D.; Vogt, B.R.; Sudol, J.J.; Theodoropulos, S.; Hedaya, E. *J. Am. Chem. Soc.* **1974**, *96*, 4673–4674.
29. Paquette, L.A.; Wyvratt, M.J. *J. Am. Chem. Soc.* **1974**, *96*, 4671–4673.
30. Paquette, L.A.; Wyvratt, M.J.; Berk, H.C.; Moerck, R.E. *J. Am. Chem. Soc.* **1978**, *100*, 5845–5855.
31. Wiberg, K.B.; Ellison, G.B.; Wendolski, J.J. *J. Am. Chem. Soc.* **1976**, *98*, 1212–1218.
32. Böhm, M.C.; Gleiter, R.; Schang, P. *Tetrahedron Lett.* **1979**, 2575–2578.
33. Wiberg, K.B.; Olli, L.K.; Golembeski, N.; Adams, R.D. *J. Am. Chem. Soc.* **1980**, *102*, 7467–7475.
34. Würthwein, E.–U.; Chandrasekhar, J.; Jemmis, E.D.; von R. Schleyer, P. *Tetrahedron Lett.* **1981**, *22*, 843–846.
35. Georgian, V.; Saltzman, M. *Tetrahedron Lett.* **1972**, 4315–4317.
36. Georgian, V., private communication, October, 1975.
37. Wolff, S.; Agosta, W.C. *J. Chem. Soc., Chem. Commun.* **1981**, 118–120.
38. Liebman, J.F.; Greenberg, A. *Chem. Rev.* **1976**, *76*, 311–365.
39. Chandrasekhar, J.; Würthwein, E.–U.; Schleyer, P.v.R. *Tetrahedron* **1981**, *37*, 921–927.
40. Schulman, J.M.; Sabio, M.L.; Disch, R.L. *J. Am. Chem. Soc.* **1983**, *105*, 743–744.
41. (a) Rao, V.B.; Wolff, S.; Agosta, W.C. *J. Chem. Soc., Chem. Commun.* **1984**, 293–294, (b) Luyten, M.; Keese, R. *Angew. Chem. Int. Ed. Engl.* **1984**, *23*, 390–391, (c) Dauben W.G.; Walker, D.M. *Tetrahedron Lett.* **1982**, *23*, 711–714, (d) Rao, V.R.; Wolff, S.; Agosta, W.C. *Abstracts of Papers, 190th National Meeting of the American Chemical Society, Chicago, IL* (American Chemical Society, Washington, DC, 1985), ORGN 80, (e) Luyten, M.; Keese, R. *Abstracts of Papers, 190th National Meeting of the American Chemical Society, Chicago, IL* (American Chemical Society, Washington DC, 1985), ORGN 81, (f) Luyten, M.; Keese, R. *Tetrahedron* **1986**, *42*, 1687–1691, (g) Rao, V.B.; George, C.F.; Wolff, S.; Agosta, W.C. *J. Am. Chem. Soc.* **1985**, *107*, 5732–5739, (h) For a review see Venepalli, B.R.; Agosta, W.C. *Chem. Rev.* **1987**, *87*, 399–410.
42. Georgian, V., private communication, March 1984. In correspondence about the term "trivial," Professor Georgian lamented that widespread usage has relegated this word to its derogatory connotation (i.e., to mean insignificant, of little value or importance, trifling). Yet that's not what chemists intend when they pick a convenient name in place of an unwieldy systematic one. So, for this latter context, Dr. Georgian yearned to replace "trivial" by something more suitable; he came up with *koinos*, a venerable Greek word that means common, standard, widely used or accepted, and so on, but that is free of demeaning implication. The pronunciation is koy-nos in ancient Greek and kee-nos in modern Greek (in each case with the second syllable slightly accented). Thus, "koinology" would become the study of common or widely used names. And the adjectival-sounding "koinal" would serve instead of "trivial" to describe unsystematic names in chemistry.
43. (a) Corbett, R.E.; Lauren, D.R.; Weavers, R.T. *J. Chem. Soc., Perkin Trans. 1*, **1979**, 1774–1790, (b) Corbett, R.E.; Couldwell, C.M.; Laurens, D.R.; Weavers, R.T. *J. Chem. Soc., Perkin Trans. 1*, **1979**, 1791–1794.

Chapter 5

CHEMISTRY IS NO FLASK IN THE PAN

Photographs of chemists typically show them surrounded by beakers, flasks, and bottles. Containers of all sorts sit on laboratory benches and shelves. So, it's not surprising that our colleagues have christened molecules after receptacles, albeit not the kind for chemicals.

Consider "barrelene," (1), named after a vessel familiar to all. The word, conceived by Howard Zimmerman in 1960 when he was on the faculty at Northwestern University, was inspired not by the shape of the carbon skeleton, but of the π-clouds.[1] The fascination here was whether the p-orbitals depicted in 2 overlap hominally (p. 178) to

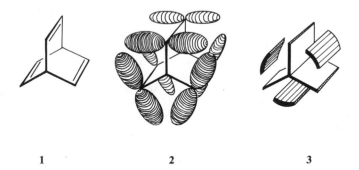

1   2   3

create three banana-shaped regions of maximum electron density (see 3). The curved clouds reminded Dr. Zimmerman of staves on a barrel outlined by the σ-bonded framework.

Barrelene was among the earliest contemporary, trivial "organic" names prompted by a familiar object, and the choice was not universally admired. John D. Roberts from the California Institute of Technology visited Zimmerman while the "barrelene" paper was in press.[2] Professor Roberts considered the term "excruciatingly inelegant" and "horribly undescriptive" and said he could think of only one name that might have been as bad, "zimmerene."[3] But the presses were rolling and it was too late for change. Shortly thereafter, Charles Hurd, a senior faculty member at Northwestern, published an article espousing systematic nomenclature and stating: "Occasionally one encoun-

58

ters such needless names as triptane,* triptycene, and barrelene, which are undecipherable unless memorized."[4] At that time, Professor Zimmerman was striving for tenure at Northwestern, and the raised eyebrows of such eminent chemists as Roberts and Hurd undoubtedly gave him second thoughts about having rolled out the barrel.

Disdain for "barrelene" did not fade entirely. Even as late as 1979 an authoritative text[5] on chemical nomenclature commented on such newly invented names as follows:

> There is, of course, also the lunatic fringe, which can be exemplified by the name barrelene... the not too difficult name bicyclo[2.2.2]octa-2, 5, 7-triene should be used, but such is the appeal of brevity that the misleading trivial name is still often employed.

In reality, the imagined orbital overlap that inspired the name barrelene does not enhance stability. In fact, the molecule is quite aggressive, as demonstrated by its passion for dicyanoacetylene at room temperature.[6] This lively behavior complies with current views from orbital symmetry[7] that p-lobes of the same phase can't interact synergetically all around the "barrel" (see representation **4**).

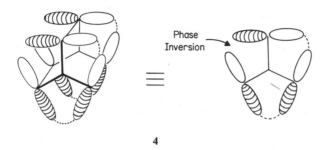

4

A barrel is fine for storage. But to consume a drink, try a chalice. And here, "calixarenes" (**5**) add a touch of elegance to our collection of molecular containers. This class of compounds was named by C. David Gutsche's connoisseurs[8] at Washington

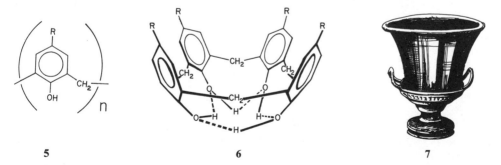

5   **6**   **7**

University, St. Louis, after *calix* (Greek for "chalice") and arene (aromatic units).[9] A molecular model of the tetramer calix[4]arene (**6**, with R = CH_3) can be cup-like when the OHs are hydrogen bonded, a shape akin to that in a type of Greek vase know as calix crater (**7**). Compounds like **5** were originally synthesized (from p-alkylphenol, formaldehyde, and alkali) by Alois Zinke and co-workers at the University of Graz, Austria.[10] They proposed tetrameric ($n = 4$) structures, but Dr. Gutsche's go-getters

---

*According to the 7th Collective Index (covering 1962–1966) of *Chemical Abstracts*, triptane was a synonym for 2, 2, 3-trimethylbutane. The name does not appear in the eighth collective index, so apparently it fell from grace in the mid-1960s. Score one point for Dr. Hurd.

showed that the calix[8]arene generally predominated.[8] Increasing the proportion of base favored a hexamer; raising the temperature improved the yield of tetramer. Some experimental conditions produced a lopsided calixarene in which one of the bridges between aromatic rings is CH_2OCH_2.[11] Calix[4]arenes have also been made by directed, stepwise routes that leave no doubt they are tetrameric.[12,13]

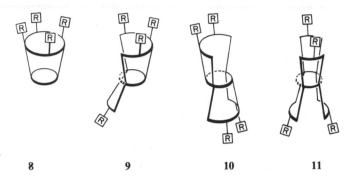

8   **9**   **10**   **11**

Dynamic ^{1}H NMR studies of several calix[4]arenes revealed that some prefer the "cone" shape (symbolized by **8**), but this form coexists at room temperature with other conformations such as **9**, **10**, and **11**. However, certain derivatives of calixarenes can maintain more rigid cavities, which the St. Louis chemists call "changeless calixes" (cf. enforced cavities in spherands, chapter 17).[14] The cone shape is strongly favored when a hydrocarbon chain links two opposite R groups in **6**.[15] Dr. Gutsche coined "pleated loop," "hinged," and "winged" to describe conformations of calixarene octamers and hexamers.[16] "Pseudocalixarenes" and "hemicalixarenes" are open-chain analogs of **5**, whose ends nevertheless hang together through stout intramolecular and intermolecular H-bonds, respectively.[17] Professor Gutsche pointed out that calixarenes can form complexes and, when suitably functionalized, might act as enzyme mimics.[11,18,19]

The chalice-like appearance of the shapely molecule **12** prompted Horst Prinzbach to dub this parent skeleton "calicene."[20] In its ionic resonance form **13**, each ring is

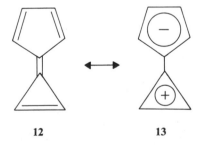

12   **13**

aromatic according to the Hückel $4n + 2$ rule. If contributor **13** has a say, the system should show a sizable dipole moment and the linkage between the rings would have a bond order considerably less than two. The experimental dipole moment of the dibenzodimethylcalicene **18** is 4.9 D, high indeed for a hydrocarbon,[21,22] and molecular orbital computations indicate a bond order of 1.44.[23] Furthermore, the methyl hydrogens of **18** absorb at $\delta = 2.63$ in the NMR,[22] whereas ordinary allylic methyls appear about $\delta = 1.8$. This extra deshielding in **18** was attributed to severe electron deficiency in the three-membered ring.[24] Professor Prinzbach's laboratory fashioned this calicene by annealing a "cup" (**14**) and a "stand" (**15**) and then removing two hydrogens (**16** → **17** → **18**). *Prosit!*

If you choose to sip sake from a calicene, try the spacious orange analog **19**.[25] It was termed "cyclic bicalicene" and was synthesized and scrutinized by collaborators from Japan's Kyoto and Osaka Universities.[26] They determined that, despite a 16π electron perimeter and an 8π inner track, this heretic is planar; it is also air-stable at room temperature, and the ring bonds don't alternate in length. Charge densities (estimated

from ^{13}C NMR and from calculations) further nurtured the notion that their non-Hückel annulene inherits stability from Kekulé-type resonance (**a** and **b**) as well as from a polar contributor **c**.[26]

Let us now look at a container for *solids*. In 1966, Satoru Masamune's artisans, then

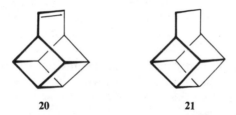

at the University of Alberta, published their synthesis of **20** and **21**.[27] During a coffee break, Masamune's faculty colleague, Dennis Tanner, suggested the name "basketene"

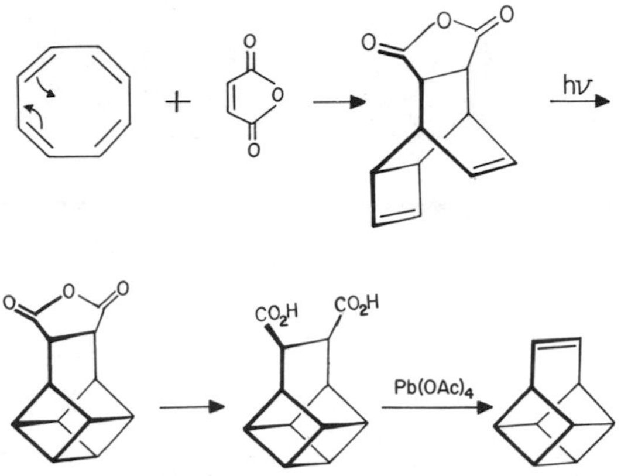

22

for olefin **20** because it resembles a basket.[28] Thus, **21** became "basketane." Masamune and company wove the basket by (1) Diels–Alder and electrocyclic reactions between cyclooctatetraene and maleic anhydride, (2) intramolecular [2 + 2] cycloaddition, (3) hydrolysis to diacid **22**, and (4) oxidative decarboxylation.

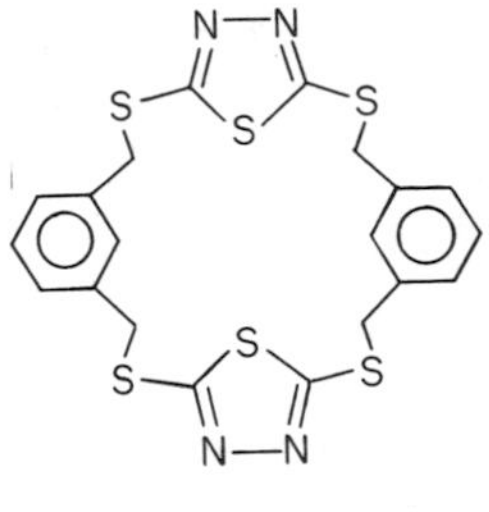

23

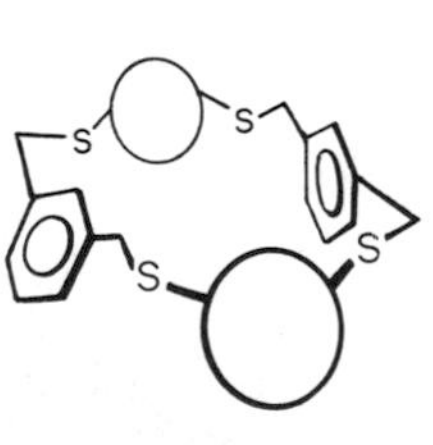

24 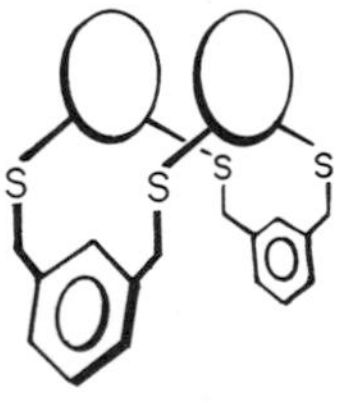   25 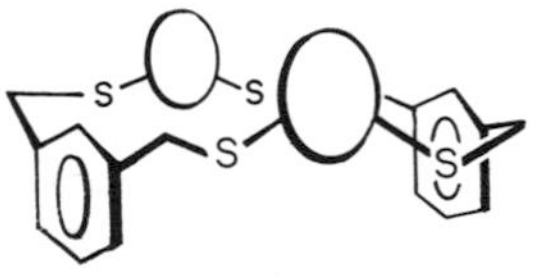   26

In 1982, Sebastiano Pappalardo and Francesco Bottino at the University of Catania, Italy, fabricated the adaptable basket **23**. Molecular models suggested it can assume conformations **24**, **25**, and **26**. These shapes look like a "basket," a "saddle," and a "crown," respectively, and were so named.[29] Thus, **26** joins Dr. Pedersen's collection of crowns (chapter 2).

Not to be outdone, inorganic chemists Tristram Chivers and John Proctor of the University of Calgary, Alberta, subsequently assembled a carbonless basket with its atoms positioned as in **27**. In this S_5N_6 sulfur nitride an $-N{=}S{=}N-$ unit bridges two sulfurs of an S_4N_4 "cradle."[30]

Jimmie Edwards of the University of Toledo predicted that yet another basket style may be possible.[31] He and graduate student Alfred Gates prepared a compound

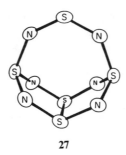

27

$H_2B_2S_5$ that, they thought, could polymerize to **28** and then to **29**.[31,32] A space-filling model of **29**, as drawn, looks like an inverted basket with no handle.[31] In a lecture, titled "Rings and Baskets of BS," Dr. Edwards further speculated that resonance in **29** might make the rings of eight sulfurs at the top and at the bottom indistinguishable and give a structure resembling a hamburger (**30**). In rejoinder, James Leitnaker of Oak Ridge National Laboratory invented the term "hamburger structure" for it.[31] What a versatile molecule **29**: it could be a picnic basket as well as part of its contents. Alas! A

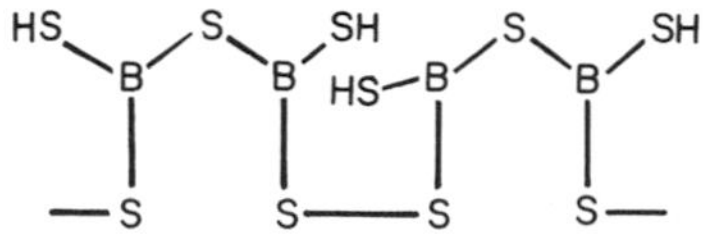

28

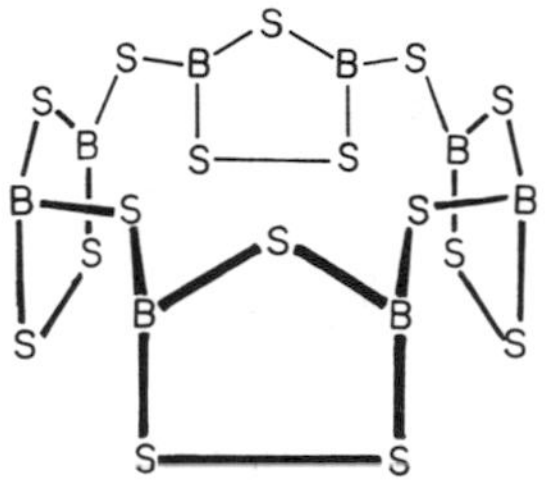

29

follow-up communication from Dr. Edwards[31] announced that X-ray diffraction[33] of polymer **29** proved it to be flat (**31**). Hence, his colleague, Dr. H. Bradford Thompson would like it renamed "pizza."[31] Culinary wizardry indeed.

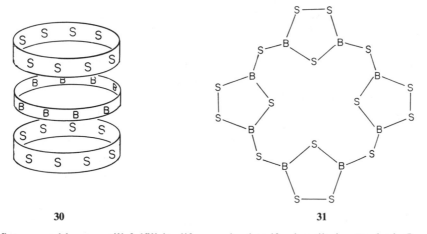

30

31

A flat ensemble can still fulfill its life as a basket if a handle is attached. Organic chemists can help here because they have been handling molecules for some time; and Mother Nature has been at it even longer. In 1942, Arthur Lüttringhaus and Heinz Gralheer at Berlin's Kaiser-Wilhelm-Institut discerned that the aliphatic chain in melecules like **32–35** loops above the aromatic unit like a handle. Consequently, they dubbed such structures "ansa-compounds" (Latin *ansa* means "handle").[34] (Dr. Lüttringhaus had synthesized several prototypes in 1937.[35]) They also resolved **35** into optical antipodes and thus demonstrated that its loop doesn't leave enough room for

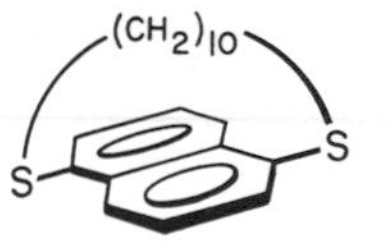

32 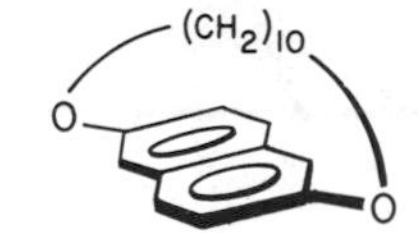   33 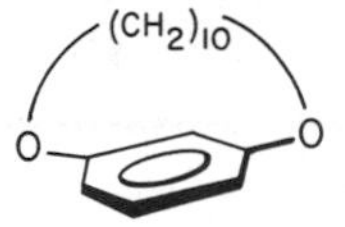   34 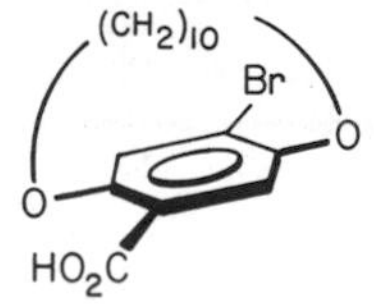   35

the substituted benzene ring to twirl.[34] In 1958, Professor Lüttringhaus and student W. Kullick prepared sandwich complexes **36** ($n = 3$, 4, and 5) and called them "ansa-ferrocenes."[36]

Mother Nature may not know Latin, but she deftly manufactures ansa molecules just the same. For example, an important family of natural products—ansamycins—

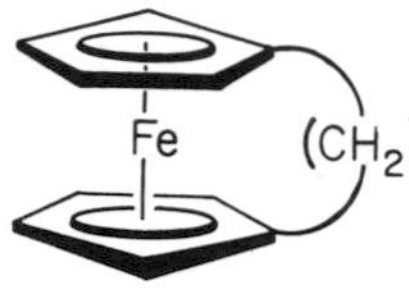

36 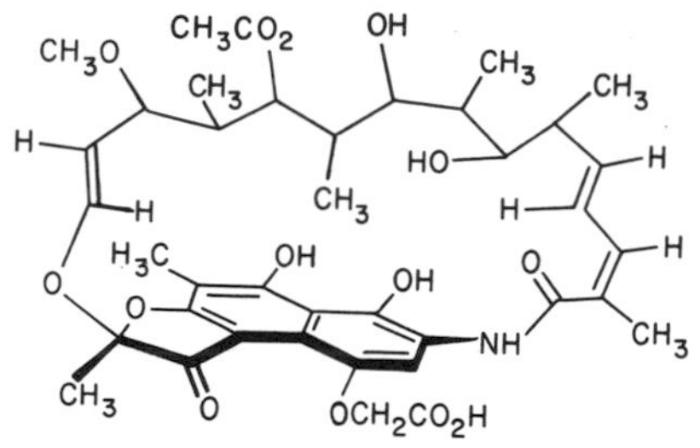   37

contain macrocyclic rings that span an aromatic segment.[37] Among these are the "rifamycins," a class of lactam antibiotics discovered at the Lepetit SpA Research Laboratories in Milan, Italy. One specific member is rifamycin B (**37**), whose structure yielded to Vladimir Prelog's team at Zurich and to Italian researchers in Milan and in Rome.[38] At Lepetit, the in-house jargon for this group of antibiotics was originally "rifomycins." This name attested to their efficiency at killing microbes and was stimulated by a gangster movie, "Rififi," popular in Europe at that time.[39] (In French underworld slang, "rififi" connotes gang war and corresponds roughly to the American "gang rumble.") The nickname eventually became official, but only after the *o* was changed to *a* to minimize confusion with a similar-sounding drug, rufomycin.[39]

The pharmaceutical industry routinely breeds colorful terminology, especially for natural products. Another case in point involved Donald Nettleton and his colleagues, who worked on an antitumor antibiotic complex at Bristol Laboratories.[40] Dr. Nettleton, an avid opera fan, titled it the "bohemic acid" complex after the Puccini opera *La Bohème*. And when the B-team eventually isolated two specific components, they named them "musettamycin" and "marcellomycin," after Musetta and Marcello, two characters in that opera.[41,42] Subsequently, Terrence Doyle and other colleagues joined the "bohemic" project, characterized several more anthracycline antibiotics, and continued the naming pattern; their publication reported "rudolphomycin," "mimimycin," "collinemycin," and "alcindoromycin."[42] (According to Doyle, they did not isolate more compound than there are characters in the libretto of *La Bohème*.)

On degradation of rudolphomycin (**38**), the researchers obtained a new unsaturated sugar (**39**), christened it "rednose," and thereby ensured that this hexose would "go down in his-tor-y."[43] Ordinarily, hard-nosed referees frown on such whimsy, but they let the name stand.[40] We note that the manuscript had a "received" date of Dec. 21, 1978 and also that Act One of *La Bohème* begins with Christmas Eve; so perhaps the referees were palliated by the Yuletide spirit.

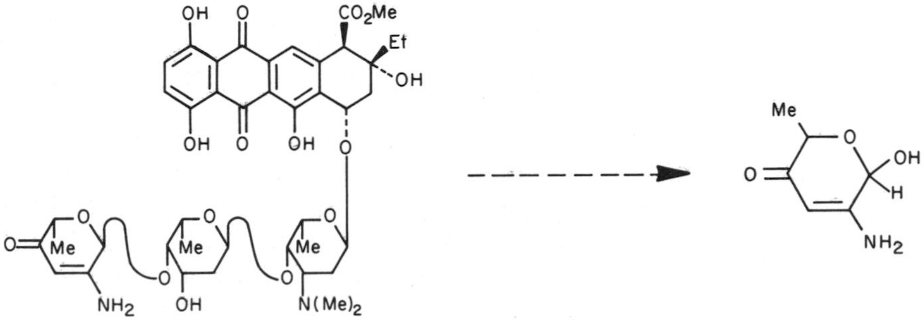

38   **39**

Although Nettleton, Doyle, and collaborators waxed operatic, other projects in the laboratory acquired their own themes. For example, Dr. Henry Schmitz named an antibiotic "hedamycin," from the first initials of his family members: Henry, Edith, David, and Anne.[40] What a nice way to get your spouse and kids excited about chemistry.

If you find such jargon offensive, and have oodles of time, try naming some of these complicated antibiotics systematically. Fearless individuals might resort to a permutational nomenclature described in 1970 by Ivar Ugi and his associates (then at the University of Southern California).[44] Their system can even cope with such behemoths as rifamycin B, which becomes (1 29 26 12 38 2 16 7 17 10 19 8 36 13 24 5 9 37) (11 21)(6 32 27)-se-⟨1_N-19_N⟩-1_N-aza-17_N-oxa-dioxo-acetoxy-methoxy-dihydroxy-hexamethyl-nonadecatriene-⟨6-8, 10-12, 30-31⟩-dipon-⟨1_N-$1'_N$, 19_N-$3''_N$, 18_N-oxa-$3'_N$⟩-(1′2″)(1″2′)(3′4″)-carbomethoxy-dihydroxy-methyl-naphthalene. If Professor Ugi's algebraic formalism[45] boggles your mind, then don't look to IUPAC for solace; what it advocates is just as frightening and definitely not for casual conversation at cocktail parties.

Our final "container," a picket fence, normally serves to keep pets and children under control. At Stanford University, James Collman and colleagues, however, designed a

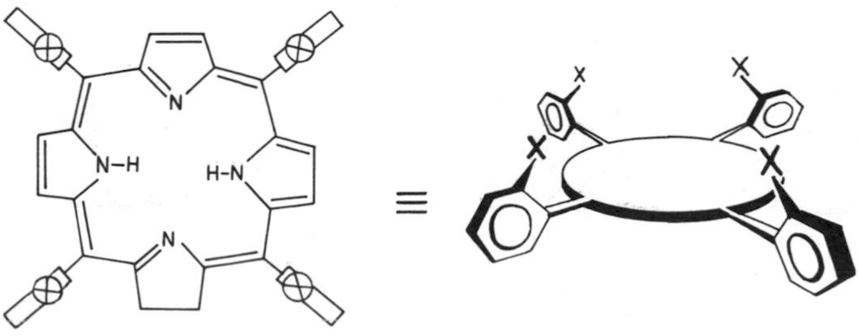

40

chemical fence to hold a frisky oxygen molecule.[46] They synthesized tetraarylporphyrins of the type **40**, which resembles a ring with four ortho "pickets." The skeleton was assembled from pyrrole and *o*-nitrobenzaldehyde units. After reduction of NO_2 groups to NH_2 an interesting isomerism emerged; four isomers of the tetraaminoporphyrin were separated. Evidently, the ortho amino group restricts rotation of the aryl ring—as in ortho-substituted biphenyls—and permits four atropisomers* in which X groups on adjacent aryls are up–up–down–down, up–up–up–down, up–down–up–down and up–up–up–up. The desired "picket fence," **40**, is the most polar of the four and travels slowest on chromatography; the other atropisomers can be reequilibrated with **40** in boiling toluene to get more of it. Each picket was then made bulky by reaction of the ortho NH_2 units with pivaloyl chloride, and a ferrous ion was attached to the porphyrin center. Dr. Collman's compound was ready to capture oxygen.[46]

Later, the Stanford chemists also erected a "tailed-picket-fence" porphyrin in which X groups are up–up–up–down (three "pickets" and a "tail").[49] The key step was thermal isomerization of an up–up–up–up isomer. David Whitten's band of researchers, then at the University of North Carolina, heightened interest in this subject by showing that irradiation does the trick also.[50] They view the process as "photo-atropisomerization," but we wager the reaction is over before you can say the word. When a "tail" terminates in an —SH, Professor Collman referred to the entity as a "mercaptan-tail" porphyrin.[51] He hoped that sulfur can shackle the metal in the middle; such complexation may be important in cytochrome P-450 enzymes.

In a nifty maneuver on **40**, and with the added talents of faculty colleague John Brauman, the Stanford tailors sewed a pivaloyl chain to one NH_2 picket and stitched a benzene triacetylchloride patch to the other three NH_2s. This sartorial coup produced a "pocket" porphyrin (**41**) with a lone picket lingering near the opening.[52] Routine minor alterations (attaching Fe(II) and N-methylimidazole) completed the job (**42**); only small units (e.g., CO and O_2) can slip inside the pocket to bind to the iron. Even so, a bound ligand would want to minimize congestion by angling itself toward the pocket entrance; and O_2 is better at this (**43**) than is CO. So, to preclude CO complexation without substantially diminishing O_2 affinity, a "picket-pocket" porphyrin is what proficient people should pick. And, thanks to input from teams at Cornell, Princeton, and Pennsylvania State universities, there are plenty of styles to choose from. These include "small," "medium" and "tall" pockets, as well as "ruffled" porphyrins.[53] For the fashion-conscious, "bis-pocket" varieties are also available.[54]

40   41

41   42   43

---

*Atropisomers are conformational isomers that can be separated at room temperature because rotation about single bonds is difficult or impossible. (Note: "a" and "trop-" come respectively from Greek etymons meaning "not" and "turning or changing.") German chemist Richard Kuhn introduced the term.[47] As this type of isomerism becomes more common,[48] his word may get shortened to "atropomer."

By the way, the Stanford article describing the tailed picket fence listed an impressive total of 10 authors.[49] This number pales when compared to the 49 collaborators on each of three publications from Harvard's chemistry department.[55] But even that task force is outnumbered by the 142(!) coauthors—from 12 research laboratories—on a 1983 paper in *Physics Letters*.[56] We always suspected it took more physicists than chemists to do a job.

While counting collaborators, we wondered what chemistry article featured the longest name for an individual author. A publication from Henry Kuivila's laboratory at State University of New York at Albany lists as one coauthor the postdoctorate associate Tiruvenkatanathapuram R. Balasubramanian (nicknamed Balu).[57] Professor Kuivila remembered that Balu was reluctant to print his full first name, but the journal preferred it that way for the sake of *Chemical Abstracts*. If you see a longer name associated with a chemistry article, please drop us a line. By the way, Balu's middle initial stands for Ramanathapuram.[58]

For the longest run of chemistry papers on a single theme, our champion is the series from Japan titled "Studies on the Syntheses of Heterocyclic Compounds and Natural Products," with patriarchal author Tetsuji Kametani. The primogeniture in Professor Kametani's series appeared in 1945 in the periodical *Yakugaku Zasshi*.[59] Publication No. 1000 showed up in the *Journal of Organic Chemistry*, 1983.[60] He spread the rest throughout more than 40 different journals. By 1983, this enduring marathon had involved 274 researchers; 85 were students who earned Ph.D. degrees. During that period, mentor Kametani served several institutions: Tokyo College of Pharmacy, Osaka University, Grelan Pharmaceutical Company, Tohoku University, and most recently, Hoshi University.[61] By the way, *doyen* Kametani also coauthored oodles of articles unconnected with that series.

Meanwhile, in the United States, Carl Djerassi (Stanford University) had also raced past a kilomarker in research productivity. A 1981 contribution in *Tetrahedron* titled "Circular Dichroism of Molecules with Isotopically Engendered Chirality" turned out to be the 1000th from his prolific pen, even though that publication gave no hint of its lofty status.[62]

The Djerassi deluge began in 1946 when he published his first paper, with Ph.D. mentor Alfred L. Wilds, University of Wisconsin.[63] The output gained momentum during C.D.'s associations with Ciba–Geigy Corporation, with Syntex (in Mexico), and with Wayne State University; then it flowed unabated from Stanford University, from Syntex in California, and from Zoecon Corporation. His 1000 publications involved about 400 collaborators worldwide[64] and enriched 80 different journals, four of which got much of the repeat business, namely *Journal of the American Chemical Society* (375 papers); *Journal of Organic Chemistry* (134); *Tetrahedron* (61); *Tetrahedron Letters* (57). The articles encompass remarkable breadth and depth. About two-thirds fell into these nine themes: "Mass Spectrometry in Structural and Stereochemical Problems" (257 papers); "Optical Rotatory Dispersion Studies" (134); "Terpenoids" (76); "Alkaloid Studies" (67); "Magnetic Circular Dichroism Studies" (59); "Applications of Artificial Intelligence for Chemical Inference" (37); "Minor and Trace Sterols in Marine Invertebrates" (27); "Macrolide Antibiotics" (15); and "Studies in Organo Sulphur Compounds" (13).[65] He also appeared as a coauthor on many of Syntex's contributions on "Steroids."

Such prodigious research accomplishment led geneticist Joshua Lederberg* to label Dr. Djerassi as one of the "wonders of nature."[66] On one occasion, a symposium

---

*Nobel Prize in medicine, 1958, shared by George W. Beadle, Edward L. Tatum, and Joshua Lederberg.

Tetraodon Letters No 37, pp 1839-1843, 1982.
Pergamon Press Ltd. Printed in Great Britain

MINOR AND TRACE STEROLS IN MARINE INVERTEBRATES. Part 2.79 x 10^2.
18,19-BISNOR-(4,5),(9,10),(8,14),(13,17)-TETRASECOPREGNANE, A NOVEL
STEROID FROM THE SPONGE <u>Aristospongia caveatemptor</u> (Meereslustig,1892)[1]

J.E.K. CRABMAN, D. DE ZOLA, A.N. KISMI, N. SCHUBERT

Collectif de Bio-Ecologie, Faculté des Sciences
Université Libre de Bruxelles - 1050 Bruxelles - Belgium.

and

Carl DJERASSI

Department of Chemistry, Stanford University, Stanford, California 94305.

(Received in UK 6 December 1979, accepted for publication 10 March 1982)

During our continuing search for novel marine sterols we have repeatedly documented [2] the presence of unusual sterol derivatives in sponges. The climax of this ex(t,p)ensive program was recently reached with the rediscovery of the title animal in New Guinea waters. The sponge *Aristospongia caveatemptor* (Porifera, Anopsispongidae), although apparently not uncommon around Laing Island, is exceedingly difficult to collect and indeed its very existence has been doubted until recently [3]. As stated in the original description [4], the colourless sponge very exactly matches the refractive index of the surrounding water and is thus unobservable for predators (and collectors) acting on simple visual cues. This remarkable adaptation is very probably due to highly specialized cell membranes microstructures, suggesting the likely presence of unusual sterol material.

An adequate methodology directly derived from the principles of optical rotatory dispersion was developed [5] in order to collect the elusive sponge. Continuous eyelid blinking induced in divers by liberal addition of stoechiometric mixtures of V.S.O.P.[6]/ S.P. [7] resulted in subliminal differential measurements between left and right eye fields at unusual wavelengths, thus permitting visualization of the sponge [8]. Since all vessels containing extracts or purified compounds of the sponge naturally appear empty *(vide supra)* the same shift reagent had to be continuously used by all members of our research teams until completion of this work.

The collection of the sponge documented a well-precedented attractive bonus, namely travel [9]. *Aristospongia caveatemptor* was gathered in 2.05 feet of water at Laing Island in the second week of September 1979 during a stiff breeze, as much as we remember. The sponge proved to be another sterol gold mine [10].

A methylene chloride extract of the sun-dried material, followed by repeated silicagel
column chromatography afforded an auspicious fraction that was shown by tlc to consist of
a mixture of essentially pure compounds. This was further fractionated by HPLC and reverse-
phase chromatography , yielding a small amount of pure, colourless and invisible compound I
(oil ?; $(\alpha)_0^{25}$ 0.002 $\pm$ 15°; UV (in hexane) : end absorption only ; IR : 2850, 2735, 1470,
1375 and 820 cm^{-1}; NMR : complex signal (35 H) centered at 1.26, ill-resolved absorption
(6H) at 0.91 ppm -several strong signals totaling 107 H were obviously due to small impurities and thus discarded). The structure was established in a remarkably short time. The most
informative information was provided by the MS, apparently consisting exclusively of metastable ions. These results were beautifully supplemented by the amazing S.M.I.P - M.S. technique [11], establishing the probable existence of a nearly-molecular quasi-ion M+1 at m/e
269 (HRSMIPMS : 269.3195 ; calc. $C_{19}H_{41}$: 269.3207). The presence of 19 carbon atoms immediately suggested the close relationship of compound 1 with 18,19-bisnorpregnane. The cor-
pus of spectral data indicated the presence of two terminal methyl groups and 17 methylene
groups. Intensive use of artificial intelligence (CONGEN—GENOA) reduced the number of possibilities to one : 18,19-bisnor-(4,5),(9,10),(8,14),(13,17)-tetrasecopregnane (I). The remarkable chemical reactivity of compound 1 (acetylation, oxidation, hydrogenation and
Bischler-Napieralski — modified according to Tchitchibabine — were all negative) is in full
agreement with the proposed structure.

The structure, stereochemistry and absolute configuration of compound 1 were unambi-
guously established by direct chemical correlation with pregnane. The epoch-marking
simultaneous discovery in Stanford and Brussels that under certain experimental conditions
a wide variety of complex marine sterols could be transformed into silicone grease and
butyl phtalate in yields of over 100% had previously led to the isolation and description
of the remarkable enzyme steroldecyclase [12]. This enzyme provided the basis for an elegant correlation between our compound and the classical steroid series.

3.756 kg of pregnane (2) [13], submitted to very mild Kuhn-Roth conditions, yielded
0.17 mg (thus justifying fully the title of this Series) of the key degradation compound
18,19-bisnorpregnane (3), accompanied by some impurities. The crude reaction mixture was
immediately incubated at 29.04 ° C, pH 5.695 with steroldecyclase [12] affording in good
yield, after purification, compound 1 , undistinguishable by all its chemical properties
from the invisible natural product.

Of considerably greater scientific relevance were the results of the treatment of
compound 3 with steroldecyclase in the presence of tritiated seawater. Total absence of
isotope incorporation completely confirmed our personal views on the reaction mechanism[13]

With our previous studies on minor and trace marine invertebrates [14] steroid chemistry
had reached an absolute summit : it has now taken a gigantic step forwards. Indeed, 18,19-
bisnor-(4,5),(9,10),(8,14),(13,17)-tetrasecopregnane is the first steroid entirely devoid
of rings and consists exclusively of an unprecedented nineteen carbons side-chain. Biogene-
tically, compound 1 is the long-awaited missing link between methane and xestosoongesterol.
The complete lack of chirality in the molecule suggests a very great biochemical plasticity
and makes it a choice candidate for interesting incorporation experiments. It is tempting
to speculate that the associated phospholipids of the cell membranes of A.caveatemptor
will also reflect the dramatic evolutionary changes evidenced in compound 1 and might even
be devoid of phosphorus.

Preliminary data indicate bright possibilities for the use of 18,19-bisnor-(4,5),(9,10)-
(8,14),(13,17)-tetrasecopregnane in human contraception.A dramatic decrease in the usual
number of pregnancies was reported by female research workers of both teams who volunteered
to hold at appropriate moments a solid pill preparation [15] of compound 1 tightly between
the knees. Further investigations along these lines are actively pursued in our laboratories.

Acknowledgements. This work was generously supported by the National Institute of Wealth
(grant JB-007) and the Belgian Institut pour l'Enragement de la Recherche Scientifique
dans l'Industrie et l'Agriculture (I.R.S.I.A.). One of us (A.N.K) gratefully aknowledges
the receipt of a Fulblast fellowship. N.S. thanks the Institut Royal des Sciences Surna-
turelles for the loan of a diving mask (n° ARS/3764-6743/NG/2098658). We are grateful to
Dr. Lois Durham who would have recorded the 360 MHz nmr spectra, had she been given any
material. We thank Prof.G.Sourison (ut eructant quirites) for lack of comments.

References.

1. This paper was presented at the 1982 convention celebrating C.D's 1000th publication. This is paper $1000 + n$, $(n \to \infty)$.

2. See this Series, parts 1.17×10^2 to 2.78×10^2 .

3. Z. von Spielvogel, *Proc.Mick.Mouse Soc.*,$\underline{78}$,382,(1980).

4. L. Meereslustig, *Dreizig Jahren unter Kannibalen*, Leipzig, 1892, p.1547.

5. J.Pierret, N.Schubert, D.Middlebrook and Carl Djerassi, *Trans.Hansa Bay Philos.Soc.*, Series K,$\underline{182}$(211),pp. 1–981,(1980).

6. Supplied by Courvoisier,Ltd.

7. Supplied by South Pacific Breweries Ltd., Port Moresby

8. N.Schubert et al., *J.Chem.Necrol.*$\underline{18}$,211,(1981).

9. Carl Djerassi, *Pure and Applied Chemistry*, $\underline{41}$,113–144,(1975) —unexpurgated version.

10. Carl Djerassi, *Pure and Applied Poetry*, in press.

11. Scrambled Metastable Ions Partition Mass Spectrometry. See : J.Smith, K.Smith, O.Jones, W.Jones, A.Dupont, W.Dupont and Carl Djerassi, *Organic Mass Spectrometry*,$\underline{12}$,432,(1980). For applications, see M.Borboleta, J.Papillon, K.Schmetterling , A.Moth and Carl Djerassi, *J.Infect Physiol.*$\underline{71}$, 123,(1981).

12. J.E.K.Crabman, D. de Zola, A.N.Kismi, N.Schubert, J. Vastifar Hadji, J.Smith, K.Smith, K.Bouter IV Jr., A. Kichi Duoduma, Kristina Katzenellenbogen née Finkelstein, R.Pinto del Pinar Martinez Rosas Pulque Bamba y Mujeres Stop and Carl Djerassi, *Spheroids*,$\underline{31}$, 231,(1979).

13. Carl Djerassi in *Marine Unnatural Products*, P.J.Scheuer, Ed., Vol.$\underline{26}$, in preparation.

14. See This Series, Part 2.77×10^2 .

15. Patents pending.

chairman introduced his friend, Carl, as the only person he knew "who has written more than he has read."

To commemorate No. 1000, friends and collaborators of the chemistry titan gathered at a party where, amid other festivities, he was presented with a fictitious publication (see pages 68–71) composed in the format of *Tetrahedron Letters*[64] by four of his past associates, J.C. Braekman, D. Daloze, M. Kaisin, and Ben Tursch. Their names show up as authors, but with letters scrambled. Professor Tursch, a Djerassi "Postdoc" at Stanford and then his "bush chemist" in Brazil, continued to collaborate with C.D. for years afterwards while Dr. Tursch pursued marine research in chemistry and biology at Belgium's Université Libre de Bruxelles. The spoof paper took shape after Professor Djerassi visited Tursch and his colleagues at their biological station in Papua New Guinea; so it contains allusions to Laing Island, Hansa Bay, etc. The pastiche pokes friendly fun at Djerassi's writing style, at his favorite expressions, and at his abstinence from alcoholic drinks. Through the courtesy of Professors Tursch[67] and Djerassi,[64] we duplicated this delightful "publication" so you can savor its wit and imbibe the spirit of that happy celebration. "Tetraodon," by the way, is a puffer fish that provides tetrodotoxin, a natural product of interest to marine chemists.

While musing about Carl Djerassi's impact on chemistry, we wondered what scientist holds the all-time record for sheer *number* of articles printed. In *Current Contents*, information entrepreneur Eugene Garfield pointed out that the bibliography of the late Nikolei Vassilevich Belov (Institute of Crystallography, Moscow) included about 1500 papers. And, entomologist Theodore D.A. Cockerell (1865–1948) published over 3904 items![66] Can anyone top that?

For those who don't like being fenced in and want to move around, how about chemical vehicles that travel like the United States Marines—"by land and air and sea." We already noted that aircraft inspired the name biplanene (chapter 1). Let us now see how land and water vehicles have enriched chemical nomenclature.

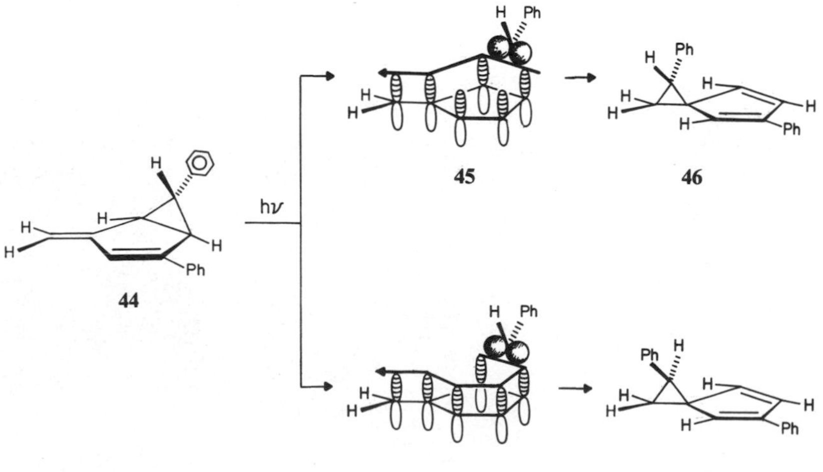

Howard Zimmerman and his students at the University of Wisconsin in Madison found that in their "Wisconsin black box" apparatus[68a] compound **44** isomerizes photically to **46** with high stereospecificity; they used "slither" to describe the journey of the PhCH moiety along the π system to its final destination.[68b] Later, Dr. Zimmerman's co-workers preferred the term "bicycle" and publicized it in subsequent papers.[69] Orbital drawings **45** and **47** show the carbenoid "bicycle" about to begin its trip. The benzylic hydrogen and the phenyl ring are handlebars, and the two attached orbitals are the wheels. Just as in a real bicycle, one handlebar points to the center of the track at all times. But, unlike the real thing, our benzylic biped could cycle forward or backward around the fulvene track (**45 → 46**, or **47 → 48**) without turning around. Molecules aren't dumb,* so all of them opt for the shorter route **45**. But, in doing so, a few neglect to leave the track at the exit carbon and instead coast past it a short distance on the curved path. Not to worry. These zealots simply back-peddle to the fork and continue rearward out of the arena. The direct cyclists end up as **46**. But the small fraction of overshooters become **48**; it's the penalty they pay for missing the exit. Zimmerman's team demonstrated that some bicycle reactions have more overshooters than this one.

Incidentally, if you itch to understand a Wisconsin-style track meet, don't turn to a "bicycle" theory developed by Nicholas Epiotis and company at Seattle's University of Washington.[71] These theoreticians proposed guidelines to predict how open-shell species (i.e., those with not all electrons paired) interact with each other and with closed-shell species (i.e., all electrons paired). Professor Epiotis lectured on this topic at Queen's University in Canada; and Saul Wolfe, in the audience, promptly baptized it the "bicycle" theory, because the hypothesis had a *bias* for *cyclic structures*.[71,72] Obviously, the cycle shops in Madison and those in Seattle peddle (pedal?) different merchandise.

Arieh Warshel (at Weizmann Institute of Science and at MRC Laboratory of Molecular Biology, Cambridge University) became expert in bicycle pedaling when he investigated polyene isomerizations in the red–purple pigment "rhodopsin" (from Greek *rhodon* for "rose" and *opsis* for "sight, appearance"). Rhodopsin consists of a multiply unsaturated aldehyde, (Z)-11-retinal linked as a Schiff base to a protein, opsin. When light enters the retina and strikes rhodopsin, the Z configuration switches to E. Not astounding, until you consider that this event is over in about 10^{-13} sec. How can such a geometric change take place so swifty in a polyene severely immobilized by protein binding?

Probing this enigma, Dr. Warshel used a computer to model the dynamics of polyene photochemistry. He found that a retinal with one Z link anywhere along the chain can end up as all-E retinal without much shift in atomic coordinates—provided that the major twist is accompanied by simultaneous partial twists elsewhere.[73]

When time came to publish the predictions, Dr. Warshel considered that a title such as "Computer Simulation of the Dynamics of the Primary Event in the Vision Process" represented accurately his work and its findings. But he needed more bait to entice readers. Grasping for a mechanical analogy, Dr. Warshel decided that either "crankshaft" or "bicycle-pedal" roughly conveyed the concerted movement of the atoms during Z → E. He chose "bicycle-pedal" because at that time almost everyone at MRC (even Nobel Prize winners) rode bicycles to get around.[74] Drawing **49** depicts the

49

---

*An exception may be the triterpenoid "semimoronic acid."[70]

rigid alkene segments as pedals. Note that Z becomes all-E with but little displacement at the polyene atoms.[73]

Later, at the University of Southern California, with postdoctorate associate Natalia Barboy, Professor Warshel adjusted his bicycle pedal but retained its key feature, namely, rotation around one bond in concert with partial movement around others.[75]

Finally, for the ultimate in "cycle" theories take a look at *The Hypercycle*, by Nobel laureate Manfred Eigen and theoretician Peter Schuster. This book presents a far-reaching hypothesis on natural self-organization at the molecular level as a basis for understanding the origin and evolution of life.[76]

Chemists who yearn for the sea have the "boat" form of cyclohexane (**50**) available. This usage of "boat" was introduced by Odd Hassel* of the University of Oslo;[83] and Derek Barton adopted it in his classic 1950 paper on conformational analysis of steroids.[84] (A picture of a ball-and-stick molecular model of cyclohexane appeared in 1919.[85]) Extending this nautical nomenclature to the boat's C–H bonds, Angyal and Mills likened some of them to parts of the vessel.[86] The two upright "flagpoles" (labeled

---

*Nobel Prize in chemistry, 1969, shared by Sir Derek H.R. Barton and Odd Hassel. Dr. Barton's award was the theme of a British commemorative postage stamp issued in 1977 on the centennial of the Royal Institute of Chemistry.[77]

Numerous Nobel laureates and their achievements have been depicted on postage. Among those most often honored were Marie and Pierre Curie. Individually or together they have appeared on about 95 stamps from 45 different countries, issued up to mid-1984. The runners-up are Albert Einstein (about 88 stamps from 31 countries); Guglielmo Marconi (30 from 19); Frederick and Irene Joliot-Curie (19 from 11); and Wilhelm K. Röntgen (18 from 11).[78] Philatelic commemoratives have also hailed the discovery and naming of many atomic elements[79] and have traced the history of other scientific achievements.[80] Interestingly, the United States has issued stamps honoring sixteen Nobel Prize winners, but no chemist is among them.[80e]

On coins, chemistry themes, or even science themes, understandably turn up much less frequently. A numismatic sampling includes Nobelists Marie Curie (coins minted by Poland in 1974 and by France in 1984), Otto Hahn (West Germany, 1979), Albert Einstein (East Germany, 1979), and organic chemist Justus Liebig (East Germany, 1978).[81] The Republic of San Marino heralded "science for the good of mankind" by striking nine coins in 1984 featuring Albert Einstein, Enrico Fermi, Guglielmo Marconi, Pierre and Marie Curie, Louis Pasteur, Alessandro Volta, Galileo Galilei, Leonardo da Vinci, and Hippocrates.[82] A Malaysian $ 1 coin of 1977 depicts the structural formula of isoprene in recognition of that country's natural rubber industry. Early astronomers Johannes Kepler and Nicolaus Copernicus are among other scientists saluted on metallic money.[81]

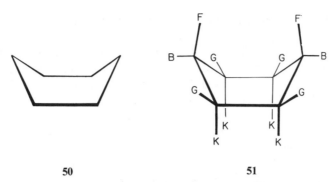

50   51

F in **51**) should satisfy patriots who like to fly banners at each end of their boats.[86] The "bowsprit" (a spar extending forward from the bow of a ship) gives its name to bonds B; again, instead of one there are two, for boats that don't know whether they are coming or going. England's William Klyne may have been the first to use "bowsprit" in this stereochemical way.[87] Richard Lawton and his University of Michigan mariners took care of the rest of the craft. They dubbed the G bonds "gunnel," after the upper edge of the sides of a ship, and the K bonds "keel," because they point down into the water (see **51**).[88]

Someone at Trinity College, Dublin, was amused or annoyed by boat, bowsprit, flagpole, and so forth. In a spoof article from that college, author "Alonzo S. Smith" proposed structure **52** for the fictitious compound Fe_6H_8 and, with tongue in cheek, set adrift the "raft" form.[89] When Smith's raft capsizes, it becomes the "hat" conformation (**53**). And, remarkably, the raft's six "stanchion" and two "keelson" bonds transform into "tassle" and "crown" positions.

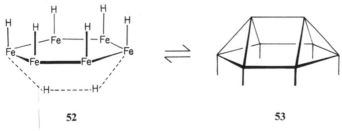

52   53

Master of this metamorphosis turns out to be Professor John T. Edward of McGill University, who was at Trinity College at that time. Believing that conformational terms were becoming unnecessarily numerous, Dr. Edward penned the raft paper one afternoon while supervising an undergraduate laboratory. He cleared it with department head Wesley Cocker, and submitted it on March 26, 1955, to *Chemistry and Industry* (*London*) under the pseudonym Alonzo S. Smith. (The previous evening Dr. Edward had attended Shakespeare's play *Twelfth Night* and it somehow moved him to couple a fancy first name with a plebian surname.) The editor enjoyed the (April Fool) joke, but printed it as a "letter" rather than a communication. Not everyone saw its humor: *Chemisches Zentralblatt* unwittingly abstracted the article.[90] In any case, we are pleased that A.S.S. has forever been exposed.[91,92]

The raft at Trinity College may have sunk, but the craft launched at Cambridge University by Jack Lewis and his mates proved seaworthy. The Cantabrigians prepared hexaosmium compounds such as $[Os_6(CO)_{17}(POMe_3)_4]$ and termed them "raft" complexes because of the near-flat arrangement of the six metal atoms.[93] The U.K. crew also uncovered activated raft clusters such as $[Os_6(CO)_{20}(CH_3CN)]$ and sailed them into unchartered waters of hexametal chemistry.[94] And they weren't joking.

By the way, letters written in jest to journal editors are not uncommon. For example, back in 1840, Justus Liebig, editor of *Annalen der Chemie und Pharmacie*, received a missive in French from "S.C.H. Windler" begging him to announce in *Annalen* "one of the most brilliant facts of organic chemistry." The writer claimed that passing chlorine through manganese acetate successively replaced the hydrogens, the oxygens, the carbons, and finally even the manganese to produce a compound of pure chlorine. Remarkably, however, the product preserved the original structural type![95] The *s(ch)windler* is thought[96] to be Friedrich Wöhler, who was attempting to ridicule a theory of that era elaborated by noted French chemist and educator Jean Baptiste André Dumas and known as the "Theory of Types."[97] Liebig didn't think much of the Dumas theory, either, and published the epistle.

In 1962, *Chemistry and Industry* (*London*) printed a letter, signed by William Graham, on the separation of enantiomers by gas chromatography. The writer proposed using mirrored surfaces as adsorbents. His idea, based on reflection symmetry, was this: a vacant site alongside an adsorbed D-molecule would appear to be occupied by an L-molecule. So, L-forms, sensing fewer L-spaces available, move along the column and elute sooner.[98]

If you can't browse through *Nachrichten aus Chemie, Technik und Laboratorium* regularly, at least try to catch their April 1st issues. Over the years, this German news magazine (roughly comparable to *Chemistry in Britain* or to *Chemical and Engineering News* and, until 1977, called *Nachrichten aus Chemie und Technik*) has often commemorated April Fools' Day by spoofing its readers. For example, in 1972 it announced that M.S. Grzimek *et al.* had synthesized the intriguing cyclopropyl compounds **54, 55,** and **56**.[99] The cumulene **54** was unusually stable to intense UV light and survived modest temperatures. But, it broke down when cooled below 0°C. Its foreshortened analog, **55**, was said to be sensitive to cold and to emit a very unpleasant odor. And **56**? The authors claimed this polycycle appeared in red, yellow, brown, and white modifications and described it as most dangerous.[100]

54   55   56

The same issue noted that "R.S. Schnauzer und Mitarbeiter" observed a completely unexpected reaction between 2, 3, 4-trimethyl-1, 3-pentadiene (**57**) and 2, 3, 4, 5-tetraethyl-3-methylbicyclo[4.1.0]heptane (**58**).[101] The encounter produced, in high yield and purity, a pair of pentamethylcyclopentanes, **59**. One more fact. The reaction was accelerated by cyclopentadienyl ion and took place only in the spring.

57   58   59

The chemistry in mock articles can be fascinating, and subtly, or not so subtly, devious. Wulf Rödder's spoof "Probleme bei der Anzucht fossiler Pflanzen, dargestellt am Palaeogelos-Beispiel," describes a new species, *Palaeogelos doleros*, cultivated from seeds found in fossil dung. It also announces discovery of two plant hormones, "hedonegamine," found only in female plants, and "nitrohedone," found only in male plants. Both hormones were thought to be responsible for the libidinous nature of this species.[102] By the way, *palaeogelos* can be translated from Greek as "ancient joke."

Sometimes the senior spoofer is revealed, as in the case of the discourse on "Temporäre Chiralität" penned by Michael J.S. Dewar, assisted by Brian Addle, Hans Krapp, Till E. Spiegel, and AA Predoctoral Fellow, Michel Ob.[103] In other instances, a well-known chemist may feign anonymity behind a cryptic pseudonym, as done in "Methine und seine Sphäromerisierungen" by A. Troischose (see p. 90).[104] A few spoofs are authorless, such as the ones titled "Elektrodynamisch induzierte chirale Synthese,"[105] "Die Potentialhyperfläche,"[106] and "The Sex Hormones of the Male Rhubarb."[107] Such delightful shenanigans, it seems, are often linked one way or another to participants in the conferences on stereochemistry held annually at Bürgenstock, Switzerland. You need to understand some German to appreciate these spoof papers.

Chemists not at ease with German can look for fun in some April issues of *CHEMTECH*. For example, in a delightful 1981 contribution from MIT titled "The generation of a highly reactive intermediate in the high-temperature pyrolysis of tetramethylsilane," Dietmar Seyferth and James Pudvin described the successful isolation of pure 3d orbitals (87% yield). Their product was free of nuclear contamination and condensed as an amber liquid. In cyclohexane solution, these orbitals became grafted onto carbons of the solvent to produce "orbital-augmented" carbons. This paper disclosed other fascinating properties of free 3d orbitals and alluded to experiments in progress on isolation of pure p and f orbitals. By the way, their research received initial support from government agencies—"until the first progress reports."[108]

Once in a while, something written with tongue-in-cheek seems credible and is accepted at face value. Such was the case of a three-page biography of Claude Émile Jean–Baptiste Litre, composed by physicist K.A. Woolner and published in an April issue of *Chem 13 News*.[109] This engaging article commemorated the 200th anniversary of the death of Monsieur Litre (1716–1778), maker of scientific glassware and namesake for the SI unit of volume. Not recognized as a jest, the biography was condensed and featured in *Chemistry International*, the official news magazine of IUPAC.[110,111] Professor M.L. McGlashan (University College, London) revealed the hoax to the magazine's editor, who dutifully relayed the exposé to its readers.[112]

Another instance of a joke item that evaded recognition, much to the chagrin of the editors, involved an abstract in the "Forensic" section of the *Analyst* (March, 1944).[113] It summarized a hoax paper, "Toxicological Significance of Laevorotatory Ice Crystals," published by Joseph Beeman in *Bulletin of the Bureau of Criminal Investigation, New York State Police* (**1943**, *8*, Dec., pp. 6–8).[114] The abstract describes how Beeman froze tap water to obtain a mixture of levorotatory monoclinic rhombs and dextrorotatory hexagonal plates, and how Beeman also observed remarkable toxicological effects of the two types of ice crystals on animals and on humans. Why this fanciful paper got abstracted is not clear. But the very next issue of *Analyst* (April, 1944) carried a terse corrigendum asking readers to ignore the abstract, which was "inserted in error."[115] According to one version of the incident, the summary was sneaked into the *Analyst* by undergraduates at Cambridge University who "got into the printing works."[16] But a more plausible version holds that a disgruntled abstractor intentionally synopsized the hoax paper after receiving a termination notice.[117]

A slip-up at the printers was, in part, the reason one April Fool spoof lasted longer than intended. It involved a 1979 article in *CHEMTECH* by Arthur Scott (Reed College) and purported to be a translation of a German paper from *Berichte* on "Thermochemical Decomposition of Water."[118] That the translation was contrived might have become apparent from the early date (1870) in the citation to the *Berichte* paper. So, the editor of *CHEMTECH* wanted to emblazon that date with color. But the color plate didn't print, the 1870 didn't appear at all, and many readers remained taken in.[119] In fact, Professor Scott received requests for more information, and his article was picked up by the abstracting service, *Technical Survey*.[120] However, at Chemical Abstracts Service, vigilant abstractors correctly labeled Scott's article "a historical account."[120,121]

If you crave a steady diet of satire, consider subscribing to *The Journal of Irreproducible Results*, established in 1955 and published four times a year.[122] This periodical prints letters, research "findings," theories, observations, and so forth that poke fun at all branches of science—from Astronomy to Zoology—or at any other scholarly discipline. But, its title notwithstanding, *JIR* is not simply an outlet for an experiment that didn't work the second time. Unless, of course, that second attempt proved your point.

REFERENCES AND NOTES

1. Zimmerman, H.E.; Paufler, R.M. *J. Am. Chem. Soc.* **1960**, *82*, 1514–1515.
2. Zimmerman, H.E., private communication, September, 1975.
3. Roberts, J.D., private communication to Zimmerman, H.E., January, 1960, conveyed to us as in reference 2.
4. Hurd, C.D. *J. Chem. Educ.* **1961**, *38*, 43–47.
5. Cahn, R.S.; Dermer, O.C. *Introduction to Chemical Nomenclature*, 5th ed. (Butterworth & Co., London, 1979), p. 89.
6. Zimmerman, H.E.; Grunewald, G.L. *J. Am. Chem. Soc.* **1964**, *86*, 1434–1436.
7. Woodward, R.B.; Hoffmann, R. *The Conservation of Orbital Symmetry* (Verlag Chemie GmbH., Weinheim/Bergstr., 1970).
8. Gutsche, C.D.; Muthukrishnan, R. *J. Org. Chem.* **1978**, *43*, 4905–4906.
9. Gutsche, C.D., private communications, September, 1975, and March, 1984.
10. (a) Zinke, A.; Ziegler, E. *Chem. Ber.* **1944**, *77*, 264–272, (b) Zinke, A.; Zigeuner, G.; Hossinger, K.; Hoffman, G. *Monatsh. Chem.* **1948**, *79*, 438–439, (c) Zinke, A.; Kretz, R.; Leggewie, E.; Hossinger, K. *Monatsh. Chem.* **1952**, *83*, 1213–1227.
11. Gutsche, C.D.; Muthukrishnan, R.; No, K.W. *Tetrahedron Lett.* **1979**, 2213–2216.
12. Hayes, B.T.; Hunter, R.F. *J. Appl. Chem.* **1958**, *8*, 743–748.
13. Kämmerer, H.; Happel, G.; Caesar, F. *Makromol. Chem.* **1972**, *162*, 179–197.
14. Gutsche, C.D.; Dhawan, B.; Levine, J.A.; No, K.H.; Bauer, L.J. *Tetrahedron*, **1983**, *39*, 409–426.
15. Böhmer, V.; Goldmann, H.; Vogt, W. *J. Chem. Soc., Chem. Commun.* **1985**, 667–668.
16. Gutsche, C.D.; Bauer, L.J. *J. Am. Chem. Soc.* **1985**, *107*, 6052–6059, 6059–6063.
17. Dhawan, B.; Gutsche, C.D. *J. Org. Chem.* **1983**, *48*, 1536–1539.
18. Gutsche, C.D.; Dhawan, B.; No, K.H.; Muthukrishnan, R. *J. Am. Chem. Soc.* **1981**, *103*, 3782–3792.
19. (a) Gutsche, C.D. *Acc. Chem. Res.* **1983**, *16*, 161–170, (b) Gutsche, C.D. *Top. Curr. Chem.* **1984**, *123*, 1–47.
20. Prinzbach, H. *Angew. Chem.* **1964**, *76*, 235–236.
21. Woischnik, E., Thesis, University of Freiburg, Germany, 1967, quoted in reference 22.
22. Prinzbach, H.; Fischer, U. *Helv. Chim. Acta* **1967**, *50*, 1669–1692.
23. Roberts, J.D.; Streitwieser, Jr., A.; Regan, C.M. *J. Am. Chem. Soc.* **1952**, *74*, 4579–4582.
24. Prinzbach, H. *Pure Appl. Chem.* **1971**, *28*, 281–292.
25. Yoshida, Z. *Pure Appl. Chem.* **1982**, *54*, 1059–1074.

26. Yoneda, S.; Shibata, M.; Kida, S.; Yoshida, Z.; Kai, Y.; Miki, K.; Kasai, N. *Angew. Chem. Int. Ed. Engl.* **1984**, *23*, 63–64.

27. Masamune, S.; Cuts, H.; Hogben, M.G. *Tetrahedron Lett.* **1966**, 1017–1021.

28. Masamune, S., private communication, October, 1977.

29. Bottino, F.; Pappalardo, S. *Tetrahedron* **1982**, *38*, 665–672.

30. Chivers, T.; Proctor, J. *J. Chem. Soc., Chem. Commun.* **1978**, 642–643.

31. Edwards, J.G., private communications, May and June, 1979, October, 1981.

32. Gates, A.S.; Edwards, J.G. *Inorg. Chem.* **1977**, *16*, 2248–2252.

33. Krebs, B.; Hürter, H.–U. *Angew. Chem.* **1980**, *92*, 479–480.

34. Lüttringhaus, A.; Gralheer, H. *Liebigs Ann. Chem.* **1942**, *550*, 67–98.

35. Lüttringhaus, A. *Liebigs Ann. Chem.* **1937**, *528*, 181–210.

36. Lüttringhaus, A.; Kullick, W. *Angew. Chem.* **1958**, *70*, 438.

37. Wehrli, W. *Top. Curr. Chem.* **1977**, *72*, 21–49.

38. (a) Oppolzer, W.; Prelog, V.; Sensi, P. *Experientia*, **1964**, *20*, 336–339, (b) Brufani, M.; Fedeli, W.; Giacomello, G.; Vaciago, A. *Experientia* **1964**, *20*, 339–342, (c) Leitich, J.; Oppolzer, W.; Prelog, V. *Experientia* **1964**, *20*, 343–344.

39. Prelog, V., private communication, June, 1977.

40. Doyle, T.W., private communication, April, 1986.

41. Nettleton, D.E. Jr.; Bradner, W.T.; Bush, J.A.; Coon, A.B.; Moseley, J.E.; Myllymaki, R.W.; O'Herron, F.A.; Schreiber, R.H.; Vulcano, A.L. *J. Antibiot.* **1977**, *30*, 525.

42. Doyle, T.W.; Nettleton, D.E.; Grulich, R.E.; Balitz, D.M.; Johnson, D.L.; Vulcano, A.L. *J. Am. Chem. Soc.* **1979**, *101*, 7041–7049.

43. From the popular Christmas song, *Rudolph The Red-Nosed Reindeer*.

44. Ugi, I.; Marquarding, D.; Klusacek, H.; Gokel, G.; Gillespie, P. *Angew. Chem. Int. Ed. Engl.* **1970**. *9*, 703–730.

45. Professor Ugi continued research on permutational nomenclature at the Technical University of Munich. For an improved version, see Ugi, I.; Dugundji, J.; Kopp, R.; Marquarding, D. *Perspectives in Theoretical Stereochemistry*, Berthier, G. *et al.* (Eds.), (Springer-Verlag, New York, 1984), vol. 36. Ugi, I., private communication, June, 1986.

46. Collman, J.P.; Gagne, R.R.; Reed, C.A.; Halbert, T.R.; Lang, G.; Robinson, W.T. *J. Am. Chem. Soc.* **1975**, *97*, 1427–1439.

47. Kuhn, R. Molekulare Asymmetrie. In *Stereochemie*, Freudenberg, K. (Ed.), (Franz Deuticke, Leipzig–Wien, 1933), pp. 803–824.

48. Michinori, Ō. Recent Advances in Atropisomerism. In *Topics in Stereochemistry*, Allinger, N.L.; Eliel, E.L.; Wilen, S.H. (Eds.), (Interscience: John Wiley & Sons, New York, 1983), vol. 14, pp. 1–81.

49. Collman, J.P.; Brauman, J.I.; Doxsee, K.M.; Halbert, T.R.; Bunnenberg, E.; Linder, R.E.; LaMar, G.N.; Del Gaudio, J.; Lang, G.; Spartalian, K. *J. Am. Chem. Soc.* **1980**, *102*, 4182–4192.

50. (a) Freitag, R.; Mercer–Smith, J.A.; Whitten, D.G. *J. Am. Chem. Soc.* **1981**, *103*, 1226–1228, (b) Whitten, D.G.; Russell, J.C.; Schmehl, R.H. *Tetrahedron* **1982**, *38*, 2455–2487.

51. Collman, J.P.; Groh, S.E. *J. Am. Chem. Soc.* **1982**, *104*, 1391–1403.

52. Collman, J.P.; Brauman, J.I.; Collins, T.J.; Iverson, B.; Sessler, J.L. *J. Am. Chem. Soc.* **1981**, *103*, 2450–2452.

53. (a) Collman, J.P.; Brauman, J.I.; Collins, T.J.; Iverson, B.L.; Lang, G.; Pettman, R.B.; Sessler, J.L.; Walters, M.A. *J. Am. Chem. Soc.* **1983**, *105*, 3038–3052, (b) Collman, J.P.; Brauman, J.I.; Iverson, B.L.; Sessler, J.L.; Morris, R.M.; Gibson, Q.H. *J. Am. Chem. Soc.* **1983**, *105*, 3052–3064.

54. (a) Suslick, K.S.; Fox, M.M. *J. Am. Chem. Soc.* **1983**, *105*, 3507–3510, (b) Gold, K.W.; Hodgson, D.J.; Gold, A.; Savrin, J.E.; Toney, G.E. *J. Chem. Soc., Chem. Commun.* **1985**, 563–564.

55. Woodward, R.B. *et al. J. Am. Chem. Soc.* **1981**, *103*, 3210–3213, 3213–3215, 3215–3217.

56. Arnison, G. *et al. Phys. Lett.* **1983** *128B*, 336–342. For a physics publication with 89 coauthors see Balea, O. *et al. Nucl. Phys.* **1973**, *B52*, 414–421.

57. (a) McWilliam, D.C.; Balasubramanian, T.R.; Kuivila, H.G. *J. Am. Chem. Soc.* **1978**, *100*, 6407–6413, (b) If we permit any number of middle names, then Gloria Berenice Chagas Tolentino de Carvalho Brazão da Silva has the edge over Balu. Tursch, B.; Tursch, E.; Harrison, I.T.; da Silva, G.B.C.T. de C.; Monteiro, H.J.; Gilbert, B.; Mors, W.B.; Djerassi, C. *J. Org. Chem.* **1963**, *28*, 2390–2394.

58. Kuivila, H.G., private communication, April, 1983.

59. Sugasawa, S.; Kametani, T. *Yakugaku Zasshi*, **1945**, *65*, 372.

60. Ihara, M.; Noguchi, K.; Ohsawa, T.; Fukumoto, K.; Kametani, T. *J. Org. Chem.* **1983**, *48*, 3150–3156.
61. Kametani, T., private communications, January, 1984 and June, 1986.
62. Barth, G.; Djerassi, C. *Tetrahedron* **1981**, *37*, 4123–4142.
63. Wilds, A.L.; Djerassi, C. *J. Am. Chem. Soc.* **1946**, *68*, 1712–1715.
64. Djerassi, C., private communications, February, March, and May, 1984.
65. Statistics were gleaned from a publication list kindly provided by Professor Djerassi.
66. Garfield, E. *Current Contents* **1982**, *13*, No. 42, 5–14.
67. Tursch, B., private communications, March, 1984, and March, 1987.
68. (a) Zimmerman, H.E. *Mol. Photochem.* **1971**, *3*, 281–292, (b) Zimmerman, H.E.; Juers, D.F.; McCall, J.M.; Schröder, B. *J. Am. Chem. Soc.* **1971**, *93*, 3662–3674.
69. (a) Zimmerman, H.E.; Cutler, P.C. *J. Org. Chem.* **1978**, *43*, 3283–3303, (b) Zimmerman, H.E.; Cutler, P.C. *J. Chem. Soc., Chem. Commun.* **1978**, 232–234, (c) For a review see Zimmerman, H.E. *Chimia* **1982**, *36*, 423–428.
70. Bagchi, A.; Sahal, M.; Sinha, S.C.; Ray, A.B.; Oshima, Y.; Hikino, H. *J. Chem. Res. (S)* **1985**, 398–399.
71. Epiotis, N.D.; Sarkanen, S.; Bjorquist, D.; Bjorquist, L.; Yates, R. *J. Am. Chem. Soc.* **1974**, *96* 4075–4085.
72. Wolfe, S., private communication, September, 1979.
73. Warshel, A. *Nature (London)* **1976**, *260*, 679–683.
74. Warshel, A., private communication, August, 1985.
75. Warshel, A.; Barboy, N. *J. Am. Chem. Soc.* **1982**, *104*, 1469–1476.
76. Eigen, M.; Schuster, P. *The Hypercycle* (Springer-Verlag, New York, 1979).
77. A postcard rendering of the stamp was kindly furnished by Sir Derek and Madam Barton.
78. (a) "Chemistry in Philately," a booklet distributed at the philatelic exhibit held at the 186th National Meeting of the American Chemical Society, Washington, DC, 1983, (b) Stierstadt, K. *Philatelia Chimica* **1984**, *6*, 20–23 (the journal of the Chemistry-on-Stamps Study Unit of the American Topical Association), (c) Professor Klaus Stierstadt (Department of Physics, University of Munich) kindly made available to us his extensive checklists of stamps dealing with physics (private communication, July, 1985).
79. Miller, F.A. *Philatelia Chimica*, **1983**, *5*, 66–73, 114–120; **1984**, *6*, 16–21 (the journal of the Chemistry-on-Stamps Study Unit of the American Topical Association). In 1986, this journal became *Philatelia Chimica et Physica*, the journal of the Chemistry and Physics On-Stamps Study Unit.
80. (a) For example, see "A Postage Stamp History of Chemistry," Miller, F.A. *Appl. Spectrosc.* **1986**, *40*, 911–924, and "The History of Spectroscopy as Illustrated on Stamps," Miller, F.A. *Appl. Spectrosc.* **1983**, *37*, 219–225, (b) Schreck, J.O.; Lang, C.M. *J. Chem. Educ.* **1985**, *62*, 1041–1042, (c) Schreck, J.O. *J. Chem. Educ.* **1986**, *63*, 283–287; (d) Chenier, P.J. *J. Chem. Educ.* **1986**, *63*, 498–500. (e) Popp, F.D. *Philatelia Chimica et Physica*, **1986**, *8*, 93–101, **1987**, *9*, 27–34.
81. (a) Krause, C.L.; Mishler, C. *Standard Catalog of World Coins*, (Krause Publications, Iola, WI, 1984), (b) Described in *Proof Collectors Corner* **1984**, Nov./Dec. issue, p. 188 (a newsletter published by the World Proof Numismatic Association, Pittsburgh, PA, 15201, USA).
82. Described in *Coin World*, **1984**, April 25, p. 48 (a numismatic newspaper published by Amos Press Inc., Sidney, OH, 45367, USA).
83. Hassel, O. *Quart. Rev.* **1953**, *7*, 221–230.
84. Barton, D.H.R. *Experientia* **1950**, *6*, 316–320.
85. Mohr, E. *J. Prakt. Chem.* [2] **1919**, *98*, 315–353.
86. Angyal, S.J.; Mills, J.A. *Rev. Pure Appl. Chem.* **1952**, *2*, 185–202.
87. Lyle, G.G., Lyle, R.E. "Letters" section, *J. Chem. Educ.* **1973**, *50*, 655–656.
88. McEuen, J.M.; Nelson, R.P.; Lawton, R.G. *J. Org. Chem.* **1970**, *35*, 690–696.
89. Smith, A.S. *Chem. Ind. (London)* **1955**, 353–354.
90. *Chem. Zentralbl.* **1956**, No. 39, 10658. For a news item see *Nachr. Chem. Techn.* **1957**, *5*, 62.
91. Wheeler, D.M.S., private communication, March, 1983. Professor Wheeler (University of Nebraska, Lincoln) was on the staff at Trinity College at the time.
92. Edward, J.T., private communications, April and June, 1983.
93. Goudsmit, R.J.; Johnson, B.F.G.; Lewis, J.; Raithby, P.R.; Whitmire, K.H. *J. Chem. Soc., Chem. Commun.* **1983**, 246–247.

94. (a) Goudsmit, R.J.; Jeffrey, J.G.; Johnson, B.F.G.; Lewis, J.; McQueen, R.C.S.; Sanders, A.J.; Liu, J.–C. *J. Chem. Soc., Chem. Commun.* **1986**, 24–26, (b) Jeffrey, J.G.; Johnson, B.F.G.; Lewis, J.; Raithby, P.R.; Welch, D.A. *ibid.* **1986**, 318–320.

95. *Liebigs Ann. Chem. Pharm.* **1840**, *33*, 308–310.

96. Friedman, H.B. *J. Chem. Educ.* **1930**, *7*, 633–636.

97. Dumas, J. *Liebigs Ann. Chem. Pharm.* **1840**, *33*, 259–300.

98. Graham, W. *Chem. Ind.* (*London*) **1962**, 1533.

99. *Nachr. Chem. Techn.* **1972**, *20*, 125.

100. Doodling with chemical formulas is, of course, a venerable pastime. For another published example, see an amusing letter depicting ketene "animals" being attacked by "Greater Hump-backed Nucleophiles." Galt, R.H.B.; Mills, S.D. *Chem. Brit.* **1965**, *1*, 230. This item was also recounted by Reese, K.M. *Chem. Eng. News* **1984**, July 9, p. 132, and by Seikaly, H.R.; Tidwell, T.T. *Tetrahedron* **1986**, *42*, 2587–2613. The whimsy in the Galt–Mills letter was inspired when a paper on ketene aminals was abstracted to be about ketene animals. Bredereck, H.; Effenberger, F.; Beyerlin, H.P. *Chem. Ber.* **1964**, *97*, 3081–3087; *Chem. Abstr.* **1965**, *62*, 1561*d*.

101. *Nachr. Chem. Techn.* **1972**, *20*, 126.

102. Rödder, W. *Pharm. Ztg.* **1976**, *121*, 479–484.

103. Addle, B.; Dewar, M.J.S.; Krapp, H.; Spiegel, T.E.; Ob, M. *Nachr. Chem. Techn.* **1974**, *22*, 135–136.

104. Troichose, A. *Nachr. Chem. Techn.* **1970**, *18*, 127–128.

105. *Nachr. Chem. Techn.* **1975**, *23*, 127–128.

106. *Nachr. Chem. Techn.* **1975**, *23*, 132.

107. *Nachr. Chem. Techn.* **1970**, *18*, 129.

108. Seyferth, D.; Pudvin, J.J. *CHEMTECH* **1981**, *11*, 230–233. We thank Dr. W.V. Metanomski (Chemical Abstracts Service) for a copy of this article.

109. Woolner, K.A. *Chem. 13 News* **1978**, No. 95, 1–3; also reprinted in *Int. Newsl. Chem. Educ.* **1979**, No. 11, 7–9.

110. *Chem. Int.* **1980**, *No. 1*, 32.

111. Gellender, M. (Editor, Chemistry International), private communication, August, 1980.

112. *Chem. Int.* **1980**, *No. 3*, 2.

113. *Analyst* (*London*) **1944**, *69*, 97–98.

114. Ramsay, O.B. *Stereochemistry* (Heyden & Son Ltd., Philadelphia, 1981), pp. 238–240.

115. *Analyst* (*London*) **1944**, *69*, 138.

116. Reese, K.M. *Chem. Eng. News* **1983**, April 18, p. 92.

117. Jones, D.E.H. *New Scientist* **1966**, *32*, 465–467.

118. (a) Scott, A.F. *CHEMTECH* **1979**, *9*, 208–209, (b) Dr. W.V. Metanomski (Chemical Abstracts Service) kindly alerted us to this episode and provided copies of relevant material.

119. (a) *CHEMTECH* **1979**, *9*, 450–451, (b) *CHEMTECH* **1979**, *9*(9) inside back cover.

120. *CHEMTECH* **1979**, *9*, 719–720.

121. *Chem. Abstr.* **1979**, *91*, 122907a.

122. (a) Address: Journal of Irreproducible Results Publishers, P.O. Box 234, Chicago Heights, IL, 60411, USA, (b) *The Best of the Journal of Irreproducible Results*, Scherr, G.H. (Ed.), (Workman Publishing Company, New York, 1982).

Chapter 6

WE AREN'T ALL SQUARES!

Cyclohexane boats and chairs evolved because early chemists (notably Hermann Sachse[1,2] and Ernst Mohr[3,4]) knew their geometry. The internal angle in a regular, planar hexagon is 120° Saturated carbons are tetrahedral (109°28'), and cyclohexane would be unhappy if flat. So it puckers to avoid angle strain. This is but one example of the importance of geometry in chemistry. As we shall now see, several compounds have been named after geometric figures.

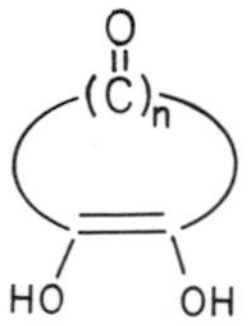

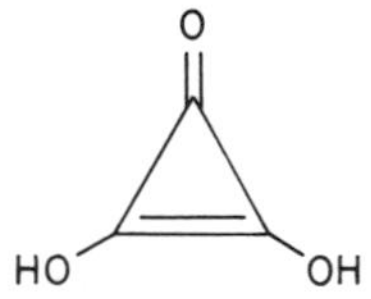

Cyclic molecules of general formula **1** were termed "oxocarbon acids" by Robert West and his students at the University of Wisconsin.[5] Such types are quite acidic, and in their dianions all the oxygens share the charge. Let's focus on the first five members of the family (**2–6**) to show how geometric shape led to their individual names.

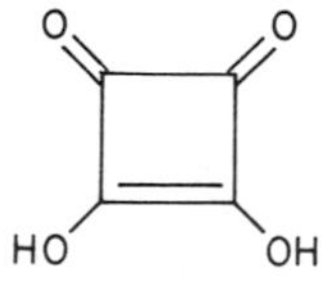

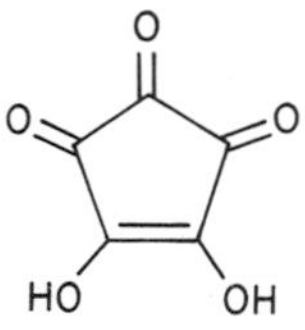

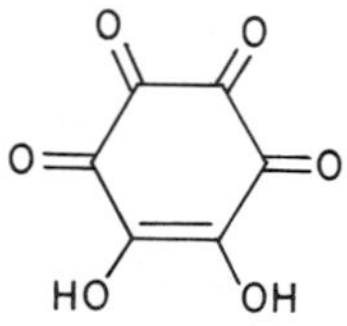

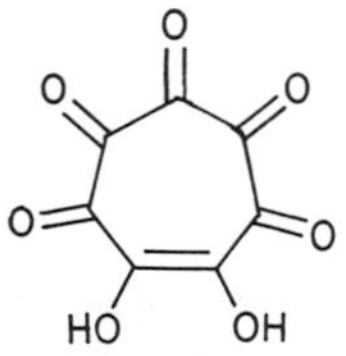

In Germany, **2** was known as "Dreiecksäure" (triangle acid). But, in 1975, the Wisconsin researchers decided that "deltic" (after the Greek capital letter delta, Δ) sounded better than "triangle."[6] So "deltic acid" is the name most commonly used.[7] The next homolog, **3**, was christened "squaric acid" around 1960 by George Van Dyke Tiers of the 3M Company, Minneapolis,[6,8] and it became "Quadratsäure" in the German literature.[9] Acids **4** and **5** and their (colored) salts have been around since 1825 and 1837, respectively, but their structures were not known in those days. Hence, Leopold Gmelin[10] and J.F. Heller[11] labeled them after the colors of their salts. Thus, **4** became "croconic acid" (from the Greek *krokos*, meaning "yellow"); and **5** is "rhodizonic acid" (from the Greek *rhodizein*, meaning to be "rose-red"). (Dr. West denied he considered renaming **4** "pentagonic acid" in appreciation of a research grant from the United States Air Force.[12]) Homolog **6** was looked upon as "heptagonic acid" in Dr. West's research group;[6] he definitely preferred this to "septic acid." No wonder! Whereas Latin *septem* means "seven," Greek *septikos* means "rot."

Along this vein we note that amine **7** is "septicine." This plant alkaloid was named by J.H. Russel, who first isolated it from *Ficus septica*.[13]

But christening plant chemicals turned out to be traumatic, at least for English, Bonner, and Haagen-Smit. In 1939, at the California Institute of Technology, they extracted vegetation that had been wounded (i.e., ground up or heated). Some of the substances isolated (referred to as wound hormones) proved capable of promoting cell enlargement and cell division in mature, unwounded plants. One such active principle is the C_{12} enedioic acid **8**, which they labeled "traumatic acid."[14]

7   8   9

Shocking, you think? Then focus on the *vic*-dimethylakanedioic acids represented by general formula **9** (the m and n chains can contain unsaturation). Investigators in Cambridge, England, obtained such acids from the lipids of bacteria (genus *Butyrivibrio*) in the rumen.[15] The first one they discovered was saturated and proved difficult to detect. For example, on GLC it eluted broadly with a retention time of about 30 hr. In fact, the U.K. team only realized its existence after accidentally having left the gas chromatograph running over the weekend.[16] It also turned out that mass spectral molecular ions from the dimethyl esters of **9** lost methanol on fragmentation. This observation initially threw the investigators off the track in their structural studies because dimethyl esters ordinarily eject a methoxy group rather than methanol. Having been thus misled, Roger Klein, Geoffrey Hazlewood, Patrick Kemp, and Rex Dawson decided to get even. They gave **9** the class name "diabolic acids," a devilish title that fit the situation because Greek *diabollo* means "to mislead."[15] Professor Klein also perceived **9** as having horns like the devil.[16] But, enough about these infernal compounds. Let's get back to oxocarbon acids and their fascinating chemistry.[17]

Resonance structures for the dianions (shown for the deltate ion) lead to the prediction that such planar, polygonal species (e.g., **10**) should have "aromatic" stability.[18,19] Aromaticity in these dianions is supported by theoretical calculations, which indicate that the occupied π molecular orbitals are completely filled.[5] Thus, potassium croconate vies with benzene as the oldest known aromatic substance; both were first isolated in 1825.[12]

10

Several syntheses of oxocarbon acids make use of nucleophilic substitution in polyhalogenated precursors. For example, the first step in the original synthesis of squaric acid in 1959 involved a threefold replacement.[20] Sidney Cohen, John Lacher, and Joseph Park, who brought off this quadrilateral coup, found that **3** was almost as

3

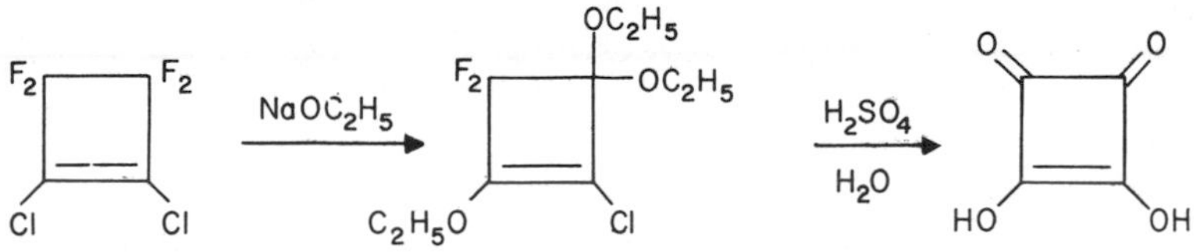

acidic as sulfuric acid! Beginners in organic chemistry might not view oxocarbons as close relatives of carboxylic acids. But they may derive some comfort from the fact that alcohols convert oxocarbon acids to esters (mono- or diesters),[9] which in turn give amides with amines (**11** → **12** → **13**).[9,21] Sound familiar?

acid **11** | diester **12** | diamide **13**

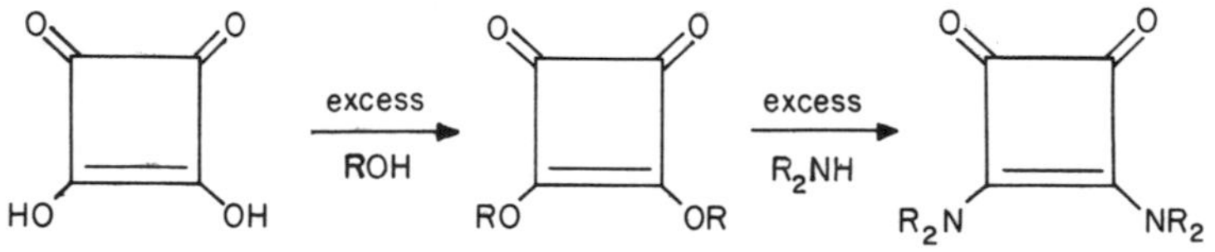

A circle can have radii emanating from its center, like the spokes of a wheel. When Edgar Heilbronner at the University of Basel sought a name for polyenes such as **14**–**17**, he conversed with John Platt at a Gordon Conference.[22] Dr. Platt recalled his elementary school geometry, noted the "radiating" methylenes in **14**–**17**, and said,

14 **15** **16** **17**

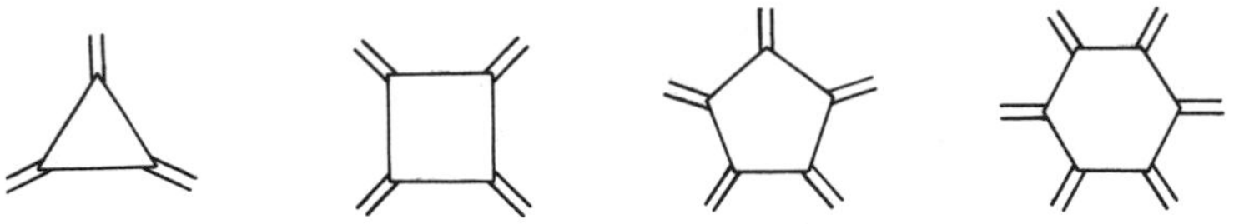

"What is wrong with 'radialenes'?" Nothing, thought Dr. Heilbronner, who adopted the term "radialene" in his paper on the absorption spectrum of the methylated analog **18**.[23] This fascinating compound had been prepared by H. Hopff and A.K. Wick at the Technische Hochschule, Zürich.[24] Later, a numeral in brackets showed up to keep track of the spokes; thus **17** is [6]radialene.

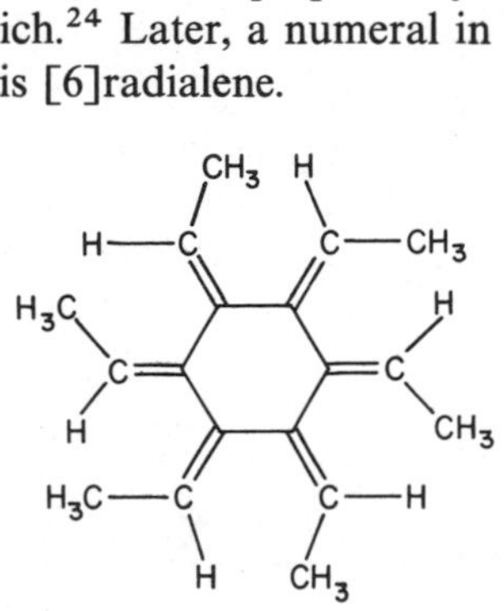

18

Since those early papers appeared, radialene chemistry has wheeled right along. The parent structure **17** is known.[25] We also have [4]radialene (**15**)[26] as well as hexamethyl- and hexacyano[3]radialene.[27,28] Not to mention radialenes in which the double bonds inhabit a quinoid system (**19**)[29] or a heterocyclic ring (**20**).[30] There is even an analog with no place for a spoke (**21**).[31] In octaphenyl[4]radialene[32] (**22**) and in heptaphenyl[4]radialene[33] the aryl rings cannot be coplanar, so these molecules may be molecular propellers (see chapter 3).

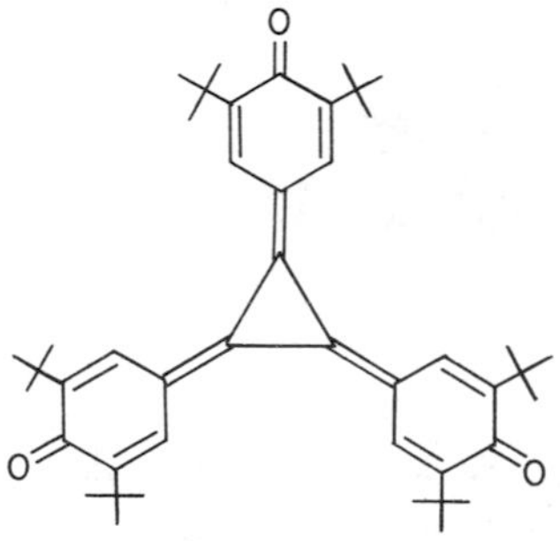

19 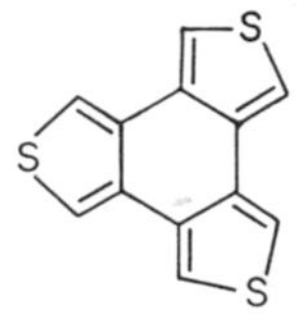   20 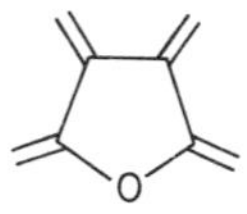   21

22

Harold Hart and co-workers at Michigan State University suggested attaching the suffix "radialene" to the name of a parent system.[34] Thus, polyene **24**, a possible intermediate in Hart's conversion of **23** to **25**, becomes "naphtharadialene."[34]

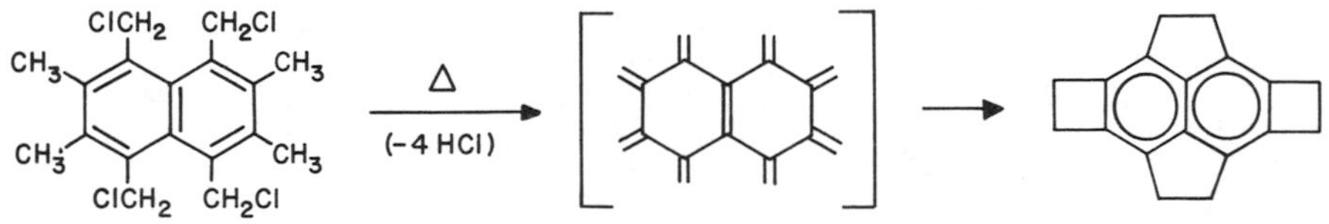

23   24   25

Bridged radialenes like **26**, **27**, and **28** fascinate chemists because the π-bonds could interact transannularly. And, a fully radialated bicyclic analog such as **29** falls, structurally, between barrelene (p. 58) and triptycene (p. 99)[35] This intriguing relationship did not go unnoticed in Switzerland. In 1980, Pierre Vogel and co-worker Olivier Pilet at the University of Lausanne synthesized **29**, and Edgar Heilbronner's spectroscopists at the University of Basel, in collaboration with Vogel et al., reported its photoelectron spectrum.[35,36]

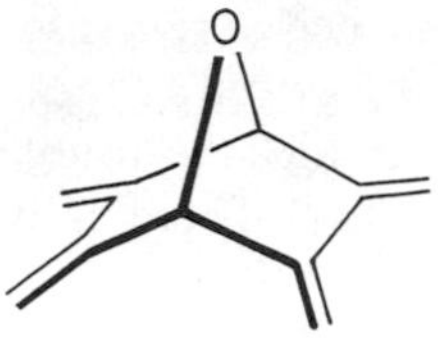

26 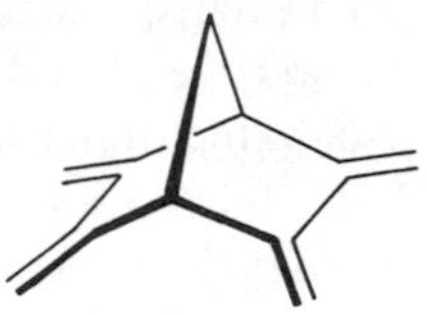   **27** 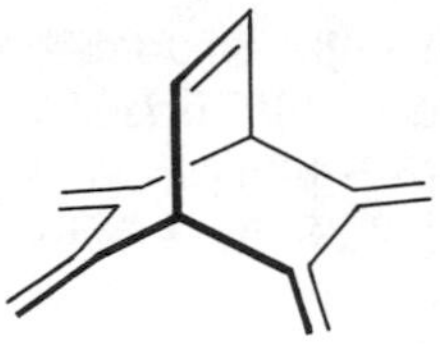   **28**

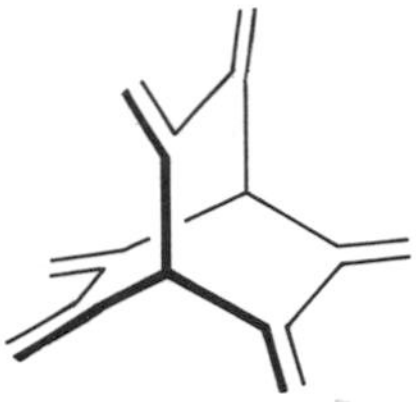

29    **30**

At coffee breaks, Dr. Vogel chatted with Hugo Wyler, a faculty colleague with a penchant for likening formulas to familiar objects or living things. For example, Professor Wyler dubbed **26** *chat écrasé* (flattened cat). And for **29** he suggested *hérisson* (hedgehog), because the spokes look like that animal's protective bristles (**30**). Dr. Vogel got the point and ultimately adopted the class name "hericene," after the Latin *hericeus* for "hedgehog."[37] As a result, the Lausanne researchers recommend [*l.m.n*]hericene as a generic shorthand for bicyclo[*l.m.n*]alkanes with $l + m + n$ methylidene units (*n* can be zero). Thus **29** carries a dual identity: [2.2.2]hericene in the west of Switzerland; and bicyclo-[2.2.2]radialene in the "Hart" of Michigan.[34]

Feast your eyes upon hexaethynylbenzene (**31**), brought into our world by Peter Vollhardt's group (University of California, Berkeley) and Roland Boese (Universität–Gesamthochschule, Essen). They fashioned this rimless wheel from hexabromobenzene in but two steps.[38] It is not a radialene, yet it radiates beauty just the same. And how intriguing to think a rim could materialize if one or both orthogonal sets of alkyne π-orbitals can diffuse through space. The original publication assigned no special name to this genre, but something so pleasing esthetically is bound to get christened sooner or later.

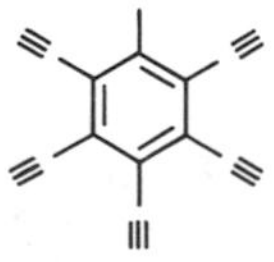

31 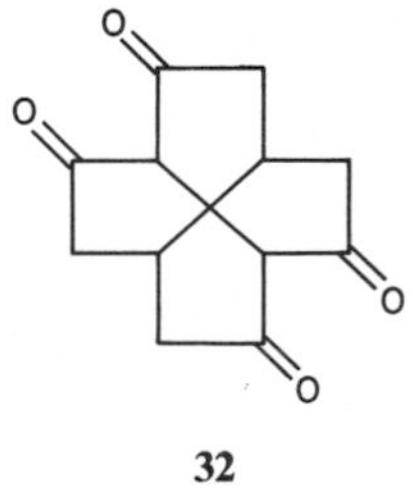   **32**    **33**

Randall Mitschka and James Cook (of the University of Wisconsin–Milwaukee), and Ulrich Weiss (of the National Institutes of Health) synthesized tetraketone **32** in 1978.[39] When Dr. Weiss mentally deleted the carbonyls, the remainder looked like a cross. Remembering, from high school, that *stauros* is Greek for "cross," he coined "staurane" for this skeleton.[40] In 1981, Reinhart Keese's team at the University of Berne suggested that a more apt descriptor would be "tetraquinacane" in analogy to triquinacene (chapter 16).[41] And naturally, Tufts University Professor Vlasios Georgian sees "rosettanes" when he looks at structures like **32** (see p. 55).

A different chemical cross came from the laboratory of André Dreiding of the University of Zürich. In 1972, he and graduate student, Marc Steinfels, studied derivatives of cyclododecane.[42] They often drew the parent skeleton as shown in **33**. This is, of course, a Swiss cross; and because the work was carried out at Zürich, Marc Steinfels referred to cyclododecane as "swisscrossane."[43]

Purdue University's Nathan Kornblum had Switzerland's flag—and not its cheese—in mind when he and his research students likewise portrayed the cyclododecane ring as **33**.[44] In their laboratory, the mononitro analog was "nitroswissane."[45]

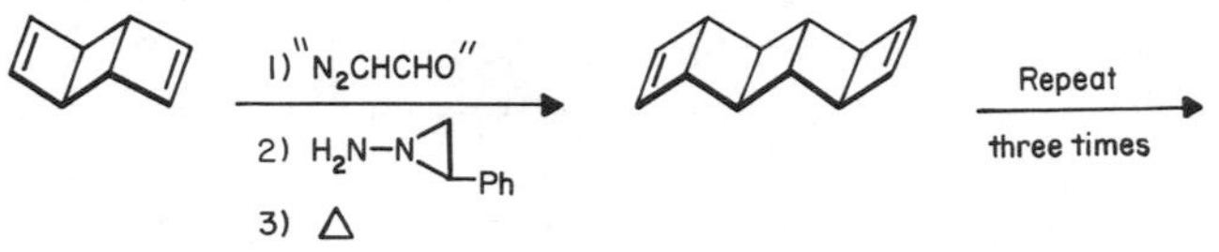

At the Eidgenösische Technische Hochschule, Albert Eschenmoser pondered the Swiss cross in depth and, in 1969, lectured on these thoughts at the Technion, Haifa, Israel.[46] For his concluding address, Professor Eschenmoser wanted to contrive a synthesis that embraced all the precepts discussed in his lecture series. So he elaborated the following route to members of the $(CH)_{24}$ class of macrocycles.[47] First, attach a ring to each end of the anti dimer (**34**) of cyclobutadiene through twofold carbenoid cyclopropanation followed by carbenoid ring expansion. This annelation at each extremity produces the homologous pentacycle **35**, which can grow to 11 rings (**36**) after three repetitions of the sequence. Finally, functionalize each double bond in **36** and join the ends to create the Star of David dodecacycle, "israelane" (**37**). There's more. The Swiss patriot then asserted that heat should transform israelane to a more stable state, "helvetane" (**38**; *Helvetia* is Latin for Switzerland). The audience was duly impressed. But to Haifa's peripatetic chemist, David Ginsburg, the notion that helvetane's durability should exceed that of israelane pinched a nationalistic nerve.[48] In fact, he and Dr. Eschenmoser privately debated that issue on several occasions. In any case, the uneasy feeling festered in his subconscious for years.

Finally, on a sultry, sleepless night in August, 1981, unable to contain his turmoil, D. Ginsburg prevailed upon his alter ego, G. Dinsburg, to take rebuttal pen in hand. It was time to set forth his own views on the stability of israelane. He composed a paper that night, let it incubate several months, then expanded it by also including Eschenmoser's original proposals. The potential international impact of Dinsburg's manuscript was not overlooked by editors. It was published in English in a French journal,[49] and a version *auf Deutsch* appeared almost simultaneously in the German literature.[50] Date of acceptance: April 1st, 1982. In this mock rebuttal, G. Dinsburg claimed to have assembled helvetane in one swoop by irradiating the right brew, and he also described a one-step synthesis of israelane by a thermal procedure. The high temperature used in the latter case was offered as preliminary proof of israelane's ability to survive. We won't spoil your fun by revealing G. Dinsburg's synthetic methodology. But if you read his delightful article, you'll go to bed chuckling—and feeling good about chemists.

A 1985 update on israelane and helvetane came from Li, Luh, and Chiu at the Chinese University, Hong Kong. Using MNDO to compute geometry and heats of formation, they surmised that israelane should be less stable than helvetane by about 383 k cal/mol.[51] CAS Registry System bestowed world citizenship on these two polycycles by assigning them passports, respectively numbered 99347-43-0 and 99396-93-7.[52]

While we're gaining perspective in chemistry, how about a lesson in solid geometry. A regular prism is a solid whose ends are parallel, regular polygons, and whose sides are rectangles. Thus, chemical structures **39–42** belong to a family known as "prismanes;" their carbons fit corners of prisms. The heat of formation and strain energy of **42** ("hexaprismane") have been predicted through calculations.[53] We saw earlier (chapter 4) that prismalogue **41** ("pentaprismane")[54] earned the alias "housane" because, when resting on a rectangular face, the structure looks like something every family should own.

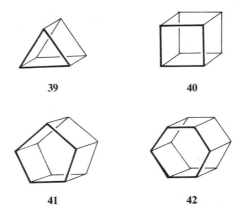

39   40

41   42

The delightful symmetry of "cubane" (**40**) has intrigued chemists for years. And it was indeed exciting when Philip Eaton's contingent at the University of Chicago announced the first synthesis in 1964.[55] With its considerable strain, cubane, not surprisingly, has more gusto than ordinary alkanes. Recall its easy isomerization to cuneane, brought about by silver perchlorate (p. 28). Another aspect is the NMR $^{13}C-^{1}H$ coupling constant of 155 Hz.[56] This high value for a saturated C—H indicates that the carbon orbital holding hydrogen has *ca.* 32% s character, made possible because the ring bonds at each corner are pulled back from ordinary tetrahedral geometry. Thus, the C—C—H angles exceed 109°28′ (they are 125°16′[57]), so the hydrogens in cubane are vinyl-like. By the way, don't confuse cubane with "cubitene" (**43**), which is not cubic at all but comes from termites known as *Cubitermes*.[58] Also, the word "polygonal" is not just an adjective for a many-sided figure; it is the name given to **44**, a norsesquiterpene aldehyde isolated from seeds of the medicinal plant *Polygonum hydropiper*.[59]

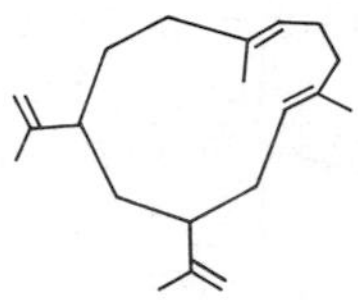

43 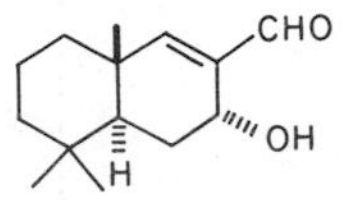   44

The C_6H_6 figure **39** ("triprismane," more commonly known as just plain "prismane") is important historically. Albert Ladenburg suggested it as the structure of benzene back in 1869.[60] Indeed, chemists have referred to **39** as "Ladenburg benzene." More than a century later, Thomas Katz and Nancy Acton at Columbia University synthesized this molecule.[61] As expected, its NMR spectra show just one ^{1}H signal and one ^{13}C signal. Prismane turned out to be surprisingly stable; it does isomerize to benzene, but the half-life is 11 hr at 90°C.[61]

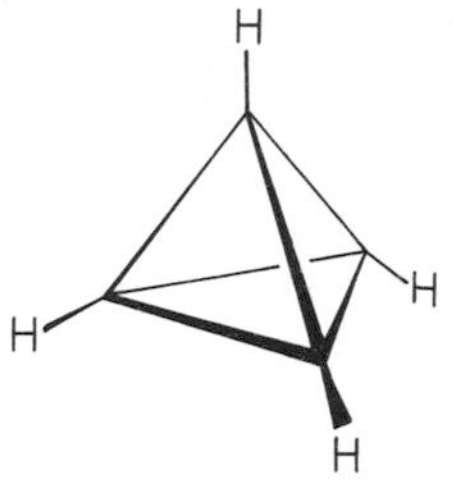

45

A beginning chemistry student soon learns about the tetrahedral nature of methane, with carbon at the center and hydrogens at the apices. Tetrahedral shapes are ubiquitous in molecules, and even two prestigious international chemistry journals are named *Tetrahedron* and *Tetrahedron Letters*. How inviting it would be to remove the center and place *carbons* at the corners to give "tetrahedrane" (**45**). Synthesis of such a contorted skeleton posed a real challenge, and chemists assailed it ingeniously.[62] In 1978, Paul Schleyer's researchers at the University of Erlangen–Nürnberg tried a tantalizing approach.[63] They irradiated dilithioacetylene (**46**; the structure shown was postulated from theoretical calculations[64]) and obtained a substance, C_4Li_4, which they thought might be tetralithiotetrahedrane.[65] Unfortunately, hydrolysis of their C_4Li_4 gave a quantitative yield of acetylene, so Dr. Schleyer at least got his "A" for effort.

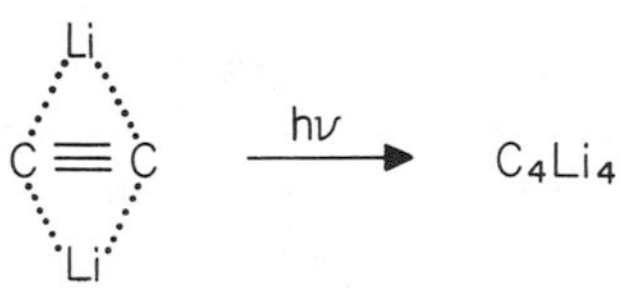

46

Shortly thereafter, Günther Maier's team at the University of Marburg synthesized tetra-*t*-butyl-tetrahedrane (**49**) by irradiating tetra-t-butylcyclopentadienone (**47**).[62] Formally, the conversion amounts to a criss-cross cycloaddition to give **48**, followed by expulsion of carbon monoxide, a common photochemical fate of ketones. Interestingly, Professor Maier's artisans found that 2, 3, 4-tri-t-butylcyclopentadienone (**50**) produced the nifty, carpeted "housenone" derivative **51** (see chapter 4), via a more conventional electrocyclic reaction.[66]

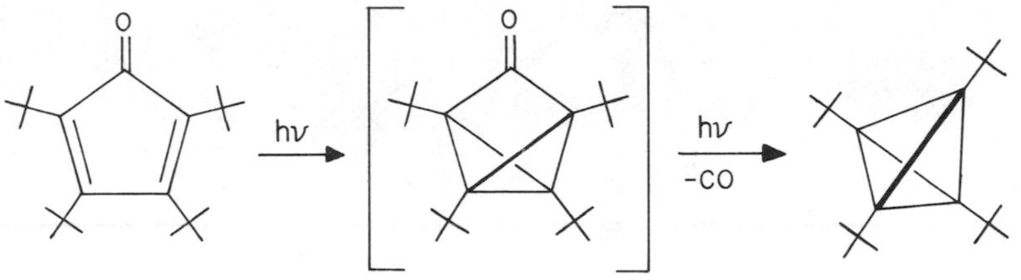

47 **48** **49**

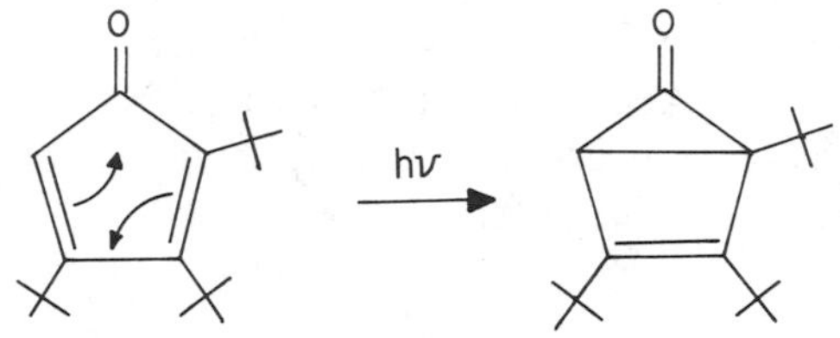

50 **51**

Despite intense interest in tetrahedrane, a 1970 announcement of its synthesis went relatively unnoticed.[67] According to that published account, Professor A. Troischose astounded the audience at the 1969 Bürgenstock stereochemistry conference with his disclosure of a successful, and simple, route to tetrahedrane. On controlled, sequential treatment with cesium and oxygen, diazomethane gave up in turn one hydrogen and molecular nitrogen to produce the fleeting "methin" (HĊ:). This frisky entity obligingly joins up to form seven $(CH)_n$ polymers. Besides acetylene (C_2H_2), benzene (C_6H_6), and cubane (C_8H_8), Dr. Troischose reported isolation and characterization of tetrahedrane (**52**; structure confirmed by X-ray crystallographer "J. Donix") and of the "sphero-compounds" **53** $(C_{12}H_{12})$ and **54** $(C_{16}H_{16})$. Equally remarkable, dodecahedrane $(C_{20}H_{20})$ was also among the products. That last polycycle had not yet been synthesized by anyone else, but it was alleged to be identical in every respect to the substance being pursued in other laboratories.

52 **53** **54**

According to the journal editor, this Bürgenstock lecture, delivered in English, commanded sufficient interest to warrant its publication in German.[67] But wait! It seems that A. Troischose credited the actual experimental work to "Fr. Ersta Prillig," which sounds suspiciously like *erst April-lig* (first of April). And the author also acknowledged help and encouragement from colleagues B. Rigoni, A. Mosenescher, H.H. Osten, and V.P. Stereolog. Sound familiar? In fact, it was at the urging of this passel of Helvetians that A. Troischose consented to deliver the lecture at Bürgenstock. And, we should have guessed that the perpetrator of this delightful hoax was noted chemist André Dreiding. After all, *drei Ding* (German) is *trois choses* in French.[68] Who said the Swiss are a staid lot?

The cube and the tetrahedron are but two of the five theoretically possible Platonic solids (i.e., convex polyhedra in which all the faces are congruent, regular polygons). These figures were discussed by Plato (*ca.* 427–347 BC) in his dialogue "Timaeus."[69] In one such solid, the icosahedron (**55**), five edges meet at each apex; so this structure cannot be constructed with *tetravalent* carbons at the corners.[70] (If you doubt this, try it with a set of "Troischose" molecular models.) In the octahedron (**56**), four edges meet at each corner, and this figure also cannot be built with *tetrahedral* carbons. We are left with the last Platonic solid, the pentagonal dodecahedron. One look at its hydrocarbon equivalent, dodecahedrane (**57**), and you understand why this molecule has been referred to as the "Mount Everest of alicyclic chemistry."[71]

55	**56**	**57**

Several research teams yearned to become the Sir Edmund Hillary and Tenzing Norgay* of dodecahedrane. Robert Woodward of Harvard University pointed out, in 1964, that triquinacene (see chapter 16), which he, Tadamichi Fukunaga, and Robert Kelly synthesized, could, in principle, mate incestuously to beget dodecahedrane (**58**).[72] We saw earlier that Leo Paquette's racemic bivalvane becomes dodecahedrane if the "half-shells" could be connected with five new bonds (see p. 9). Also, Philip Eaton's peristylane needs only a five-membered "roof" (see p. 51). In a remarkable achievement, Paquette's troupe at the Ohio State University synthesized 1, 16-dimethyldodecahedrane (**59**), the first example of this class.[73] Its structure, confirmed by X-ray crystallography, graced the cover of the February 6, 1981, issue of *Science*, where the event was first proclaimed. A mere 18 months more of ascent brought the Ohio State's expedition to methyldodecahedrane and also to the ultimate summit, dodecahedrane![74†]

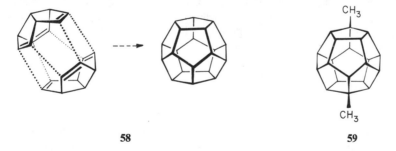

58	**59**

The high melting points of these 12-faced behemoths reflect their symmetry. For example, the dimethyl analog melts at 410°C; dodecahedrane itself has the expected symmetry[76] and shows no sign of liquifying even at 450°C. Such extraordinary lattice stability shattered the record for saturated hydrocarbons held since 1980 by 2-methyltriamantane (mp *ca.* 314°C)[77] and long before that by venerable ex-champion,

---

*Edmund Hillary, mountaineer and explorer from New Zealand, and Tenzing Norgay, his Sherpa guide, were the first persons to reach the summit of Mount Everest (May 29, 1953).

†If you like reaching great heights too, the Aldrich Chemical Co. offers help through its trademarked reducing agents, "Alpine-Borane" and "Alpine-Hydride."[75] These optically active boranes contain the chiral isopinocampheyl unit and convert carbonyl groups to alcohols with high enantioselectivity.

adamantane (mp *ca.* 269°C).[78] Now let's leave edges and corners and touch on something smoother.

The helix is a geometric figure that obsessed the research of several illustrious chemists. Linus Pauling* proved that α-keratin and certain other proteins are helical;[79] and James Watson† and Francis Crick built on the long, patient investigations by Maurice Wilkins and Rosalind Franklin[80] to evolve their double-helix structure for DNA.[81] These helices result from the secondary structure of the chains (i.e., from the way chains fold because of hydrogen bonding and other attractive forces).

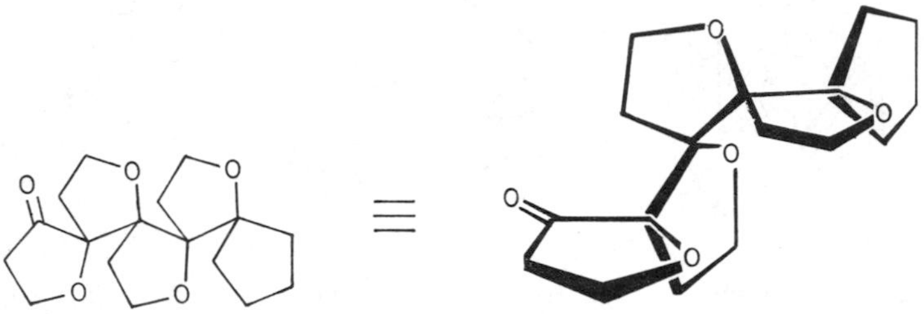

60

A joint research effort headed by Philip Magnus (then at Ohio State) and crystallographer Jon Clardy of Cornell has turned out molecules typified by **60**, which has a helical *primary* structure.[82] In other words, the relationship of the quaternary carbons forces the scaffold to spiral. The workers built their "helixanes" one ring at a time, so the potential length may be limited only by the size of their research grants.

When molecules get really long, polymer chemists can play the name game. The University of Toronto's Mitchell Winnick referred to a ring closure that joins the ends of an open-chain polymer (**62 → 61**), as a "polygon cyclization."[83] The open form (**62**) is a "self-avoiding walk." (See chapter 18 for other chemical walks.) When an end clamps on to an interior atom in the chain, we get "tadpole cyclization" (**64**), which qualifies for our chemical zoo (chapter 1). These terms originated in graph theory.[84] Finally, if two interiors couple up (**63**), we have "Ω graph cyclization," because the closed structure resembles the Greek capital letter Ω (omega).[83]

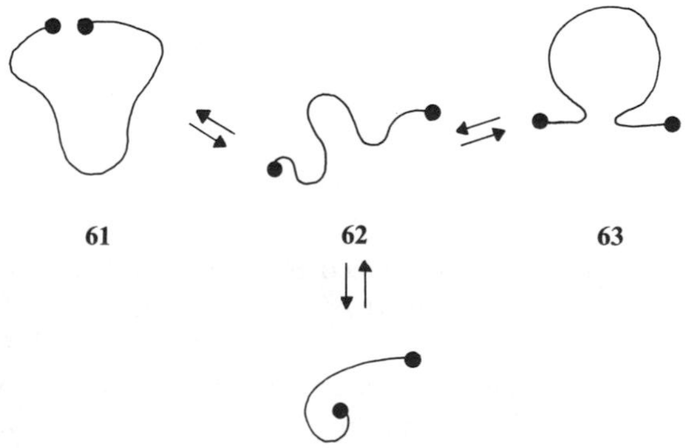

61   **62**   **63**

64

*Nobel Prize in chemistry, 1954; Nobel Peace Prize, 1962.

†Nobel Prize in medicine, 1962, shared by Francis H. Crick, James D. Watson, and Maurice H.F. Wilkins.

And while we are in a joining mood let us not overlook "oligomer," a name that probably first saw print in a laboratory manual on resins and plastics written in 1943 by G.F. D'Alelio at General Electric Co.[85] He sought a simple word for "low-molecular-weight (or number) polymer." One of his colleagues, L.V. Larsen, recommended "oligomer," from Greek *oligos* ("few") and *meros* ("part"). D'Alelio happily approved, and thereafter these oligopeople pushed the term. Not much push was necessary because chemists had already met "oligo-" nomenclature through the courtesy of Burckhardt Helferich. In 1930, at the University of Greifswald, he and co-workers Eckart Bohn and Siegfried Winkler coined "oligo-saccharides" for carbohydrates composed of a few monose units.[86] And in 1940, at Leipzig University, he and Horst Grünert proposed "oligo-peptides" for substances made up of a small number of amino acids.[87]

In the polymer world, molecules are born in an initiation, they grow by propagation, and finally they die in a termination process that is inherent or contrived. When a monomer pool is exhausted and the system does not have a way to terminate, the macromolecules can remain alive.[88] These virile bodies resume growth spontaneously when a fresh supply of monomer arrives. In 1956, Michael Szwarc, State University of New York, Syracuse, designated them "living polymers."[89] Postponement of natural death does not necessarily mean immortality, because the experimenter can add an agent to "kill" the growth. For instance, suitable anionic polymerization of styrene affords a living polymer that can be slain by a proton source like H_2O. Researchers strive to achieve "immortal polymerization" (i.e., to develop living polymers that defy death even when accosted by killer agents).[90] It may sound like molecular warfare, but actually it's just good science.

Anyone who gazes at stars thinks of them as esthetically symmetrical figures. Small wonder that these heavenly bodies inspire theologists, philosophers, songwriters, and even chemists. Hans Musso certainly saw stars at the University of Marburg when he examined molecular models of **65**, **66** and **67**, and coined the euphonyms* tri-, tetra-,

65   66   67

and pentaasterane, respectively.[91] Asterane derives from Latin *aster*, meaning "star."[92] Dr. Musso's space gazers converted acid **68** to the diazoketone **69**. Heating with powdered copper gave ketone **70**, probably via a carbenoid addition to the double

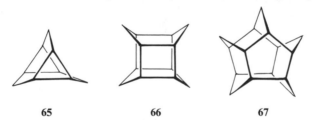

68   69

---

*A euphonym is a name with a pleasant sound.

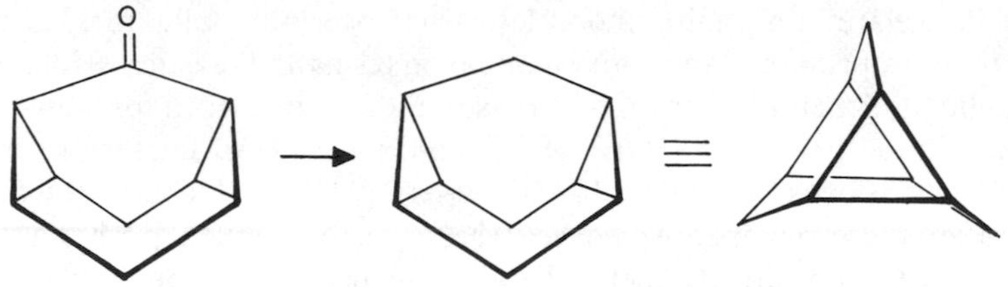

70   **65**

bond. After Wolff–Kishner reduction, "A Star was Born."[93] And others were to follow.[94]

As for "diasteranes," you are more likely to encounter some while drilling on earth than cruising in space. Petroleum chemists apply the homograph* "diasteranes" (pronounced dia-steranes) to steroidal hydrocarbons ("steranes") that have angular methyls attached to positions 5 and 14, rather than to 10 and 13. Steranes, diasteranes, and other skeletal types found in petroleum serve as informative "biomarkers" that give clues about the age and geologic history of crude oils.[95]

In 1968, University of Glasgow stellar chemists headed by Eric Clar introduced the title "starphene" for molecules such as **71**, in which benzene rings emanate in three directions from a central one (**72**).[96,97] Numbers within a bracket identify how many rings extend in each direction; thus, **71** is [3, 2, 2]starphene.[98] This short notation replaced one suggested earlier by Glasgow's Thomas Goodwin and D.A. Morton–Blake.[99]

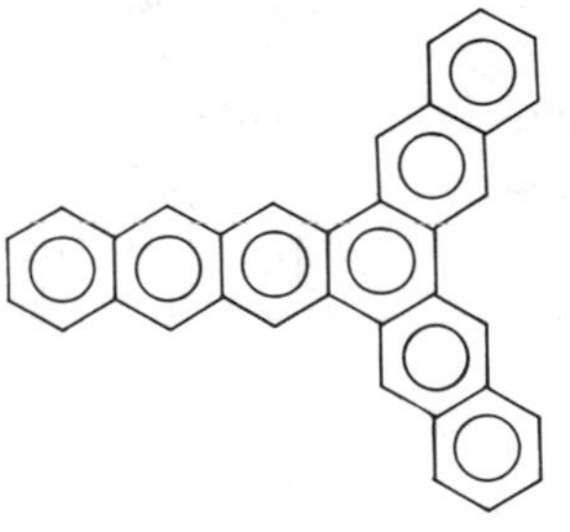

71 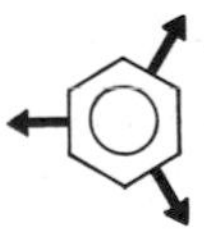   **72**

In 1972, a squad at Rutgers University led by Ronald Sauers made tricyclo[3.3.0.0^{3,7}]octan-2-one (**73**),[100] a structure destined for stardom. Later, Jan Windhorst prepared it at the University of Leiden and, in his Ph.D. thesis in 1975,

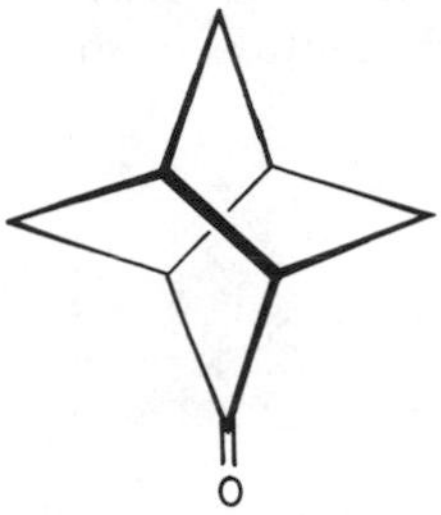

73   **74**    **75**

*Homograph: a word identical with another in spelling but differing in pronunciation and in meaning, as in bass (fish) and bass (voice).

dubbed it "natone"[101] because its carbon skeleton resembles the four-pointed star **74**, symbol of the North Atlantic Treaty Organization (NATO).[102] Although **73** is chiral (C_2 symmetry), its parent hydrocarbon has D_{2d} symmetry and is achiral. Nakazaki and co-workers synthesized the corresponding achiral dione **75** to probe for possible interactions between the carbonyl tips of the star.[103]

REFERENCES AND NOTES

1. Sachse, H. *Chem. Ber.* **1890**, *23*, 1363–1370.
2. Sachse, H. *Z. Physik. Chem.* **1892**, *10*, 203–241.
3. Mohr, E. *J. Prakt. Chem.* [2] **1919**, *98*, 315–353.
4. Mohr, E. *Chem. Ber.* **1922**, *55*, 230–231.
5. West, R.; Powell, D.L. *J. Am. Chem. Soc.* **1963**, *85*, 2577–2584.
6. West, R., private communication, October, 1975.
7. Eggerding, D.; West, R. *J. Am. Chem. Soc.* **1975**, *97*, 207–208.
8. Park, J.D.; Cohen, S.; Lacher, J.R. *J. Am. Chem. Soc.* **1962**, *84*, 2919–2922.
9. Maahs, G. *Liebigs Ann. Chem.* **1965**, *686*, 55–63.
10. Gmelin, L. *Ann. Phys.* **1825**, *4*, 31–62.
11. Heller, J.F. *Liebigs Ann. Chem.* **1837**, *24*, 1.
12. West, R. *Aldrichimica Acta* **1968**, *1*, (No. 2), 3–6, a publication of the Aldrich Chemical Co., Milwaukee, WI.
13. (a) Russel, J.H. *Naturwissenschaften* **1963**, *50*, 443–444, (b) Russel, J.H.; Hunziker, H. *Tetrahedron Lett.* **1969**, 4035–4036.
14. English, J. Jr.; Bonner, J.; Haagen–Smit, A.J. *J. Am. Chem. Soc.* **1939**, *61*, 3434–3436.
15. Klein, R.A.; Hazlewood, G.P.; Kemp, P.; Dawson, R.M.C. *Biochem. J.* **1979**, *183*, 691–700.
16. Klein, R.A., private communication, November, 1985.
17. For reviews see the following. West, R.; Nia, J. In *The Chemistry of the Carbonyl Group*, Zabicky, J. (Ed.), (Interscience Publishers, New York, 1970), vol. 2, pp. 241–275, Maahs, G.; Hegenburg, P. *Angew. Chem. Int. Ed. Engl.* **1966**, *5*, 888 893, *Oxocarbons*, West, R. (Ed.), (Academic Press, Inc., New York, 1980).
18. Yamada, K.; Mizuno, N.; Hirata, Y. *Bull. Chem. Soc. Jpn.* **1958**, *31*, 543–549.
19. Washino, A.; Yamada, K.; Kurita, Y. *Bull. Chem. Soc. Jpn.* **1958**, *31*, 552–555.
20. Cohen, S.; Lacher, J.R.; Park, J.D. *J. Am. Chem. Soc.* **1959**, *81*, 3480.
21. Cohen, S.; Cohen, S.G. *J. Am. Chem. Soc.* **1966**, *88*, 1533–1536.
22. Heilbronner, E., private communication, July, 1977.
23. Weltin, E.; Gerson, F.; Murrell, J.N.; Heilbronner, E. *Helv. Chim. Acta* **1961**, *44*, 1400–1413.
24. Hopff, H.; Wick, A.K. *Helv. Chim. Acta* **1961**, *44*, 19–24.
25. Barkovitch, A.J.; Strauss, E.S.; Vollhardt, K.P.C. *J. Am. Chem. Soc.* **1977**, *99*, 8321–8322.
26. Griffin, G.W.; Peterson, L.I. *J. Am. Chem. Soc.* **1962**, *84*, 3398–3400.
27. Köbrich, G.; Heinemann, H. *Angew. Chem. Int. Ed. Engl.* **1965**, *4*, 594–595.
28. (a) Fukunaga, T. *J. Am. Chem. Soc.* **1976**, *98*, 610–611, (b) Fukunaga, T.; Gordon, M.D.; Krusic, P.J. *J. Am. Chem. Soc.* **1976**, *98*, 611–613.
29. West, R.; Zeicher, D.C. *J. Am. Chem. Soc.* **1970**, *92*, 155–161.
30. Hart, H.; Sasaoka, M. *J. Am. Chem. Soc.* **1978**, *100*, 4326–4327.
31. Jullien, J.; Pechine, J.M.; Perez, F.; Piade, J.J. *Tetrahedron Lett.* **1980**, 611–612.
32. Tanaka, K.; Toda, F. *Tetrahedron Lett.* **1980**, *21*, 2713–2716.
33. Hart, H.; Ward, D.L.; Tanaka, K.; Toda, F. *Tetrahedron Lett.* **1982**, *23*, 2125–2128.
34. Hart, H.; Teuerstein, A.; Jeffares, M.; Kung, W.–J. H.; Ward, D.L. *J. Org. Chem.* **1980**, *45*, 3731–3735.
35. (a) Pilet, O.; Vogel, P. *Angew Chem. Int. Ed. Engl.* **1980**, *19*, 1003–1004, (b) Pilet, O.; Birbaum, J.–L.; Vogel, P. *Helv. Chim. Acta* **1983**, *66*, 19–34.
36. Mohraz, M.; Jian-qi, W.; Heilbronner, E.; Vogel, P.; Pilet, O. *Helv. Chim. Acta* **1980**, *63*, 568–570.
37. (a) de Picciotto, L.; Carrupt, P.–A.; Vogel, P. *J. Org. Chem.* **1982**, *47*, 3796–3799, (b) Vogel, P., private communication, March, 1983.
38. Diercks, R.; Armstrong, J.C.; Boese, R.; Vollhardt, K.P.C. *Angew. Chem. Int. Ed. Engl.* **1986**, *25*, 268–269.
39. Mitschka, R.; Cook, J.M.; Weiss, U. *J. Am. Chem. Soc.* **1978**, *100*, 3973–3974.

40. Weiss, U., private communication, July, 1978.
41. Schori, H.; Patil, B.B.; Keese, R. *Tetrahedron* **1981**, *37*, 4457–4463.
42. Steinfels, M.A.; Dreiding, A.S. *Helv. Chim. Acta* **1972**, *55*, 702–739.
43. Dreiding, A., private communication, October, 1978.
44. Kornblum, N.; Singh, H.K.; Kelly, W.J. *J. Org. Chem.* **1983**, *48*, 332–337.
45. Kornblum, N., discussion by phone, May, 1983.
46. Eschenmoser, A. *Organic Synthesis, Lectures at Technion, Haifa, March–April 1969*, collated by Ben-Bassat, J.; Shatzmiller, S. pp. 69–72 (in Hebrew).
47. Eschenmoser, A., private communication, April, 1983.
48. Ginsburg, D., private communication, August, 1982.
49. Dinsburg, G. *Nouveau J. Chim.* **1982**, *6*, 175–177.
50. Dinsburg, G. *Nachr. Chem. Tech. Lab.* **1982**, *30*, 289–291.
51. Li, W.K.; Luh, T.Y.; Chiu, S.W. *Croat. Chem. Acta* **1985**, *58*, 1–3; *Chem. Abstr.* **1986**, *104*, 19194d.
52. Metanomski, W.V. (Chemical Abstracts service), private communication, June, 1986.
53. Engler, E.M.; Andose, J.D.; Schleyer, P.v.R. *J. Am. Chem. Soc.* **1973**, *95*, 8005–8025.
54. Eaton, P.E.; Or, Y.S.; Branca, S.J.; Shankar, B.K.R. *Tetrahedron*, **1986**, *42*, 1621–1631.
55. Eaton, P.E.; Cole, T.W. *J. Am. Chem. Soc.* **1964**, *86*, 962–964; *ibid.* **1964**, *86*, 3157–3158.
56. Luh, T.–Y.; Stock, L.M. *J. Am. Chem. Soc.* **1974**, *96*, 3712–3713.
57. Schultz, H.P. *J. Org. Chem.* **1965**, *30*, 1361–1364.
58. Prestwich, G.D.; Wiemer, D.F.; Meinwald, J.; Clardy, J. *J. Am. Chem. Soc.* **1978**, *100*, 2560–2561.
59. Asakawa, Y.; Takemoto, T. *Experientia* **1979**, *35*, 1420–1421.
60. Ladenburg, A. *Chem. Ber.* **1869**, *2*, 140–142.
61. Katz, T.J.; Acton, N. *J. Am. Chem. Soc.* **1973**, *95*, 2738–2739.
62. Maier, G.; Pfriem, S.; Schäfer, U.; Matusch, R. *Angew. Chem. Int. Ed. Engl.* **1978**, *17*, 520–521.
63. Rauscher, G.; Clark, T.; Poppinger, D.; Schleyer, P.v.R. *Angew. Chem. Int. Ed. Engl.* **1978**, *17*, 276–278.
64. Apeloig, Y.; Schleyer, P.v.R.; Binkley, J.S.; Pople, J.A.; Jorgensen, W.L. *Tetrahedron Lett.* **1976**, 3923–3926.
65. For calculations of various C_4Li_4 structures, see Disch, R.L.; Schulman, J.M.; Ritchie, J.P. *J. Am. Chem. Soc.* **1984**, *106*, 6246–6249.
66. Maier, G.; Schäfer, U.; Sauer, W.; Hartan, H.; Matusch, R.; Oth, J.F.M. *Tetrahedron Lett.* **1978**, 1837–1840.
67. *Nachr. Chem. Techn.* **1970**, *18*, 127–128.
68. Dreiding, A., private communication, October, 1978. We thank Professor Dreiding for his confession and Professor G. Meier for providing clues.
69. *Encyclopedia Britannica* (William Benton, Publisher, London, 1972).
70. Eaton, P.E. *Tetrahedron* **1979**, *35*, 2189–2223.
71. *Nachr. Chem. Techn.* **1977**, *25*, 59–70.
72. Woodward, R.B.; Fukunaga, T.; Kelly, R.C. *J. Am. Chem. Soc.* **1964**, *86*, 3162–3164.
73. Paquette, L.A.; Balogh, D.W.; Usha, R.; Kountz, D.; Christoph, G.G. *Science* **1981**, *211*, 575–576.
74. (a) Paquette, L.A.; Ternansky, R.J.; Balogh, D.W. *J. Am. Chem. Soc.* **1982**, *104*, 4502–4503, (b) Ternansky, R.J.; Balogh, D.W.; Paquette, L.A. *J. Am. Chem. Soc.* **1982**, *104*, 4503–4504.
75. Aldrich Chemical Co., advertisement, *J. Org. Chem.* **1983**, *48*, Feb. 11 issue, back cover. For leading references on applications, see Midland, M.M.; McLoughlin, J.I. *J. Org. Chem.* **1984**, *49*, 1316–1317; Brown, H.C.; Pai, G.G.; Jadhav, P.K. *J. Am. Chem. Soc.* **1984**, *106*, 1531–1533.
76. Gallucci, J.C.; Doecke, C.W.; Paquette, L.A. *J. Am. Chem. Soc.* **1986**, *108*, 1343–1344.
77. Hollowood, F.S.; McKervey, M.A.; Hamilton, R.; Rooney, J.J. *J. Org. Chem.* **1980**, *45*, 4954–4958.
78. Fort, Jr., R.C.; Schleyer, P.v.R. *Chem. Rev.* **1964**, *64*, 277–300.
79. Pauling, L.; Corey, R.B.; Branson, H.R. *Proc. Natl. Acad. Sci. U.S.A.* **1951**, *37*, 205–211.
80. Julian, M.M. *J. Chem. Educ.* **1983**, *60*, 660–662.
81. Watson, J.D.; Crick, F.H.C. *Nature (London)* **1953**, *171*, 737–738.
82. Gange, D.; Magnus, P.; Bass, L.; Arnold, E.V.; Clardy, J. *J. Am. Chem. Soc.* **1980**, *102*, 2134–2135.
83. Winnick, M.A. *Chem. Rev.* **1981**, *81*, 491–524.

84. Trueman, R.E.; Whittington, S.G. *J. Phys.* **A1972**, *5*, 1664–1668.
85. (a) Larsen, L.V. *Chem. Eng. News* **1984**, January 23, p. 58, (b) Mallavarapu, L.X. *Chem. Eng. News* **1984**, March 12, p. 45, (c) Hendry R.A. *Chem. Eng. News* **1984**, March 12, p. 45.
86. Helferich, B.; Bohn, E.; Winkler, S. *Chem. Ber.* **1930**, *63B*, 989–998.
87. Helferich, B.; Grünert, H. *Naturwissenschaften* **1940**, *28*, 411.
88. For a review see Szwarc, M. *Adv. Polym. Sci.* **1983**, *49*, 1–177.
89. Szwarc, M. *Nature (London)* **1956**, *178*, 1168–1169.
90. Asano, S.; Aida, T.; Inoue, S. *J. Chem. Soc., Chem. Commun.* **1985**, 1148–1149 and references cited.
91. Biethan, U.; Gizycki, U.v.; Musso, H. *Tetrahedron Lett.* **1965**, 1477–1482.
92. Musso, H., private communication, July, 1977.
93. Musso, H.; Biethan, U. *Chem. Ber.* **1964**, *97*, 2282–2288.
94. Fritz, H.–G.; Hutmacher, H.–M.; Musso, H.; Ahlgren, G.; Åkermark, B.; Karlsson, R. *Chem. Ber.* **1976**, *109*, 3781–3792.
95. Philp, R.P. *Chem. Eng. News* **1986**, February 10, pp. 28–43.
96. Clar, E.; Mullen, A. *Tetrahedron* **1968**, *24*, 6719–6724.
97. Clar, E., private communication, August, 1982.
98. Biermann, D.; Schmidt, W. *J. Am. Chem. Soc.* **1980**, *102*, 3173–3181.
99. Goodwin, T.H.; Morton–Blake, D.A. *Theor. Chim. Acta (Berl.)* **1964**, *2*, 75–83.
100. Sauers, R.R.; Kelly, K.W.; Sickles, B.R. *J. Org. Chem.* **1972**, *37*, 537–543.
101. Windhorst, J.C.A., Ph.D. Dissertation, University of Leiden, 1975. Information kindly furnished by Professor Leo Paquette, Ohio State University, private communication, September, 1980.
102. Barraclough, C.B.E., Crampton, W.G. *Flags of the World* (Frederick Warne, Ltd., London, 1978), p. 219.
103. Nakazaki, M.; Naemura, K.; Harada, H.; Narutaki, H. *J. Org. Chem.* **1982**, *47*, 3470–3474.

Chapter 7

THE WELL-FURNISHED
CHEMICAL DOMICILE

Chemists are really quite versatile. They not only build molecular houses and churches (chapter 4), but they can even supply some of the furniture. To begin with, the "chair" form of cyclohexane was on sale as early as 1919[1] and gained fashion through Derek Barton's landmark paper on conformational analysis.[2] These chemical chairs are now as ubiquitous as real ones. And if you need flexible furniture, designed for reclining as well as sitting, shop for a six-membered ring with a double bond. Most often, a cyclohexene system adopts a "half-chair" posture (**1**) (i.e., four carbons in a plane and two carbons on opposite sides of that plane).[3] But, in 1958, Thomas S. Wheeler and

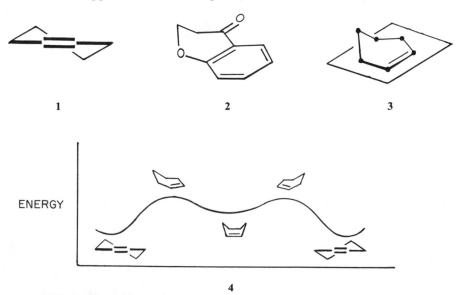

colleague Eva M. Philbin at University College, Dublin, used infrared carbonyl bands to infer that the enone ring of flavanones (e.g., **2**) favors a slightly different shape: five ring atoms planar, and only one atom out of plane (**3**). In jest, they dubbed **3** the "sofa" conformation because they had been amused by a mock article about a "raft" form published earlier by "Alonzo S. Smith" (p. 75).[4,5] Wheeler and Philbin also suggested that other oxygen-containing cyclohexenes might find comfort in this less-puckered form.[4] In cyclohexene itself, the sofa is only *ca.* 5 kcal/mol (21 kJ/mol) less stable than half-chair and corresponds to the crests of the energy wave (**4**) for ring-inversion *via* a half-boat.[6] You might think a raft could ride the wave even better; but, in fact, the raft has gone under, while the sofa still floats.

Furniture comes in different styles, and so do half-chairs. For example, cyclopentane is not flat, and its slightly puckered ring undulates. During this continuous rippling, the carbons in swift succession take turns poking out of plane, so that the blips appear to rotate location (hence the designation, pseudorotation).[7] Two of the nonplanar stages possess some symmetry, namely, C_2 (**5**) and C_s (**6**); and in certain substituted cyclopentanes, or in heterocyclopentanes, one of these shapes can become a true energy minimum. In C_2, three carbons define a plane, and the remaining two lie on opposite sides of this surface. University of Pennsylvania chemist Fred V. Brutcher, Jr., and his students termed the C_2 form "half-chair," by analogy to cyclohexene. But in the C_s form, with four carbons level and the fifth out of plane, they did not see a sofa they liked. Instead, they envisioned an envelope with its flap open; so **6** became the "envelope" form.[8]

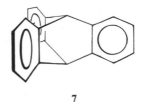

a three-panelled device called a "triptych," often used in antiquity to display art or writing. The three panes were hinged centrally (**8**) like the three aromatic rings in **7**, which thus became "triptycene."[10] Modern triptychs have panels joined differently, as in **9**, so it is indeed fortunate that Drs. Bartlett and Hammond met that day.

At that time, triptycene fascinated organic chemists. A bridgehead benzylic hydrogen could not be removed under normal conditions to give an anion or a radical.[9]

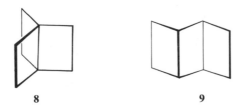

Evidently, the anion or radical lacks sufficient resonance stabilization because its orbital can't mesh effectively with the aromatic π-system (**10**). Furthermore, the three

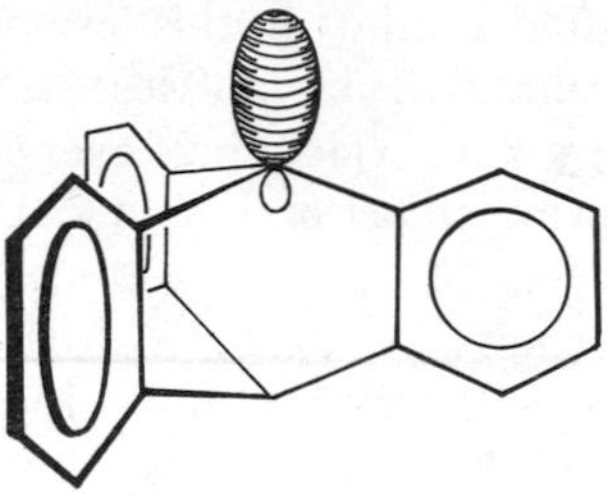

10

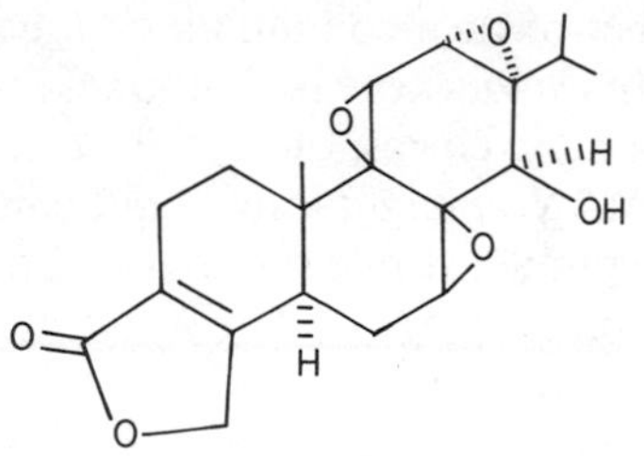

11

bonds to the bridgehead carbon cannot be coplanar; and this geometric constraint may also inhibit production of a tertiary radical. Incidentally, please do not confuse triptycene with the highly oxygenated natural products triptolide (**11**) or its ketone triptonide, whose names arise from their botanical source (*Tripterygium wilfordii*).[11]

"Triptindan"[12] (**12**; *tript*ycene with *indan* panels) and perhydrotriptycene are also on the market. Triptindan is actually a propellane (p. 30), because the bridgehead carbons are directly connected. (Keep triptindan distinct from its clover-shaped cousin "trindane," **13**.[13]) Nine stereoisomers are possible for perhydrotriptycene, because each cyclohexane can be *cis* or *trans* on the central bicyclo[2, 2, 2]octane unit. Isomers

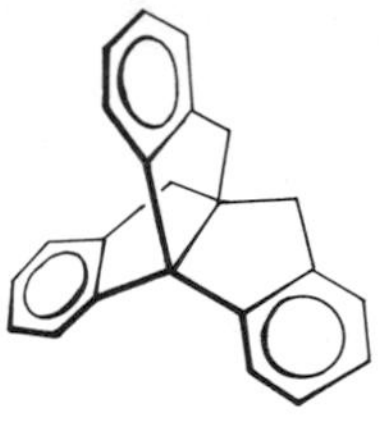

12

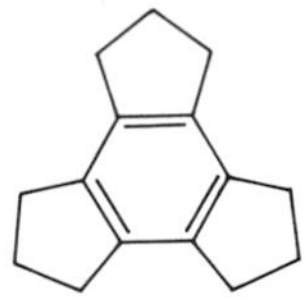

13

14–18, synthesized in Italy, are shown viewed along an imaginary axis through both bridgeheads.[14] Look at these structures sequentially and you may see an animated maestro conducting a Verdi opera. "Bravo" to our orchestral Italian colleagues! (You will hear more chemical music later in this chapter.)

14

15

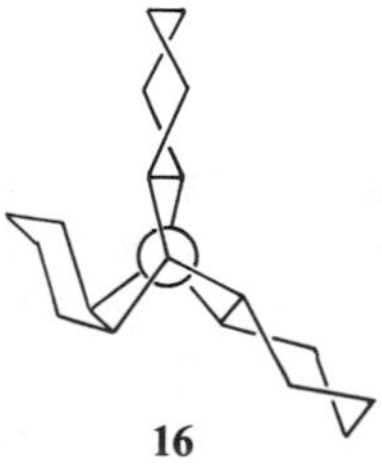

16

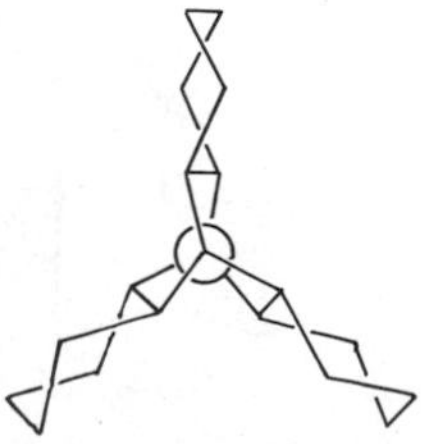

17

18

Harold Hart's researchers at Michigan State University conglomerated triptycene

moieties, as in **19**.[15] Because of the prototypal precedent set by triptycene, Joel Liebman (University of Maryland—Baltimore County) suggested the general term

19	**20**

"iptycenes" to Dr. Hart while they were having lunch.[16] More specifically, **19**, which has arene units in five planes, is a pentiptycene.[15] Before the Hart team adopted "iptycene" nomenclature, they had considered calling **19** "X-fightene" because the end view (**20**) resembles the X-fighters in the film *Star Wars*.[16] But finally, Hart and company settled for the more general term. Then they went on to synthesize extended iptycenes such as **21** (called tritriptycene and conveniently depicted by the end-on view **22**)[17] and to conceptualize other analogs, such as **23** and the cycloiptycenes **24** and **25**.[18] Given also that naphthalene, anthracene, and so forth could replace benzene units in such molecules, no one can deny that Hart's crew had ventured into a virtual mini-universe of fascinating "cenes."

21	**22**

23	**24**	**25**

For furnishings with an oriental touch, try an old-fashioned Japanese lantern known as a *chochin*.* When Masao Nakazaki and co-workers at Osaka University synthesized multilayered [2, 2]-cyclophanes such as **26**, they noted a resemblance to the lantern (**27**) and named this family of compounds *chochins*.[19] (See chapter 16 for a discussion of cyclophanes.) The helical array of ethylene bridges in **26** even simulates the helical twist of bamboo fiber used to construct a (lantern) chochin. Professor Nakazaki first suggested his euonym at a 1972 meeting of the Chemical Society of Japan, and the proposal was greeted with hearty laughter.[20] An interesting aspect of chochins is that

---

*Not to be confused with the English word "cochin," a type of large domestic fowl.

they are chiral molecules without chiral atoms. Some allenes (28) also have this property.

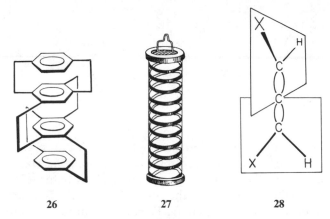

26   27   28

For the back yard, caterers Roald Hoffmann and David Thorn have given the term "picnic tables" to molecules such as 29.[21] The top of the table need not be azulene, as in 29; many other ligands will do.

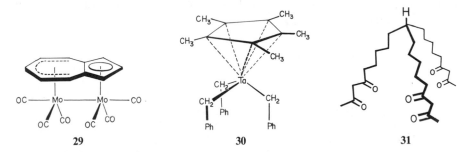

29   30   31

Teams led by Richard Schrock at MIT and Galen Stucky at the University of Illinois helped seat the picnic crowd with their "three-legged stool" (30).[22] This product of chemical carpentry features a pentamethylcyclopentadienide "seat" and three benzyl "legs." And, you should find it more comfortable than the "tripod" hexaketone 31 fabricated at the General Electric Company—unless you happen to be a Mn(III) ion. In that case you will complex strongly with the three enolic β-diketone units in 31 and stay put even after some reductant decides to convert you to Mn(II).[23] So, sit where you like and enjoy the hamburgers and pizza (chapter 5).

After the meal let us stroll back into our housane for a musicale. After all, chemistry and music are linked. Aleksandr Porfir' evich Borodin (1833–1887) demonstrated this link convincingly. He found time to earn a doctorate in medicine, to become professor of chemistry at the St. Petersburg Medical–Surgical Academy in his native Russia, to have a reaction named after him,[24] and to compose (among other works) two symphonies and the opera *Prince Igor*.[25] His arrangements provided the music for the 1953 operetta, *Kismet*, and his country honored him on a 1951 postage stamp. The "Borodin reaction" is more commonly known as the "Hunsdiecker reaction" and involves combination of a silver carboxylate with bromine:

$$RCOOAg + Br_2 \longrightarrow RBr + CO_2 + AgBr$$

Borodin carried out the conversion with silver acetate (R = CH_3); in 1861 he described the product as an "odorous, bromine-containing gas" but did not prove that it was methyl bromide.[24] Later, Hunsdiecker and Hunsdiecker developed and generalized the transformation[26] and thus share the credit with Borodin.

The name "Hunsdiecker reaction" does not disclose that two Hunsdieckers, Heinz and wife Cläre, collaborated in that venture. And this brings to mind that "Reimer–Tiemann reaction" gives no inkling of some rather remarkable coincidences surrounding Reimer. In a carefully researched article on this matter,[27] Alan Rocke and Aaron Ihde reminded readers that the well-known formylation of phenols was developed by Karl Ludwig Reimer and Friedrich Tiemann in A.W. Hofmann's Berlin laboratory. Now, would you believe that *two* Karl Ludwig Reimers worked on this reaction with Tiemann in the mid-1870s?

One of them (call him "K.L.R. 1"; 1845–1883) was son of Karl August Reimer, a publisher and bookseller; the other one, "K.L.R. 2" (1856–1921) was son of Dietrich Reimer, who also was a publisher and bookseller. At the time, K.L.R. 1 was affiliated with industry and was also collaborating with Tiemann (a *Privatdozent*). In 1876, K.L.R. 1 announced discovery of a method that easily generates salicylaldehyde from phenol. He and Tiemann explored the reaction and jointly published a few papers, on which K.L.R. 1 is listed as "K. Reimer." Because of preoccupation with his industrial job, K.L.R. 1 could not pursue the studies, so Tiemann took on other collaborators, one of whom was K.L.R. 2, a research student. K.L.R. 2 continued work on the formylation reaction and appeared on publications as "K.L. Reimer." But his effort seems to have been confined to polishing of experimental details. Therefore, conclude Rocke and Ihde, K.L.R. 1 and not K.L.R. 2 deserves essential credit for what we know as the Reimer–Tiemann reaction.[27]

You need not play a piano to make use of a piano stool—at least at Texas A & M University. There, in 1974, Cotton, Hunter and Lahuerta examined cyclooctatetra-enemolybdenum tricarbonyl, $C_8H_8Mo(CO)_3$, by ^{13}C NMR.[28] In solution below 0°C this complex displayed four signals for the ring carbons, in tune with the known X-ray crystal structure **32**.[29] At higher temperatures the four ring signals broadened and

32   33

collapsed, all at the same rate. To account for this type of signal averaging, the Cotton crew imagined a transition state with the metal positioned over the center of a flattened octagon (**33**). Fittingly, they dubbed this symmetric state the "piano stool." Since then, chemists have extended the term to metal complexes having three or four legs and a variety of seats.[30]

Love of music may have motivated Lothar Knothe, a co-worker of Horst Prinzbach at the University of Freiburg, to name **34** "fidecene."[31] This structure resembles a string instrument, the Latin word for which is *fides*. (The English "fiddle" and German *Fiedel* also derive from *fides*.)[32] Like calicene (p. 60), fidecene can be described by ionic resonance structures (**35**) in which both rings own $4n + 2$ π electrons and are aromatic. Drs. Prinzbach, Knothe *et al.* have expanded their orchestra by preparing several derivatives of fidecene and have identified them by spectroscopy.[31,33,34] They even

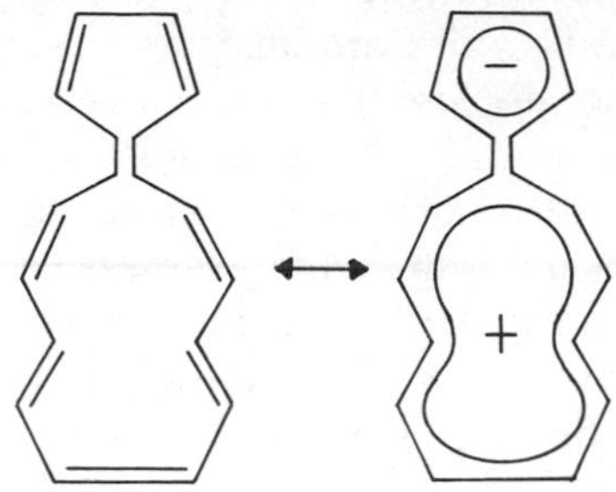

34　　　　　　　　35 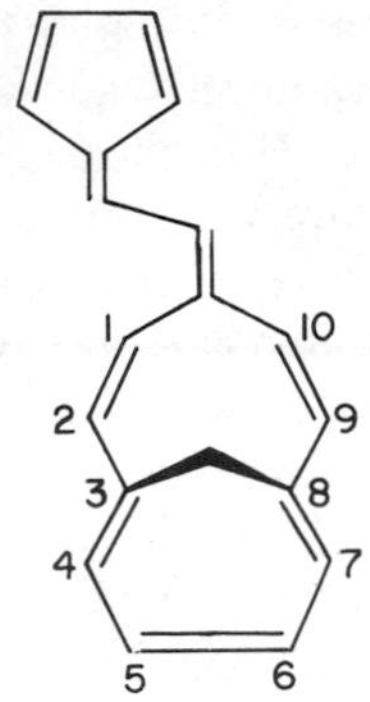　　　　　　　　　　36

enlarged the fiddle itself (and added a bridge between carbons 3 and 8) when they made vinylogous fidecene **36**.[35]

Church bells usually do not belong in a symphony hall, but their pleasant chimes qualify them as musical instruments, at least for James C. Martin (then at the University of Illinois) and graduate student Robert Basalay. They synthesized sulfonium salts **37** and **38** (X = trifluoroacetate).[36] The proton NMR spectrum of **37** at room temperature showed, for the methyl units, two separate peaks, which did not coalesce completely even at 200°C. Evidently, one of the methyls remains *syn* to the

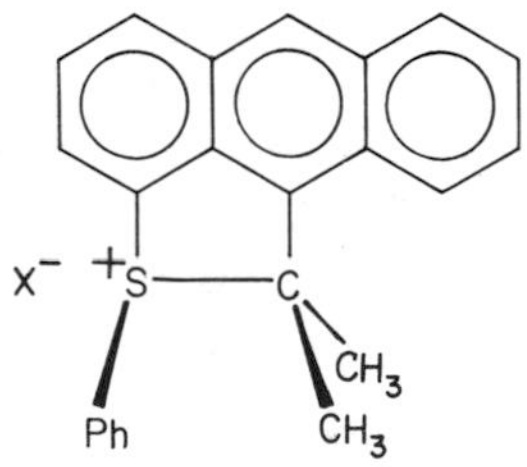

37

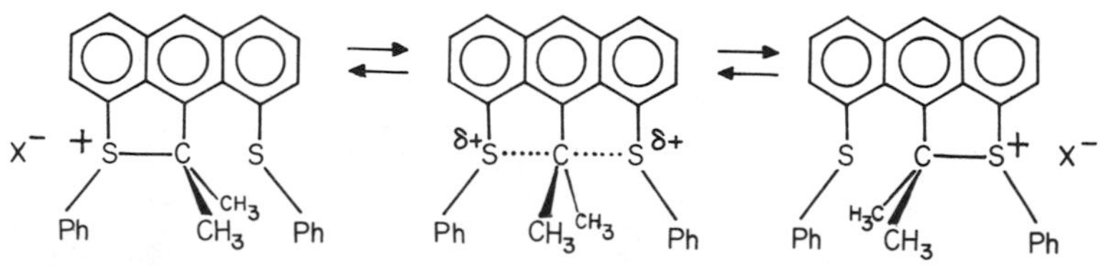

38　　　　　　　　39　　　　　　　　40

phenyl group on sulfur, the other *anti*. In contrast, the bis-sulfur analog **38** displayed only one methyl signal at, or above, room temperature. Martin *et al.* believe the divalent sulfur performs as an internal nucleophile. Displacement on carbon with Walden inversion produces **40**, in which the two sulfurs have switched roles. The phenyl on the bivalent sulfur can hover above the ring plane in some molecules and below it in others. Rapid transit between **38** and **40** averages the environment for both methyls. The resultant back-and-forth motion of the methyl groups brought to mind the swinging of the clapper inside a bell, so Martin and Basalay dubbed this the "bell-clapper mechanism."[36,37] The clapper's tune in **38** changes when the temperature drops. For example, at −10°C, *two* methyl peaks appear in the ^{1}H NMR. This temperature behavior probably rules out the long-term existence of species **39**, in which carbon is pentacoordinate akin to that of an S_N2 transition state. Bells may yet become sound members of the chemical orchestra.

REFERENCES AND NOTES

1. Mohr, E. *J. Prakt. Chem.* [2] **1919**, *98*, 315–353.
2. Barton, D.H.R. *Experientia* **1950**, *6*, 316–320.
3. Barton, D.H.R.; Cookson, R.C.; Klyne, W.; Shoppee, C.W. *Chem. Ind. (London)* **1954**, 21–22.
4. Philbin, E.M.; Wheeler, T.S. *Proc. Chem. Soc.* **1958**, 167–168.
5. Wheeler, D.M.S., private communication, March, 1983. We thank D.M.S.W. for recalling this event about his father, T.S. Wheeler.
6. Anet, F.A.L.; Haq, M.Z. *J. Am. Chem. Soc.* **1965**, *87*, 3147–3150.
7. Kilpatrick, J.E.; Pitzer, K.S.; Spitzer, R. *J. Am. Chem. Soc.* **1947**, *69*, 2483–2488.
8. Brutcher, F.V. Jr.; Roberts, T.; Barr, S.J.; Pearson, N. *J. Am. Chem. Soc.* **1959**, *81*, 4915–4920.
9. Bartlett, P.D.; Ryan, M.J.; Cohen, S.G. *J. Am. Chem. Soc.* **1942**, *64*, 2649–2653.
10. Bartlett, P.D., private communication, January, 1978.
11. van Tamelen, E.E.; Leiden, T.M. *J. Am. Chem. Soc.* **1982**, *104*, 1785—1786.
12. Thompson, H.W. *J. Org. Chem.* **1968**, *33*, 621–625.
13. Katz, T.J.; Slusarek, W. *J. Am. Chem. Soc.* **1980**, *102*, 1058–1063.
14. Farma, M.; Morandi, C.; Mantica, E.; Botta, D. *J. Org. Chem.* **1977**, *42*, 2399–2407.
15. (a) Hart, H.; Shamouilian, S.; Takehira, Y. *J. Am. Chem. Soc.* **1981**, *103*, 4427–4432, (b) Hart, H.; Raju, N.; Meador, M.A.; Ward, D.L. *J. Org. Chem.* **1983**, *48*, 4357–4360.
16. Hart, H., private communication, December, 1981.
17. Hart, H.; Bashir–Hashemi, A. *Abstracts of Papers, 190th National Meeting of the American Chemical Society, Chicago, IL*, (American Chemical Society, Washington, DC, **1985**), ORGN 86.
18. Hart, H.; Bashir–Hashemi, A.; Luo, J.; Meador, M.A. *Tetrahedron* **1986**, *42*, 1641–1654.
19. Nakazaki, M.; Yamamoto, K.; Tanaka, S.; Kametani, H. *J. Org. Chem.* **1977**, *42*, 287–291.
20. Nakazaki, M., private communication, December, 1977.
21. Thorn, D.L.; Hoffmann, R. *Inorg. Chem.* **1978**, *17*, 126–140.
22. Messerle, L.W.; Jennische, P.; Schrock, R.R.; Stucky, G. *J. Am. Chem. Soc.* **1980**, *102*, 6744–6752.
23. Hallgren, J.E.; Lucas, G.M. *Tetrahedron Lett.* **1980**, *21*, 3951–3954.
24. Borodine, A. *Liebigs Ann. Chem.* **1861**, *119*, 121–123.
25. (a) *Encyclopedia Britannica* (William Benton, Publisher, London, 1972), (b) Kauffman, G.B.; Rae, D.; Solov'ev, I.; Steinberg, C. *Chem. Eng. News*, February 16, 1987, pp. 28–35, (c) White, A.D. *J. Chem. Educ.* **1987**, *64*, 326–327.
26. Hunsdiecker, H.; Hunsdiecker, C. *Chem. Ber.* **1942**, *75*, 291–297. For patents thereof, see *Chem. Abstr.* **1937**, *31*, 2233, 2616; *ibid.* **1940**, *34*, 1685.
27. Rocke, A.J.; Ihde, A.J. *J. Chem. Educ.* **1986**, *63*, 309–310.
28. (a) Cotton, F.A.; Hunter, D.L.; Lahuerta, P. *J. Am. Chem. Soc.* **1974**, *96*, 4723–4724, (b) *ibid.*, **1974**, *96*, 7926–7930.
29. McKechnie, J.S.; Paul, I.C. *J. Am. Chem. Soc.* **1966**, *88*, 5927–5928.
30. Some examples: (a) Catheline, D.; Astruc, D. *J. Organomet. Chem.* **1984**, *272*, 417–426, (b) Ambrosius, H.P.M.M.; Bosman, W.P.; Cras, J.A. *J. Organomet. Chem.* **1981**, *215*, 201–213, (c) Pfeiffer, E.; Vrieze, K.; McCleverty, J.A. *J. Organomet. Chem.* **1979**, *174*, 183–189, (d) Dreyer, E.B.; Lam, C.T.; Lippard, S.J. *Inorg. Chem.* **1979**, *18*, 1904–1908, (e) Cotton, F.A.; Kolb, J.R. *J. Organomet. Chem.* **1976**, *107*, 113–119.
31. Prinzbach, H.; Knothe, L.; Dieffenbacher, A. *Tetrahedron Lett.* **1969**, 2093–2096.
32. Knothe, L., private communication, July, 1977.
33. Prinzbach, H.; Knothe, L. *Angew. Chem.* **1967**, *79*, 620–622.
34. Prinzbach, H.; Knothe, L. *Angew. Chem. Int. Ed. Engl.* **1968**, *7*, 729–730.
35. Beck, A.; Knothe, L.; Hunkler, D.; Prinzbach, H. *Tetrahedron Lett.* **1982**, *23*, 2431–2434.
36. Martin, J.C.; Basalay, R.J. *J. Am. Chem. Soc.* **1973**, *95*, 2572–2578.
37. Martin, J.C., private communication, October, 1975.

Chapter 8

FOOD FOR THE PALATE AND THE MIND

In 1951–1952 the research teams of Peter Pauson[1] and of John Tremaine[2] first prepared and documented the substance we now know as "ferrocene." In a conversation with Dr. Pauson, William von E. Doering suggested the correct structure for this remarkable iron compound, but neither one wanted to publish the idea without X-ray data.[3] Soon thereafter, Geoffrey Wilkinson,* Myron Rosenblum, Mark Whiting, and Robert Woodward,*,[4] as well as Ernst Fischer* and W. Pfab,[5] published evidence for the unique constitution **1**; and Woodward's group proclaimed it "ferrocene."[6] Thereupon, Jack Dunitz and Leslie Orgel coined the wonderfully descriptive term "sandwich compounds"†[7] for such structures.[7] These events launched a new type of chemical delicatessen. And since then, chemists have varied the "filling" (metal atom) and the "bread slices" (ligands) extensively to create a menu that most sandwich shops would envy.[11-13]

1

We now have "hot" sandwiches with a uranium filling ("uranocene").[14] Roald Hoffmann *et al.* suggested that a completely organic sandwich, with carbon filling, may

---

*Nobel Prize in chemistry, 1973, shared by Ernst O. Fischer and Geoffrey Wilkinson. Nobel Prize in chemistry, 1965, won by Robert B. Woodward.

†Iron is often the center of a sandwich but also it can be the center of activity—especially in catalysis. A case in point concerns the enzyme aconitase, which mediates the biological interconversion of citrate, isocitrate, and *cis*-aconitate ions in the Krebs cycle. At Fox Chase Cancer Center, Jenny Glusker pondered the stereospecificity of aconitase action. Based on several considerations, which included X-ray analyses of substrate ions, she came up with a dynamic model in which an enzyme-bound ferrous ion rotates its coordination octahedron *ca.* 90° while the substrate revolves 90° in the opposite sense. Excitedly, she discussed these details with colleague Albert Mildvan.[8] He thought the mechanism deserved a name and suggested "ferrous wheel." Dr. Glusker used the phrase in her 1968 publication,[9] and it caught on. More recently, Professor Mildvan, at Johns Hopkins University, implied that the ferrous wheel needed repair.[10]

be possible.[15] Chemical chefs have assembled "half-sandwiches" (**2**) with two dissimilar types of bread.[16] Structure **2** also illustrates ionic sandwiches and toppings made of several small ligands (usually CO or NO) rather than a single, larger one. (Have you ever eaten bologna on rye with three saltines on top?) Hungry technicians may order triple-decker concoctions with three identical bread slices (**3**)[17] or with the middle slice different from the others (**4**).[18]

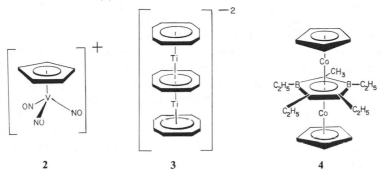

2   **3**   **4**

Delicatessens and fast-food shops use "double-decker" for sandwiches with three layers of bread and two fillings (e.g., hamburger patties or corned beef). This usage differs from "triple-decker" designations for **3** and **4**. Evidently, food people look at the number of *fillings*, whereas chemists focus on what holds these fillings. To each his own. For really big appetites, we have "tetra-decker sandwiches" (**5**),[19] "penta-decker sandwiches" (**6**),[20] and "supersandwiches" (**7**).[21] The huge spirocyclic filling in the last morsel qualifies it as a chemical "Dagwood sandwich," in honor of the American comic strip character, Dagwood Bumstead, whose classic sandwiches held "everything but the kitchen sink."[22]

5   **6**   **7**

If (for cultural or gastronomical reasons) you prefer snacks made with bagels (sometimes spelled "beigels"), don't despair. Franz Sondheimer's artisans synthesized 1, 3, 5, 7, 9, 11, 13, 15, 17-cyclooctadecanonene (**8**).[23] A space-filling molecular model with its π-cloud resembles a donut or bagel, so they addressed **8** as "beigelene." Appropriately, these chefs baked it at the Weizmann Institute in Israel. This name got around privately but was not advertised in print.[24] For public consumption, Sondheimer *et al.* coined "annulene," derived from the Latin *annulus* (ring), to describe **8** and other cyclic polyenes of type **9**.[25] A numeral in brackets discloses the number of ring atoms; so **8** is [18]annulene.

8   **9**

Annulenes are intriguing because they allow a test of Hückel's $4n + 2$ rule, which enjoins us to expect aromaticity only when **9** has an odd number of olefinic bonds. The aromatic character of [6]annulene (benzene, **10**) and lack of it in [8]annulene (1, 3, 5, 7-cyclooctatetraene, **11**) are well documented.[26] The smallest annulene, 1, 3-cyclobutadiene (**12**), is a famous chemical "holy grail"; for many years crusaders tried unsuccessfully to get their hands on it. It may exist fleetingly in the synthesis of pterodactyladiene (p. 13); and its tetrakis (trifluoromethyl) derivative[27] and a few other family members have been made.[28,29] Thus, if stability and aromaticity go hand in hand, 1, 3-cyclobutadiene is certainly not aromatic.

10   **11**   **12**

Stability, however, is a relative term; a more modern criterion for aromaticity is the ability to sustain a ring current, as assessed by proton NMR spectroscopy.[30] The magnetic field induced by circular electron flow deshields hydrogens outside the perimeter and shields any that find themselves in the infield.[31] Small rings like benzene don't have "inner" hydrogens; but large rings can, because of isomerism about individual bonds. By this criterion, [14]annulene (**13**) *is* aromatic (as predicted by Hückel's $4n + 2$ rule); at $-60°$C, its ten outer H's absorb near $\delta = 7.6$ and its four inner ones at $\delta = 0.0$.[32] Nevertheless, at room temperature, the compound succumbs to light and air. Presumably, congestion in the middle warps the plane and dampens π-delocalization; so the molecule yearns to react. Remove this interference and you

13   **14**

restore stability. For example, the bridged [14]annulene **14** survives at its melting point of 119–120°C.[33] And large crystals of the [18]annulene **8**, whose big hole welcomes the inner hydrogens, rested contentedly in a bottle for 14 days at room temperature even without a sun screen.[23]

[16]Annulene, with 16 π-electrons, illustrates a nonaromatic cyclic polyene. In solution, the ring buckles and the double bonds dispose themselves as in **15**, and not as in **16**.[34] Above 30°C all H's are equivalent (singlet at $\delta = 6.71$) as a result of rapid conformational and configurational averaging. At $-130°$C the NMR signals of the

four inner hydrogens ($\delta = 10.43$) and the 12 outer ones ($\delta = 5.40$) show up distant from where they would if the macrocycle had much ring current. Its dearth of aromaticity is also indicated by alternation in the carbon–carbon bond lengths (i.e., 1.44–1.47 Å for the single bonds and 1.31–1.35 Å for the double bonds).[35] Evidently, bagels agree with Dr. Hückel.

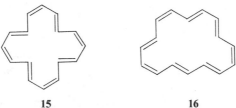

15   16

Lawrence Scott and his students at the University of Nevada, Reno, took a gamble on a different sort of Hückel game. First, they pondered cyclic homoconjugation (i.e., the possibility that π-bonds could meld across space). Then, the Nevada researchers conceptualized "[N]pericyclynes"—cyclic hydrocarbons with N acetylene units and N methylene groups alternating around the ring, as in 17, 18, and 19.[36] Their coined term is apt. "Cyclynes" stands for "cyclic alkynes"; and peri (a Greek prefix meaning "all around") indicates that a triple bond lies along every side and also suggests the possibility of cyclic delocalization.[37]

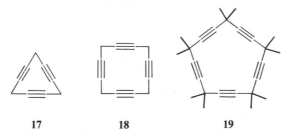

17   18   19

Pericyclynes hold many fascinations. Valence isomerization in the [3] analog might produce tautomers 20 or 21, or even the bond-convergent hybrid 22.[38] A homolog such as [6]pericyclyne could pucker into chair (23), boat, and twist boat shapes. The alkyne units hold the saturated corners far apart, so such higher homologs have been termed "exploded cycloalkanes."[39] The same spacers might serve as π-ligands for metals (e.g., 24).[38]

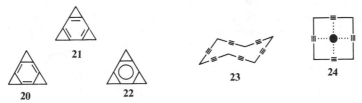

21

20   22   23   24

In 1983, Professor Scott's dealers played their hand by assembling the first prototype of this appealing class, decamethyl[5]pericyclyne (19),[36] followed soon by the [6], [7], and [8] counterparts.[38] The payoff? Spectroscopy on the permethylated [5] and [6] analogs suggested sizeable interaction among the π-units.[39] A Tohoku University squad led by Hideki Sakurai synthesized silicon analogs, including the hexamethylated corner-silated counterpart of 17.[40] Theoreticians are understandably intrigued by pericyclynes and have placed their bets.[39,41,42]

John McMurry at Cornell and Roger Alder at University of Bristol also envisioned rings, orbitals, and cavities when they dreamed about polyalkenes. We can appreciate their dreams by slicing the tricyclic diene **25** two ways. The first, as shown by dashed line a, produces a hypothetical fragment that can join up with others of its kind to create

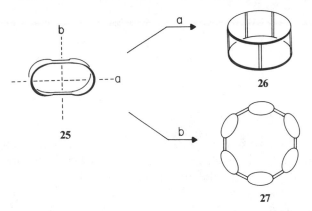

structures like **26**. For years, Professor Alder thought of such oligomers as molecular belts;[43] so, in 1985, his publication of calculated strain energies in those hypothetical compounds referred to them as "beltenes."[44] They are torus-shaped and semirigid; and the large ones contain good-sized cavities. In chapter 12 you will meet a saturated relative of [3]beltene under the name "iceane."

Bisection of **25** through the alkene bonds (line b) and oligomerization leads to arrays like **27**. In 1984, McMurry *et al.* reported the synthesis of trimer **28** as well as hexamer

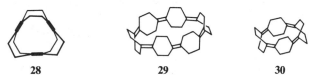

29.[45] Professor McMurry had the notion to coin a simple name for this class of compounds but refrained from doing so, presumably to avoid hassles with editors and referees. However, in his laboratory such structures were called "circus" compounds, because each comprises a ring of rings.[46] In 1986, McMurry's roustabouts synthesized the four-ring circus **30** and found that it produced a stable, square-planar complex with Ag[+] ion.[47] It's a safe way to store silver.

With our taste buds tingling for torus-like edibles, how about a snack of "donut-shaped cyclophanes" (e.g., **31** and **32**), baked by the husband–wife team of Howard and

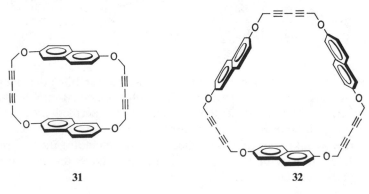

Barbara Whitlock and Esa Jarvi at the University of Wisconsin? The large voids in these acetylenic donuts allow the naphthalenes to swivel freely around their C–O axes and may be spacious enough to trap a benzenoid molecule.[48] (We shall come back to cyclophanes in more detail in chapter 16.)

To develop an appetite for all these tasty goodies, there's nothing better than a brisk workout. So, Nikolay Zefirov of Moscow State University in Russia designed athletic orbitals for this purpose. His research group examined the ^{1}H NMR spectra of several trans-3, 3, 6, 6-tetradeuterio-2-chlorocyclohexyl alkyl sulfides (33).[49] The chair–chair

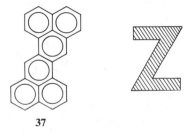

33   **34**

equilibria featured goodly amounts (up to 42%) of the diaxial conformer **34**. Rather than lose sleep over the matter, the Muscovites decided that lack of strong equatorial preference is due, at least in part, to repulsion between unshared electrons that fill chlorine and sulfur orbitals (**35**).[49] The right-hand portion of **35**, rotated, is shown

35   **36**

separately in **36**. To Professor Zefirov, a hockey enthusiast in a hockey-loving country, **36** looked like two hockey sticks, poised for a faceoff. Thus, he coined "hockey-sticks effect" for this type of orbital destabilization.[49–52] Fortunately, iceane (p. 153) is now available, so the game can proceed!

37

Eric Clar and co-workers at the University of Glasgow synthesized the polynuclear hydrocarbon **37**, noted its resemblance to the letter "Z", and dubbed it "zethrene."[53] A fitting end to our preceding chapters inspired by molecular shape.

REFERENCES AND NOTES

1. Kealy, T.J.; Pauson, P.L. *Nature (London)* **1951**, *168*, 1039–1040.
2. Miller, S.A.; Tebboth, J.A.; Tremaine, J.F. *J. Chem. Soc.* **1952**, 632–635.
3. Kauffman, G.B. *J. Chem. Educ.* **1983**, *60*, 185–186.
4. Wilkinson, G.; Rosenblum, M.; Whiting, M.C.; Woodward, R.B. *J. Am. Chem. Soc.* **1952**, *74*, 2125–2126.
5. Fischer, E.O.; Pfab, W. *Z. Naturforsch. B* **1952**, *7*, 377–379.
6. Woodward, R.B.; Rosenblum, M.; Whiting, M.C. *J. Am. Chem. Soc.* **1952**, *74*, 3458–3459.
7. For the first chemical use of the generalized term "sandwich" see Dunitz, J.P.; Orgel, L.E. *Nature (London)* **1953**, *171*, 121–122.
8. Glusker, J.P., private communication, February, 1986.
9. Glusker, J.P. *J. Mol. Biol.* **1968**, *38*, 149–162.
10. Mildvan, A., personal discussion, May, 1986.
11. Wilkinson, G., Cotton, F.A. In *Progress in Inorganic Chemistry* (Interscience Publishers, New York, 1959), vol. 1, pp. 1– 124.
12. Bublitz, D.E.; Rinehart, K.L., Jr. In *Organic Reactions* (John Wiley and Sons, New York, 1969), vol. 17, pp. 1–154.
13. Rausch, M.D. *Pure Appl. Chem.* **1972**, *30*, 523–538.
14. Butcher, J.A., Jr.; Chambers, J.Q.; Pagni, R.M. *J. Am. Chem. Soc.* **1978**, *100*, 1012–1013.
15. Tatsumi, K.; Hoffmann, R.; Whangbo, M.-H. *J. Chem. Soc., Chem. Commun.* **1980**, 509–511.
16. Herberhold, M.; Klein, R.; Smith, P.D. *Angew. Chem. Int. Ed. Engl.* **1979**, *18*, 220–221.
17. Kolesnikov, S.P.; Dobson, J.E.; Skell, P.S. *J. Am. Chem. Soc.* **1978**, *100*, 999.
18. Siebert, W.; Edwin, J.; Bochmann, M. *Angew Chem. Int. Ed. Engl.* **1978**, *17*, 868–869.
19. Siebert, W.; Böhle, C.; Krüger, C.; Tsay, Y.-H. *Angew. Chem. Int. Ed. Engl.* **1978**, *17*, 527–528.
20. Whiteley, M.W.; Pritzkow, H.; Zenneck, U.; Siebert, W. *Angew Chem. Int. Ed. Engl.* **1982**, *21*, 453–454.
21. Werner, H.; Khac, T.N. *Angew. Chem. Int. Ed. Engl.* **1977**, *16*, 324–325.
22. Young, C., then Raymond, "Blondie," syndicated comic strip in many U.S. newspapers, e.g., *Washington Post* and *Chicago Tribune*.
23. Sondheimer, F.; Wolovsky, R.; Amiel, Y. *J. Am. Chem. Soc.* **1962**, *84*, 274–284.
24. Sondheimer, F., private communication, November, 1975.
25. Sondheimer, F.; Wolovsky, R. *J. Am. Chem. Soc.* **1962**, *84*, 260–269.
26. March, J. *Advanced Organic Chemistry*, 3rd ed. (John Wiley & Sons, New York, 1985), pp. 48–55.
27. Masamune, S.; Machiguchi, T.; Aratani, M. *J. Am. Chem. Soc.* **1977**, *99*, 3524–3526.
28. Masamune, S.; Nakamura, N.; Suda, M.; Ona, H. *J. Am. Chem. Soc.* **1973**, *95*, 8481–8483.
29. Kimling, H.; Krebs, A. *Angew. Chem. Intl. Ed. Engl.* **1972**, *11*, 932–933.
30. Garratt, P.J. *Aromaticity* (John Wiley & Sons, New York, 1986).
31. Haddon, R.C.; Haddon, V.R.; Jackman, L.M. *Fortschr. Chem. Forsch.* **1971**, *16*, 103–220.
32. Gaomi, Y.; Melera, A.; Sondheimer, F.; Wolovsky, R. *Proc. Chem. Soc.* **1964**, 397–398.
33. Boekelheide, V.; Phillips, J.B. *J. Am. Chem. Soc.* **1967**, *89*, 1695–1704.
34. Schröder, G.; Oth, J.F.M. *Tetrahedron Lett.* **1966**, 4083–4088.
35. Johnson, S.M.; Paul, I.C.; King, G.S.D. *J. Chem. Soc. B*, **1970**, 643–649.
36. Scott, L.T.; DeCicco, G.J.; Hyun, J.L.; Reinhardt, G. *J. Am. Chem. Soc.* **1983**, *105*, 7760–7761.
37. Scott, L.T., private communication, June, 1986.
38. Scott, L.T.; DeCicco, G.J.; Hyun, J.L.; Reinhardt, G. *J. Am. Chem. Soc.* **1985**, *107*, 6546–6555.
39. Houk, K.N. and 16 coauthors. *J. Am. Chem. Soc.* **1985**, *107*, 6556–6562.
40. Sakurai, H.; Eriyama, Y.; Hosomi, A.; Nakadaira, Y.; Kabuto, C. *Chem. Lett.* **1984**, 595–598.
41. Gleiter, R.; Schäfer, W.; Sakurai, H. *J. Am. Chem. Soc.* **1985**, *107*, 3046–3050.
42. Dewar, M.J.S.; Holloway, M.K. *J. Chem. Soc., Chem. Commun.* **1984**, 1188–1191.
43. Alder, R.W., private communication, February, 1986.
44. Alder, R.W.; Sessions, R.B. *J. Chem. Soc., Perkin Trans. 2* **1985**, 1849–1854.
45. (a) McMurry, J.E.; Haley, G.J.; Matz, J.R.; Clardy, J.C.; Van Duyne, G.; Gleiter, R.; Schäfer, W.; White, D.H. *J. Am. Chem. Soc.* **1984**, *106*, 5018–5019, (b) *Ibid.* **1986**, *108*, 2932–2938.
46. McMurry, J.E., discussion by phone, June, 1986.
47. McMurry, J.E.; Haley, G.J.; Matz, J.R.; Clardy, J.C.; Mitchell, J. *J. Am. Chem. Soc.* **1986**, *108*, 515–516.
48. Whitlock, B.J.; Jarvi, E.T.; Whitlock, H.W. *J. Org. Chem.* **1981**, *46*, 1832–1835.

49. Zefirov, N.S. *J. Org. Chem. U.S.S.R.* (*Engl. Transl.*) **1970**, *6*, 1768–1771.
50. Zefirov, N.S.; Blagoveshchensky, V.S.; Kazimirchik, I.V.; Surova, N.S. *Tetrahedron* **1971**, *27*, 3111–3118.
51. Zefirov, N.S.; Gurvich, L.G.; Shashkov, A.S.; Krimer, M.Z.; Vorob'eva, E.A. *Tetrahedron* **1976**, *32*, 1211–1219.
52. Zefirov, N.S., private communication, September, 1979.
53. Clar, E.; Lang, K.F.; Schulz–Kiesow, A. *Chem. Ber.* **1955**, *88*, 1520–1527.

Chapter 9

JOIN ORGANIC AND SEE THE WORLD

Chemistry is truly international, and its practitioners often travel far and wide to pursue their profession. So it's not surprising they sometimes coin names after geographic locations. Let's "beta-hop"* on a biplanene (chapter 1) and take a chemistry tour around the world.

We journey first to Beirut, Lebanon, in the mysterious Middle East. In the mid-1960s, at the American University of Beirut (founded by missionaries 100 years earlier), Costas Issidorides and Makhluf Haddadin made good use of a grant from the Research Corporation. They invented a new way to prepare quinoxaline 1, 4-dioxides (e.g., **3**) from enamines (e.g., **1**) and benzofurazan 1-oxide (**2**).[2] Dioxides such as **3**, as well as the monooxides and parent amines accessible by reduction, are commercially useful antibactericides but were hard to synthesize by other means.[3] Drs. Haddadin and Issidorides reached an agreement with the Pfizer Company. While Pfizer further

---

*Collaborators Albert Stoessl and Jake Stothers in London, Ontario, coined "beta-hop" for a deuterium 1, 2-migration that ultimately perturbs a ^{13}C NMR chemical shift as a result of a β-isotope effect.[1]

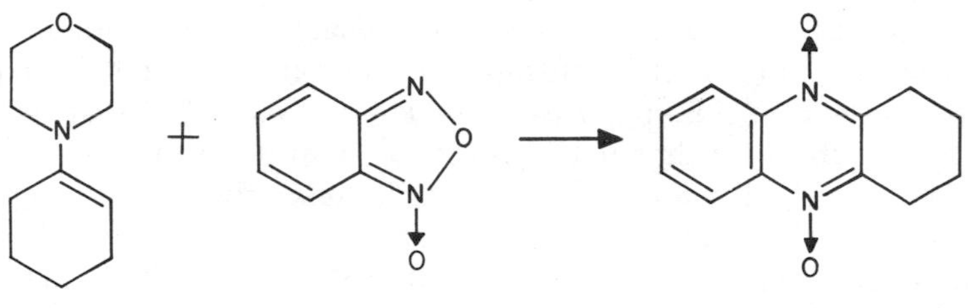

developed and scaled up the process, its chemists referred to it as the "Beirut reaction."[4,5] One of the compounds produced this way was marketed by Pfizer under the registered trademark Mecadox as a drug to combat dysentery and salmonellosis in swine. The "Beirut reaction," which brought substantial royalties to the university and to the Research Corporation,[5] has been extended to include β-diketones[6,7] and phenolic compounds[3,7] in place of the enamines.[8] At first glance, the chemical change seems like Middle East magic. However, Professors Issidorides and Haddadin presented a very reasonable mechanism.[6,8]

Our next stop is the island nation of Sri Lanka, known as Ceylon until 1972 and located near the southern tip of India. *Sri Lanka* is Sinhalese for "the resplendent land"; small wonder that natural products chemist Maurice Shamma of Pennsylvania State University chose to go there on one of his plant-gathering journeys. When his research students back home later extracted a new compound from one of those plants and identified it as **4**,[9] they named it "srilankine" in honor of its native land.

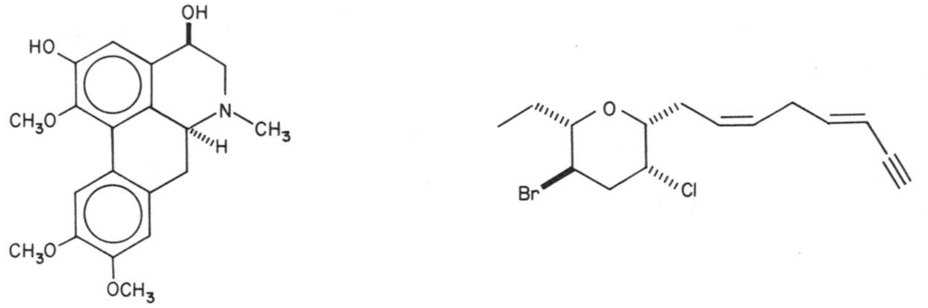

A pleasant bit of international cooperation is associated with this substance.[10] Dr. Shamma and a local plant taxonomist, S. Balasubramanian, were motoring through southern Sri Lanka looking for plants of the *Alseodaphne* species that might provide new isoquinoline alkaloids. Suddenly, the taxonomist stopped the vehicle at a cottage with a tree in its yard and exclaimed: "This is it!" The two scientists found the owners very friendly and bought one quarter of the tree from them for the princely sum of 50 cents. Back in the United States, this plant (*Alseodaphne semicarpifolia Nees*) surrendered its alkaloid **4**; and the rest is a matter of chemical record.[9]

By the way, Professor Shamma's researchers have commemorated other countries and regions in their pursuit of plant alkaloids. Some examples: "puntarenine" (plant collected near the town of Punta Arenas, Chile);[11] "istanbulamine" (collected in western Anatolia, Turkey);[12] "kalashine" (from the Kalash Valley, Pakistan, near the Afghan border);[13] "pakistanine" (gathered in the northern regions of West Pakistan).[14] So, if you want your locality emblazoned in the chemical literature, invite Shamma's explorers down to scout for new alkaloids.

And if the countryside is short on flora and fauna, try to persuade University of Hawaii marine chemist Paul Scheuer to dip into your waters. He and his collaborators isolated a new metabolite, **5**, from the sea hare, *Aplysia oculifera*.[15] They named this C_{15} acetylene "srilankenyne" in honor of co-worker Dilip de Silva, who collected the animals in Sri Lanka, his native country. At the time, the marine chemists were unaware of Dr. Shamma's similar-sounding alkaloid, srilankine (**4**); otherwise they could have paid tribute to de Silva and his home land differently.[16,17] After all, before Ceylon became Sri Lanka, its old Arabic name was Serendib (source of the word serendipity[18]); and its Sanskrit name was Sinhala. So, **5** might have become serendibyne or sinhalyne.*

A nine-hour flight southeast from Sri Lanka carries us to Sydney, Australia. There, in 1935, J. Campbell Earl and Alan Mackey treated N-nitrosophenylglycine (**6**) with acetic anhydride. They obtained a substance $C_8H_6N_2O_2$ and bestowed on it the unusual structure **7**.[20] Dr. Earl realized their supporting evidence was not overwhelming. Nevertheless, he and Ronald Eade later dubbed this and similar compounds

| 6 | 7 |

"sydnones," in honor of the city in which the molecules first saw the light of day.[21] Wilson Baker and his student W. David Ollis, of the University of Bristol, proposed that "sydnones" cannot be represented by any single covalent formula. They are resonance hybrids of ionic canonical forms such as **8–10**, conveniently symbolized as **11**.[22,23] For such compounds, the British chemists proffered the expression "meso-ionic" to emphasize their highly polarized hybrid character.[23] Since then, many more meso-ionic substances have joined the club.[24]

| 8 | 9 | 10 | 11 |

An overnight jet above the Pacific Ocean and halfway across the United States brings us to Knoxville, Tennessee, host city to the 1982 World's Fair. There, in 1973, at the University of Tennessee, Richard Pagni and graduate student Charles Watson synthesized **12**.[25] In the laboratory Charles referred to it as "knoxvalene" in acclaim of Knoxville.[26] The name degenerated to "obnoxvalene" whenever his research did not go well. (In this vein, we wonder if investigators at Scripps Institution of Oceanography had trouble dealing with "obtusadiol," a dibromoditerpenoid they isolated from a Mediterranean alga.[27]) We think the toponym[†]"osakavalene" might also be appropriate for **12** because Japan's Ichiro Murata and Kazuhiro Nakasuji of the University of Osaka reported its synthesis in the same issue of *Tetrahedron Letters* that carried the Pagni–Watson publication.[28]

The "valene" suffix comes from the word introduced by Heinz Viehe for molecules that partake in valence isomerization.[29,30] (For discussions of "valence isomerization"

---

*The minerals serendibite and sinhalite also take their names from these origins.[19]

†Toponym: a name derived from the name of a place.

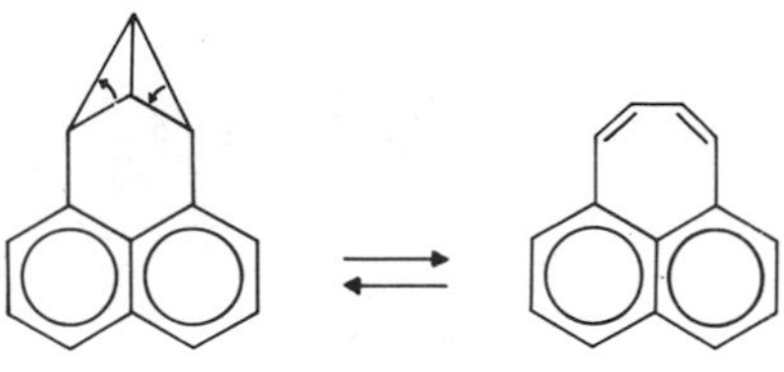

12   **13**

and "valene," see chapter 20.) Hence, **12** is a valene because it shuffles its valencies to form pleiadiene (**13**; p. 172).[31]

We depart Knoxville for a brief visit to Panacea, Florida. Residents there tell of a bevy of scientists from Cornell and Columbia Universities who came to fish the coastal waters and who fled with eight full-grown marine mollusks, *Aplysia brasiliana* (a so-called sea hare). The abductors transported the booty across state lines and back to their laboratories. Probing the digestive glands of these mollusks, they extracted an optically active substance that proved to be a novel aromatic bromoallene, **14**. Citizens of Panacea normally do not get excited over allenes; but they should like this one because the researchers named it "panacene" to remind the world where it came from.[32]

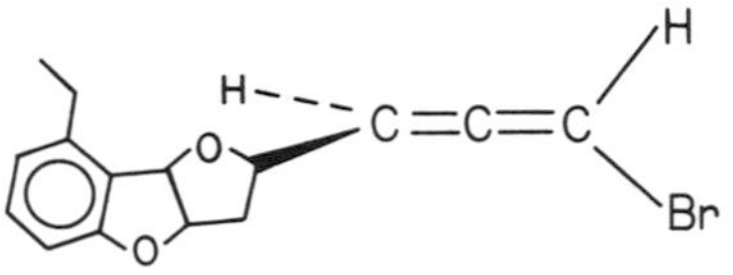

14

After a short hop from Florida, we touch down at Bethesda, Maryland, home of the National Institutes of Health (NIH), a division of the Department of Health and Human Services (H$_2$S). At NIH in 1961, Bernhard Witkop and John Daly prepared some *p*-tritiophenylalanine (**15**, X = T), which colleague Sidney Udenfriend planned to use to assay the enzyme phenylalanine hydroxylase. This enzyme catalyzes the biological para-hydroxylation of phenylalanine to tyrosine. Imagine everyone's chagrin when 90% of the tritium showed up in the tyrosine (**16**, X = T).[33] Surely, they thought, someone had goofed. However, the same phenomenon occurred with *p*-deuteriophenylalanine (**15**, X = D); proton NMR showed that much deuterium had moved next door (**16**). Even halogens slid over, so the process appeared quite general.[33] By this time, the number of researchers engaged in the project was so large it was not feasible to name this transformation after a specific discoverer, so Dr. Udenfriend

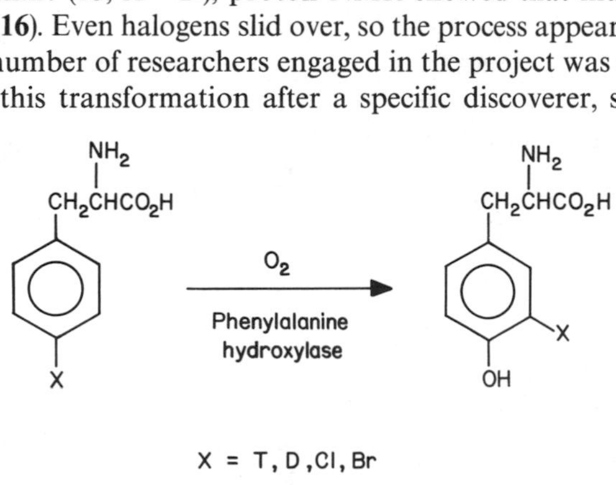

X = T, D, Cl, Br

15   **16**

suggested "NIH Shift."[33,34] The mechanism shown on p. 118 was proposed in the original NIH shift paper.[33]

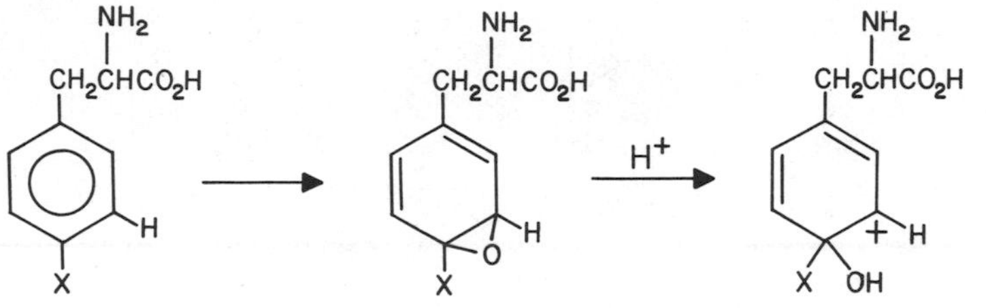

15

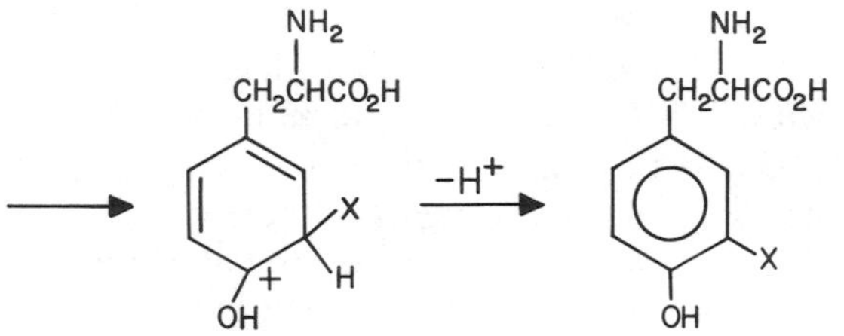

16

In 1968, the Bethesda chemists proved that naphthalene-1,2-oxide is a labile metabolite when naphthalene converts to α-naphthol in the liver.[35] When 1-deuterionaphthalene (**17**) was used, the label took the NIH shift.[36] The demonstrated intermediacy of an arene oxide (**18**) in this study strengthened their speculation that the phenylalanine → tyrosine conversion also involves a similar intermediate. Witkop recently recounted the serendipitous discovery of the NIH shift.[37]

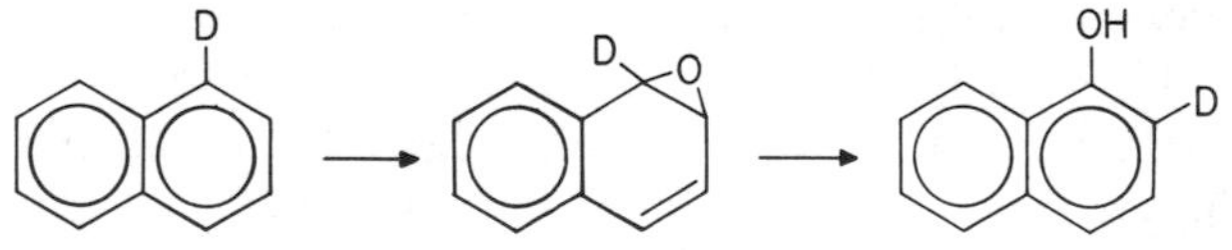

17   **18**

A pleasant 30 minute drive through rolling Maryland countryside brings us to the city of Frederick, site of the Frederick Cancer Research Facility. It was there in 1982 that Ramesh Pandey and Renuka Misra, in collaboration with NIH crystallographer James Silverton, elucidated the structure of an antitumor antibiotic (**19**) produced by a soil bacterium, *Streptomyces griseus*. The oncologists paid homage to their host city by dubbing this L-shaped spirocycle "fredericamycin A,"[38] a permanent reminder of

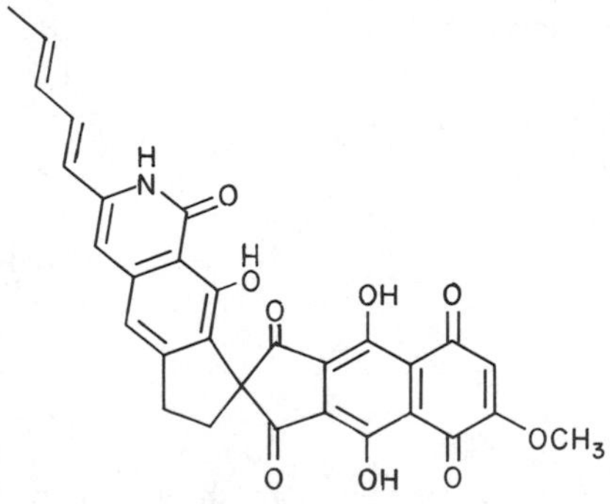

19

where the work was carried out. Not a bad ploy, especially when the institute is ready to apply for continued research funding. (In similar vein we note that the proclaimed "Gif system for oxidation of saturated hydrocarbons" leaves no doubt that this radical chemistry sprang from the renowned CNRS institute at Gif-sur-Yvette, France.[39])

By the way, when NIH researchers Harry Gelboin, Donald Jerina, and their colleagues talked about the "bay region" you might suspect they had in mind a weekend of boating and fishing on Maryland's famous upper Chesapeake Bay. Perhaps so; but more likely they were discussing supermutagens.[37] To convince you of this, let us first board a trans-Atlantic flight to England and drop in on The University (formerly the Institute of Technology) at Bradford, Yorkshire. There in 1966 we could have joined seafarers Keith Bartle, Harry Heaney, Derry Jones, and Peter Lees, for a cruise around polynuclear aromatic hydrocarbons. Theoreticians and experimentalists took such trips many times—particularly in the early days of ^{1}H NMR—to discover how benzenoid ring currents and how magnetic anisotropies of bonds influence chemical shifts. Polynuclear aromatics were nice islands to explore because their contours provided a variety of environments for ring hydrogens.

Circumnavigating triphenylene (20), the British crew sighted two essentially different kinds of H's; and Derry Jones spontaneously called one type "bay" and the other "peninsular." They published these descriptive "self-explanatory" names,[40] but not without some apprehension. In fact, Professor Jones confessed to having dreamed that referees rejected their manuscript on account of its "bay" and "peninsular" nomenclature.[41] In structure 20, we have labeled these two types as "b" and "p," respectively.

20   21

22   23

Note that peninsulars are not angled toward any other nearby hydrogen, whereas a bay hydrogen inclines its bond toward another peripheral one. Phenanthrene (21) has five dissimilar sets: One bay, and four different kinds of peninsulars (two of which are also peri). By the same token, all the H's in benzene, in naphthalene, and in anthracene are peninsular.

Not all angling takes place in bays. For example, in benzo[c]phenanthrene (22) the hydrogen labelled H_f lies near the entrance of a deeper coastal inlet; so the Bradford researchers dubbed it a "fjord" hydrogen[42] and thereby defined a "fjord region."[43] And the angled H shown in pentacycle 23 became a "double-bay" type.[42]*

---

*Britishers are not alone in their love for angling. At the Université Libre of Brussels, Belgium, Richard H. Martin probed ^{1}H NMR of substituted polynuclear aromatic hydrocarbons. In 1964, Professor Martin classified hydrogens as nonangular and angular; and he compiled a list of symbols to distinguish more precisely various types within each class.[44]

In the 1970s, evidence quickly accumulated to implicate certain arene oxides as key metabolites in the carcinogenic action of polycyclic aromatics. One example is 9, 10-oxido-7, 8, 9, 10-tetrahydrobenzo[a]pyrene (**24**). This mutagenic oxide, and others, involved a bay carbon and so became "bay-region" epoxides.[45]

Tagging names onto segments of aromatic hydrocarbons was not new. It has been in vogue with theoreticians since the mid 1940s. And—typical of the world of fashion—the style was set in Paris, specifically by wife–husband team Alberte and Bernard Pullman. Pullman fashions were actually originated by Alberte; Bernard married into the business but soon became an equal partner. This prolific coterie of quantum mechanicians blessed the world with designer labels K, L, and M. The trend began in 1945 when Alberte first published on the electronic structure of some carcinogenic polynuclear aromatic hydrocarbons.[46] She used "meso-phenanthrenic" (m–p) to refer to the 9, 10-bond in phenanthrene and to portions of other polycycles that correspond to this bond (e.g., see **25** and **26**). In 1945–46, Madame simplified this mouthful to "K region."[47] Later (1953), the 9, 10-"mesoanthracenic" positions in anthracene became the "L region",[48] and in 1954 the House of Pullman came out with "M region" for sites particularly active toward initial metabolic attack.[49] The presence of a K region together with absence of an L region seemed characteristic of potent carcinogens.

KLM phraseology became popular, but the origins of these terms were not always remembered. Some investigators thought K stood for "Krebs," the German word for

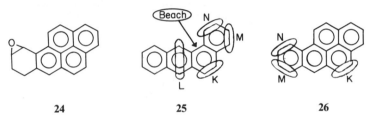

24	**25**	**26**

"cancer." One compatriot stated K meant "kill"; and presumably he would think L stood for "live." The KLM triad also has naught to do with the Dutch airline, Koninklijke Luchtvaart Maatschappij (Royal Airline Company) or with electron shells around a nucleus. In fact, the Pullmans assured us, the initial choice of K was completely arbitrary; and the other letters just followed the alphabet. (They had the notion to begin with "P," but modesty won out.)[50]

A newer item in the Pullman line is "N region," which hit the retail market in 1979.[51] This label identifies a peripheral ring bond containing a bay carbon (see **25** and **26**). Meanwhile, in England, the Bradford bunch had not been idle. They sailed into a bay, landed at the inner carbon–carbon bond, and crystallographically found it had an extended length. Henceforth, that site became the "beach bond" (e.g., see **25**).[52] So, dear patron of polynuclear aromatics, you can go with either fashion: British nautical or French alphabetical.

From Paris we wing back over the English Channel to London, where research teams at two nearby institutions, Wye College and Oxford University, isolated the antifungal ketone **27** from the broad bean.[53] They named it "wyerone" in recognition of one of the collaborating schools. Wye not? David Knight's team at Nottingham University synthesized wyerone in 1982.[54]

Now let's cross the North Sea to Norway, whose many attractions include the fjords, beautiful arms of water that penetrate inland between steep cliffs. This sight so inspired Emory University's Fred Menger that he borrowed "fjord" to describe a chemical phenomenon.[55]

In water, soap molecules form clusters known as micelles, in which the ionic ends extend outward from an inner core of hydrophobic groups (e.g., **28**).[56] In 1973, Scandinavian chemists Bengt Svens and Björn Rosenhohn of Åbo Academy, Åbo, Finland, suggested that water penetrates deeply into this core.[57] Dr. Menger called this

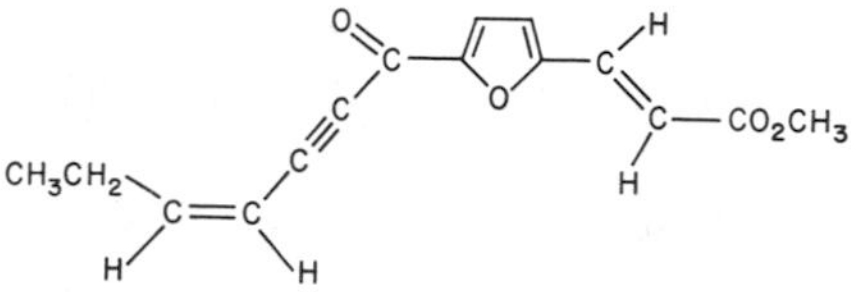

27 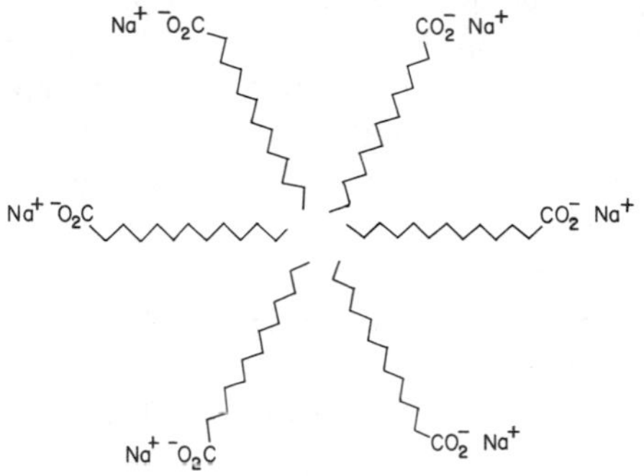   **28**

viewpoint of micellar structure the "fjord" model.[55,58] Alas, Dirk Stigter at the Western Regional Research Center, Berkeley, California, saw no fjord in his future because he believed water does not penetrate the core at all.[59] Professor Menger dubbed Stigter's notion the "reef" model, since aqueous inlets do not penetrate reefs.[55]

An "inverted" micelle points its hydrophilic (ionic) groups inward, and this hydrophilic region holds quite a bit of water in its interior.[60] So, Menger and mates coined the hydronym "water pools" for these highly hydrated micelles.[60,61] Some workers did not like this description because "pool" can be regarded to mean "common source," whereas Menger's intended meaning was "pond." So, he regrets not having dubbed these micelles "water blobs."[58] Clearly, the micellists at Emory knew how to keep their research afloat, even if it meant getting wet occasionally.[62]

The properties of multiply charged electrolytes are obviously important for an understanding of micellar structure. In 1951 at Yale University, Raymond Fuoss and doctoral student David Edelson coined the expression "bolaform" electrolytes for ions with two (or more) positive charges separated by chains of atoms.[63] The term stems from *bola*, a throwing weapon used to entangle the legs of animals. It consists of two or three balls, usually of iron or stone, fixed to the ends of a cord (e.g., **29**). The positive

29 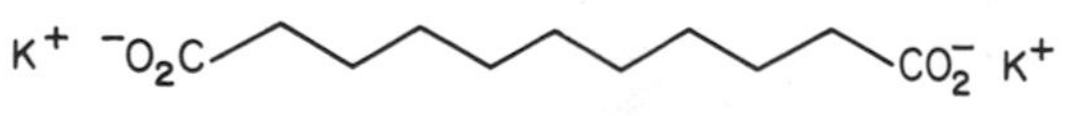   **30**

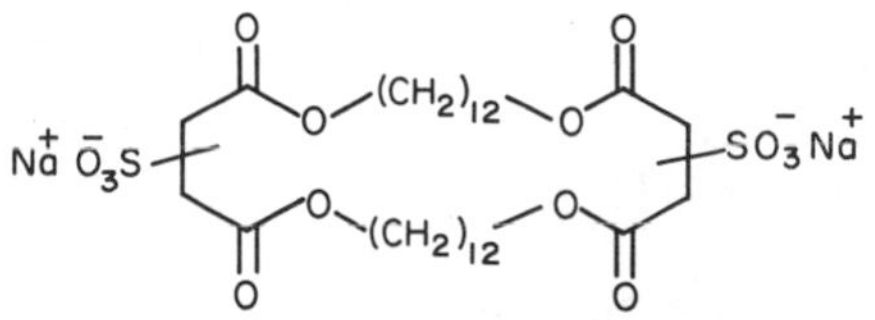

31 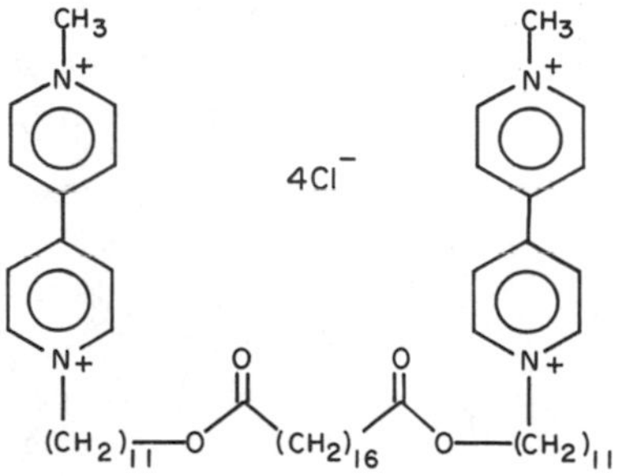   **32**

charges on bolaform electrolytes were initially those of quaternary ammonium ions; but the concept evolved to embrace any molecule with multiple ionic sites tethered by a long nonpolar chain (e.g., **30, 31, 32**).[64] Bolaform electrolytes have also been called bolaform amphiphiles, bolaamphiphiles, bolions, and bolytes.[64c,65]

While on the subject of micelles, we should note they can consist of more than one component. Ryuichi Ueoka of Kumamoto Institute of Technology and Katsutoshi Ohkubo of Kumamoto University prepared "comicelles" using a hydroxamic acid (e.g., $CH_3(CH_2)_8CONHOH$) and a quaternary salt (e.g., $CH_3(CH_2)_{17}\overset{+}{N}(CH_3)_3Cl^-$).[66] These micelles proved to be good catalysts for ester hydrolysis; and because catalysis required *co*operative action of *micellar* components, the Japanese collaborators regard them as "comicellar catalysts."[66,67]

Near the middle of the Irish Sea, surrounded by Ireland, Scotland, and England, lies the Isle of Man, our next stop. Most of the inhabitants of the isle, as well as the dialect spoken there, are Manx. And, the renowned, tailless Manx cats are said to have

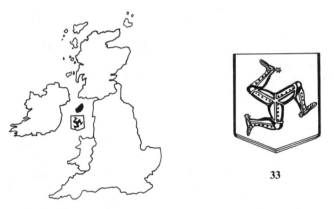

33

originated there. The official coat of arms (**33**) of this tiny, independent country consists of three armored legs, which seem to be "kicking at Scotland, ignoring Ireland, and kneeling to England."[68] The emblem, known as a triskelion, also appears on many Manx coins. Interestingly, some of the Isle's coinage shows opposite 'footedness" (i.e., the toes point clockwise rather than counterclockwise). Manx officials give equal recognition to each enantiomer! But what has this country to do with chemical names?

In 1970, William Parker, head of a joint research team from the Universities of Ulster and Glasgow, reported the synthesis of bicyclo[3.3.3]undecane.[69] The proton NMR spectrum of this achiral hydrocarbon, taken at 35°C, indicated rapid interconversion of conformations **34a** and **34b**. The resemblance of these structures to the Isle of Man's coat of arms led Parker and colleagues Doyle, Gunn, Martin, and McNicol to christen the compound "manxane."

At about the same time, Nelson Leonard's research group at the University of Illinois independently prepared **34** as well as an analog with a bridgehead nitrogen (**35**).[70] Among other things, they were interested in the amine's basicity and in the bond angles

34a   **34b**

around nitrogen. Adopting Parker's nomenclature, the Illinois chemists dubbed this *manx*ane-like am*ine*, "manxine." But Professor Leonard thought this amine should

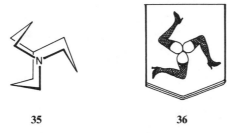

35   36

have its own coat of arms and, since "manxine" sounds like a woman's name, the design **36** was concocted.[68] The properties of manxane and manxine suggest they differ in shape around the bridgehead atoms.[71] Vive la différence!

By the way, a dictionary defines "triskelion" as a symbolic figure consisting of three curved branches, bent legs, or arms, radiating from a center. The word stems from Greek *tri* (three) and *skelos* (leg). In 1966, Michael Cava borrowed seven-tenths of "triskelion" when his research squads (then at Ohio State and Wayne State Universities) converted 9, 10-dihydro-9, 10-*o*-xyleneneanthracene (**37**) to the interesting structure **38**. They called this new hydrocarbon. "triskelene."[72] And why not? It's handier than 5, 10-(*o*-benzeno)-4b, 5, 9b, 10-tetrahydro-7H-indeno-[2, 1-a]indene.

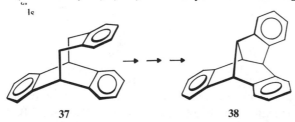

37   38

A short trip from the Isle of Man brings us to Germany and specifically to München (Munich), famed not only for its Oktoberfest but also for excellent organic chemistry. Rolf Huisgen and his colleagues at the Universität München saw to that. Professor Huisgen's research school thoroughly examined cycloadditions in which a 1, 3-dipole (**39**) combines with a dipolarophile (**40**) to create a five-membered ring.[73] (Dipolarophiles with triple bonds also work; and the a–b bond can be single.) Naturally, they paid attention to the sydnones (p. 116), whose canonical forms include structures that are 1, 3-dipolar ions.

$$\begin{array}{ccc} \overset{c^-}{\underset{b}{\diagup}} & + & \overset{e}{\underset{d}{\parallel}} & \longrightarrow & \overset{c-e}{\underset{a-d}{b}} \\ a^+ & & & \end{array}$$

39   40

Indeed, sydnone **41** combines with dipolarophile **42** to produce **43**, which loses CO_2 and ends up as **44**.[74] To extend the scope of meso-ionic substrates for their work, Dr. Huisgen's co-workers treated **45** with acetic anhydride and, in a reaction that parallels the first sydnone synthesis,[20] obtained **46**.[75] In keeping with sydnones and city pride, they were tempted to call **46** and similar substances "munchnones."[76a] Intentional or not, the term appears to have caught on.[76b-d]

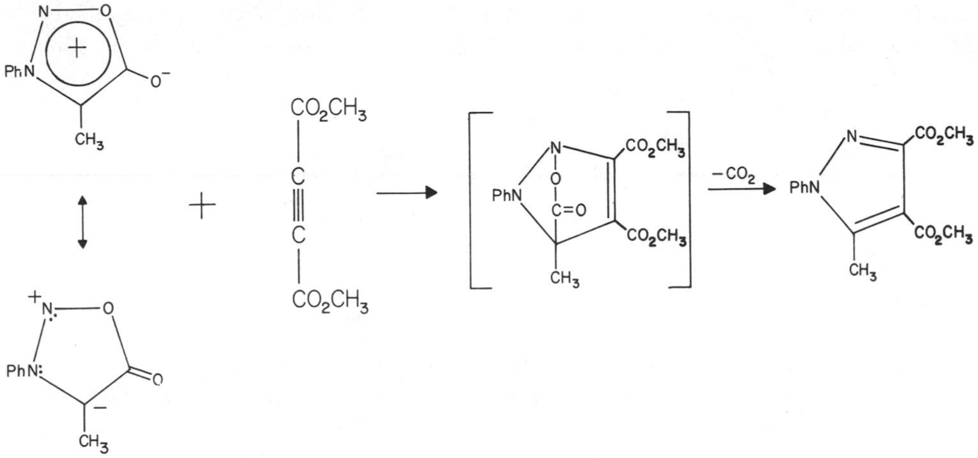

41   **42**   **43**   **44**

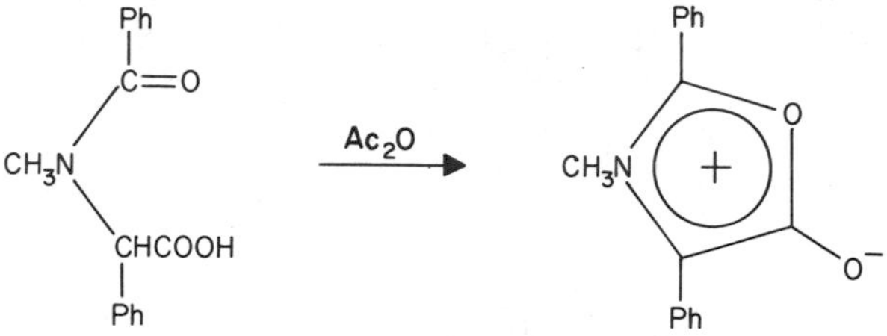

45   **46**

The next stop on our journey, Hesperia, does not appear on a modern globe. It means "the Western Land" in Greek.* Grecians used the term to refer to Italy, whereas Romans used it to mean Spain. In any case, for Harvard's Robert B. Woodward and his visionaries it became the basis of the name "hesperimine," which they gave to structure **47**. And therein lies a tale of how the West was won.

A total synthesis of vitamin B_{12} was achieved jointly by battalions led by Robert Woodward at Harvard and Albert Eschenmoser at Zurich's ETH. Their masterful route, forged during 11 years with 99 co-workers from 19 countries,[77] ultimately involved marriage of two major fragments related to **47** and **48**.[78] Component **48** features the B/C rings of the final molecule and was crafted in Zurich. Eschenmoser's

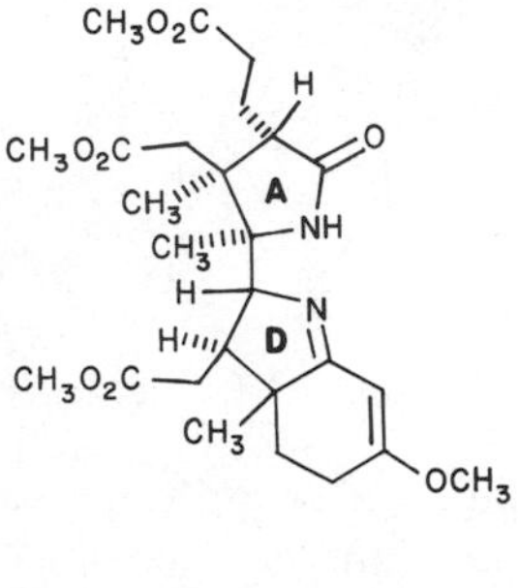

47 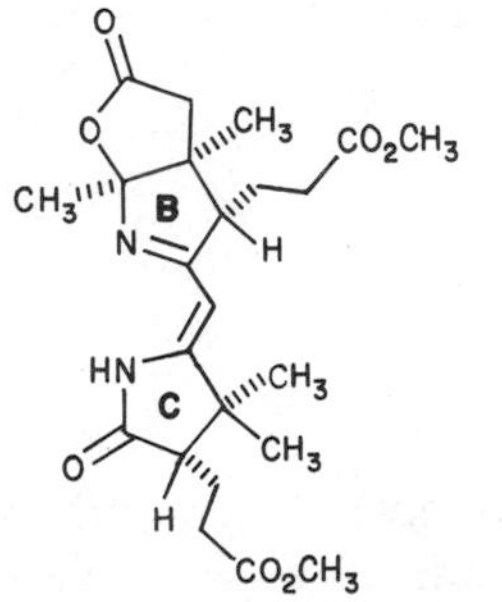   **48**

*Derived from Greek *hesperios* for "western," or "towards the evening sun." The same root occurs in "hesperidin" (a flavinoid glucoside in lemons and sweet oranges) and hesperetin (the aglucone of hesperidin).

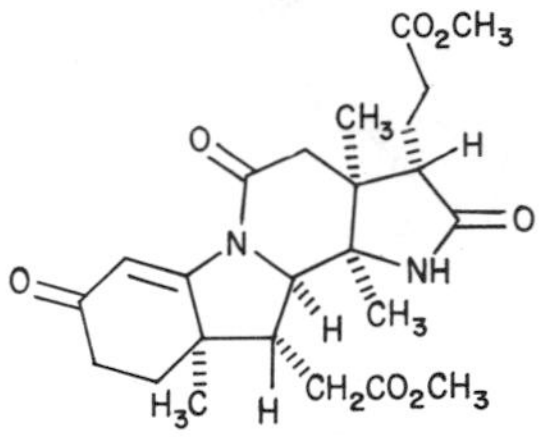

49

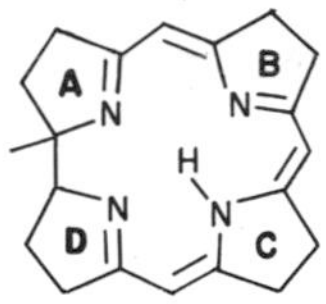

50

people called it "dextrolin" (from the Latin *dexter*, meaning "right"), because it was to become the right half of their target compound and because their Harvard allies used "right" and "left" to refer to the two segments. Although the name dextrolin was not used in publications, it facilitated the many trans-Atlantic phone discussions between Robert and Albert during their intense collaboration.[79] When the Harvard juggernaut later synthesized fragment **47** (i.e., the left, or A/D, portion), Woodward sought a convenient name. Logically, he first considered "sinistraline" (from the Latin *sinister*, meaning "left side"). But "sinister" has odious connotations (e.g., evil, base, adverse). Furthermore, he knew that the Swiss stalwarts commonly spoke of "eastern" and "western" parts of the macrocycle. Since maps normally show the west toward the left, Woodward coined "hesperimine." So, onomastically, each research group paid a subtle compliment to the other.[78]

The B_{12} odyssey produced another trivial tidbit in the guise of tetracycle **49**, jocularly labeled "β-corrnorsterone" by Woodward.[78] The "corr" in this appellation represented hope that they could eventually transform **49** to a corrinoid (i.e., a molecule with the corrin ligand system in **50**);[80] and the "norsterone" devolved because ketone **49** has a norsteroid skeleton, if we ignore the nitrogens. And finally, noted Woodward, if the name is pronounced in "Slurvian," it becomes "cornerstone."[78] The pivotal title proved apt, because β-corrnorsterone did, in fact, generate hesperimine (**47**) on treatment with methanolic hydrogen chloride.

Robert Woodward's penchant for wordplay was displayed with subtlety when he lectured and wrote about the synthesis of chlorophyll, a milestone reached earlier at his Harvard precinct. For example, a 1961 account described the chemistry of a series of tetrapyrrolic substances having one sp^3 carbon bridge (as in **51**); and he introduced for these the class name "phlorins," a term suggested by postdoctorate associate Eugene

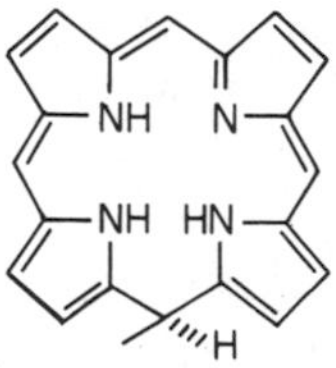

51

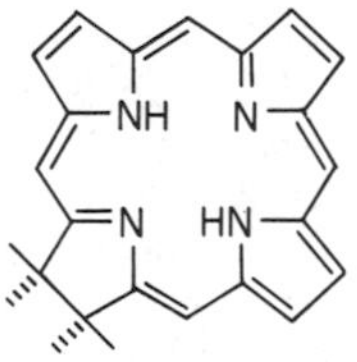

52

LeGoff.[81] At the time, chemists were already acquainted with "chlorins," a group of green compounds of general type **52** having two tetrahedral carbons in one ring. According to Michael King at George Washington University, R.B.W. was amused by the thought that when these structures are lined up the series would read: phlorin, chlorin, ...[82]

If you skim papers in the general area of porphyrin chemistry, it's handy to know that a "corphin" means the parent ligand system drawn in **53**; and a "pyrrocorphin" is **54**. Professor Eschenmoser's researchers prepared such hexahydroporphyrins and regis-

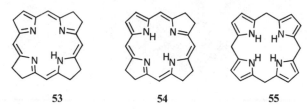

53   54   55

tered these two names in 1968[83] and 1982,[84] respectively. For comparison, we show the long-known "porphyrinogen" system in **55**.

Finally, to spotlight the nitrogen in these macrocycles, we visit Libya to track down the naming of a vital building block—the molecule NH_3. It ranks second only to sulfuric acid in importance as a chemical raw material for industrial use.[85] The word "ammonia" derives from "sal ammoniac" (NH_4Cl) or "salt of Ammon." This salt in turn, was named after the chief god of the ancient Egyptians, because it was supposedly first obtained from camel dung near the temple of Jupiter Ammon, in Libya, around 332 BC.[86] On this historic note we terminate our global journey.

REFERENCES AND NOTES

1. Stoessl, A.; Stothers, J.B. *J. Chem. Soc., Chem. Commun.* **1982**, 880–881.
2. Haddadin, M.J.; Issidorides, C.H. *Tetrahedron Lett.* **1965**, 3253–3256.
3. Abu El–Haj, M.J.; Dominy, B.W.; Johnston, J.D.; Haddadin, M.J.; Issidorides, C.H. *J. Org. Chem.* **1972**, *37*, 589–593.
4. Issidorides, C.H., private communication, February, 1976.
5. *Research Corporation Quarterly Bulletin*, (Foundation for the Advancement of Science, Tucson, AZ, Summer, 1975), pp. 1–2.
6. Issidorides, C.H.; Haddadin, M.J. *J. Org. Chem.* **1966**, *31*, 4067–4068.
7. Johnston, J.D.; Abu El–Haj, M.J.; Dominy, B.W.; McFarland, J.W.; Issidorides, C.H.; Haddadin M.J. *Abstracts of Papers*, 156th National Meeting of the American Chemical Society, Atlantic City, NJ, (American Chemical Society, Washington, DC, 1968), MEDI-15.
8. Haddadin, M.J.; Issidorides, C.H. *Heterocycles* **1976**, *4*, 767–816.
9. Smolnycki, W.D.; Moniot, J.L.; Hindenlang, D.M.; Miana, G.A.; Shamma, M. *Tetrahedron Lett.* **1978**, 4617–4620.
10. Shamma, M., private communication, May, 1981.
11. Fajardo, V.; Elango, V.; Chattopadhyay, S.; Jackman, L.M.; Shamma, M. *Tetrahedron Lett.* **1983**, *24*, 155–158.
12. Guinaudeau, H.; Freyer, A.J.; Minard, R.D.; Shamma, M.; Baser, K.H.C. *Tetrahedron Lett.* **1982**, *23*, 2523–2526.
13. Hussain, S.F.; Shamma, M. *Tetrahedron Lett.* **1980**, *21*, 3315–3318.
14. Shamma, M.; Moniot, J.L.; Yao, S.Y.; Miana, G.A.; Ikram, M. *J. Am. Chem. Soc.* **1972**, *94*, 1381–1382; *J. Am. Chem. Soc.* **1973**, *95*, 5742–5747.
15. de Silva, E.D.; Schwartz, R.E.; Scheuer, P.J.; Shoolery, J.N. *J. Org. Chem.* **1983**, *48*, 395–396.
16. Scheuer, P.J., private communication, March, 1983.
17. Professor Scheuer's team also originated the name "kalihinol-A" for a diterpenoid antibiotic isolated from a Guam marine sponge. Kalihi, a residential neighborhood in Honolulu, was the early home of research collaborator Clifford Chang. Chang, C.W.J.; Patra, A.; Roll, D.M.; Scheuer, P.J.; Matsumoto, G.K.; Clardy, J. *J. Am. Chem. Soc.* **1984**, *106*, 4644–4646.
18. Morris, W.; Morris, M. *Dictionary of Word and Phrase Origins* (Harper & Row, New York, 1962), p. 310.
19. Mitchell, R.S. *Mineral Names: What Do They Mean?* (Van Nostrand Reinhold, New York, 1979).
20. Earl, J.C.; Mackney, A.W. *J. Chem. Soc.* **1935**, 899–900.
21. Eade, R.A.; Earl, J.C. *J. Chem. Soc.* **1946**, 591–593.
22. Baker, W.; Ollis, W.D. *Nature (London)* **1946**, *158*, 703.
23. Baker, W.; Ollis, W.D.; Poole, V.D. *J. Chem. Soc.* **1949**, 307–314.
24. Ramsden, C.A. *Tetrahedron* **1977**, *33*, 3203–3232.

25. Pagni, R.M.; Watson, Jr., C.R. *Tetrahedron Lett.* **1973**, 59–60.
26. Pagni, R., private communication, January, 1976.
27. Howard, B.M.; Fenical, W. *Tetrahedron Lett.* **1978**, 2453–2456.
28. Murata, I.; Nakasuji, K. *Tetrahedron Lett.* **1973**, 47–50.
29. Viehe, H.G. *Angew. Chem. Int. Ed. Engl.* **1965**, *4*, 746–751.
30. Viehe, H.G., private communication, October, 1977.
31. *Nachr. Chem. Techn.* **1973**, *21*, 49.
32. Kinnel, R.; Duggan, A.J.; Eisner, T.; Meinwald, J.; Miura, I. *Tetrahedron Lett.* **1977**, 3913–3916.
33. Guroff, G.; Daly, J.W.; Jerina, D.M.; Renson, J.; Witkop, B.; Udenfriend, S. *Science* **1967**, *157*, 1524–1530.
34. Witkop, B., private communications, August, 1977 and March, 1986.
35. Jerina, D.M.; Daly, J.W.; Witkop, B.; Saltzman–Nirenberg, P.; Udenfriend, S. *J. Am. Chem. Soc.* **1968**, *90*, 6525–6527; *Biochemistry* **1970**, *9*, 147–156.
36. Boyd, D.R.; Daly, J.W.; Jerina, D.M. *Biochemistry* **1972**, *11*, 1961–1966.
37. Witkop, B. In *NIH: An Account of Research in its Laboratories and Clinics*, Stetten, D., Jr.; Carrigan, W.T. (Eds.), (Academic Press, New York, 1985), chapter 11, pp. 213–216.
38. Misra, R.; Pandey, R.C.; Silverton, J.V. *J. Am. Chem. Soc.* **1982**, *104*, 4478–4479.
39. (a) Barton, D.H.R.; Boivin, J.; Ozbalik, N.; Schwartzentruber, K.M.; Jankowski, K. *Tetrahedron Lett.* **1985**, *26*, 447–450, (b) Barton, D.H.R.; Göktürk, A.K.; Morzycki, J.W.; Motherwell, W.B. *J. Chem. Soc., Perkin Trans 1* **1985**, 583–585.
40. Bartle, K.D.; Heaney, H.; Jones, D.W.; Lees, P. *Spectrochim. Acta* **1966**, *22*, 941–951; Bartle, K.D.; Jones, D.W. *Trans. Farad. Soc.* **1967**, *63*, 2868–2873. Actually, the authors used "bay," without comment, in a publication (*Tetrahedron* **1965**, *21*, 3289–3295) that preceded these.
41. Jones, D.W., private communication, July, 1983.
42. Bartle, K.D.; Jones, D.W. *Adv. Org. Chem.* **1972**, *8*, 317–423.
43. For an example of usage see Silverman, D.B.; LaPlaca, S.J. *J. Chem. Soc., Perkin Trans. 2* **1982,** 415–417
44. Martin, R.H. *Tetrahedron* **1964**, *20*, 897–902.
45. Jerina, D.M.; Lehr, R.E.; Yagi, H.; Hernandez, O.; Dansette, P.M.; Wislocki, P.G.; Wood, A.W.; Chang, R.L.; Levin, W.; Conney, A.H. *In Vitro Metabolic Activation in Mutagenesis Testing*, de Serres, F.J.; Fouts, J.R.; Bend, J.R.; Philpot, R.M. (Eds.), (Elsevier/North–Holland Biomedical Press, Amsterdam, 1976), pp. 159–177.
46. Pullman, A. *Compt. Rend.* **1945**, *221*, 140–142.
47. Pullman, A. *Bull. du Cancer* **1946**, *33*, 120–130; Compt. Rend. Soc. Bio. **1945**, *139*, 1056–1058.
48. Pullman, A. *Comp. Rend.* **1953**, *236* Part 2, 2318–2320.
49. Pullman, A.; Pullman, B. *Bull. Soc. Chim. Fr.* **1954**, *21*, 1097–1104.
50. (a) Pullman, B. *Int. J. Quantum Chem., Quantum Biol. Symp.* **1979**, *6*, 33–45, (b) Pullman, A., private communication, September, 1983.
51. Pullman, B. *Int. J. Quantum Chem.* **1979**, *16*, 669–689.
52. (a) Jones, D.W.; Sowden, J.M. *Cancer Biochem. Biophys.* **1979**, *4*, 43–47, (b) Briant, C.E.; Jones, D.W. In *Polynuclear Aromatic Hydrocarbons: Formation, Metabolism, and Measurement*, Cooke, M.; Dennis, A.J. (Eds.), (Battelle Press, Columbus, OH 1983), pp. 191–200, (c) Jones, D.W., private communications, July, 1983 and March, 1984.
53. Fawcett, C.H.; Spencer, D.M.; Wain, R.L.; Fallis, A.G.; Jones, E.R.H.; Le Quan, M.; Page, C.B.; Thaller, V.; Shubrook, D.C.; Whitham, P.M. *J. Chem. Soc. (C)* **1968**, 2455–2462.
54. Knight, D.W.; Nott, A.P. *J. Chem. Soc., Perkin Trans. 1* **1982**, 623–625.
55. Menger, F.M.; Jerkunica, J.M.; Johnston, J.C. *J. Am. Chem. Soc.* **1978**, *100*, 4676–4678.
56. Hess, K.; Gundermann, J. *Chem. Ber.* **1937**, *70*, 1800–1808.
57. Svens, B.; Rosenholm, B. *J. Colloid Interface Sci.* **1973**, *44*, 495–504.
58. Menger, F.M., private communication, September, 1978.
59. Stigter, D. *J. Phys. Chem.* **1974**, *78*, 2480–2485.
60. Menger, F.M.; Donohue, J.A.; Williams, R.J. *J. Am. Chem. Soc.* **1973**, *95*, 286–288.
61. Menger, F.M.; Saito, G. *J. Am. Chem. Soc.* **1978**, *100*, 4376–4379.
62. Menger, F.M.; Doll, D.W. *J. Am. Chem. Soc.* **1984**, *106*, 1109–1113.
63. Fuoss, R.; Edelson, D. *J. Am. Chem. Soc.* **1951**, *73*, 269–273.
64. (a) Menger, F.M.; Carnahan, D.W. *J. Am. Chem. Soc.* **1986**, *108*, 1297–1298, (b) Fuhrhop, J–H.; David, H–H.; Mathieu, J.; Liman, U.; Winter, H–J.; Boekema, E. *J. Am. Chem. Soc.* **1986**, *108*, 1785–1791, (c) Fuhrhop, J–H.; Mathieu, J. *Angew. Chem. Int. Ed. Engl.* **1984**, *23*, 100–113.

65. Morawetz, H.; Kandanian, A.Y. *J. Phys. Chem.* **1966**, *70*, 2995–3000.
66. Ueoka, R.; Ohkubo, K. *Tetrahedron Lett.* **1978**, 4131–4134.
67. Ueoka, R., private communication, March, 1981.
68. Nelson, N.J., private communications, September, 1975; March and May, 1984.
69. Doyle, M.; Parker, W.; Gunn, P.A.; Martin, J.; MacNicol, D.D. *Tetrahedron Lett.* **1970**, 3619–3622.
70. Leonard, N.J.; Coll, J.C. *J. Am. Chem. Soc.* **1970**, *92*, 6685–6686.
71. Leonard, N.J.; Coll, J.C.; Wang, A. H–J.; Missavage, R.J.; Paul, I.C. *J. Am. Chem. Soc.* **1971**, *93*, 4628–4630.
72. Cava, M.P.; Krieger, M.; Pohlke, R.; Mangold, D. *J. Am. Chem. Soc.* **1966**, *88*, 2615–2616.
73. Huisgen, R. *Angew. Chem. Intl. Ed. Engl.* **1963**, *2*, 565–598.
74. Huisgen, R.; Grashey, R.; Gotthardt, H.; Schmidt, R. *Angew. Chem. Intl. Ed. Engl.* **1962**, *1*, 48–49.
75. Huisgen, R.; Gotthardt, H.; Bayer, H.O.; Schaefer, F.C. *Angew. Chem. Int. Ed. Engl.* **1964**, *3*, 136–137.
76. (a) Huisgen, R. *Special Publication No. 21* (The Chemical Society, London, England, 1967), pp. 51–73; (b) Padwa, A.; Gingrich, H.L.; Lim, R. *Tetrahedron Lett.* **1980**, 3419–3422, (c) Padwa, A.; Burgess, E.M.; Gingrich, H.L.; Roush, D.M. *J. Org. Chem.* **1982**, *47*, 786–791; (d) Newton, C.G.; Ramsden, C.A. *Tetrahedron* **1982**, *38*, 2965–3011.
77. Ollis, W.D. *Chem. Brit.* **1980**, *16*, 210–216.
78. Woodward, R.B. *Pure Appl. Chem.* **1968**, *17*, 519–547.
79. Eschenmoser, A., private communication, April, 1983.
80. Eschenmoser, A. *Quart. Revs.* **1970**, *24*, 366–415.
81. (a) Woodward, R.B. *Pure Appl. Chem.* **1961**, *2*, 383–404, (b) LeGoff, E., discussion by phone, July, 1986.
82. King, M.M., private communication, March, 1984. Professor King's Ph.D. Dissertation (Harvard, 1970, under R.B.W.) concerned pentapyrrolic macrocycles and related compounds.
83. Johnson, A.P.; Wehrli, P.; Fletcher, R.; Eschenmoser, A. *Angew. Chem. Intl. Ed. Engl.* **1968**, *7*, 623–625.
84. Schwesinger, R.; Waditschatka, R.; Rigby, J.; Nordmann, R.; Schweizer, W.B.; Zass, E.; Eschenmoser, A. *Helv. Chim. Acta* **1982**, *65*, 600–610.
85. Chenier, P.J. *J. Chem. Educ.* **1983**, *60*, 411–413.
86. Felty, W.L. *J. Chem. Educ.* **1982**, *59*, 170.

Chapter 10

OUR COLLEAGUES AND SOME OTHER FOLKS

Even though chemists work a great deal with inanimate things (instruments, glassware, and chemicals), they can be sentimental when it comes to personal relationships. So, they frequently identify chemical compounds with individuals. Some names spring from religion and mythology; one (barbituric acid) has as many alleged origins as people who have tried to track them down.

The name August Kekulé is linked historically to an important event in chemistry: elucidation (in 1865 at the University of Bonn) of the structure of benzene.[1] Students hear the story about a dozing Dr. Kekulé who imagined chains of atoms going through snake-like contortions when, suddenly, one "snake" grabbed its own tail.[2] The cyclic structure for benzene was born (or reborn, see p. 3)! Since then, the words "Kekulé" and "benzene" have been practically synonymous.

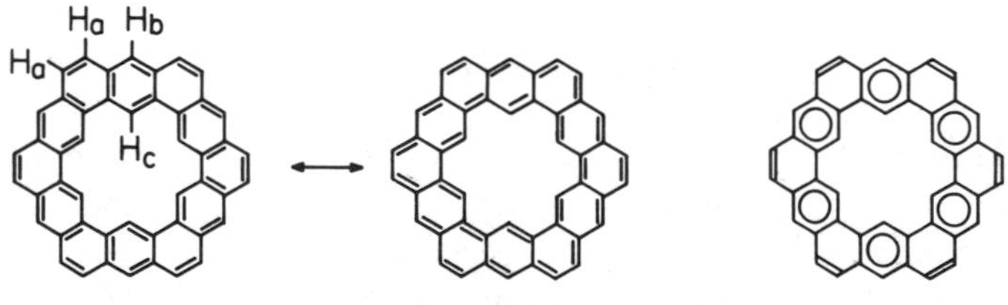

1a   1b   1c

During the 1965 Kekulé centennial celebration of the Gesellschaft Deutschen Chemiker, Heinz Staab of the Max-Planck-Institute, Heidelberg, announced preliminary work[3] on his pursuit of **1**, a polynuclear aromatic hydrocarbon for which 200(!) canonical structures, including **1a** and **1b**, are possible.[4] Since **1a** is a "benzene ring of benzene rings" and Kekulé was being honored, Professor Staab suggested "kekulene" for this compound.[5] The non-Kekulé canonical form **1b** has special interest, as it consists of a [30]annulene (see chapter 8) outer track and an [18]annulene inner track, both aromatic according to the Hückel $4n + 2$ rule. These concentric loops are connected by 12 sigma bonds.[6] If **1b**, rather than **1a**, adequately describes kekulene, the six inner hydrogens would be strongly shielded in the NMR.

Jenny, Baumgartner, and Paioni at CIBA A.-G., Basel, Switzerland, claimed to have synthesized **1** in 1970 but gave no details.[7] Staab and Vögtle may have prepared a tiny amount of it in 1968;[8] their product gave a mass spectral peak at m/e 600 (the mass of kekulene) but also showed an ion of about equal intensity at m/e 604, suggestive of a tetrahydrokekulene. And in 1978, Diederich and Staab reported the first unequivocal synthesis.[6] The high symmetry of kekulene results in a melting point above 620°C and extremely low solubility in organic solvents. The latter feature presented real problems for ^{1}H NMR, the ultimate "referee" in the contest between structures **1a** and **1b**. The spectrum required 50,000 scans of a saturated solution in 1, 3, 5-trichlorobenzene-d_3 at 215°C! It consisted of peaks at $\delta = 7.94$, 8.37 and 10.45 in an area ratio of 2:1:1. Dr. Staab allocated these absorptions to hydrogens labeled a, b and c, respectively, in the Kekulé structure **1a**.[6] No signal appeared far upfield, where the inner aromatic H's of **1b** should absorb; so collect your money if you wagered on structure **1a**. Bond lengths from X-ray analysis also fit **1a** better than **1b**. In fact, Professor Staab and his ringmasters favor the sextet formulation **1c** as the best single representation of π-cliques in kekulene.[9] We think August Kekulé would have preferred it that way also.

Teams spearheaded by John Barton (University of Bristol) and by Peter Vollhardt (University of California, Berkeley) found other ways to please Dr. Kekulé. They built linearly fused π-corrals, **2**, which Vollhardt *et al.* called "multiphenylenes," or "[*N*]phenylenes" (*N* = number of benzene rings).[10] In such a series, the alternating benzene and cyclobutadiene units make opposing (e.g., stabilizing and destabilizing) contributions to the overall resonance energy.

2   **3**   **4**

Professor Barton's squad forged [3]phenylene (**2**, $N = 3$) by a pyrolytic route;[11] and independently the Berkeley bunch made this red hydrocarbon by cobalt-catalyzed cycloadditions.[10] Previously, both groups had also applied their individual methodologies to assemble the angular prototype **3**.[12,13] Professor Vollhardt and co-worker Rainer Diercks noted that a series of angular fusions would culminate in the corrugated analog **4**, which they termed an "anti-kekulene" because its inner and outer π-circuits are both $4n$, unlike in kekulene.[13]

Diercks and Vollhardt deftly applied their cobalt-mediated reactions to convert hexaethynylbenzene (**5**, see p. 86) in two steps to the heptacycle **6**.[14] In this type of triangular multiphenylene, the central ring has its bonds strongly fixed, like that in the hypothetical "1, 3, 5-cyclohexatriene." Interestingly, flash vacuum pyrolysis transformed **6** to **7**. This benzene → trialkyne retrocyclization demonstrated that multiphenylenes may have inherited some pericyclyne bloodline (see p. 109).

two steps Δ 700°C

5   **6**   **7**

We hinted earlier (chapter 1) that the "horse" in equinene (**8**) (*equus* is Latin for "horse") and the "bull" in "bullvalene" (**9**) are of the two-legged varieties. How so? The "horse" turns out to be Harry Wasserman of Yale University. While a graduate student

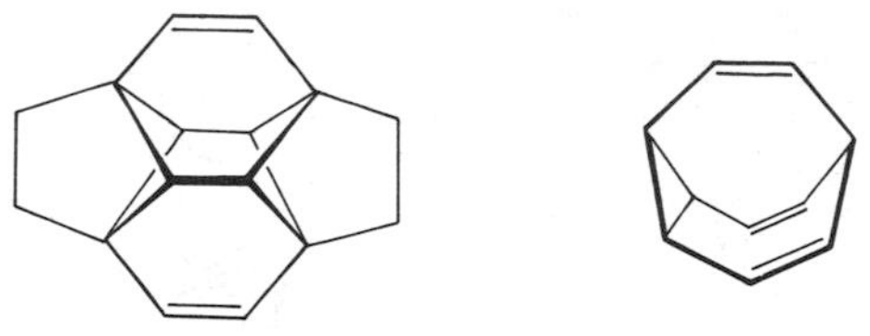

| 8 | 9 |

at Harvard, Wasserman worked tirelessly and so was nicknamed "Harry the Horse" by his doctoral supervisor, Robert B. Woodward.[15] (In return, the Horse dubbed his mentor "Robert the Ram.") This ribbing did not end at Harvard. Later, mail clerks at Yale had to contend with letters from Woodward addressed "Harry the Horse Wasserman." In research at Yale, Professor Wasserman and graduate student Philip Keehn, in 1967, synthesized **12** by beaming light onto **10**.[16] They interpret the change as two successive internal Diels–Alder reactions via **11**. Protégé Keehn was aware of Dr. Wasserman's old nickname, so he called **12** "dibenzoequinene" to honor him. Thus, the

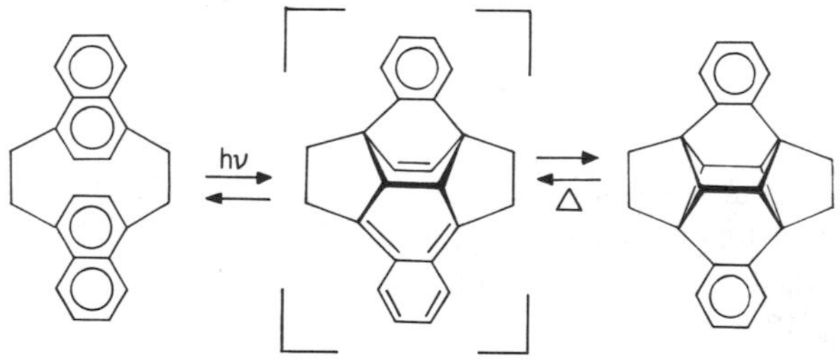

| 10 | 11 | 12 |

parent hydrocarbon **8** became "equinene." When heated, dibenzoequinene galloped back to **10**.[16]

An air of mystery surrounds the naming of "bullvalene" (**9**), and cryptic statements appear in the literature.[17,18] Princeton's Maitland Jones, Jr. and the University of Tennessee's Ronald Magid helped solve part of the mystery;[19,20] structure **9** was named to acclaim William Doering who, with Wolfgang Roth, predicted some of its remarkable properties before it was synthesized.[21] It seems that during Professor Doering's early days on the faculty at Columbia University, his research students nicknamed him "The Bull" for reasons that are not clear. Perhaps it was because "Bill Doering" sounds like "Bull Durham," a popular brand of tobacco;[19] but other explanations can be found. Dr. Doering and his nickname subsequently transferred to Yale University, and it was there at a group seminar on October 9, 1961, that he disclosed his predictions on degenerate rearrangements of the then unknown $(CH)_{10}$ structure. Two graduate students, Ronald M. Magid and Maitland Jones, Jr., sat near the back of the room; one of them turned to the other and blurted out "bullvalene."[20] (The "valene" suffix may have spun off from "fulvalene" (p. 288), a compound synthesized earlier by Dr. Doering.[22]) Which of the two students was the blurter and which was the blurtee remains unsettled; but their term caught on within the research group. And everyone was surprised when Doering and Roth formally adopted it in their seminal publications in 1963.[21]

Professor Doering predicted that through Cope rearrangements (see arrows in drawings **13**) a molecule constituted as in **9** would be "fluxional," namely, characterized "by the existence of two or more identical chemical structures which are

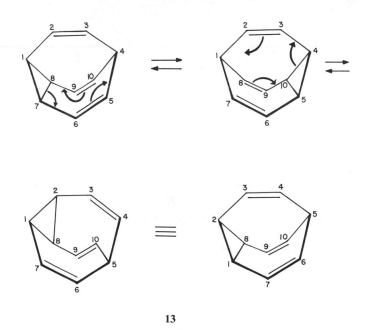

13

interconverted by an autogenous intramolecular rearrangement... ."[17] (In ordinary tautomerism, the participants are not identical.) In 1978, Stephen Stobart and coworkers at Canada's University of Victoria labeled fluxion among *non*identical structures as "quasifluxional."[23]

Bullvalene was first prepared (somewhat serendipitously) by Gerhard Schröder in 1963 at Union Carbide European Research Associates, Brussels. He had been working on the structure of a cyclooctatetraene dimer (**14**) and found that irradiation cleaved the cyclobutane ring and delivered bullvalene.[24] And, just as Professor Doering presaged,[21] the proton NMR spectrum of this intriguing hydrocarbon (at 100°C) consisted of only one signal![24] Therefore, each hydrogen spends 10% of its time at the bridgehead, 30% at a cyclopropyl position, 30% at one olefinic site, and 30% at the other olefinic site. To describe fully the itinerary of all the atoms at this temperature, you would need to draw 1,209,600 (i.e., 10!/3) degenerate structures![17] (Below 100°C, the spectrum of bullvalene becomes more complex as the Cope convolutions slow down.) Shortly after Schröder's report, Doering's team completed its own rational synthesis as shown.

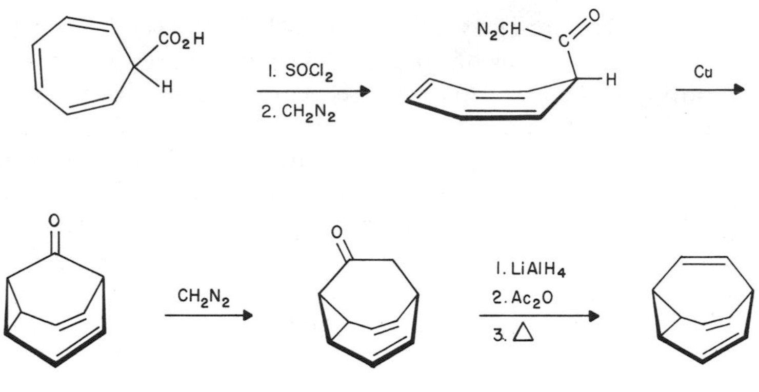

15

A crucial ketone (**15**) in Doering's route became known as "barbaralone" to acknowledge Dr. Barbara M. Ferrier, a team member instrumental in its synthesis.[17,21] Who came up with that name is not certain, but everyone in the group used it except Dr. Ferrier.[25] Barbaralone is also fluxional; in fact, it "Copes" better than bullvalene. At only 25°C, the proton NMR spectrum consists of three signals in an area ratio of 2:1:1. Evidently, the equilibrium **15a** $\rightleftarrows$ **15b** is rapid; and the signals correspond to the three sets of hydrogens labeled a, b, and c.*

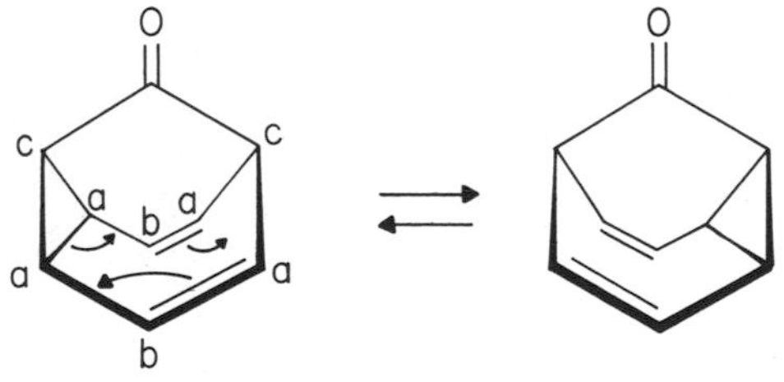

15a   **15b**

Barbaralone was definitely named for a Barbara. But barbituric acid (**16**), which is related to the barbiturates (**17**) we hear so much about, is another matter. Adolf von Baeyer first prepared and named barbituric acid in 1863.[29] The "-uric" presents no

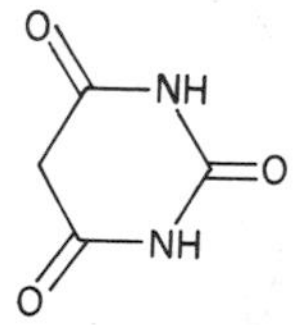

16 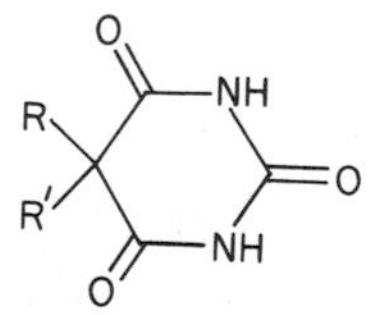   **17** 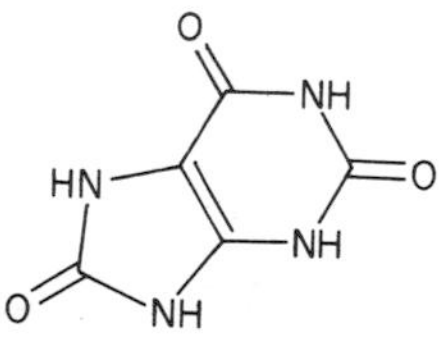   **18**

puzzle, because **16** resembles uric acid (**18**) structurally. But Dr. von Baeyer never revealed what the "barb" means. Romantics may accept the legend that the good doctor courted a lady named Barbara.[30-32] Nonromantics may prefer one of the following theories, or may want to invent their own: (1) The compound was discovered

---

*In 1960, Balaban and Farcasiu introduced the word "automerization" for a process (such as the fluxion of **9** or **15**) in which the structural formula of a molecule remains intact.[26,27] Binsch, Eliel, and Kessler in 1971 recommended that such a change be termed "topomerization."[28] If you venture into the world of "topomers," be prepared to confront "stereoheterotopomerization," its offspring "diastereotopomerization" and "enantiotopomerization," and some cousins. But don't turn back. These creatures are actually friendly and can help steer perplexed souls through NMR intricacies of conformationally mobile systems.

on or just before Saint Barbara's Day (December 4).[33-35] (2) It is a key compound; the German word *Schlüsselbart* means "the beard of a key" (i.e., the serrated bit), and *barba* is Latin for "beard."[36] Where are you, Sherlock Holmes, now that we need you?*

The story behind the euonym "smissmanones" for bicyclic compounds such as 19 has a sad note. Researchers at the University of Kansas headed by Edward Smissman began the synthesis of the 2, 4-oxazolidinedione 19 ($n = 1$), but Professor Smissman died in 1974 before they reached this goal.[39] Dr. Smissman's student Wayne Brouillette, along with Warren Quillin, continued these efforts under Professor Gary Grunewald's direction; but they did not succeed. Several years later at the University of Alabama in Birmingham, Dr. Brouillette independently prepared 19 with $n = 2$ and 3.[40] He named compounds of type 19 "smissmanones" in memory of his departed mentor.[41]

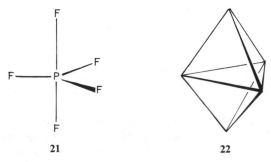

Smissmanones illustrate how chemists strive to improve physiological activity by modifying chemical structure. 2, 4-Oxazolidinediones (20) were known to be anticonvulsants before the investigators at Kansas began their work. However, these drugs did not help some epileptics and promoted unpleasant side effects, such as sedation. Dr. Smissman and co-workers had reasoned that the trimethylene bridge in 19 would block one face of the heterocyclic ring and permit the molecule to bind a receptor site from only the other face. Such selective binding might improve the drug's specificity and thus decrease side effects. Also, if nucleophilicity of nitrogen is vital for binding, 19 might be more potent than are simple oxazolidinediones (20); the alkane bridge discourages electron seepage such as in 19b by preventing planarity at the bridgehead (Bredt's rule).[39,40]

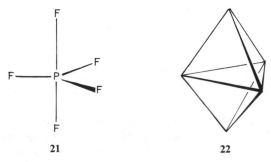

---

*Chemists have been known to turn to legendary supersleuth Sherlock Holmes for assistance even on loftier matters of science. For example, to elucidate crystallographically the arrangement of boron atoms in a B_9 hydride, Richard Dickerson, Peter Wheatley, Peter Howell, and William Lipscomb (all then at the University of Minnesota) and Riley Schaeffer (then at Iowa State College) said they used a computer and resorted to "the method of Holmes," to wit: "...when all other contingencies fail, whatever remains, however, improbable, must be the truth."[37] In 1971, Julian Heicklen and trainee William Wood at Pennsylvania State University also called upon the "Holmes method" to draw mechanistic conclusions about carbon disulfide–oxygen explosions.[38]

A mechanism for shuffling ligands at pentacovalent phosphorus proposed in 1960[42] by Stephen Berry (then at the University of Michigan) has become known as "Berry pseudorotation."[43] The fluorines in phosphorus pentafluoride (**21**) sit at the corners of a trigonal bipyramid (**22**). The three bonds in the trigonal plane are referred to as equatorial, and the two perpendicular bonds are called apical. Despite the two different ligand sites, the ^{19}F NMR spectrum of neat **21** at $-90°$C consists only of one signal.[44]

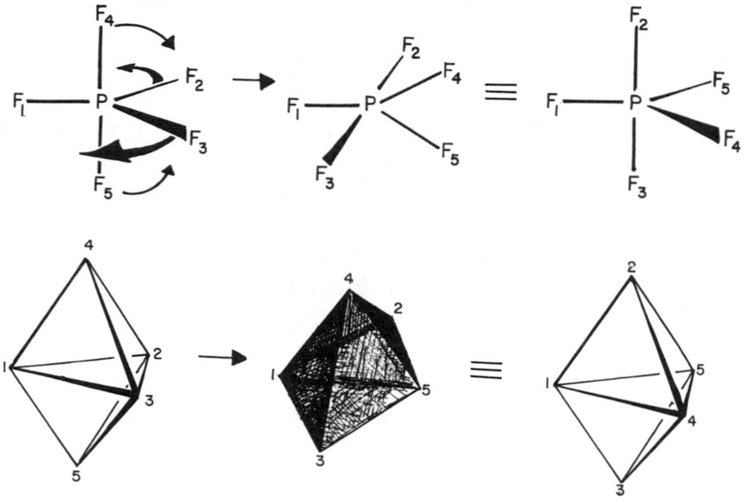

23

All five halogens are equivalent. Enter Dr. Berry. He proposed that two equatorial fluorines (F_2 and F_3) can easily spread apart while the apical ones (F_4 and F_5) swing toward each other (see diagram **23**). Presto, chango! Two erstwhile equatorial fluorines become apical and *vice versa*; and the solid trigonal bipyramid appears to have rotated in space (hence "pseudorotation"). Repeat performances with F_1 getting into the act lead eventually to time-averaged equivalence of the five ligands. George Whitesides and his band (then at the Massachusetts Institute of Technology) provided supporting evidence.[45]

Two research groups, headed by Fausto Ramirez (State University of New York at Stony Brook)[43] and by Ivar Ugi (then at University of Southern California),[46] unburied an alternative pathway with bisphosphorus compounds such as **24**. From proton and ^{19}F NMR spectra of **24**, these investigators found all 6 hydrogens

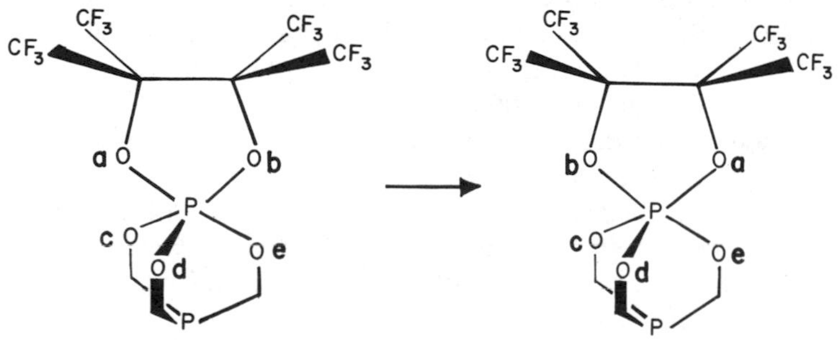

24   **25**

equivalent and all 12 fluorines equivalent. But the Ugi and Ramirez teams argued that Berry pseudorotation is highly unlikely here, because the ligands on the *pentavalent* phosphorus are held back by the rings.[43,46] They suggest that the upper and lower portions of **24** can rotate around an axis through the two phosphorus atoms to give degenerate form **25**. This motion reminded Professor Ugi and colleagues of turnstiles, so they dubbed the process "turnstile mechanism."[46] For certain types of structures, they believe that turnstile rotation is more likely than Berry pseudorotation.[47] By the way, the word ligand (Latin *ligare*, to bind) was coined in 1917 by German chemist Alfred Stock. A historical account is available tracing the origin and dissemination of this term.[48]

"Free radicals are unstable." Students in organic chemistry often hear that generalization. And it certainly is true for many simple ones, such as $CH_3 \cdot$ and $C_2H_5 \cdot$. However, radicals become more comfortable if the odd electron spreads around by resonance; and they also can persist if bulky groups prevent dimerization, a favorite annihilation path. Galvin Coppinger of the Shell Development Co., Emeryville, California, used both of these tacks when he first synthesized radical **26**.[49] The odd electron can diffuse to the starred atoms, and the tert-butyl groups keep this beast from mating. Indeed, Paul Bartlett and Toshio Funahashi, then at Harvard University, found that only 10% of **26** decomposed during six summer weeks. (Summers can be hot in Cambridge, Massachusetts!) Dr. Bartlett christened radical **26** "galvinoxyl" after the chemist who gave it birth; the -oxyl suffix proclaims it as an aryloxy free radical.[50]

Charles Friedel, famous nineteenth-century French chemist and mineralogist, is best known as codiscoverer of the Friedel–Crafts reaction. Thus he joins the enviable cadre of chemists who have had reactions named after them. But Friedel has the added distinctions of having a natural product and a mineral labeled in his honor. Thus, in

26   27

1899, Constantin Istrati and A. Ostrogovitch of the University of Bucharest isolated a compound from cork and named it "friedelin."[51] More than half a century later, researchers led by Leopold Ruzicka[52] in Switzerland and by Elias J. Corey[53] in the United States established its structure as **27**. Friedelin has since been found in many plants other than cork.[54] The mineral that pays homage to Dr. Friedel is "friedelite"; it consists mostly of oxides of silicon and manganese.[55]

By the way, some other minerals that salute well-known chemists are berzelianite and berzeliite (Jöns J. Berzelius, Swedish); bunsenite (Robert W. Bunsen, German); curite (Pierre Curie, French); davyne (Sir Humphry Davy, British); gaylussite (Joseph L. Gay–Lussac, French); gmelinite (Christian G. Gmelin, German); heyrovskyite (Jaroslav Heyrovsky, Czechoslovak, Nobel Prize in chemistry 1959); lechatelierite (Henri Le Chetalier, French); liebigite (Justus von Liebig, German); lipscombite (William N. Lipscomb, Jr., American, Nobel Prize in chemistry, 1976); paulingite (Linus C. Pauling, American, Nobel Prize in chemistry, 1954); scheelite (Karl W. Scheele, Swedish); sklodowskite (Marie Sklodowska–Curie, Polish–French); ullmannite (Johann C. Ullmann, German–Hessian); ureyite (Harold C. Urey, American,

Nobel Prize in chemistry, 1934); vanthoffite (Jacobus H. van't Hoff, Dutch, Nobel Prize in chemistry, 1901); wohlerite (Friedrich Wöhler, German); wurtzite (Charles A. Wurtz, French).

Lest you think one must achieve fame before becoming a mineral, please consult Mitchell's monograph on the subject.[56] You will learn that the 2500 or so minerals therein have common names connected with persons, places, or things, as well as with physical or chemical properties. When persons are honored, the list suggests that whom you know may sometimes count more than what you know. But, for the most part, propriety is the order of the day in mineral neologism. For example, at Leeds University, England, Derry Jones and John Smith investigated ^1H NMR spectra of crystalline $CaHPO_4 \cdot 2H_2O$,[57] a mineral known as brushite and named after Yale mineralogist George Jarvis Brush (1831–1912).[56] Later, the British duo determined the crystal structure of a related calcium orthophosphate, $Ca(H_2PO_4)_2 \cdot H_2O$, and tended to refer to it as "scrubite."[58] But, when it came time to publish, whimsy gave in to restraint; they kept the literature cleaner by not using the term "scrubite" in their paper.[59]

During World War I, renowned chemist Roger Adams of the University of Illinois led a research regiment that contributed importantly to the arsenial arsenal of the U.S. Government. For example, they developed **28** as a chemical warfare agent. This compound was later named "adamsite" in his honor.[60] Be sure to stress the first syllable.

28

29

In 1966, a nonaromatic sextet headed by Monroe Wall of Research Triangle Institute in North Carolina isolated **29** from *Camptotheca acuminata*.[61] They named it comptothecin because of its source. Ten years later, Chinese chemists in Shanghai described a preparation of this compound;[62,63] and, more recently, Dr. Wall and co-workers also reported a similar synthesis.[64] The combination of the nationality of one research group and the leader of the other inspired Richard Hutchinson of the University of Wisconsin to dub that method the "Chinese/Wall synthesis" of camptothecin.[65] But don't get the wrong idea. The route is neither long nor tortuous.

Chemists who study natural products often depend on botanists to gather exotic vegetation. For this reason, W. David Ollis and his collaborators at the University of Sheffield named **30** "duartin" in appreciation of Brazilian botanist Apparicia Pereira Duarte, who collected many of the plants explored in Professor Ollis's laboratory.[66] Who said that chemists only honor other chemists?

Another nonchemist was remembered when James Moore and Stanley Eng at the University of Delaware isolated five new substances from Jamaica Dogwood and gave one of them the paedonym* "lisetin," after Professor Moore's daughter, Lise.[67] The structure of lisetin turned out to be **31**.[68] Alas, Dr. Moore's gesture did not influence Lise to become a chemist.[69]

---

*Paedonym: a name derived from one's child.

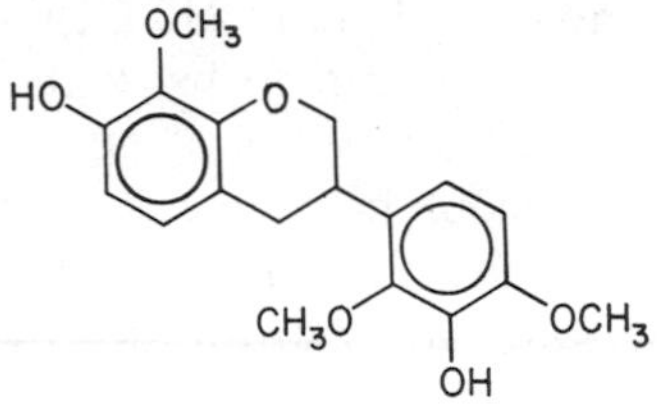

30 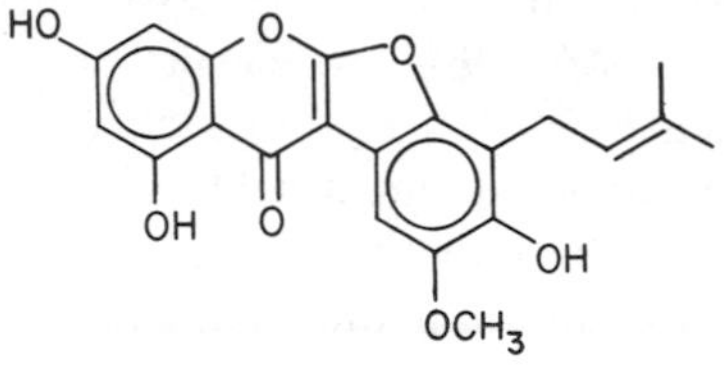　　　**31**

If a daughter can be honored, why not a wife? Indeed. When Heinz Balli and graduate student Martin Zeller at the University of Basel's Institute für Farbenchemie synthesized the π-excess heteroarene **32**, they titled it "ullazine."[70] Molecules with this shapely network could win shares in the aromatic marketplace by polarization that develops a 14 π-electron perimeter.[71] But whether **32** is aromatic or not, Professor Balli's name for this N-heterocycle pleased his wife, Ulla.[72]

And while we're into family relations, let us not neglect fathers. When Dietmar Seyferth's research group at MIT studied complexes of the type $RCCo_3(CO)_9$, graduate student Ralph Spohn felt that it took longer to say "alkylidenetricobalt nonacarbonyl" than to make one.[73,74] So, to simplify life, he called this breed of compounds "Fred," after his father, and started a trend in Professor Seyferth's laboratory. Soon thereafter, the related complexes $(RC)_2Co_2(CO)_6$ became "Ralph"

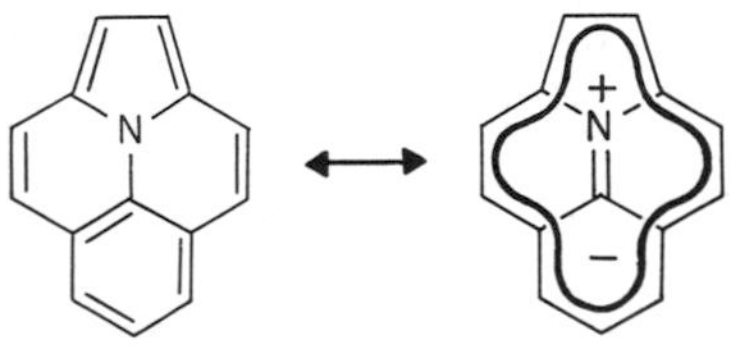

32 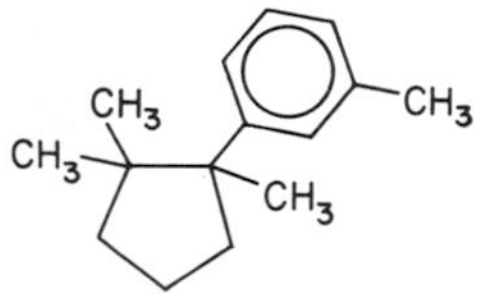　　　**33**

(after Ralph Spohn); and $RCAsCo_2(CO)_6$ became "Arsa–Joe" (after Joseph S. Merola, a graduate student who first prepared it).[74] On this basis you might expect "herbertene" to be an inorganic complex named for a youth in Dr. Seyferth's studio. Not so. Actually, it is an aromatic sesquiterpene (**33**) isolated by Japanese researchers from a leafy liverwort, *Herberta adunca* Dicks.[75] We didn't follow up this last name, but the simple Seyferth system may have its counterpart in the world of botany.

Subramania Ranganathan at the Indian Institute of Technology at Kanpur reached far back into the history of his native India when he read that Charles Wilcox's group at

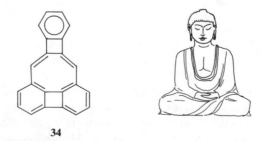

34

Cornell University had synthesized **34**.[76] The structure reminded Professor Ranganathan of Buddha (a religious philosopher and teacher in India about 563–483 BC and founder of Buddhism) in meditation; he pointed this out in *Current Organic*

Chemistry Highlights, a publication of abstracts edited by him and his chemist–wife Darshan and circulated by their institute.[77]

In the United States, pioneer steroid researcher Russell E. Marker published prodigiously on sapogenin chemistry between 1935 and 1949. While processing over 40,000 kg of plants, comprising more than 400 species collected mainly in Mexico and the Southern United States, he and his collaborators characterized many novel sapogenins. Running out of terms that reflected their botanical sources, he labeled some of his new compounds after people and places. For example, Marker coined "pennogenin" (after Pennsylvania State College, where he conducted research from 1935 to 1943); "kammogenin" (after Dr. Oliver Kamm, then Director of Research at Parke, Davis and Co.); "rockogenin" (after chemist Frank Whitmore, who became Dean of Chemistry at Penn State and was nicknamed "Rocky"); "nologenin" (after Stanford University chemist Carl R. Noller); and "fesogenin" (after Harvard chemist Louis F. Fieser). Marker's colleagues returned these tributes by naming three steroids in his honor; the literature knows these as "Marker acid," "Marker–Lawson acid," and "marcogenin."[78] And, the famous "Marker degradation" for conversion of diosgenin to progesterone remains a classic, efficient way to shear the sapogenin side chain. Russell Marker was a cofounder of Syntex S.A., a Mexican company whose commercial production of steroid hormones strongly influenced the direction of steroid research in Mexico. Incidentally, Mr. Marker completed his graduate research at the University of Maryland (under Morris S. Kharasch) but was never granted the Ph.D. degree because he refused to take certain required courses in physical chemistry. A short biography of Marker's career is available.[79]

Mexican chemist Xorge Domínguez continued the trend of christening natural products after persons and institutions.[80] For example, when he and his collaborators isolated and characterized flavanone derivative **35**, they called it "louisfieserone."[81] This title justly complimented Professor Louis Fieser, with whom Dr. Domínguez had received research training. Diterpene **36**, extracted from an arid-region Mexican shrub, became "riolozatrione"[82] to salute Leopoldo Rio de la Loza, a nineteenth century Mexican chemist (1813–1874).[80] On occasion, Professor Domínguez chose tongue-

35   36

twisting titles like "netzahualcoyone,"[83] "cuauthemone,"[84] and "ixtlixochiline"[85]— not to slow your reading but to honor Aztec kings in Mexican Indian history.[80]

In a few instances, we wonder if the Domínguez crew spiced its nomenclature with a pinch of politics. Consider for example, "itesmol" (**37**),[86] named for Dr. Domínguez's institute, ITESM (Instituto Tecnologico de Estudios Superiores de Monterrey); "eisacol" (**38**),[87] named for EISAC (a Mexican granting agency, similar to the U.S. National Science Foundation), and "luiselizondin" (a flavonol of formula $C_{20}H_{20}O_9$),[88] named for Luis Elizondo, a local scientific benefactor. Need we add that Professor Domínguez's research was always well funded?

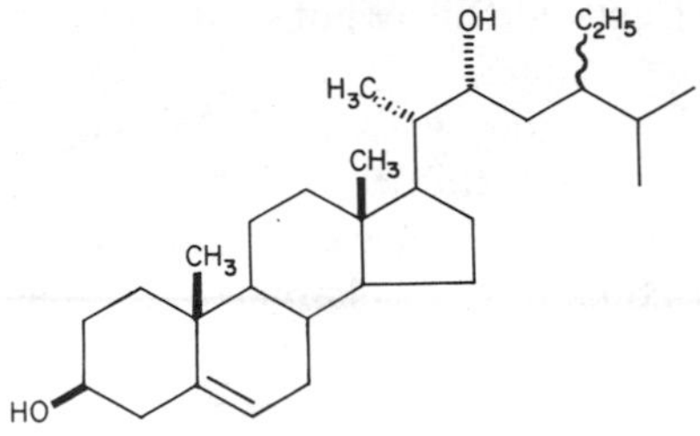

37

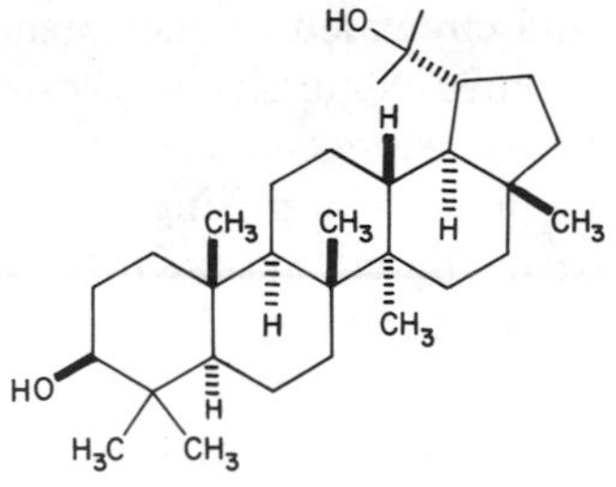

38

In any case, chemical neology is enticing and deserves fuller exploitation. After all, who knows what benefits could befall those who name a compound after (say) a congressman, senator, or other influential politician. A similar gesture would surely brighten the day for the president of your firm, the head of your division, or the dean of your faculty. Plenty of souls out there yearn to be immortalized; and chemists have the means to fulfill those desires.*

Of course, coining a term is one thing; getting it accepted by your scientific brethren is another. But here, we can learn much about strategy and tactics in the name game from Cornell physicist N. David Mermin. In a delightful account in *Physics Today*, Professor Mermin described his resolve to introduce the word "boojum" into the physics literature. He recounted his struggles with editors and referees, his temporary setbacks, and his ultimate success in getting boojum accepted internationally.[90] Read his entertaining article. You will chuckle throughout even if you don't understand the physics.

So far, we have dealt only with compounds and phenomena named after real people. Let us look at some linked to persons that may be real or imaginary, depending on your belief.

According to scripture, the Biblical strongman, Samson, "took hold of the two middle pillars upon which the house stood and on which it was borne up, of the one

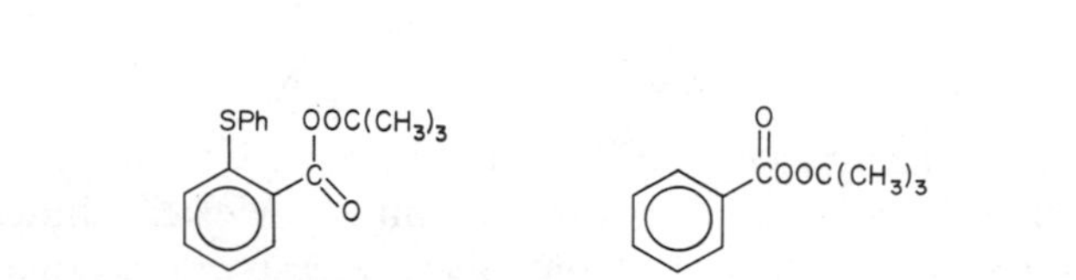

39

40

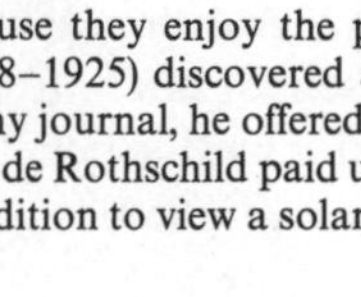

41

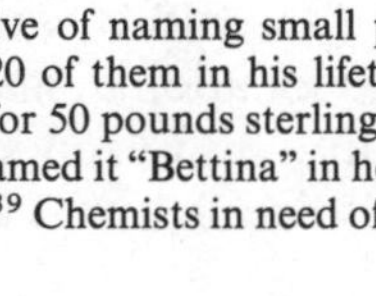

42

---

*So do astronomers, because they enjoy the prerogative of naming small planets they find. Austrian stargazer Johann Palisa (1848–1925) discovered about 120 of them in his lifetime; and once, through an advertisement in an astronomy journal, he offered to sell (for 50 pounds sterling) the right to title one of his new planetoids. Baron Albert de Rothschild paid up and named it "Bettina" in honor of his wife. The money helped Palisa finance an expedition to view a solar eclipse.[89] Chemists in need of research dollars take heed.

with his right hand and of the other with his left." He then "bowed himself with all his might and the house fell."[91] Thousands of years later (in 1974, to be exact), James C. Martin and his students at the University of Illinois found their own Samson—the "S" atom in the double peroxyester **39**. They discovered that **39**, when heated, spurts forth tert-butoxy radicals 80 times faster than does the monoperester **41**,[92] which in turn generates them a million times quicker than does **42**.[93,94] Evidently, sulfur assists the decomposition anchimerically and does so best when it pushes simultaneously in opposite directions.[92] (Indeed, **39** produces the stable hypervalent sulfur compound **40** when it reacts.[92]) The sulfur behaves like the biblical Samson, so Professor Martin dubbed its action "the Samson effect."[95]

In 1967, Stanley Cristol of the University of Colorado and David Lewis synthesized **43**[96] and looked to mythology for onomastic help.[97] In this structure, half the molecule

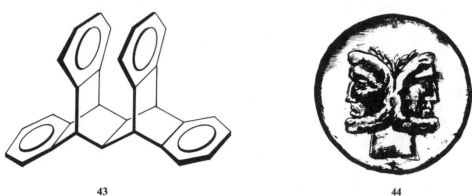

43   44

seems to be looking at its own reflection, just like Narcissus, handsome son of the Greek god Cephissus.[98] It seems that Narcissus loved a beautiful nymph named Echo, but when she lost her voice Narcissus spurned her. Later, when he looked into a pond, he fell in love with his own image—a poor substitute, indeed! Eventually, Narcissus turned into a daffodil, and so this lovely flower is also known as a narcissus. Several alkaloids extracted from this plant have narcissistic titles, for example, narcissamine and narcissidine.[99] Professor Cristol wanted to avoid confusion, so he shunned any similar-sounding name for **43**. No problem. He just pointed the faces outward and switched to Roman mythology.

The Roman god Janus had a head with two faces (one at the front, the other at the back) and guarded outer doors and gates from invaders. The two faces made it impossible for anyone to come or go without his knowledge. (Some ancient Roman coins have hailed Janus; and one bronze issue, *ca.* 211 BC, is shown in **44**.)[100] Accordingly, colleague Dr. Walter M. Macintyre suggested the name "janusene" for **43**, and Dr. Cristol happily followed his advice.[97] Extending the analogy, the Colorado chemists refer to the two benzene rings at the top of **43** as face (F) rings and the two at the sides as lateral (L) rings.[96]

Janus never heard of magnetic resonance, but his namesake shows an interesting NMR phenomenon. The hydrogens of each F ring invade the shielding zone of the other F ring. As a result these H's absorb upfield (by *ca.* 0.4 δ unit) from those on the L rings.[96]

Narcissus did receive chemical recognition, however, when Lionel Salem's theoreticians at the University of Paris coined "narcissistic reactions" for processes in which the product is the mirror image of the reactant.[101–103] An example is **45 → 46**, a symmetry-allowed suprafacial 1, 3-sigmatropic shift with inversion of configuration at the migrating center.[104] We can almost see **45** falling in love with **46**; they have much in common, like Narcissus and his reflection.

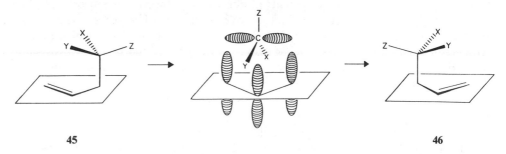

45   **46**

REFERENCES AND NOTES

1. Kekulé A. (a) *Bull. Soc. Chim. France* **1865**, *3*, 98–110, (b) *Bull. Acad. Roy. Belgique* (2) **1865**. *19*, 551, (c) *Liebigs Ann. Chem.* **1866**, *137*, 129–196.
2. According to various sources, Kekulé had this vision in front of a fireplace in his bachelor apartment in Ghent, Belgium, early in 1861. See *Chem. Eng. News*, **1984**, June 11, p. 108 and July 16, p. 52. How much of this folklore is fact and how much fiction has been debated. See Wotiz, J.H.; Rudofsky, S. *Chem. Brit.* **1984**, *20*, 720–723; Ramsay, O.B.; Rocke, A.J. *Chem. Brit.* **1984**, *20*, 1093–1094; Smethurst, P.C. *Chem. Brit.* **1985**, *21*, 347; Rudofsky, S.; Wotiz, J.H. *Chem. Brit.* **1985**, *21*, 347; Simpson, P. *Chem. Brit.* **1985**, *21*, 349; Rocke, A.J.; Ramsay, O.B. *Chem. Brit.* **1985**, *21*, 633; Seltzer, R.J. *Chem. Eng. News* **1985**, Nov. 4, pp. 22–23.
3. Staab, H.A. Plenary lecture at the annual meeting of the Gesellschaft Deutscher Chemiker (Kekulé Centennial), Bonn, September 14, 1965.
4. Aichara, J. *Bull. Chem. Soc. Jpn.* **1976**, *49*, 1429–1430.
5. Staab, H., private communication, May, 1978.
6. (a) Diederich, F.; Staab, H.A. *Angew. Chem. Int. Ed. Engl.* **1978**, *17*, 372–374, (b) Staab, H.A.; Diederich, F. *Chem. Ber.* **1983**, *116*, 3487–3503.
7. Jenny, W.; Baumgartner, P.; Paioni, R. *ISN A-Proc.*, Sendai, Japan, August, 1970. (Cited by Krieger, C.; Diederich, F.; Schweitzer, D.; Staab, H.A. *Angew. Chem. Int. Ed. Engl.* **1979**, *18*, 699–701.
8. Vögtle, F.; Staab, H.A. *Chem. Ber.* **1968**, *101*, 2709–2716.
9. Staab, H.A.; Diederich, F.; Krieger, C.; Schweitzer, D. *Chem. Ber.* **1983**, *116*, 3504–3512.
10. (a) Berris, B.C.; Hovakeemian, G.H.; Lai, Y.–H.; Mestdagh, H.; Vollhardt, K.P.C. *J. Am. Chem. Soc.* **1985**, *107*, 5670–5687, (b) Hirthammer, M.; Vollhardt, K.P.C. *J. Am. Chem. Soc.* **1986**, *108*, 2481–2482.
11. Barton, J.W.; Rowe, D.J. *Tetrahedron Lett.* **1983**, *24*, 299–302.
12. Barton, J.W.; Walker, R.B. *Tetrahedron Lett.* **1978**, 1005–1008.
13. Diercks, R.; Vollhardt, K.P.C. *Angew. Chem. Int. Ed. Engl.* **1986**, *25*, 266–268.
14. Diercks, R.; Vollhardt, K.P.C. *J. Am. Chem. Soc.* **1986**, *108*, 3150–3152.
15. Wasserman, H.H., private communications, June and July, 1977.
16. Wasserman, H.H.; Keehn, P.M. *J. Am. Chem. Soc.* **1967**, *89*, 2770–2772.
17. See footnote 15 in Doering, W. von E.; Ferrier, B.M.; Fossel, E.T.; Hartenstein, J.H.; Jones, M., Jr.; Klumpp, G.; Rubin, R.M.; Saunders, M. *Tetrahedron* **1967**, *23*, 3943–3963.
18. See footnote 8 in Jones, M., Jr.; Reich, S.D.; Scott, L.T. *J. Am. Chem. Soc.* **1970**, *92*, 3118–3126.
19. Jones, M., Jr., private communications, June, 1976 and March, 1984.
20. Magid, R.M., private communication, April, 1984.
21. Doering, W. von E.; Roth, W.R. *Angew Chem. Int. Ed. Engl.* **1963**, *2*, 115–122; *Tetrahedron* **1963**, *19*, 715–737.
22. Doering, W. von E. *Theoret. Org. Chem., Papers Kekulé Symposium, London*, **1958**, 35–48 (pub. 1959); *Chem. Abstr.* **1959**, *53*, 21704i. See also *Chem. Abstr.* **1963**, *59*, 3828.
23. Bonny, A.; Stobart, S.R.; Angus, P.C. *J. Chem. Soc., Dalton Trans.* **1978**, 938–943.
24. Schröder, G. *Angew. Chem. Int. Ed. Engl.* **1963**, *2*, 481–482.

25. Ferrier, B.M., private communication, January, 1985.
26. Balaban, A.T.; Farcasiu, D. *J. Am. Chem. Soc.* **1967**, *89*, 1958–1960.
27. Farcasiu, D., private communication, July, 1977.
28. Binsch, G.; Eliel, E.L.; Kessler, H. *Angew. Chem. Int. Ed. Engl.* **1971**, *10*, 570–572.
29. von Baeyer, A. *Liebigs Ann. Chem.* **1983**, *127*, 199–236.
30. Carter, M.K. *J. Chem. Educ.* **1951**, *28*, 524–526.
31. (a) Willstätter, R. Memories of Adolf von Baeyer. In *From My Life: The Memoirs of Richard Willstätter*, translated by L.S. Hornig (W.A. Benjamin, Inc., New York, 1965), p. 119, (b) Huisgen, R. *Angew. Chem. Int. Ed. Engl.* **1986**, *25*, 297–311.
32. Patterson, A.M. *Chem. Eng. News* **1952**, *30*, p. 1455.
33. Adams, E. *Sci. Amer.* **1958**, *198*, January, pp. 60–64.
34. Oesper, R.E. *The Human Side of Scientists* (University of Cincinnati Press, Cincinnati, OH, 1975), p. 8.
35. Noller, C.R. *Chemistry of Organic Compounds*, 3rd ed. (W.B. Saunders Co., Philadelphia, 1965), p. 691.
36. McOmie, J.F.W. *J. Chem. Educ.* **1980**, *57*, 878.
37. Dickerson, R.E.; Wheatley, P.J.; Howell, P.A.; Lipscomb, W.N.; Schaeffer, R. *J. Chem. Phys.* **1956**, *25*, 606–607.
38. Wood, W.P.; Heicklen, J. *J. Phys. Chem.* **1971**, *75*, 861–866.
39. Grunewald, G.L., private communication, December, 1975.
40. (a) Brouillette, W.J. *Abstracts of Papers, 184th National Meeting of the American Chemical Society, Kansas City, MO* (American Chemical Society, Washington, DC, 1982), ORGN 215, (b) Brouillette, W.J.; Einspahr, H.M. *J. Org. Chem.* **1984**, *49*, 5113–5116.
41. Brouillette, W.J., private communication, March, 1984.
42. Berry, R.S. *J. Chem. Phys.* **1960**, *32*, 933–938.
43. Ramirez, F.; Pfohl, S.; Tsolis, E.A.; Pilot, J.F.; Smith, C.P.; Ugi, I.; Marquarding, D.; Gillespie, P.; Hoffmann, P. *Phosphorus* **1971**, *1*, 1–16.
44. Gutowsky, H.S.; McCall, D.W.; Slichter, C.P. *J. Chem. Phys.* **1953**, *21*, 279–292.
45. Whitesides, G.M.; Mitchell, H.L. *J. Am. Chem. Soc.* **1969**, *91*, 5384–5386.
46. Ugi, I.; Marquarding, D.; Klusacek, H.; Gokel, G.; Gillespie, P. *Angew Chem. Int. Ed. Engl.* **1970**, *9*, 703–730.
47. (a) Ugi, I., private communication, June, 1986, (b) Ugi, I.; Dugundji, J.; Kopp, R.; Marquarding, D. Perspectives in Theoretical Chemistry, In *Lecture Notes in Chemistry*, Berthier, G. *et al.* (Eds.), (Springer-Verlag, Berlin, 1984), vol. 36.
48. Brock, W.H.; Jensen, K.A.; Jørgensen, C.K.; Kauffman, G.B. *Polyhedron* **1983**, *2*, 1–7.
49. Coppinger, G.M. *J. Am. Chem. Soc.* **1957**, *79*, 501–502.
50. (a) Bartlett, P.D.; Funahashi, T. *J. Am. Chem. Soc.* **1962**, *84*, 2596–2601. (b) Bartlett, P.D., private communication, October, 1977.
51. Istrati, C.; Ostrogovitch, A. *Compt. Rend.* **1899**, *128*, 1581–1584.
52. Dutler, H.; Jeger, O.; Ruzicka, L. *Helv. Chim. Acta* **1955**, *38*, 1268–1273.
53. (a) Corey, E.J.; Ursprung, J.J. *J. Am. Chem. Soc.* **1955**, *77*, 3668–3669, (b) *Ibid.* **1956**, *78*, 5041–5051.
54. Chandler, R.F.; Hooper, S.N. *Phytochemistry* **1979**, *18*, 711–724.
55. Bauer, L.H.; Berman, H. *Amer. Mineral* **1928**, *13*, 341–348.
56. Mitchell, R.S. *Mineral Names: What Do They Mean?* (Van Nostrand Reinhold, New York, 1979).
57. Jones, D.W.; Smith, J.A.S. *Trans. Farad. Soc.* **1960**, *56*, 638–647.
58. Jones, D.W., private communication, July, 1983.
59. Jones, D.W.; Smith, J.A.S. *Nature (London)* **1962**, *195*, 1090–1091.
60. Tarbell, D.S.; Tarbell, A.T. *Roger Adams: Scientist and Statesman* (American Chemical Society, Washington, DC, 1981), p. 63.
61. Wall, M.E.; Wani, M.C.; Cook, C.E.; Palmer, K.H.; McPhail, A.T.; Sim, G.A. *J. Am. Chem. Soc.* **1966**, *88*, 3888–3890.
62. *K'o Hsuch Tung Pao* **1976**, *21*, 40–42; *Chem. Abstr.* **1976**, *84*, 122100n.
63. *Scientia Sinica* **1978**, *21*, 87–98. (Contribution from Pharmaceutical Plants and Institutes in Shanghai; authors unspecified.)
64. Wani, M.C.; Ronman, P.E.; Lindley, J.T.; Wall, M.E. *J. Med. Chem.* **1980**, *23*, 554–560.
65. Hutchinson, C.R. *Tetrahedron* **1981**, *37*, 1047–1065.
66. (a) Ollis, W.D.; Sutherland, I.O.; Alves, H.M.; Gottlieb, O.R. *Phytochemistry* **1978**, *17*, 1401–1403, (b) Ollis, W.D., personal discussion, June, 1979.

67. Moore, J.A.; Eng, S. *J. Am. Chem. Soc.* **1956**, *78*, 395–398.
68. Falshaw, C.P.; Ollis, W.D.; Moore, J.A.; Magnus, K. *Tetrahedron, Supplement No. 7*, **1966**, 333–348.
69. Moore, J.A., private communication, June, 1979.
70. Balli, H.; Zeller, M. *Helv. Chim. Acta* **1983**, *66*, 2135–2139.
71. Cunningham, R.P.; Farquhar, D.; Gibson, W.K.; Leaver, D. *J. Chem. Soc. C* **1969**, 239–243.
72. Balli, H., private communication, February, 1984.
73. Seyferth, D. *Adv. Organomet. Chem.* **1976**, *14*, 97–144.
74. Seyferth, D., private communication, August, 1979.
75. (a) Matsuo, A.; Yuki, S.; Nakayama, M.; Hayashi, S. *J. Chem. Soc., Chem. Commun.* **1981**, 864–865, (b) Matsuo, A.; Yuki, S.; Nakayama, M. *J. Chem.Soc., Perkin Trans. 1* **1986**, 701–710.
76. Wilcox, C.F., Jr.; Grantham, G.D. *Tetrahedron* **1975**, *31*, 2889–2895.
77. Ranganathan, S. *Current Organic Chemistry Highlights*, **1976**, March. (Published at one time by the Indian Institute of Technology, Kanpur).
78. Fieser, L.F.; Fieser, M. *Steroids* (Reinhold Publishing Corp., New York, 1959), chapters 19 and 21.
79. Lehmann, F.P.A.; Bolivar, G.A.; Quintero, R.R. *J. Chem. Educ.* **1973**, *50*, 195–199.
80. Domínguez, X.A., private communications, January, 1976; March and April, 1984.
81. Domínguez, X.A.; Martinez, C.; Calero, A.; Domínguez, X.A., Jr.; Hinojosa, M.; Zamudio, A.; Zabel, V.; Smith, W.B.; Watson, W.H. *Tetrahedron Lett.* **1978**, 429–432.
82. Domínguez, X.A.; Cano, G.; Franco, R.; Villarreal, A.M.; Watson, W.H.; Zabel, V. *Phytochemistry*, **1980**, *19*, 2478.
83. González, A.G.; Fraga, B.M.; González, C.M.; Ravelo, A.G.; Ferro, E.; Domínguez, X.A.; Martinez, M.A.; Fayos, J.; Perales, A.; Rodriguez, M.L. *Tetrahedron Lett.* **1983**, *24*, 3033–3036.
84. Nakanishi, K.; Crouch, R.; Miura, I.; Domínguez, X.; Zamudio, Z.; Villarreal, R. *J. Am. Chem. Soc.* **1974**, *96*, 609–611.
85. Domínguez, X.A.; Franco, R.; Cano, G.; Villarreal, R.; Bapuji, M.; Bohlmann, F. *Phytochemistry*, **1981**, *20*, 2297–2298.
86. Zanno, P.R.; Nakanishi, K.; Morales, J.G.; Domínguez, X.A. *Steroids* **1973**, 829–833.
87. Domínguez, X.A.; Barragán, V.A.; León, J.; Morales, J.; Watson, W.H. *Revista Latinoamerica de Quimica* **1971**, 35–37.
88. Domínguez, X.A.; Torre, B. *Phytochemistry* **1974**, *13*, 1624–1625.
89. Weber, R.L. *More Random Walks in Science* (The Institute of Physics, London, England, 1982), p. 141.
90. Mermin, N.D. *Physics Today* **1981**, April, pp. 46–53. Follow-up correspondence about this article appeared in *Physics Today* **1981**, September, pp. 11, 13, 15.
91. *The Holy Bible*, Judges 16:29–30.
92. Martin, J.C.; Chau, M.M. *J. Am. Chem. Soc.* **1974**, *96*, 3319–3321.
93. Martin, J.C.; Bentrude, W.G. *Chem. and Ind. (London)* **1959**, 192–193.
94. Bentrude, W.G.; Martin, J.C. *J. Am. Chem. Soc.* **1962**, *84*, 1561–1571.
95. Martin, J.C.; Perozzi, E.F. *Science* **1976**, *191*, 154–159.
96. Cristol, S.J.; Lewis, D.C. *J. Am. Chem. Soc.* **1967**, *89*, 1476–1483.
97. Cristol, S.J., private communication, January, 1976.
98. *Encyclopedia Britannica* (William Benton, Publisher, London, 1972), vol. 16, p. 31.
99. Glasby, J.S. *Encyclopedia of the Alkaloids* (Plenum Press, New York, 1975), pp. 987–988.
100. Doty, R.G. *The Macmillan Encyclopedic Dictionary of Numismatics* ((Macmillan, New York, 1982), p. 85.
101. Salem, L., private communication, June, 1981.
102. Salem, L.; Durup, J.; Bergeron, C.; Cazes, D.; Chapuisat, X.; Kagan, L. *J. Am. Chem. Soc.* **1970**, *92*, 4472–4474.
103. Salem, L. *Accts. Chem. Res.* **1971**, *4*, 322–328.
104. Woodward, R.B.; Hoffmann, R. *J. Am. Chem. Soc.* **1965**, *87*, 2511–2513.

Chapter 11

ELEMENTS, MY DEAR WATSON

When hydrogen played oxygen,
And the game had just begun,
Hydrogen racked up two fast points,
But oxygen had none.

Then oxygen scored a single goal
     And thus it did remain.
Hydrogen 2 and oxygen, 1,
Called off because of rain.

—Anonymous[1]

Atomic elements are the cornerstones of chemistry, and researchers sometimes name molecules after the elements they contain.

In 1964, Vladimir Prelog* and his research personnel at the Eidgenössische Technische Hochschule in Zürich called red–brown metabolites such as **1**, "siderochromes."[2,3] In Greek, *sideros* is the word for "iron" and *chroma* means "color." At the University of California at Berkeley in 1973, John Neilands suggested that "siderochromes" should not be confined to hydroxamate structures but ought to embrace other iron compounds with similar biological functions.[4] During the same year, Charles Lankford of the University of Texas felt that a class name based on color isn't such a

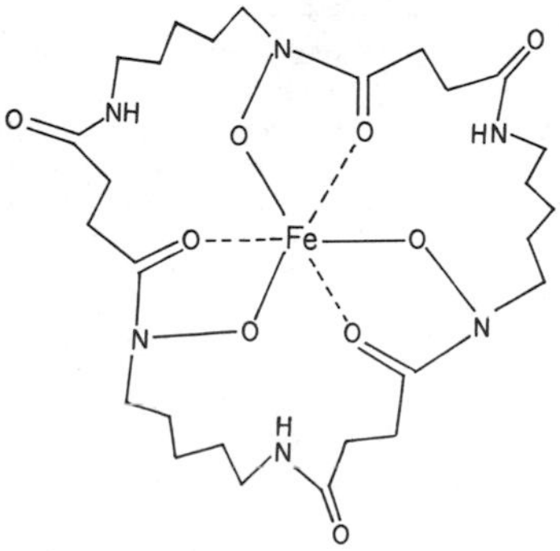

1

---

*Nobel Prize in chemistry, 1975, shared by John W. Cornforth and Vladimir Prelog.

red-hot idea; what if someone discovered a colorless example? He prefers "sidero-phore";[3,5] "*phorein*" is from a Greek verb meaning "to carry." A 1984 review about siderophores told us this onomastic difference was ironed out.[6]

A Greek word for copper or brass, namely *chalkos*,[7] gives rise to the names of several minerals containing copper, such as chalcocite (Cu_2S) and chalcopyrite ($CuFeS_2$). Furthermore, the art of engraving on copper or brass is known as chalcography. "Chalcogens" are elements in main group VI of the periodic table (O, S, Se, Te, Po), because these atoms are often combined with copper in nature.[7] Chalcogens united with things other than copper (e.g., $KFeS_2$) become "chalcogenides."[8] In fact, PhSSPh might be flattered to learn it has been called a "diphenyl dichalcogenide."[9] By the way, don't confuse chalcogens with chalcone (1, 3-diphenyl-2-propen-1-one)[10] or with chalones (substances vital in the control of tissue growth).[11]

The name "pnicogen" for the main group V family of elements (N, P, As, Sb, Bi)[12] may stem from Greek *pniktos*, meaning "strangled, stifled." According to Professor Matthew Schlecht, pniktos (sometimes pronounced niktos in English) may be a Greekification of the German word for nitrogen, *Stickstoff*, which is a "substance that strangulates or asphyxiates."[13] Now take a deep breath, relax, and make a note in your address book. Our old friends from the table of elements are to have new house numbers. In 1983, the Committee on Nomenclature of the American Chemical Society decreed that the society's official format for the periodic table henceforth will consist of 18 groups and no A, B subgroups. So the venerable pnicogens and chalcogens now live in columns 15 and 16, respectively.[14] Not everyone rejoiced at the news,[15] but IUPAC hoped to follow suit.[16]

The first and last letters of the English alphabet, when combined to give "az," mean "nitrogen" in many contexts (e.g., azobenzene, diazonium salts). It seems that the early French literature used Az as the chemical symbol for the element. The French word *azote* for "nitrogen " is derived from the Greek *a* (not) and *zoe* (life). Nitrogen, unlike oxygen, does not sustain life. In the Hantzsch–Widman nomenclature system the letters "oc" (from "octo," which is derived from Greek *okto* meaning eight) signify an eight-membered ring.[17] So, when Joseph Lambert of Northwestern University constructed heterocycle **2** for conformational studies, he dutifully assembled the proper prefix and stem to call it "azocane."[18,19] Table 11.1 lists affixes for heteromono-

TABLE 11.1 Affixes for Heteromonocycles (IUPAC 1983 recommendations)

Heteroatom	Prefix[a]	Number of Atoms in Ring	Suffix if Ring is	
			saturated	unsaturated[b]
O	oxa	3	irane	irene
S	thia	4	etane	ete
Se	selena	5	olane	ole
N	aza	6	ane or inane[c]	ine or inine[c]
P	phospha	7	epane	epine
As	arsa	8	ocane	ocine
Sb	stiba	9	onane	onine
Si	sila	10	ecane	ecine
B	bora			

[a]Drop the "a" when a vowel comes next.
[b]With the maximum number of noncumulative double bonds.
[c]The preferred suffix depends on the heteroatoms. See ref. 20.

cycles as recommended by IUPAC in 1983.[20] Quick quiz: Draw structures for azinine and 1, 2, 4, 3-triazasilolane before peeking at **3** and **4**. You may need to look up the rules for numbering a ring with two or more different heteroatoms, as in **4**.

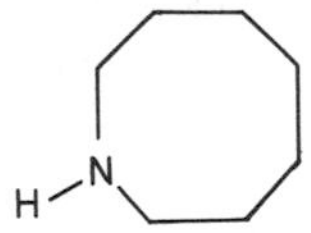

2 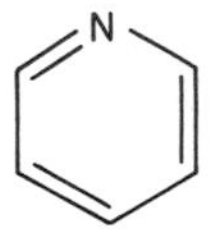   3 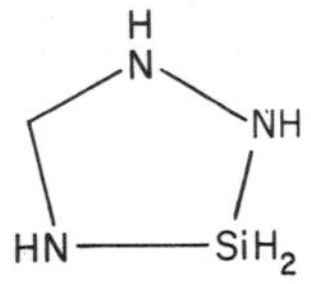   4

Let us focus again on nitrogen and reflect on the origin of the prefix "nor-," commonly used to denote the lack of one or more carbon units. (For example, **5** is camphor and **6** is often called norcamphor.) Austin M. Patterson, a noted authority on chemical nomenclature, thought that nor- in this context simply came from the word "normal".[21] But, according to one source, nor- arose from early alkaloid chemistry, where degradation of R_2NCH_3 to R_2NH was sometimes used in structure elucidation.[22] As in illustration, consider the amines **7** and **8**. With respect to **7**, secondary amine **8** has its "*N ohne Radikal*" (German for "N without radical"). Nor- is just an acronym of this German phrase. The *Radikal** was often CH_3, so nor- usually means

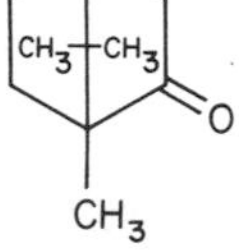

5 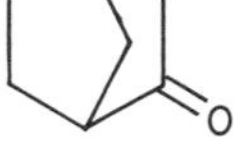   6 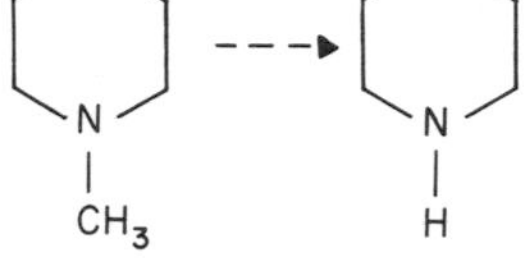   7   8

an analog with *one* less carbon. Although purists might prefer trisnorcamphor or trinorcamphor for **6**, the extra prefix is not mandatory.[17]

We saw earlier that homo- refers to one *more* carbon, and now we know nor- indicates one *less*. Zdenko Majerski at Rudjer Boskovic Institute, Zagreb, joined both prefixes when he and his co-workers referred to **10** as "9-homonoradamantane."[24] We can relate this interesting tricycle to adamantane (**9**) by mentally removing one methylene carbon (e.g., C10) (nor) and inserting it between C1 and C9 (homo). It seems that minus one, plus one, does not bring us back to zero.

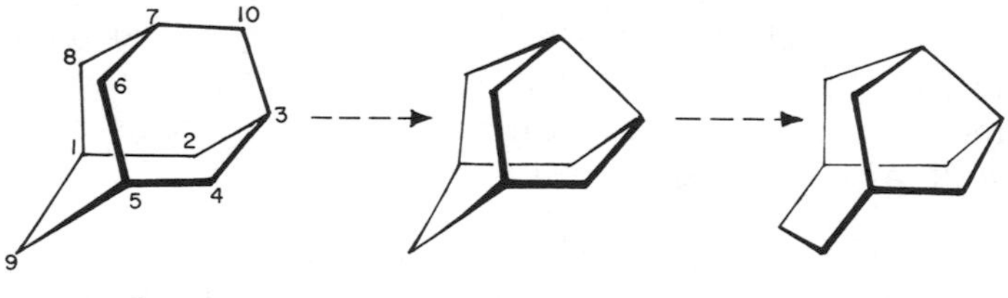

9   10

Phosphorus and arsenic, both chemical relatives of nitrogen, have also enriched organic terminology. For example, a set of chemists in France referred to **11** as "cyclodiop" because its *cyc*lic structure includes two (*di*) oxygens and two *p*hosphorus atoms.[25] Kyba and Chou at the University of Texas (Austin) addressed the molecular

---

*Organic chemists routinely used R for any group or *Radikal*. In physical chemistry, *R* ordinarily denotes the gas constant. Its use in this latter context has been traced to a publication by E. Clapeyron in 1834.[23]

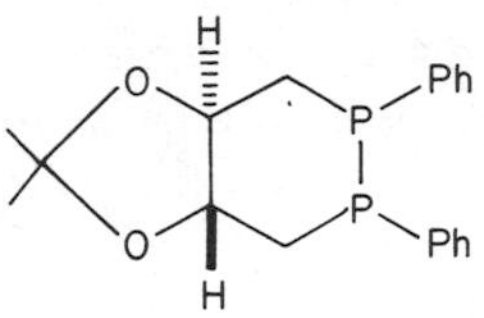

11

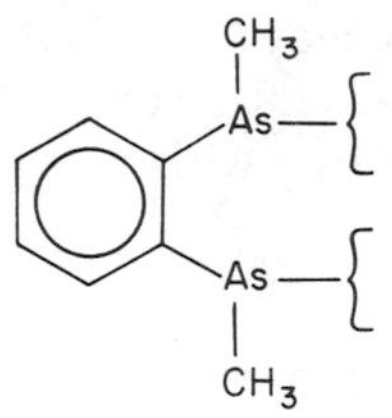

12

portion **12**, with its two *ar*senics, as the "diars moiety."[26] Earlier, Elmer Alyea had assigned the abbreviation "diars" specifically to *o*-phenylenebis(dimethylarsine), that is, to structure **12** having two Me groups on each arsenic. In a 1973 survey of complexation between metals and ditertiary arsines he also adopted contractions for other arsine ligands, (e.g., "edas," "vdiars," "dam," "ffars").[27] Perhaps chemists should look into the reaction between **12** and "old lace."

In this age when spaceships circle our globe in a mere 90 minutes, we also observe chemists in a hurry. Andrew Streitwieser and his band at the Univeristy of California, Berkeley, were in tune with the times when they found that ethyl trifluoromethanesulfonate (**13**) is really anxious to react. It solvolyzes in acetic acid 30,000 times faster than does ethyl methanesulfonate (i.e., **13** with CH_3 instead of CF_3); in fact,

$$CH_3CH_2OSO_2CF_3 + CH_3CO_2H \longrightarrow CH_3CO_2-CH_2CH_3 + {}^-OSO_2CF_3 + H^+$$

13

faster than you can say "*tri*fluoromethanesulfon*ate*." Professor Streitwieser took care of that mouthful by shortening the title of this speedster to "triflate."[28] James Hendrickson and his students at Brandeis University lopped off an oxygen and contributed "trifyl" for the *tri*fluoromethanesulfon*yl* (CF_3SO_2-) group, "triflinate" for the corresponding sulfinate ion ($CF_3SO_2^-$), and "triflone" for the related sulfone (CF_3SO_2R).[29] Michael Hanack and collaborators at the University of Tübingen added a trifle more with "nonaflate" for *nona*fluorobutane-sulfon*ate* ($CF_3CF_2CF_2CF_2SO_2O^-$) and "nonaflon" for *nona*fluorobutylsulfonyl ($CF_3CF_2CF_2CF_2SO_2-$).[30] Fittingly, these names single out fluorine, since it is this electronegative element that makes triflate and nonaflate superb leaving groups.

Not to be outdone, Lester Kuhn and Masahiro Inatome at Aberdeen Proving Ground, Maryland, hailed three elements at once when they coined "BON–BON" for heterocycles like **14**.[31] The government scientists simply mixed the borinic acid, R_2BOH, with the hydroxylamine, $R'R''NOH$, to get $R_2BONR'R''$; then conjugal union between boron and nitrogen created the dimer **14**. The ring atoms, in sequence, spell out BON–BON. A sweet project indeed.

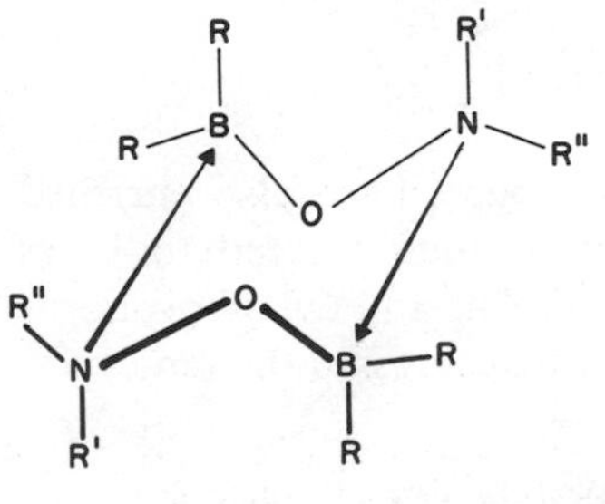

14

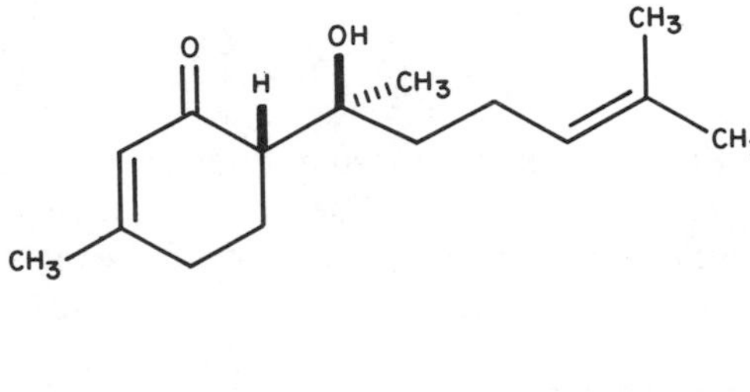

15

If you crave sweetness, then the (+)-enantiomer of **15** may be to your liking. This sesquiterpene of the bisabolane class is more than 1000 times sweeter than sucrose and might serve as a model for synthesis of sugar substitutes. A. Douglas Kinghorn and colleagues (University of Illinois at Chicago) isolated this ketol from a Mexican Plant, *Lipia dulcis*.[32] They named it "hernandulcin" after a Spanish physician, Francisco Hernández, who described and illustrated the dulcet plant (known to the Aztecs as "sweet herb") in a monograph he wrote in the 1570s. The Chicago squad synthesized racemic hernandulcin by a directed aldol reaction.

Have you ever wondered what chemical contains the most elements? At Chemical Abstracts Service, W. Val Metanomski thought about such things and decided to scan the vast information stored in the CAS chemical database.[33] Out popped the compound with most elements (10) as well as the highest hypothetical element studied (atomic number 226). But that's not all; the computer search also turned up the longest name in the Chemical Abstracts Index (1578 characters); the longest Wiswesser Line Notation (749 characters); the longest aliphatic chain (384 carbons); the largest single-ring parent (288 carbons); and the largest ring system (61 rings).

THE COMPOUND WITH MOST ELEMENTS (10)

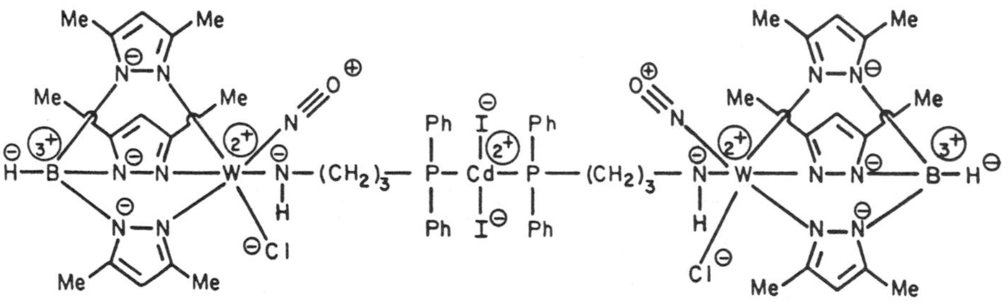

Tungsten, dichloro(diiodocadmium)bis[μ-[3-(diphenylphosphino)-1-propanaminato-*N*:*P*]] dinitrosylbis[tris(3,5-dimethyl-1*H*-pyrazolato-*N¹*)hydroborato(1-)-*N²*,*N²*,*N²*]di-

$C_{60}H_{78}Br_2CdCl_2I_2N_{16}O_2P_2W_2$

1 2 3 4 5 6 7 8 9 10

CA 99:224075k (1983)

THE HIGHEST HYPOTHETICAL ELEMENT STUDIED

Bibihexium
(Element of atomic number 226)

Bbh

CA 92:65093b (1980)

THE LONGEST CA INDEX NAME
(1,578 characters)

Adenosine, N-[4-(dimethylethyl) benzoyl]-2'-O-(tetrahydromethoxypyranyl) adenylyl-(3'→5')-4-
deamino-4-(2,4-dimethylphenoxy)-2'-O-(tetrahydromethoxypyranyl) cytidylyl-(3'→5')-4-deamino-4-
(2,4-dimethylphenoxy)-2'-O-(tetrahydromethoxypyranyl) cytidylyl-(3'→5')-N-[4-(dimethylethyl) benzoyl]-
2'-O-(tetrahydromethoxypyranyl) cytidylyl-(3'→5')-N-[4-(dimethylethyl) benzoyl]-2'-O-
(tetrahydromethoxypyranyl) cytidylyl-(3'→5')-N-[4-(dimethylethyl) phenyl] acetyl]-2'-O-
(tetrahydromethoxypyranyl) guanylyl-(3'→5')-N-[4-(dimethylethyl) phenyl] acetyl]-2'-O-
(tetrahydromethoxypyranyl) guanylyl-(3'→5')-N-[4-(dimethylethyl) benzoyl]-2'-O-
(tetrahydromethoxypyranyl) adenylyl-(3'→5')-N-[4-(dimethylethyl) benzoyl]-2'-O-(tetrahydromethoxy-
pyranyl) cytidylyl-(3'→5')-4-deamino-4-(2,4-dimethylphenoxy)-2'-O-(tetrahydromethoxypyranyl)-
cytidylyl-(3'→5')-4-deamino-4-(2,4-dimethylphenoxy)-2'-O-(tetrahydromethoxypyranyl) cytidylyl-
(3'→5')-N-[[4-(dimethylethyl) phenyl] acetyl]-2'-O-(tetrahydromethoxypyranyl) guanylyl-(3'→5')-4-
deamino-4-(2,4-dimethylphenoxy)-2'-O-(tetrahydromethoxypyranyl) cytidylyl-(3'→5')-N-[4-
(dimethylethyl) benzoyl]-2'-O-(tetrahydromethoxypyranyl) cytidylyl-(3'→5')-N-[4-(dimethylethyl)-
benzoyl]-2'-O-(tetrahydromethoxypyranyl) cytidylyl-(3'→5')-N-[4-(dimethylethyl) benzoyl]-2'-O-
(tetrahydromethoxypyranyl) adenylyl-(3'→5')-N-[4-(dimethylethyl) benzoyl]-2'-O-
(tetrahydromethoxypyranyl) cytidylyl-(3'→5')-N-[4-(dimethylethyl) benzoyl]-2'-O-
(tetrahydromethoxypyranyl) cytidylyl-(3'→5')-N-[4-(dimethylethyl) benzoyl]-2',3'-O-
(methoxymethylene)-, octadecakis (2-chlorophenyl) ester, 5'-[2 (dibromomethyl) benzoate]

$C_{601}H_{678}Br_2Cl_{18}N_{66}O_{184}P_{18}$
CA 99:38770h (1983)

THE LONGEST WISWESSER LINE NOTATION
(749 characters)

T E6 J6 O6 T6 B&6 G&6-35-6 U&6 E&&6 J&&6 C-44-
O---------6 T&&6 B&&&6 G&&&6 U&-35-/E--H/J--
M/O--R/T--W/B&--E&/G&--J&/CN&/E&&--H&&/J&&--
M&&/O----------O&&/T&&--W&&/B&&&--E&&&/
G&&&--J&&&/W&K&&& A B D E- E- E- E- E-- I J-
J- J-- J-- J-- N O- O- O---------- O---------- T- T-- B&- B&--
G&- G&-- V& D&& E&&- E&&- E&&-- E&&-- I&& J&&-
J&&- J&&-- J&&-- N&& T&&- T&&-- B&&&- B&&&--
G&&&- G&&&-- -15-CEHJMOO--O----------
U&C&&E&&H&&J&&M&&O&& K&&& DO E-O IO J-O
NO O-O O----N O--------N O----------O SO T-O A&O
B&-O F&O G&-O K&O M&O P&N S&N V&O D&&O
E&&-O I&&O J&&-O N&&O S&&O T&&-O A&&&O
B&&&-O F&&&O G&&&-O K&&&O O--U O------
U O---------U T--U W B&--U- E& G&--U- J& N&U P&U
R&U T&U T&&--U- W&& B&&&--U- E&&& G&&&--U-
J&&& EH JH E&&--H J&&--H&&&TTTT
&&&&T&TTTTJ

$C_{78}H_{68}N_4O_{28}$
CA 91: 20922j (1979)

THE LONGEST ALIPHATIC CHAIN (384 carbons)

$CH_3—(CH_2)_{382}—CH_3$ Tetraoctacontatrictane

$C_{384}H_{770}$
CA 102:185585k (1985)

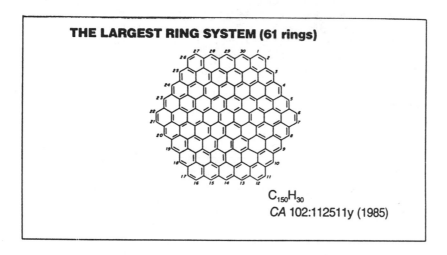

THE LARGEST SINGLE-RING PARENT (288 carbons)

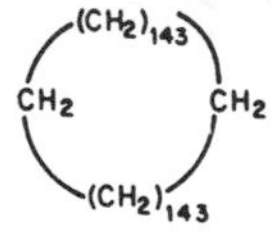

Cyclooctaoctacontadictane

$C_{288}H_{576}$
CA 89:179585g (1978)

THE LARGEST RING SYSTEM (61 rings)

$C_{150}H_{30}$
CA 102:112511y (1985)

To everyone's delight, Dr. Metanomski revealed these record-breaking features in a poster session at a national meeting in April, 1986.[34] His collection, shown here,[35] attracted considerable attention, as did a companion one he presented on "Unusual Names Assigned to Chemical Substances."[34b] And, *The New York Times*[36] and *Chemical and Engineering News* [37] even carried articles about his posters. Dr. Metanomski pointed out that published items not in the CAS database would necessarily be excluded in his survey.[33,38] So, drop him a line if you know of candidates that topple these record holders.

REFERENCES AND NOTES

1. Quoted by Weber, R.L. *More Random Walks in Science* (The Institute of Physics, London, England, 1982), p. 64.
2. Keller–Schierlein, W.; Prelog, V.; Zähner, H. *Fortsch. Chem. Org. Natur.* **1964**, *22*, 279–322.
3. Neilands, J.B., private communication, February, 1976.
4. Neilands, J.B. In *Inorganic Biochemistry* Eichhorn, G., (Ed.), (Elsevier, New York, 1973), vol. 1, p. 168.
5. Lankford, C.E. *Crit. Rev. Microbiol.* **1973**, *2*, 273–331.
6. Raymond, K.N.; Müller, G.; Matzanke, F. *Top Curr. Chem.* **1984**, *123*, 49–102.
7. Ellis, J.E. *J. Chem. Educ.* **1978**, *55*, 781.
8. Davis, E.A. *Endeavor* **1977**, *1*, 103–106.
9. Baldo, M.; Forchioni, A.; Irgolic, K.J.; Pappalardo, G.C. *J. Am. Chem. Soc.* **1978**, *100*, 97–100.
10. (a) Kostanecki, St. v.; Tambor, J. *Chem. Ber.* **1899**, *32*, 1921–1926, (b) See Appendix A.

11. Houck, J.C.; Daugherty, W.F., Jr. *Chalones: A Tissue-Specific Approach to Mitotic Control* (Medcom Press, New York, 1974).
12. Ellis, J.E. *J. Chem. Educ.* **1976**, *53*, 2–6.
13. Schlecht, M.F., Polytechnic Institute of New York, private communication, June, 1985.
14. Loening, K.L. *J. Chem. Educ.* **1984**, *61*, 136.
15. For example, see: (a) *Chem. Eng. News.* **1985**, Aug. 12, pp. 2–3, (b) Nelson, P. *Chem. Brit.* **1985**, *21*, 1077–1079, (c) *Chem. Eng. News* **1986**, Jan. 27, pp. 22–24.
16. *Chem. Brit.* **1985**, *21*, 751.
17. (a) International Union of Pure and Applied Chemistry, *Nomenclature of Organic Chemistry*, Sections A and B, 2nd ed. (Butterworths, London, 1966), p. 51, (b) *Pure Appl. Chem.* **1965**, *11*, 42–43; *Ibid.* **1979**, *51*, 1995–2003.
18. Lambert, J.B., private communication, October, 1978.
19. Lambert, J.B.; Khan, S.A. *J. Org. Chem.* **1975**, *40*, 369–374.
20. *Pure Appl. Chem.* **1983**, *55*, 409–416.
21. Patterson, A.M. *Chem. Eng. News* **1954**, *32*, p. 1818.
22. *The Merck Index* 9th ed. (Merck and Co., Rahway, NJ, 1976), p. xiv.
23. Rothenberger, O.S.; Leeds, A. *Abstracts of Papers, 191st National Meeting of the American Chemical Society, New York, NY* (American Chemical Society, Washington, DC, 1986), HIST 33.
24. Majerski, Z.; Djigas, S.; Vinkovic, V. *J. Org. Chem.* **1979**, *44*, 4064–4069.
25. Zhang, S.Y.; Yemul, S.; Kagan, H.B.; Stern, R.; Commereuc, D.; Chauvin, Y. *Tetrahedron Lett.* **1981**, *22*, 3955–3958.
26. Kyba, E.P.; Chou, S.–S. P. *J. Am. Chem. Soc.* **1980**, *102*, 7012–7014.
27. Alyea, E.C. In *Transition Metal Complexes of Phosphorous, Arsenic, and Antimony Ligands* McAuliffe, C.A. (Ed.), (Halsted Press, New York, 1973), pp. 311–373, 387–389.
28. Streitwieser, A., Jr.; Wilkins, C.L.; Kiehlmann, E. *J. Am. Chem. Soc.* **1968**, *90*, 1598–1601.
29. (a) Hendrickson, J.B.; Sternbach, D.D.; Bair, K.W. *Acc. Chem. Res.* **1977**, *10*, 306–312, (b) Hendrickson, J.B.; Bergeron, R.; Giga, A.; Sternbach, D. *J. Am. Chem. Soc.* **1973**, *95*, 3412–3413, (c) Hendrickson, J.B.; Giga, A.; Wareing, J. *J. Am. Chem. Soc.* **1974**, *96*, 2275–2276.
30. Hanack, M.; Massa, F. *Tetrahedron Lett.* **1977**, 661–664.
31. Kuhn, L.P.; Inatome, M. *J. Am. Chem. Soc.* **1963**, *85*, 1206–1207.
32. Compadre, C.M.; Pezzuto, J.M.; Kinghorn, A.D.; Kamath, S.K. *Science* **1985**, *227*, 417–419.
33. W.V. Metanomski, private communications, March, April, and October, 1986.
34. (a) Metanomski, W.V. *Abstracts of Papers, 191st National Meeting of the American Chemical Society, New York, NY* (American Chemical Society, Washington, DC, 1986), HIST 20, (b) *Ibid.*, HIST 21.
35. Copyright 1986 by the American Chemical Society, used with permission.
36. The New York Times, April 22, 1986, p. C3.
37. *Chem. Eng. News*, April 28, 1986, p. 66.
38. Dr. Metanomski (reference 33) noted that according to the 1987 Guinness Book of World Records the longest scientific name is the systematic name for deoxyribonucleic acid of the human mitochondrion, which contains 16,569 nucleotide residues, and is *ca.* 207,000 letters long (Anderson, S., *et al. Nature* **1981**, *290*, 457–465). However, the full name was not printed but was abbreviated in terms of nucleotide symbols on a computer printout.

Chapter 12

OUR MOLECULES ARE REALLY BUILT!

Much of what we know about physical and chemical properties can be understood in terms of molecular structure. In this chapter we meet compounds and unstable intermediates with names that describe their makeup.

First, consider the interesting tetracycle **1**, which was conceived by E. Muller in 1940.[1] It consists of two cyclohexane chairs linked at alternate carbons by axial bonds. These connections create three boat rings, namely, carbons 1, 2, 12, 6, 7, 11; 1, 2, 3, 4, 9,

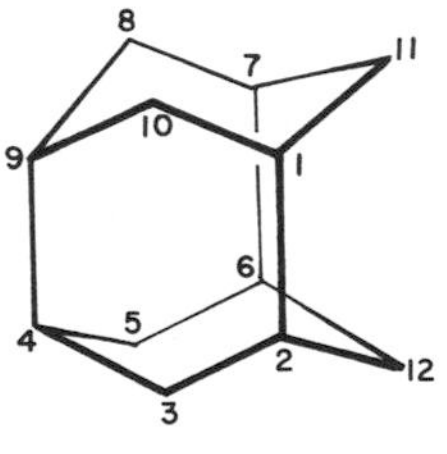

1	**2**

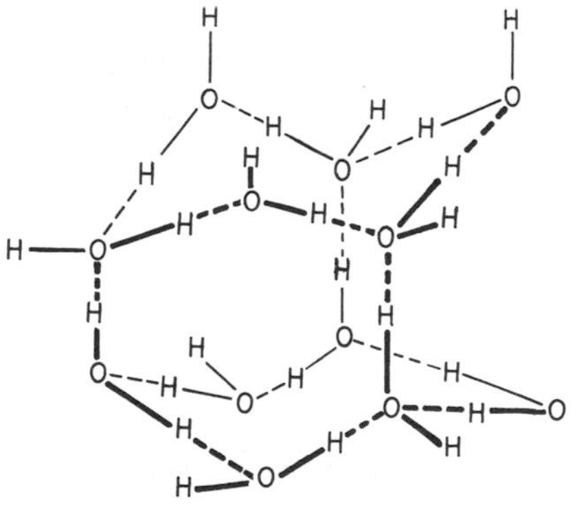

10; and 4, 5, 6, 7, 8, 9. Independently, Louis Fieser of Harvard University noted that this skeletal shape resembles that of ice (**2**).* Thus, in 1965, before **1** had been synthesized, he called it "iceane."[3] In structure **2**, the solid and dashed lines stand for covalent and hydrogen bonds, respectively. In ice, this lattice array repeats indefinitely.

Chris Cupas's explorers at Case Western Reserve University achieved the first synthesis of iceane in 1974.[4] A Michael addition of **4** to tropone (**3**) furnished an equilibrium mixture of **5** and two isomers with the diene double bonds moved. When heated, the mixture produced **6**, by an internal Diels–Alder reaction in **5**. Wolff–

---

*Structure **2** represents "ice I," the form that prevails under ordinary conditions. Several other forms of ice exist under high pressure.[2]

Kishner dispatch of the carbonyl was followed by treatment with hydrobromic acid, which led to 8-bromoiceane (7) via two Wagner–Meerwein shifts. Finally, the bromine gave way to hydrogen with $LiAlH_4$. Cool chemistry indeed.

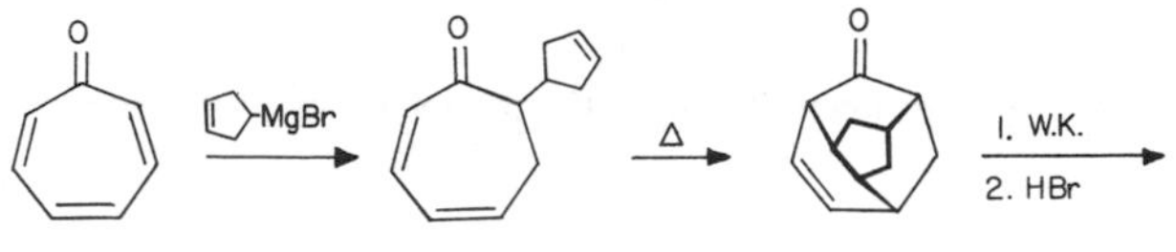

A year later, David Hamon's team at the University of Adelaide, Australia, rose from down under to report a different iceane synthesis;[5] and several oxaiceanes (one CH_2 group replaced by an oxygen) have also been prepared.[6–8] Professor Fieser may have started another ice age!

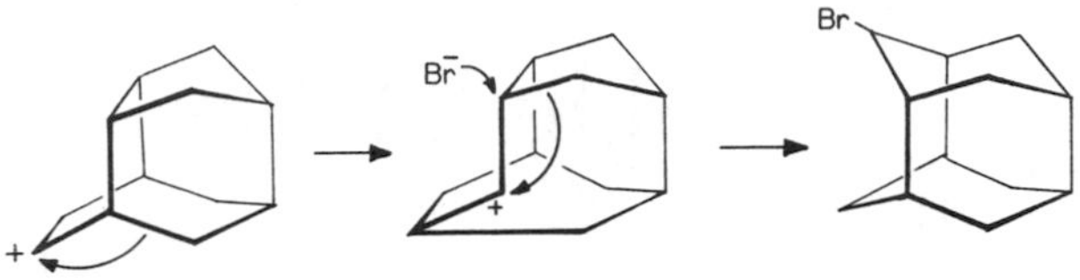

Camille Ganter and co-workers at the Technische Hochschule in Zürich felt that "Wurtzitane" is a better title for 1 than is "iceane."[6] In their opinion, 1 mimics the structure of wurtzite* (a crystal form of zinc sulfide) more closely than that of ice I; and "Wurtzitane" translates into other languages more easily than "iceane."

In the early 1950s William von Eggers Doering at Yale University and Lawrence Knox of the Hickrill Chemical Research Laboratory (Katonah, NY) irradiated ethyl diazoacetate (8) in cyclohexane and obtained nitrogen and ethyl cyclohexylacetate (9). They suggested that 8 first expels N_2 to form species 10, which then inserts itself between a carbon and a hydrogen in the alkane.[10] We can track the electrons by

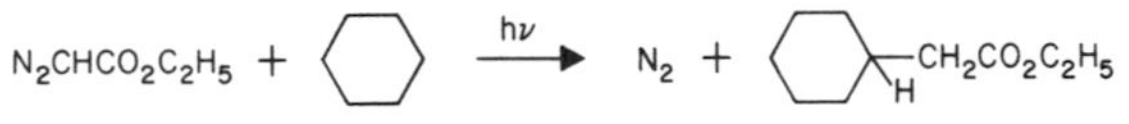

looking at resonance form 8a. Dr. Doering wanted a descriptive name for divalent carbon structures—like 10—with "two singly covalently bonded substituents and two unshared electrons."[10] He discussed the matter with Saul Winstein (UCLA) and Robert Woodward (Harvard) during a late-night taxi ride in Chicago, and the word "carbene" was born before these three eminent midwives reached their destination.[†]

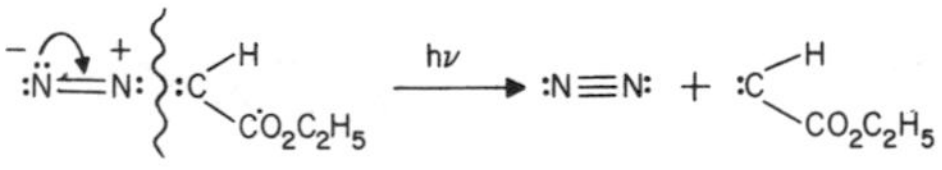

---

*The mineral wurtzite honors French organic chemist Charles Adolphe Wurtz, inventor of the classic Wurtz reaction.[9]

†Actually the name was reborn, having been used in different contexts before. It has been defined as "a bitumen soluble in carbon disulfide but insolube in carbon tetrachloride" in *Webster's New International Dictionary*, 2nd ed. (G. and C. Merriam Co., Springfield, MA., 1952), p. 402. And "carbene" appeared in *Chemical Abstracts* 15 times before Doering and Knox published their findings. (See reference 10, footnote 9.)

Professor Doering thought the name delightfully fits a *carb*on with the same degree of "unsaturation" as an alk*ene*; indeed, a carbene can be viewed as half of an alkene. Drs. Doering and Knox first announced this term at a meeting of the American Chemical

$$R—C\colon \qquad \boxed{R—C\colon}\colon C—R$$

11   **12**

Society in 1951,[11] and now it is entrenched in our chemical language. The nomenclature has since stretched to include carbyne (**11**),[12] which can be thought of as half of an alkyne (**12**).

Now don't become too logical and expect chemists to call $R_3C\cdot$ a "carbane" just because a radical is half of an alkane. But you will find "carbane" in the book *The HIRN System—Nomenclature of Organic Chemistry*, where author K. Hirayama proposed it as a class name for unbranched, acyclic hydrocarbons. Dr. Hirayama's monograph describes a radial nomenclature system designed for man-to-man as well as man-to-machine communication. The word "HIRN" is an acronym for *HI*rayamasche *R*adiale *N*omenklatur—and also happens to be German for "intelligence."[13]

Before leaving carbenes, you should know that the nitrogen analog of a carbene unofficially became "nitrene" courtesy of Gerald Smolinsky at Bell Telephone Laboratories.[14] At first, the word nitrene appeared in a dissertation by Karl Miescher at Zurich's ETH in 1918 and was applied by Staudinger and Miescher to nitrogen compounds they prepared and thought were of the type $R_2C{=}N(R){=}CR_2$.[15] When such structures later proved to be erroneous[16] that usage of "nitrene" became forsaken.[17]

In 1949, Donald Cram of the University of California at Los Angeles studied the acetolysis of optically active threo-3-phenyl-2-butyl *p*-toluenesulfonate (**13**) and also the erythro diastereomer.[18-20] To account for the stereochemical outcome in his acetate products, Professor Cram postulated the intermediacy of bridged ion **14**. He

13   **14**

envisioned it could arise by neighborly assistance of the phenyl ring as the OTs leaves. Dr. Cram sought an appropriate name for cations like **14**. The suffix "-onium" often signifies positive ions, as in ammonium and phosphonium. Since a benzene carbon forms the bridge in **14**, the UCLA investigator first thought of "benzonium ion." But John D. Roberts (then at MIT) pointed out that the prefix "benz-" is used for benzene with another carbon attached, as in "*benzyl*" ($C_6H_5CH_2$—). Dr. Cram decided he would rather switch than fight, and settled for "phenonium ion".[20,21]

We ordinarily think of carbon–carbon triple bonds as quite reactive; for example, many reagents add easily to acetylenes. But what happens if alkyl chains engulf the unsaturation? Will this "shield" keep reagents at bay? Roger Macomber of the

University of Cincinnati thought these questions might be answered with alkynes like
15, in which the π-cylinder seems well concealed.[22] In conversation with a colleague,
Professor Macomber learned that Greek for "to conceal" is κευθειν, which he anglicized

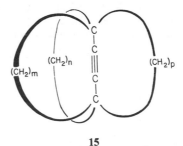

15

to "keuthyne."[23] Thus, **15** may be termed [m, n, p]keuthyne.[24] Macomber and his
students began a search for such ensconced acetylenes. In chapter 17 we will meet this
shielding concept again with olefins (e.g., "betweenanenes,") and with other structures
("capped" porphyrins).

In 1980, Howard Zimmerman's zealots studied the ability of one functional group to
relay electronic energy to another in the same molecule. To keep the two chromophores
separated a fixed distance they prepared **16** ($n = 1$ or 2).[25] The bicyclo[2, 2, 2]octane
units are rigid, so in effect the naphthyl and acetyl parts are held apart by a "rod," whose
length depends on n. The workers used a digit to specify the length of the rod; for
example, **16** with $n = 2$ was acetyl-α-naphthyl-2-rod.[25]

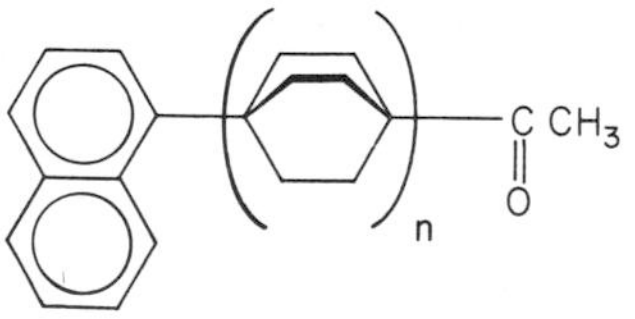

16

Naphthyl groups absorb 270 nm radiation but acetyl does not. When **16** ($n = 1$) was
bathed with 270 nm light, fluorescence occurred at 335 nm (characteristic of naphthyl)
and also at 405 nm (characteristic of acetyl). They suggest that some of the excitation
energy absorbed by naphthyl was transferred to the carbonyl intramolecularly through
the rod. However, no significant 405 nm emission appeared when the homolog with n
= 2 was irradiated; Dr. Zimmerman had lengthened the rod and spoiled the transfer.

Chemists have long been intrigued by the possibility of interlocking two or more
rings to produce a chemical chain (**17**). Cornell's Jerrold Meinwald reminded us[26] that
ca. 1950 in Harvard's research laboratories students called such structures "ballantine"
compounds, after the logo **18** of Ballantine Brewery of Newark, New Jersey (a beer
company no longer in business). Emblem **18**, as depicted here, does not define how the
circles are linked.[27] But the symbol likely evolved from the Borromean rings **19**, found

17   **18**    **19**

in the coat of arms of the famous Renaissance Italian family of Borromeo.[28] Note that even though all three hoops in **19** interlock, no two are linked. Break one of the circles and you end up with a line and two unconnected rings.[29] In 1960, Edel Wasserman, then at Bell Laboratories, reported the first synthesis of type **17**; and for such molecules he coined the term "catenanes" (from the Latin *catena* meaning "chain").[30,31]

Dr. Wasserman reasoned that rather large loops would be needed; small ones do not have a big enough space to admit another link. His strategy: conduct an acyloin ring closure on the ester **20** of a C_{34} diacid in the presence of a 34-membered cycloalkane that contained five deuterium markers (**21**). During translational motion, a small fraction of the long molecules became threaded through the cycloalkane, so at any instant the mixture consisted of free diester (**20**) free d_5-cycloalkane (**21**), and some threaded units (**22**). After the acyloin event, the flask contained initial **21**, monocyclic acyloin (**23**), and *ca.* 1% of catenane **24**, which he separated. Just to be certain, ringmaster Wasserman oxidatively split the ketol coil in **24**; he got back his **21** and also the diacid related to **20**.

A year later, Dr. Wasserman and a colleague, Dr. Harry Frisch, described this new kind of topological isomerism in detail.[32] From their viewpoint, the unlinked *pair* **23** and **21** can be considered an isomer of **24**. Similarly, a rotaxane (chapter 3), in which a wheel slides on an axle, is viewed as isomeric with its duo of components, namely, the separated wheel and axle. Another topological duo would include a knotted loop (**25**) and its unknotted counterpart (**26**). Harvard's James Wang pointed out that DNA loops are sometimes catenated—or even contorted into knots.[33]

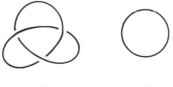

25   26

Synthetic chemist David Walba, crystallographer R. Curtis Haltiwanger, and graduate student Rodney Richards, all of the University of Colorado, were determined to twist rings. They prepared and characterized a new triene of type **27** (a molecular "cylinder") as well as its novel topoisomer **28** (a "molecular Möbius strip").[34] Snip the three double bonds in **27** and you get two monocycles. Doing the same in **28** produces a single, double-sized circle. Look up Professor Walba's engaging review on how to convolute rings—it could warp your thinking forever.[29]

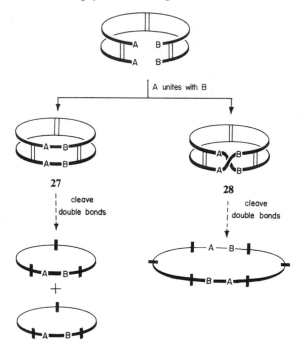

Professor Schill and his fellow smiths fabricated derivatives of a [2]catenane in 1964[35] and of [3]catenanes (**29**) in 1977[36] and 1981.[37] The bracketed numeral tells how many rings make up the linked chain. They synthesized the first all-hydrocarbon analog in 1983; it was a [2]catenane with one loop C_{28} and the other C_{46}.[38] Their 1981 paper on [3]catenanes also introduced a new type of isomerism.[37] One of the three

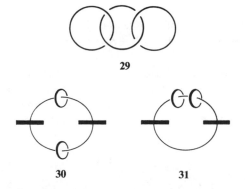

rings carried two bulky substituents (represented by shaded blocks in **30** and **31**); these act like the cotter pins in rotaxanes. Hence, the other two cycles can find themselves separated by obstructions or not. The Schill team viewed **30** and **31** as "translational isomers" because they could interconvert only by transport of a ring past a barrier group.[37]

Jean–Pierre Sauvage and colleagues at Strasbourg, France, have applied template magic to create "metallo-catenanes." This wizardry called upon the ability of Cu(I) to coordinate four nitrogens tetrahedrally. For example, the metal ion can gather a cyclic diamine (**32**) and an open chain diamine (**33**) to a threaded condition (**34**) ready for chemical construction of the second ring (**35**). With this template strategy, the

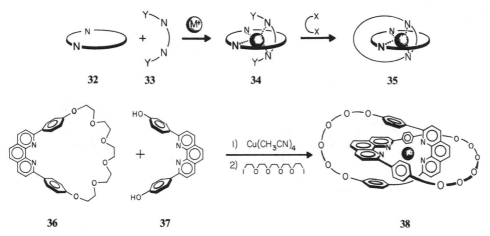

Strasbourg stalwarts assembled **36** and **37**, and then joined the phenolic tips with an α, ω-dihalide to fashion the mammoth [2]catenane **38**, complete with ensconced copper.[39] They could pull the ion out and slip it (or others like Ag$^+$ and Li$^+$) back in easily. Without a caged inmate, these giant interlocked ligands go by the title "catenands."[40]

When component **36** carried phenyl substituents at the 4, 7-positions of the phenanthroline, their template scheme provided a catenane (symbolized by **39**) in which the phenylated macrocycle can't slip around.[41] In such structures, a topographical isomer (**40**) is possible if chains m and n differ. The restricted freedom in cases like **39** and **40** is akin to that in rotaxanes (chapter 3) and in the catenanes **30** and **31** conceived by Schill and co-workers.

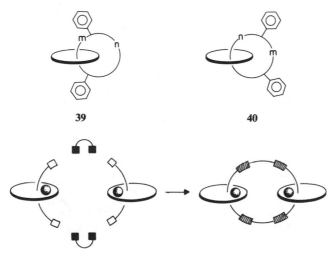

41

By choosing dihalide chains too short for intramolecular looping, the team at Strasbourg managed to dimerize a pair of metallo-complexes by means of two linking fragments and, thereby, to construct impressive dimetallic [3]catenates like **41**; these were demetalated to the corresponding [3]catenands.[42]

Incidentally, the well-known symbol of the International Olympics (**42**) is a [5]catenane; and the grant of arms (**43**) of the Ciba Foundation displays a [5]catenane with its rings portrayed slightly differently. Neither of these concatenated emblems has chemical significance. Rather, both symbols connote universal brotherhood among the five continents—the former in athletics and the latter in education and knowledge.[43] (However, the regular *hexagons* in the Ciba Foundation shield may well stand for benzene and thereby allude specifically to chemical knowledge.)

42   43

You may have met other athletic catenanes. For example, [3]catenane (**29**) is the emblem of the Mediterranean Games, begun in 1951 and held every four years. The three connected hoops represent the continents bordering the Mediterranean Sea. And the sexicatenated bracelet **44** officially symbolizes SEAG, the South East Asia Games. Chemists with an eye for esthetics and athletics should find this sports logo ("seagane?") a formidable synthetic target.

Finally, if you are not superstitious, if your job is secure, and if you control a hefty research budget, how about tackling **45**, the ultimate in manacled circles? This wreath of 13 ringlets is linked to American history and traces its origin to Benjamin Franklin, eminent statesman of colonial America. He designed emblem **45** for use in Continental currency. The pattern first went public on paper money issued in 1776; and, after confederation, it also appeared on the first coin—known as the "Fugio" cent—authorized by the United States of America.[44] Each circle stood for one of the 13 colonies; and the center of the emblem was inscribed with the legend, also created by Benjamin Franklin, "AMERICAN CONGRESS. WE ARE ONE." The chain design

44   45

and its motto prompted Tory poet Joseph Stansbury of Philadelphia to include the following stanza in a numismatic verse he composed in 1776:[45]

> But the last fashion'd Money
>   we all must commend,
> Where a Circle of Rings
>   join in Rings without end,
> Each Ring is a State,
>   and (the Motto explains)
> They all are a Congress—
>   a Congress in Chains!

So, fellow chemist, if you like to work in circles and can synthesize the looped [13]catenane **45** ("franklinane"?), your feat will forever be shackled to a piece of Americana. But for the moment, catenanes at least link us to the next chapter.

REFERENCES AND NOTES

1. Muller, E. *Neuere Anschauungen der Organischen Chemie* (Springer, Berlin, 1940), p. 30. Professor D.P.G. Hamon kindly informed us of this reference.
2. Pauling, L.; Hayward, R. *The Architecture of Molecules* (W.H. Freeman and Co., San Francisco, CA, 1964).
3. Fieser, L.F. *J. Chem. Educ.* **1965**, *42*, 408–412.
4. Cupas, C.A.; Hodakowski, L. *J. Am. Chem. Soc.* **1974**, *96*, 4668–4669.
5. Hamon, D.P.G.; Taylor, G.F. *Tetrahedron Lett.* **1975**, 155–158; *Aust. J. Chem.* **1976**, *29*, 1721–1734.
6. Klaus, R.O.; Tobler, H.; Ganter, C. *Helv. Chim. Acta* **1974**, *57*, 2517–2519.
7. Hamon, D.P.G.; Taylor, G.F.; Young, R.N. *Tetrahedron Lett.* **1975**, 1623–1626.
8. Hamon, D.P.G.; Taylor, G.F.; Young, R.N. *Aust. J. Chem.* **1977**, *30*, 589–598.
9. Mitchell, R.S. *Mineral Names: What Do They Mean?* (Van Nostrand Reinhold Co., New York, 1979).
10. Doering, W. von E.; Knox, L.H. *J. Am. Chem. Soc.* **1956**, *78*, 4947–4950.
11. Doering, W.E.; Knox, L.H., *Abstracts of Papers, 119th Meeting of the American Chemical Society, Boston, MA*, (American Chemical Society, Washington, DC, 1951), p. 2M.
12. (a) Strausz, O.P.; DoMinh, T.; Fout, J. *J. Am. Chem. Soc.* **1968**, *90*, 1930–1931, (b) March, J. *Advanced Organic Chemistry*, 3rd ed. (John Wiley & Sons, New York, 1985), p. 173.
13. Hirayama, K. *The HIRN System—Nomenclature of Organic Chemistry* (Maruzen Co. Ltd., Tokyo, and Springer-Verlag, New York, co-pub., 1984).
14. Smolinsky, G. *J. Am. Chem. Soc.* **1960**, *82*, 4717–4719.
15. Staudinger, H.; Miescher, K. *Helv. Chim. Acta* **1919**, *2*, 554–582.
16. Hassall, C.H.; Lippman, A.E. *J. Chem. Soc. (London)* **1953**, 1059–1063.
17. Horner, L.; Christmann, A. *Angew. Chem. Int. Engl.* **1963**, *2*, 599–608.
18. Cram, D.J. *J. Am. Chem. Soc.* **1949**, *71*, 3863–3870.
19. Cram, D. J. *J. Am. Chem. Soc.* **1949**, *71*, 3883–3889.
20. Cram, D.J. *J. Am. Chem. Soc.* **1952**, *74*, 2129–2137.
21. Cram, D.J., private communication, July, 1977.
22. Bauer, D.P.; Macomber, R.S. *J. Chem. Res. (S)* **1978**, 194–195; *ibid. (M)* **1978**, 2469–2495.
23. Macomber, R.S., private communication, May, 1979.
24. Greenberg, A.; Liebman, J.F. *Strained Organic Molecules* (Academic Press, New York, 1978), p. 368.
25. Zimmerman, H.E.; Goldman, T.D.; Hirzel, T.K.; Schmidt, S.P. *J. Org. Chem.* **1980**, *45*, 3933–3951.
26. Meinwald, J., private communication, June, 1977.
27. In a private communication, T. Toland (P. Ballentine and Sons) informed Norman van Gulick (University of Oregon) that since about 1955 all crossing-point detail in the Ballentine trademark was omitted in the interest of simplification. Dr. van Gulick cited this tidbit in a manuscript, "Theoretical Aspects of the Linked Ring Problem," which he wrote in 1960 but never published. Professor D. Walba kindly conveyed this information to us in November, 1984.

28. Gardner, M. *The Scientific American Book of Mathematical Puzzles & Diversions* (Simon & Schuster, New York, 1959), p. 70.
29. Walba, D.W. *Tetrahedron* **1985**, *41*, 3161–3212.
30. Wasserman, E. *J. Am. Chem. Soc.* **1960**, *82*, 4433–4434.
31. Wasserman, E., private communication, July, 1977.
32. Frisch, H.L.; Wasserman, E. *J. Am. Chem. Soc.* **1961**, *83*, 3789–3795.
33. Wang, J.C. *Scient. Amer.* **1982**, *247*, July, pp. 94–109.
34. Walba, D.M.; Richards, R.M.; Haltiwanger, R.C. *J. Am. Chem. Soc.* **1982**, *104*, 3219–3221.
35. Schill, G.; Lüttringhaus, A. *Angew. Chem.* **1964**, *76*, 567–568.
36. Schill, G.; Zürcher, C. *Chem. Ber.* **1977**, *110*, 2046–2066.
37. Schill, G.; Risler, K.; Fritz, H.; Vetter, W. *Angew. Chem. Int. Ed. Engl.* **1981**, *20*, 187–189.
38. Schill, G.; Schweickert, N.; Fritz, H.; Vetter, W. *Angew. Chem. Int. Ed. Engl.* **1983**, *22*, 889–891.
39. Dietrich–Buchecker, C.O.; Sauvage, J.–P.; Kintzinger, J.P. *Tetrahedron Lett.* **1983**, *24*, 5095–5098.
40. Dietrich–Buchecker, C.O.; Sauvage, J.–P.; Kern, J.–M. *J. Am. Chem. Soc.* **1984**, *106*, 3043–3045.
41. Dietrich–Buchecker, C.O.; Sauvage, J.–P.; Weiss, J. *Tetrahedron Lett.* **1986**, *27*, 2257–2260.
42. Sauvage, J.–P.; Weiss, J. *J. Am. Chem. Soc.* **1985**, *107*, 6108–6110.
43. Evered, D.C. (Ciba Foundation, London), private communication, July, 1982. We also thank Dr. Philip B. Flagler, Ciba Pharmaceutical Co., Summit, NJ, for help in this matter.
44. Newman, E.P. *The Numismatist*, **1983**, *96*, 2271–2281.
45. Newman, E.P. *The Numismatist*, **1983**, *96*, 2282–2284.

Chapter 13

DON'T JUST STAND THERE. DO SOMETHING!

Plenty of molecules are lovely to look at. But looks aren't everything. After all, chemistry has its penicillins as well as its dodecahedranes, and its Alexander Graham Bells as well as its Michelangelos. So let's peek at some names that describe what molecules can do.

Regulating acidity is very important in industrial processes as well as in living systems. Hundreds of acids and bases are known that can scavenge hydroxide ions and hydrogen ions, respectively, and thus give chemists pH control. An interesting example is 1, 8-bis(dimethylamino)naphthalene, **1**, which literally soaks up a proton and clasps it tightly between two nitrogens (**2**).[1] For this reason, the Aldrich Chemical Company marketed **1** under the trademark "Proton Sponge."[2] It is several pK units more basic than expected for aromatic amines. All four N-methyl groups need to be there to make

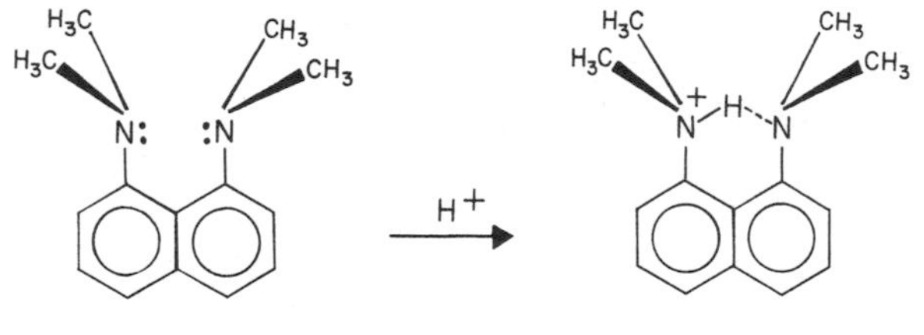

| 1 | 2 |

1 an effective sponge; with fewer alkyls the diamines are far less potent. Structure **1** is strained. The methyls cannot fit comfortably between the peri-nitrogens, nor can the methyl carbons lie in the naphthalene plane. So the lone pairs of electrons can't ooze into the rings as in most arylamines, and they may even destabilize the system by repelling each other. Protonation, however, gives a cation (**2**) in which hydrogen bonding replaces this repulsion. The hydrogen ion clings even tighter when methoxy groups reside at the 2, 7-sites[3] or when the dimethylamino units are attached to fluorene at the 4, 5 positions.[4] There's no denying it: "Proton Sponges" have absorbing jobs. Incidentally, if any boron atoms want work in the acid–base field, they should declare their skills at reversing polarity and send applications to Howard Katz. In 1984 at AT&T Bell Laboratories, he replaced N by B in structure **1** and began to spawn a shoal of "hydride sponges."[5]

Roy Olofson's students at the Pennsylvania State University faced a different proton disposal problem. They longed to generate carbenoids (**4**) from benzylic chlorides (**3**) by α-elimination of HCl.[6] A strong abstractor was called for because benzylic hydrogens

$$ArCH_2Cl \xrightarrow[-HCl]{base} ArCH:$$

$$\textbf{3} \qquad\qquad\qquad \textbf{4}$$

are only slightly acidic. But the base had to be relatively nonnucleophilic to preclude direct S_N displacement of halogen. They found that lithium 2, 2, 6, 6-tetramethylpiperidide (**5**) filled the bill nicely; it generated the carbenes, which were trapped by olefins to give excellent yields of arylcyclopropanes.[6] Other hindered lithium amides work also.[7] Evidently, these amide bases can remove a proton selectively without assailing carbon. This ability to snatch H^+ out of a molecule may have reminded the Olofson group of Captain Ahab's effort to pull Moby Dick out of the ocean,* so they named this collection of bases "H^+arpoons."[7] They also coined in-house titles for each specific H^+arpoon; for example, amide **5** is "Protorooter" and amide **6** is "H^+ooker."[8] Had **6** been available a century earlier, Captain Ahab's men

5   **6**

might have enjoyed the hunt more! Professor Olofson's box of H^+ lures also contained "H^+arpy," "H^+atchet," and "Protocide," among others. For a full inventory, drop him a line.

Jean–Marc Lalancette and his followers at the University of Sherbrooke, Quebec, strived for a different type of chemical selectivity.[9] A "textbook" route to aldehydes by oxidation of primary alcohols with chromic oxide (or acidic dichromate) is not always clean. Overoxidation to the carboxylic acid is common; and secondary alcohols and alkenes are also attacked.[10] In 1972, the Lalancette team found that combining the CrO_3 with graphite tamed the chromium and thus made it much more selective. Primary alcohols become aldehydes without overoxidation; and other functional groups survive. For example, cinnamyl alcohol (**7**) gave cinnamaldehyde in 100% yield; the olefinic group was unscathed.

$$PhCH{=}CHCH_2OH \xrightarrow[graphite]{CrO_3} PhCH{=}CHC{\nwarrow}^{O}_{H}$$

7

In this nifty oxidant, the CrO_3 is intercalated[11] into the graphite structure such that CrO_3 molecules lie between layers of graphite.[12] Edward Sullivan, technical manager of Ventron Corporation, which marketed this reagent,[13] dubbed it "Seloxcette," for "*sel*ective *ox*idation according to Lalan*cette*."[14] Shakespeare might have called it "The Taming of the CrOOO." Incidentally, intercalation compounds are not new. They were

---

*In the novel *Moby Dick*, by Herman Melville, a ship's crew, under Captain Ahab, tried repeatedly to harpoon a whale named Moby Dick. Ultimately, they failed.

discovered by the Chinese nearly 3000 years ago. But only in the past decade or two has intercalation methodology come into its own. Chemists and material scientists find that slipping "foreign" molecules or atoms between layers in a preexisting compound without completely disrupting its structure can also lead to valuable applications as superconductors, catalysts, lubricants, membranes, *etc.*[11]

Thiols, or mercaptans, are notorious for their disagreeable odors. Anyone who inhales a few molecules, whether from a reagent bottle or from a skunk, is unlikely to forget the experience. Interestingly, however, the name "mercaptan" derives from a beneficial service these molecules render. The word comes from the Latin "*mercurium captans*," which means "seizing (or capturing) mercury,"[15] and reflects the ability of —SH groups to tightly complex ions of mercury and of other toxic heavy metals. Two proximate —SH units in one molecule are even more effective, since an extremely stable chelate results. An example is 2,3-dimercapto-1-propanol (**8**), which detoxifies mercuric ions through the chelate **9**.[16] This bimercaptan sells as an antidote for mercury, arsenic, and lead poisoning under tradenames such as British Anti-Lewisite (Lewisite is $ClHC{=}CHAsCl_2$), Dimercaprol, and Dicaptol.[17] It may not smell like violets, but it gets the lead out!

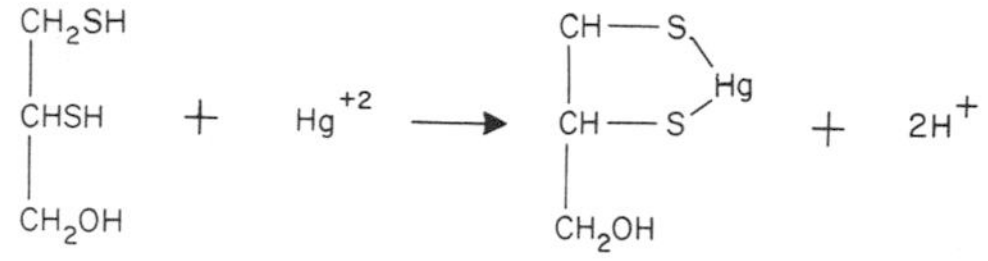

$$\begin{array}{ccc} \textbf{8} & & \textbf{9} \end{array}$$

We know that triflate ($CF_3SO_2O^-$) ejects readily because electron withdrawal by fluorines helps disperse the negative charge. Fluorosulfonate (FSO_2O^-) is another ion that departs as briskly as employees at quitting time. Consequently, methyl fluorosulfonate (**10**) reigns as a powerful methylating agent. For example, it converts most ethers to oxonium salts, a feat methyl iodide cannot match.[18] This talent has earned **10** the trademarked dionym* Magic Methyl.[19]

James F. King and his court at the University of Western Ontario let Magic Methyl do its thing on N,N-diethylmethanesulfonamide (**11**).[20] They obtained the onium species **12**, which proved to be a fine source of a methanesulfonyl (mesyl) group. For example, it transforms alcohols easily to their mesylate esters. Professor King dubbed **12** "Easy Mesyl," which is euphonious and describes its talent. He considered pluralizing the name to cover this family of compounds but decided that "easy mesyls" sounded like something contagious.[21]

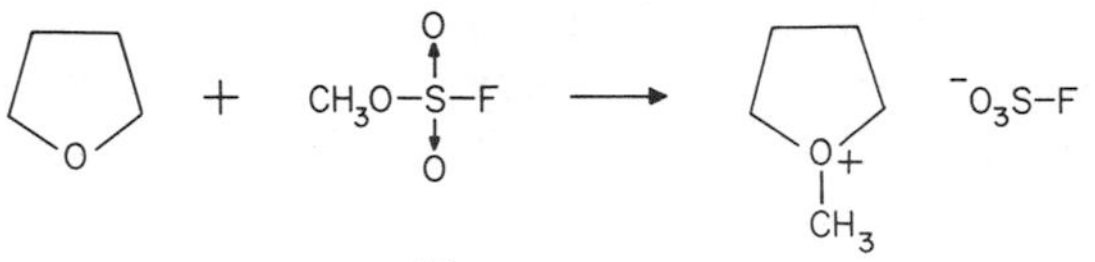

10

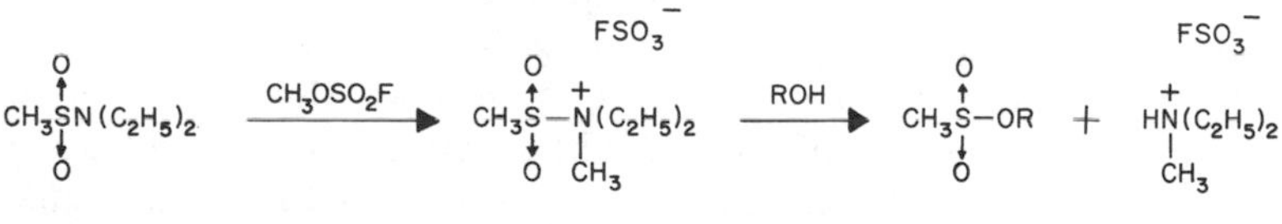

11   **12**

---

*Dionym: a name that consists of two terms.

Barry Trost's trios at the University of Wisconsin in Madison found that thionium ions (**13**) react more smartly than do their carbonyl analogs (**14**); so they think of them as "super carbonyls."[22,23] They attribute this behavior to the higher polarity (because of the positive charge) and lower π-bond strength in **13**. These super carbonyls came in

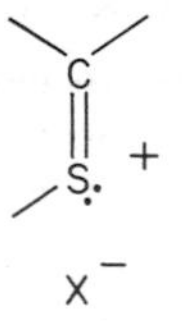

13 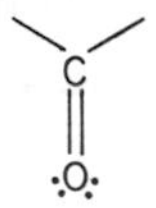   **14**

handy when Professor Trost's troupe ran Friedel–Crafts reactions with acid-sensitive systems such as pyrrole rings.[22] For example, **15** gave a tarry mixture with *p*-toluenesulfonic acid (a strong acid often used to protonate aldehydes and ketones). However, with *p*-toluenesulfinic acid (**16**, a weaker acid, pKa ~ 1.7[24]) **18** turned up in high yield. Evidently, **15** ionized to the thionium species **17**, whose "super" reactivity allowed it to cyclize under mild conditions that preserved the acid-sensitive pyrrole ring. Finally, *p*-toluenesulfinate ion unseated the CH_3S— group.

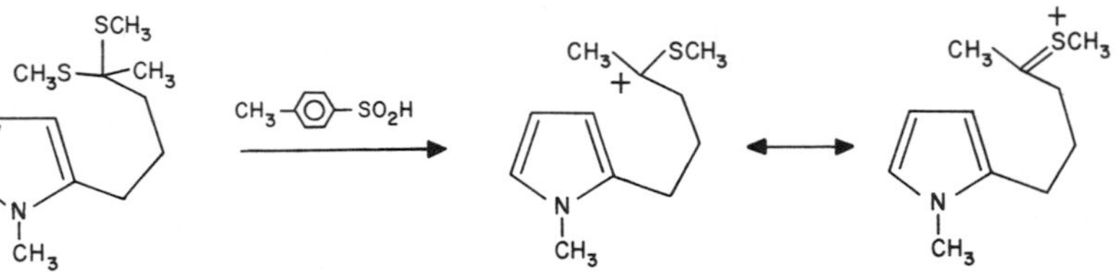

15   **16**   **17**

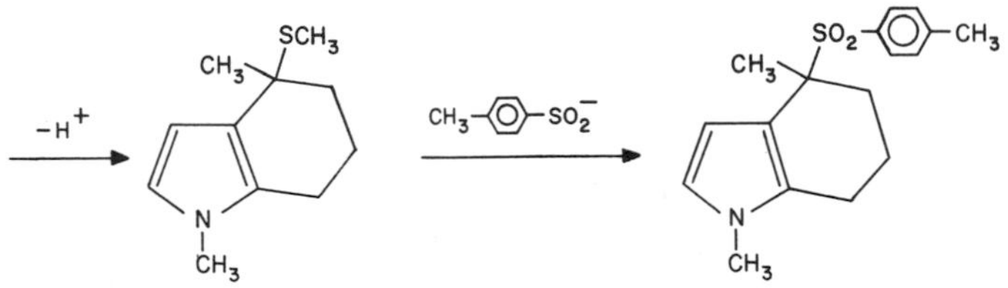

18

By the way, the descriptors "super" and "magic" are not newcomers to chemical jargon. In 1927, Harvard's James B. Conant and co-worker Norris F. Hall sowed the "super" seeds in publications titled "A Study of Superacid Solutions."[25] They were extending views on basicity and acidity adumbrated by Brönsted[26] and by Hantzsch.[27] The Harvard investigators demonstrated, by titration, that in certain nonaqueous media (e.g., HOAc) acids such as sulfuric or perchloric form salts virtually completely with bases even as weak as amides. They called such acid media "superacid solutions."

The 1960s saw several laboratories, including those of George Olah (then at Case Western Reserve University) and Ronald Gillespie (McMaster University), probe molecular behavior in mixtures of FSO_3H–SbF_5, with or without added SO_3. Olah's team generated numerous carbocations readily in FSO_3H–SbF_5. In the winter of 1966 something prompted his postdoctorate Joachim Lukas to drop into this acid medium a piece of a Christmas candle, left over from a laboratory party. It dissolved nicely; and the seasonal spirit moved him further to scan an NMR of the solution. Lo and behold, to everyone's amazement he obtained a very clear spectrum of the tert-butyl cation. From then on, he and others in the laboratory looked upon the FSO_3H-SbF_5 system as a "magic acid."[28] This nickname took hold and soon cast its spell over visitors and friends. At first, Professor Olah was reluctant to use this term in print and eschewed it in a 1967 review;[29] but it materialized in a news release about calorimetric research conducted by Edward Arnett (then at University of Pittsburgh and Mellon Institute) and John Larson at University of Pittsburgh.[30] A 1968 communication by Olah and Schlosberg[31] formally adopted the phrase "magic acid" for FSO_3H-SbF_5 (particularly the 1:1 mixture.).[32] Gillespie and co-workers did not think FSO_3H-SbF_5 should be singled out with its magic title. They liken this pair to other "superacids,"[33] such as $FSO_3H-SbF_5-nSO_3$ and $FSO_3H-AsF_5-nSO_3$, whose acidities also exceed that of anhydrous sulfuric acid or FSO_3H alone.[34] In any case, buy Magic Acid[35] under that trade name and you too can make molecules do super things.[36] But if bases turn you on more than acids do, check with Manfred Schlosser's laboratory at the University, Lausanne for a "superbase" comprised of butyllithium and a potassium alcoholate. They call such things "LICKOR-reagents" from the formula notation $LiC-$ $+ KOR$.[37] Then drop in on Edward Arnett at Duke University to learn how "superbases" sustain their drive.[38]

What about a "superproton?" Logically, you might expect this to be a sizzling H^+ from a superacid. But in a 1981 review, Cambridge University's Ian Fleming pointed out that a superproton need not be a proton[39] at all, at least not in silicon chemistry. For example, bases and various nucleophiles handily remove a carbon-bound trialkylsilyl unit when it is vicinal to a nucleofuge or to a carbocation (see **19**). It's a sterling way to make alkenes. In such cases, R_3Si simulates a highly abstractable H^+ and Fleming termed it a "superproton."[40]

19

When Si is connected to oxygen (for example as in a trimethylsilyl ether), you can also pull it off like a proton, although the Si—O bond breaks slower than a corresponding H—O bond. Here Professor Fleming regarded R_3Si as behaving like a "feeble proton." At Johns Hopkins University, Jih Ru Hwu and student John Wetzel perceived in these feeble fellows a useful virtue: bulk. And they demonstrated that reagents serving as formal sources of "R_3Si^+" can promote selective ketalizations better than do their H^+ counterparts. So, the JHU researchers looked upon R_3Si^+ as a "bulky proton."[41]

Chemiluminescence (emission of light from a chemical reaction) is a phenomenon that always evokes "oohs" and "aahs" from an audience. A well-known example is the

20

oxidation of luminol (**20**).[42] Fireflies and glowworms were carrying out chemilumine-scence long before **20** was known, of course.

Chemists can induce compounds to give off light by other stimuli, which lead to several descriptive words ending in "-luminescence." For example, when some carbohydrates or amino acids are irradiated and then rapidly dissolved in a suitable solvent, light is emitted.[43-45] The phenomenon is called "lyoluminescence," from the Greek *lyein* ("to loose or set free"). This stem also pops up in the term "lysis of cells," for a process in which a cell spills its contents and essentially disintegrates in the surroundings. (We are all familiar with -lysis as a suffix in hydrolysis, photolysis, etc.)

Sugar and certain candies emit light when put under sudden mechanical stress. This oddity is known as "triboluminescence"; the prefix "tribo-" derives from Greek *tribein* ("to rub").[46] You can demonstrate it by crushing a *Lifesaver* candy between the jaws of a pair of pliers, preferably in a dark room.[47] It's a clever way to generate light and cut your sugar intake at the same time.

Eye-catching lecture demonstrations, such as those that produce light, have a certain appeal or charm about them. So, Richard Ramette of Carleton College coined "exocharmic" for reactions having an endearing quality.[48] (Presumably, a de-monstration that flops becomes endocharmic). Australia's R.G. Amiet described a laboratory experiment with lucigenin that has lots of charm.[49]

By the way, nuclear physicists can also exude endearment if they work at *C*ERN (in Geneva), at *H*amburg, at *A*msterdam, at *R*ome, or at *M*oscow. A joint research effort from these five distinguished institutes of physics is known as the "CHARM Collaboration."[50] And anyone brave enough to burst into the nucleus of an atom will find bundles of charm in the form of "charmed quarks" and "charmed antiquarks." But that's not all. High-energy physicists have also cornered the "onium" market with discoveries of states they describe as "charmonium," "quarkonium," "gluonium," "bottomonium"—all quite the rage.[51,52] Undeniably, the subnuclear world has its "strong" and "weak" aspects, as well as "ups" and "downs." And now, by showing "naked charm" and even "naked bottom,"[53] physics—in an era of revelation—is finally exposing its raw beauty in all colors ("chromons") and flavors ("flavons").[54,55] But if "oniums" continue to proliferate, physicists should do what chemists did with their onium salts: eliminate them. A.W. Hofmann has bequeathed details.[56]

Trevor McMorris's research group at the University of California, San Diego, dubbed steroids of structure **21** "oogoniols" because these compounds induced

21

formation of oogonia (female sex organs) in a mold.[57] The "oo-" arises from Greek *oion*, meaning "egg"; and "-gonia" stems from *gonos*, Greek for "procreation." Dr. McMorris and his students proved structure **21** in 1978.[58]

A protein in red blood cells has also been heralded for its biological function. In 1979,

Vann Bennett and mates, then at Wellcome Research Laboratories, Research Triangle Park, North Carolina, realized that spectrin (a major protein in the cytoskeleton of human erythrocytes) can become bound to the inner walls of these cells.[59,60] They proved that spectrin anchors there by interacting with a particular protein in the cell wall. Since *ankyra* is Greek for "anchor," the workers called this wall protein "ankyrin."[60] We wonder what it would do to a cyclohexane boat? By the way, Dr. Bennett initially probed erythrocyte research at Wellcome but shipped out and continued it at the Johns Hopkins University.

"Pheromone" is a euonym that describes the biological functions of a class of materials. The term, introduced in 1959 by German biochemist Peter Karlson and Swiss entomologist Martin Luscher,[61] derives from the Greek *phérein* ("to carry or transfer") and *horman* ("to excite or stimulate"). ("Hormone" comes from horman, also.) A pheromone is secreted into the environment by an animal and evokes an adaptive response from a receiving member of the *same* species.[62] It thus serves as a messenger among animals. Great interest surrounds the use of pheromones to control harmful insects without resorting to insecticides.

William L. Brown, Jr., and collaborators at Cornell University have introduced two words for chemicals that convey messages between members of *different* species. An "allomone" (from Greek *allos* meaning "other") is one that favors the emitting species;[63] a defensive secretion is an example.[62] (The next time you smell a skunk, take comfort in knowing the mercaptan has become an allomone.) A "kairomone" (from Greek *kairos*, meaning "opportune" or "exploitable") favors the receiving species;[64] it could lead a predator to its unwitting prey.[62] If you don't know whether a chemical missive is a pheromone, allomone, or kairomone, just follow the suggestion of biochemists John Law and Fred Regnier and call it a "semiochemical." In 1971 they introduced this handy handle (from Greek *semeion*, meaning "a mark or signal") for any chemical substance that conveys a message between organisms.[65]

The struggle for survival by means of chemical warfare is not unique to animal species. Plants also wage battle against animals and other plants by releasing metabolites known as "allelopathic" chemicals. The term allelopathy literally means "reciprocal harm" and was coined by Austrian botanist Hans Molisch in 1937 to include toxicities exerted by microorganisms (bacteria, fungi, etc.) as well as by higher plants. Nature's allelochemicals[66] can range from simple gases (ammonia, hydrogen cyanide, ethylene, etc.) to complex organic molecules (terpenoids, coumarins, etc.). A plant species seems to use its natural herbicides as an offensive weapon for adaptive advantage, while the receiving species is harmed.[67] Seemingly, in the plant world it's better to give than to receive.

REFERENCES AND NOTES

1. (a) Alder, R.W.; Bowman, P.S.; Steele, W.R.S.; Winterman, D.R. *J. Chem. Soc., Chem. Commun.* **1968**, 723–724, (b) Alder, R.W.; Bryce, M.R.; Goode, N.C.; Miller, N.; Owen, J. *J. Chem. Soc., Perkin Trans 1* **1981**, 2840–2847.
2. Aldrich Chemical Company, Milwaukee, WI, Technical Information Bulletin, Product No. 15,849–6, January, 1971.
3. Hibbert, F.; Hunte, K.P.P. *J. Chem. Soc., Perkin Trans 2* **1983**, 1895–1899.
4. Staab, H.A.; Saupe, T.; Krieger, C. *Angew. Chem. Int. Ed. Engl.* **1983**, *22*, 731–732.
5. Katz, H.E. *J. Am. Chem. Soc.* **1985**, *107*, 1420–1421.
6. Olofson, R.A.; Dougherty, C.M. *J. Am. Chem. Soc.* **1973**, *95*, 581–582.
7. Olofson, R.A.; Dougherty, C.M. *J. Am. Chem. Soc.* **1973**, *95*, 582–584.
8. Olofson, R.A., private communication, September, 1976.
9. Lalancette, J.–M.; Rollin, G.; Dumas, P. *Can. J. Chem.* **1972**, *50*, 3058–3062.

10. March, J. *Advanced Organic Chemistry*, 3rd ed. (John Wiley & Sons, New York, 1985), pp. 1057 and 1070.
11. (a) *Chem. Eng. News*, 1982, Oct. 4, pp. 25–27, (b) Whittingham, M.S.; Jacobson, A.J. Eds. *Intercalation Chemistry* (Academic Press, New York, 1982).
12. Croft, R.C. *Aust. J. Chem.* **1956**, *9*, 201–205.
13. (a) Advertisement, Ventron Corporation, *J. Org. Chem.* **1974**, *39*, 6A, (b) Sullivan, E.A., private communication, March, 1984.
14. Lalancette, J.–M., private communication, June, 1977.
15. Reese, K.M. *Chem. Eng. News* **1980**, March 3, p. 64.
16. Zuman, P.; Zumanová, R.; Teisinger, J. *Collect. Czech. Chem. Commun.* **1955**, *20*, 139–146.
17. *The Merck Index*, 9th ed. (Merck and Company, Rahway, NJ, 1976), p. 426.
18. Aldrich Chemical Company, Inc., Milwaukee, WI, Technical Information Bulletin, Product No. 16,048–2, July, 1974.
19. The sale of Magic Methyl was eventually discontinued because of its toxicity., Bader, A.R., Aldrich Chemical Company, private communications, January, 1976 and March, 1984.
20. King, J.F.; du Manoir, J.R. *J. Am. Chem. Soc.* **1975**, *97*, 2566–2567.
21. King, J.F., private communication, December, 1975.
22. Trost, B.M.; Reiffen, M.; Crimmin, M.T. *J. Am. Chem. Soc.* **1979**, *101*, 257–259.
23. Trost, B.M.; Vaultier, M.; Santiago, M.L. *J. Am. Chem. Soc.* **1980**, *102*, 7929–7932.
24. (a) Lovén, J.M. *Z. Phys. Chem.* **1896**, *19*, 456–464, (b) Coats, R.R.; Gibson, D.T. *J. Chem. Soc.* **1940**, 442–446.
25. (a) Hall, N.F.; Conant, J.B. *J. Am. Chem. Soc.* **1927**, *49*, 3047–3061, (b) Conant, J.B.; Hall, N.F. *J. Am. Chem. Soc.* **1927**, *49*,. 3062–3070.
26. Brönsted, J.N.; Guggenheim, E.A. *J. Am. Chem. Soc.* **1927**, *49*, 2554–2584.
27. Hantzsch, A. *Chem. Ber.* **1927**, *60*, 1933–1950.
28. Olah, G., private communication, April, 1983.
29. Olah, G. *Chem. Eng. News* **1967**, March 27, pp. 77–78.
30. *Chem. Eng. News* **1968**, Feb. 26, pp. 36–37.
31. Olah, G.A.; Schlosberg, R.H. *J. Am. Chem. Soc.* **1968**, *90*, 2726–2727.
32. Commeyras, A.; Olah, G.A. *J. Am. Chem. Soc.* **1969**, *91*, 2929–2942.
33. Gillespie, R.J. *Acc. Chem. Res.* **1968**, *1*, 202–209.
34. Gillespie, R.J.; Ouchi, K.; Pez, G.P. *Inorg. Chem.* **1969**, *8*, 63–65.
35. Aldrich Chemical Co., Milwaukee, WI.
36. Olah, G.A.; Prakash, G.K.S.; Sommer, J. *Superacids* (John Wiley & Sons, New York, 1985).
37. Schlosser, M.; Strunk, S. *Tetrahedron Lett.* **1984**, *25*, 741–744.
38. Arnett, E.M.; Venkatasubramaniam, K.G. *J. Org. Chem.* **1983**, *48*, 1569–1578.
39. Just how the term "proton" (from Greek *protos*, meaning "first") became associated with the positively charged hydrogen atom is not clear. Moore, C.E.; Jaselskis, B.; Smolinsky, A.v. *J. Chem. Educ.* **1985**, *62*, 859–860.
40. Fleming, I. *Chem. Soc. Rev.* **1981**, *10*, 83–111.
41. Hwu, J.R.; Wetzel, J.M. *J. Org. Chem.* **1985**, *50*, 3946–3948.
42. Landgrebe, J.A. *Theory and Practice in the Organic Laboratory*, 2nd ed. (D.C. Heath and Co., Lexington, MA, 1977), p. 392.
43. Laflin, P.; Baugh, P.J. *J. Chem. Soc., Chem. Commun.* **1979**, 239–240.
44. Baugh, P.J.; Mahjani, M.G.; Ellis, S.C.; Rees Evans D. *Internat. J. Radiation Phys. Chem.* **1977**, *10*, 21.
45. Ettinger, K.V.; Mallard, J.R.; Srirath, S.; Takavar, A. *Phys. Med. Biol.* **1977**, *22*, 481.
46. Angelos, R.; Zink, J.I.; Hardy, G.E. *J. Chem. Educ.* **1979**, *56*, 413–414.
47. Walker, J. *Scient. Amer.* **1982**, *247*, July, pp. 146–153.
48. Ramette, R.W. *J. Chem. Educ.* **1980**, *57*, 68–69.
49. Amiet, R.G. *J. Chem. Educ.* **1982**, *59*, 163–164.
50. Jonker, M. *et al. Nucl. Instrum. Methods* **1982**, *200*, 183–193.
51. Bloom, E.D.; Feldman, G.J. *Scient. Amer.* **1982**, *246*, May, pp. 66–77.
52. Harari, H. *Scient. Amer.* **1983**, *248*, April, pp. 56–68.
53. *Chem. Eng. News* **1982**, May 10, p. 68.
54. Mistry, N.B.; Poling, R.A.; Thorndike, E.H. *Scient. Amer.* **1983**, *249*, July, pp. 106–115.
55. Quigg, C. *Scient. Amer.* **1985**, *252*, April, pp. 84–95.
56. (a) Hofmann, A.W. *Liebigs Ann. Chem. Pharm.* **1851**, *78*, 253–286, (b) *Ibid.* **1851**, *79*, 11–39.
57. McMorris, T.C.; Seshadri, R.; Weihe, G.R.; Arsenault, G.P.; Barksdale, A.W. *J. Am. Chem. Soc.* **1975**, *97*, 2544–2545.

58. McMorris, T.C.; Schow, S.R.; Weihe, G.R. *Tetrahedron Lett.* **1978**, 335–338.
59. Bennett, V., private communication, September, 1979.
60. Bennett, V.; Stenbuck, P.J. *J. Biol. Chem.* **1979**, *254*, 2533–2541.
61. Karlson, P.; Luscher, M. *Nature (London)* **1959**, *183*, 55–56; see also Karlson, P.; Butenandt, A. *Ann. Rev. Entomol.* **1959**, *4*, 39–58.
62. (a) Albone, E. *Chem. Brit.* **1977**, *13*, 92–99, (b) Wood, W.F. *J. Chem. Educ.* **1983**, *60*, 531–539.
63. Brown, W.L., Jr. *Amer. Natur.* **1968**, *102*, 188–191.
64. Brown, W.L., Jr.; Eisner, T.; Whittaker, R.H. *Bioscience* **1970**, *20*, 21–22.
65. Law, J.H.; Regnier, F.E. *Ann. Rev. Biochem.* **1971**, *40*, 533–548.
66. Rosenthal, G.A. *Scient. Amer.* **1986**, *254*, Jan., pp. 94–99.
67. Putnam, A.R. *Chem. Eng. News* **1983**, April 4, pp. 34–45 and references cited therein.

Chapter 14

WE'VE GOT YOUR NUMBER— AND YOUR LETTER

Much of our science is a numbers and letters game. In chemical equations digits tell us how many atoms each molecule has and in what ratios the molecules combine. They also reveal the location of substituents and the per cent efficiency of reactions. In this chapter we glimpse how our colleagues use Greek and Latin prefixes for numbers to construct names and also how chemists pillage our alphabet.

We already know that the prefixes tri- (Latin or Greek) and nona- (Latin) represent the numbers of fluorines in triflate (CF_3SO_2O—) and nonaflate ($C_4F_9SO_2O$—) units. In a similar vein, William Barnett (then at the University of Georgia) and his research team used "tritylone" to describe the moiety **1**.[1a] The abbreviation "trityl" for

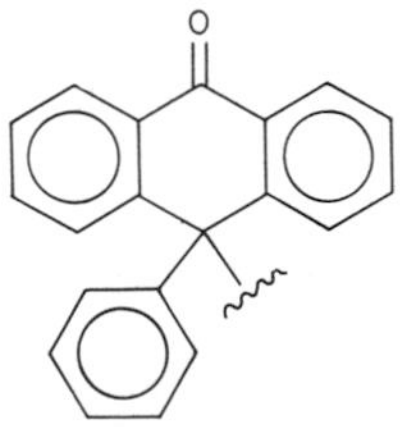

1 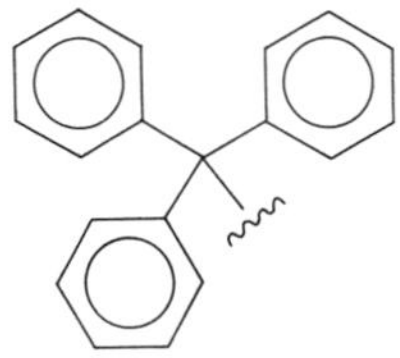   2

triphenylmethyl (**2**) is well known and comes from "tritane," an old short name for triphenylmethane.[2] So Dr. Barnett merely attached "-one" to remind us **1** is a ketone. Joseph Lambert and collaborators were in a position to trifle when they prepared the triphenylsilyl cation in 1986. "With some trepidation" they dubbed it "sityl" (rhymes with trityl).[1b]

Earlier we met Louis Fieser, who named iceane (chapter 12). He also contributed the stem of the term "pleiadiene" for **5**, which is a valence tautomer of knoxvalene (p. 116). How are the name and structure of pleiadiene related? In 1933, Professor Fieser and his chemist–wife Mary introduced the prefix "pleiad-" for several compounds with seven-membered rings.[3] For example, they called **3** "pleiadene." (The word *pleiad* means "a group of seven illustrious or brilliant persons.")

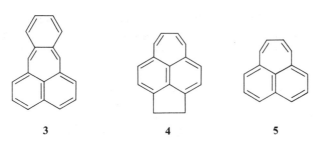

3	**4**	**5**

Eighteen years later, Virgil Boekelheide's bunch at the University of Oregon prepared **4** and suggested that its title and that of **3** be based on the unknown parent structure **5**. He christened **5** "pleiadiene"; **3** and **4** became "benzopleiadiene" and "acepleiadiene," respectively.[4] Then, in 1956, Professor Boekelheide and Gerald Vick succeeded in the first synthesis of **5**.[5] It took only two illustrious and brilliant persons, not seven, to do the job!

This type of parlance is not wedded to whole numbers. John Barton and Robert Walker of the University of Bristol, England, adopted the Latin prefix "sesqui-," which means "one and one-half," when they dubbed **6** and **7** "sesquibiphenylenes."[6] Both

6	**7**	**8**

structures are one and one-half biphenylene (**8**) molecules. (We met **6** and **7** as multiphenylenes in chapter 10.) If this sort of pastime perks you up, have a go with "sester-," from Latin *sesterius*, meaning "two and one-half times as great." Natural products chemists find the prefix handy when their terpenes have 25 carbons. Because a monoterpene is a C_{10} isoprenoid, a "sesterterpene" is a C_{25} counterpart.[7]

Outside the realm of organic chemistry, numerals might take over the nomenclature of elements completely.[8] The International Union of Pure and Applied Chemistry (IUPAC) recommended that, starting with number 104, elements should no longer be named for people (e.g., lawrencium, curium), for places (e.g., californium, americium, francium, europium), for celestial bodies (e.g., helium, uranium, plutonium), and the like (e.g., indium after indigo[9] and rhenium after Rhine). Instead, atomic number will serve as the basis according to the following code:

nil	= 0	pent	= 5
un	= 1	hex	= 6
bi	= 2	sept	= 7
tri	= 3	oct	= 8
quad	= 4	enn	= 9

The suffix is to be "-ium" and makes the name sound "element-like."[8] For example, element 104 becomes *unnilquad*ium; the symbol is Unq.

It was culprit 104 that forced the IUPAC into this historic action. It seems that researchers in the Soviet Union and in the United States discovered this element independently at about the same time; and each team proposed its own name. The

IUPAC committee could not determine which country deserved priority. Hence, it rejected both suggestions and, like a twentieth-century Solomon, devised the politically neutral system described above.[10] Teachers can now have errant students list all undiscovered elements through ennennennium! But, first, they should know that when enn precedes nil, one of the three "n's" is dropped. Thus, 290 would become biennilium. Scientists created unnilseptium (107) in 1981 and unnilennium (109) in 1982, so the "undiscovered" list is shrinking.[11]

Actually, IUPAC does not deny the discoverer of a new element the right to propose a simple name. But its Commission on Inorganic Chemistry will not sanction the proposal until scientists have accepted the discovery and agree that the coined name is satisfactory. In effect, IUPAC's system provides an interim title and symbol for a new member of the periodic table while the scientific community deliberates on what to call it permanently.[12]

The discovery of new elements is usually trumpeted with fanfare at prestigious scientific gatherings, and with care to prevent leaks beforehand. But in 1945, Glenn T. Seaborg first disclosed elements 95 and 96 on a children's radio show in the United States. Actually, Professor Seaborg was slated to break the news at a meeting of the American Chemical Society on November 16, 1945. But it so happened that he was a guest on a radio program, "The Quiz Kids," on November 11 (Armistice Day) to talk about the atomic bomb. Near the end of the commercially sponsored event, contestants put questions to the guest. One youngster asked if elements beyond plutonium (94) would be found. Professor Seaborg blurted out that, indeed, members 95 and 96 had already been discovered! It is not clear what his superiors thought about this premature disclosure. But Dr. Seaborg, who shared a Nobel Prize with Edwin McMillan in 1951, often chuckled over the incident. After all, it's not every day that an important scientific finding is first announced under the aegis of a sponsor such as AlkaSeltzer.[13]

Chemists look to numbers to describe events as well as substances. Manfred Reetz of the University of Marburg sought a handle for reactions in which two σ-bonded groups switch sites intramolecularly (e.g., $9 \rightarrow 10$).[14] Dr. Reetz's wife, who had studied Greek,

9   10

pointed out that *dyo* means "two,"[15] so he dubbed these conversions "dyotropic rearrangements."[14] Vinylogous counterparts can involve more than two sites, and both substituents need not journey the same distance. In such cases the "order" of the rearrangement $9 \rightarrow 10$ is $[m, n]$, where the letters tell how many atoms the two migrating groups traverse. In 1974, Sir Derek Barton's team at Imperial College

11   12

(London) independently labeled such changes "α, ω-rearrangements."[16] The case with $m = n = 2$ would be an "α, β-rearrangement"; if $m = n = 4$ we have "α, δ" and so on. Mechanistic studies[17] suggest that the long-known mutarotation of *trans*-5, 6-dibromocholestane[18] involves a [2, 2]-dyotropic event ($11 \rightleftharpoons 12$).

A different [2, 2]-dyotropic reaction may be occurring in your body right now. Vitamin B_{12} mediates several interconversions, including that of methylmalonyl-CoA (13) and succinyl-CoA (14).[14,19] ("HSCoA" is the abbreviation for "coenzyme A.") Experimental evidence indicates that the cobalt in the vitamin replaces a methyl hydrogen to give 15. The cobalt moiety and —COSCoA group then exchange places ($15 \rightarrow 16$)[19] and, seemingly, fit Dr. Reetz's definition of a dyotropic switch.

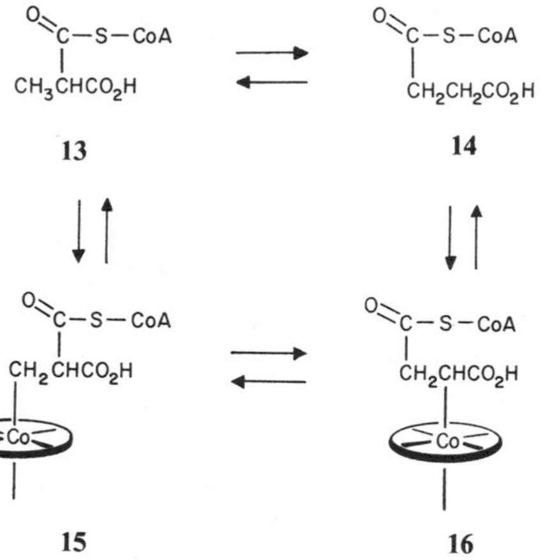

The Greek word *dis* means "twice," so "di-" frequently shows up in chemistry to mean "two." T. Mastryukova's researchers at the U.S.S.R. Academy of Science, Moscow, had this in mind in 1980 when they coined "diadic tautomerism" for processes such as $17 \rightleftharpoons 18$.[20] The hydrogen simply moves between two adjacent atoms, so the tautomeric system just involves the "diad," phosphorus and carbon. (By contrast, the more common keto–enol tautomerism involves, a C—C—O "triad.")

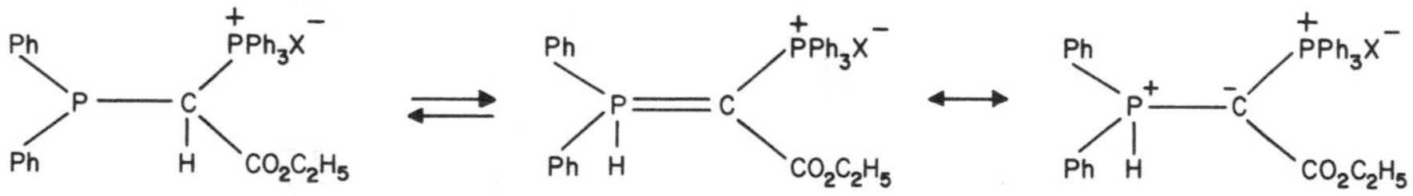

A triad is three; but in 1955 Nathan Kornblum, Robert Smiley, Robert Blackwood, and Don Iffland convinced everyone it can sometimes boil down to two—at least when it comes time to settle for a partner. The Purdue University chemists focussed on anions (SCN^-, NO_2^-, enolate, etc.) that possess two different reaction sites, and they studied factors that determine where alkylation occurs.[21] In the course of this research, Professor Kornblum decided a simple designation was needed for anions that can bond covalently at one or the other of two positions. "Bidentate" would not do—that term was already in use for ligands fastened at two places. A learned female colleage from Purdue's Classics department preferred some wisdom, but it was rather esoteric. So Dr. Kornblum began his own *aufbau*. He merely linked "ambi" from ambidexterous with "dent" from dental to fashion "ambident" (literally, "either tooth"). The term is apt. Ambident anions can bare one or the other set of teeth to form a covalent bond.

Alkylation of nitrite ion (**19**) and of the 2-hydroxypyridine anion **20** illustrate the dichotomy. The Purdue professor confesses "ambident" did not strike him as being euphonious; and he thought the word might serve better as the name of toothpaste.[22] But it took hold and became part of our chemical lore.[23a] In a broader sense, ambidents need not be nucleophiles or even ions. Appropriate cations, radicals, and neutrals can also manifest "either/or" schizophrenia.[23b]

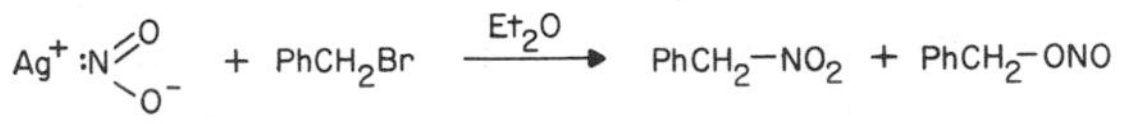

19

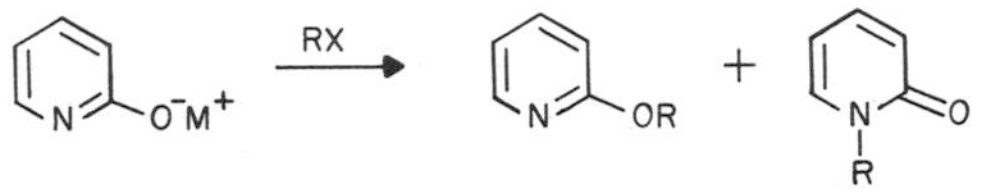

20

Herbert C. Brown of Purdue University has contributed mightily to chemistry (over 1000 publications as of 1986).[24] Besides their Nobel-prize work on hydroboration, he and his prolific platoons—at Wayne State University and then at Purdue—studied steric effects. And they were among the earliest to demonstrate that congestion can bring relief as well as misery to a transition state. Initially these researchers measured the stability of coordination between amines and boranes, but later they included other types of reactions. From their findings, H.C.B. conceptualized three different sorts of strain that can influence reaction rates and equilibria. He called them "F-strain," "B-strain," and "I-strain."[25-27] Let us illustrate each.

When an amine (**21**) binds to a trialkylboron (**22**), the complex can suffer from the steric interference between substituents on nitrogen and those on boron. This stress develops on the leading side (i.e., the front) of both reactants as they approach to produce the new bond (see **23**); hence the designation "F-strain."[25]

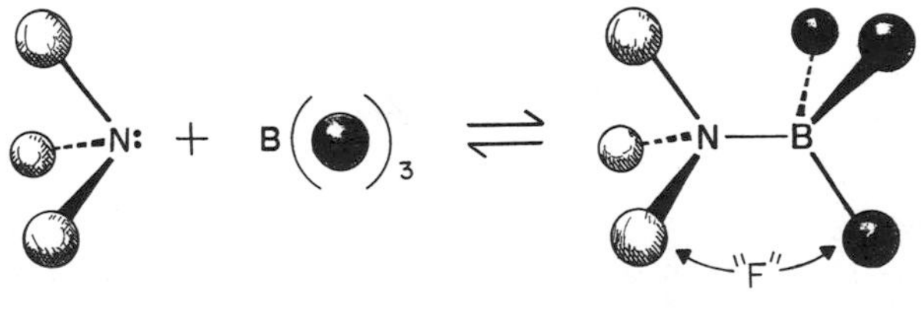

21   **22**   **23**

But sometimes things get more crowded in the back than up front. For example, when a tertiary halide (**24**) ionizes, its remaining ligands spread apart because the sp^3 carbon becomes trigonal as the halide ion bids farewell. The expanding angles can alleviate compression among the three groups, particularly if they are bulky. This relief

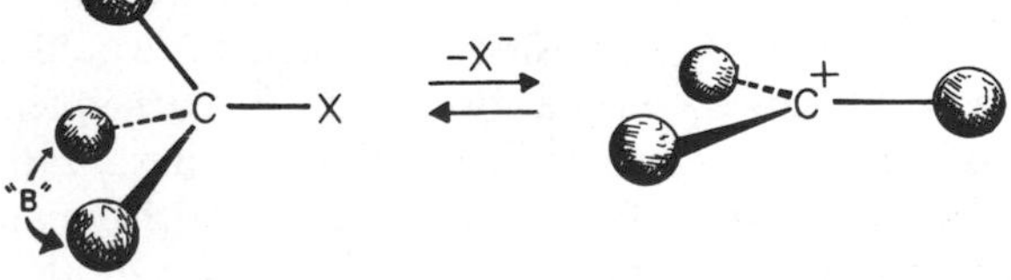

24   **25**

of strain at the *back* (hence "B")[25] could help X depart from the front. In the reverse reaction (**25 → 24**), B-strain would increase when the covalent bond forms.[27]

Those who work closely with cyclic substrates encounter "I-strain" sooner or later. The affliction sets in when a ring atom alters its hybridization ($sp^3 \to sp^2$ or vice versa) during a reaction. The new bond angles bring about changes in ring pucker and in nonbonded interactions. Such transfiguration can increase or decrease the net *internal* (hence "I") strain in the ring and thereby influence the transition-state energy.[26] On this basis we are able to appreciate why reagents (e.g., HOH) add so readily to cyclopropyl ketones (**26 → 27**). The I-strain is less troublesome in the sp^3 addition product than in the sp^2 ketone, because the fixed ring angle (*ca.* 60°) deviates less from ideality in **27** than in **26**.[26,27]

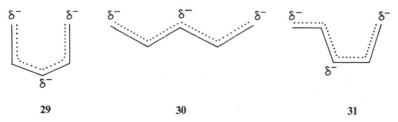

$$\text{26} \qquad\qquad\qquad\qquad\qquad \text{27}$$

If you are beginning to feel the effects of Front-strain, Back-strain, and I-strain, then have pity on heme–imidazole chelates studied by Teddy Traylor at the University of California, San Diego. He and his athletic assistants Jon Geibel, John Cannon, and Dwane Campbell put such complexes (see stylized drawing **28**) through strenuous workouts. They modified the length of the connecting chain (*n*) and the substituents A

28

and B. The aim of these exercises was to induce different types of strains that force changes in Fe—N distance and bond angle. In turn, these changes alter the iron's ability to bind another ligand L (e.g., O_2 or CO). The California training program included "ring strain," when *n* is small and the spanning chain can't reach around comfortably; "face strain," when A is large enough to press the face of the porphyrin; and "springboard strain," when the steric effect of B near the perimeter pulls on the chain and its attached imidazole.[28] Coach Traylor's obedient model compounds may come away with sore muscles, but in the end our comprehension of hemoglobin action should be in better shape.

We can sometimes cut linguistic labor with suitable implements. Robert Bates did his bit by implanting an agricultural tool into chemical jargon. At the University of Arizona his team was concerned with conformational integrity in carbanions derived

29   **30**   **31**

from 1, 3-pentadienes by removal of an allylic proton.[29] He considered the three flat shapes (**29, 30** and **31**) for the mesomeric anion and coined the descriptive terms "U" and "W" for the first two. But no alphabetical letter looks quite like conformation **31**. At first Professor Bates considered the appellation "dipper," but the high interest in matters Russian swung his decision to "sickle."[30] (Perhaps he anticipated the détente that the United States and Russia began in the 1970s.) Dr. Bates cleverly compelled such anions to retain specific shapes by incorporating them into ring systems. For example, base-catalyzed equilibration of **32** and **34** logically involves anion **33**, with its "sickle" delocalization. We will not keep you in suspense; the "U" pattern appears to be more stable than the other two.[29]

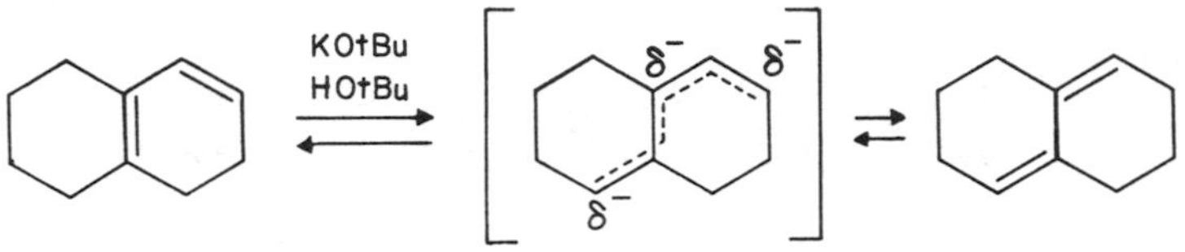

            32   **33**   **34**

The terms "U", "W", and "sickle" also conveniently describe arrangements of five-atom segments involved in concerted 1, 3-eliminations to give cyclopropanes.[31] If A and B represent the two groups to be lost, they can adopt "U" (**35**), "sickle" (**36**) and

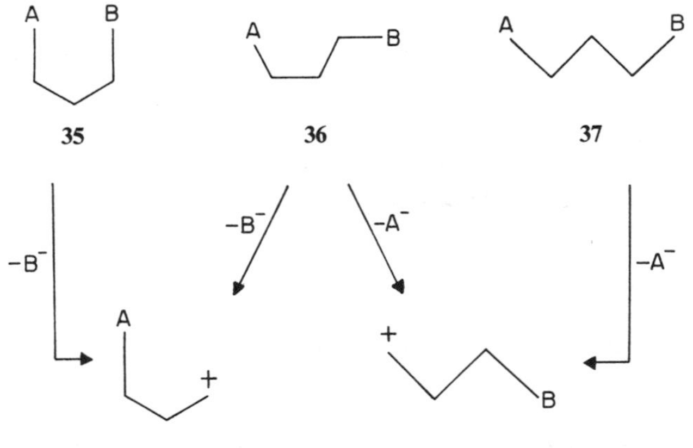

      35   **36**   **37**

            38   **39**

"W" (**37**) geometries. For a nonconcerted mechanism, departure at one end leaves a reactive intermediate (e.g., cation) that was termed "semi-U" (**38**) or "semi-W" (**39**). We can find U, W, and S (sickle) arrays in a norbornyl skeleton (see bold lines in **40**). Thus,

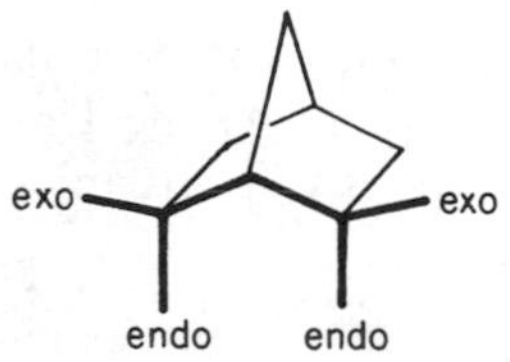

            40

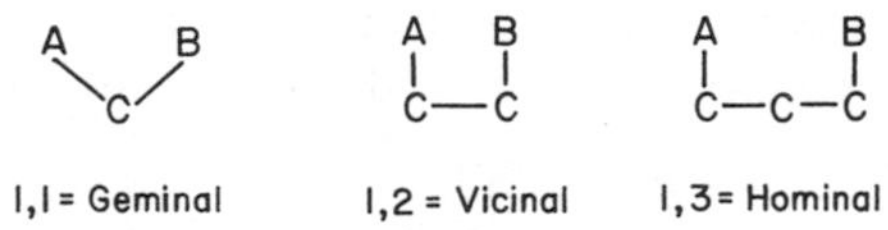

U includes two endo bonds, W involves two exo bonds, and S uses one of each. Incidentally, Johns Hopkins chemists coined the word "hominal" to describe a 1, 3 relationship between two entities.[32] This term extends the sequence "geminal" and

"vicinal" and stems from "homo-," a prefix widely used to denote a homologous structure. The ending -inal, rather than -onal, maintains the spelling as in geminal and vicinal.

In the 1970s at New Mexico State University, Dennis D. Davis and his students were knee deep in cyclopropanes, particularly those from 1, 3-eliminations in sulfonates carrying a tin or silicon substituent.[33] In one phase of this research, Professor Davis collaborated on molecular-orbital calculations with a colleague whose outlook was highly theoretical. For example, she looked upon hybrid orbitals as an unfruitful concept and considered anything less than *ab initio* calculations akin to black magic. On the other hand, D.D.D. was a physical–organic chemist steeped in the tradition of overlapping orbitals (even "empty" ones), so the two partners communicated with some difficulty.[34]

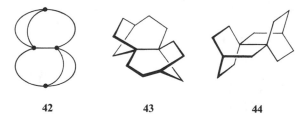

41

When experiments on stereochemistry and kinetics of ring closure led Dr. Davis to consider back-lobe orbital interaction in a W conformation (**41**), he tweaked his colleague's sensibilities by referring to it as overlap "through the tails." She was decidedly not happy with that terminology and finally insisted it be replaced. So D.D.D. called upon his education in the classics and came up with "percaudal homoconjugation." (Latin *caudalis* means "of, pertaining to, or of the nature of a tail.") The term showed up in his publications[35] and has since been wagged by others.[36] Meanwhile, Professor Davis's colleague maintained that new terms forever being invented by organic chemists should be taken "percaudally."[34]

The Latin *geminus* means "twin," so it's understandable that Leo Paquette and students Hokoon Park and Patrick King at Ohio State University coined "geminane" for tetracycles of general type **42**.[37] These structures contain two identical bicyclic segments fused, Siamese-like, by a bridgehead bond common to each segment. The Ohio laboratory gave birth to specific analogs **43** and **44**, whose trim titles, *syn-*

42   **43**   **44**

$[3.2.1]^2$geminane and $[2.2.2]^2$geminane, are easy to decipher if you know basic von Baeyer.[38] Glance back at p. 7 and you should recognize that lepidopterene became a charter member of the geminane club *ipso facto*; it is after all, a tetrabenzo derivative of **44**.

In 1948, at the University of California at Los Angeles, Saul Winstein and graduate student Ernest Grunwald offered chemists "*Y*" to designate solvent polarity.[39] Let us pause a moment to ponder polarity. When someone says water is more polar than methanol, and methanol is more polar than pentane, almost everyone would nod agreement. But what should be the criterion for polarity? Dipole moment? Dielectric constant? Capacity to dissolve salts? How would the criterion rank two solvents such as acetone and dimethylformamide? Or mixed media such as 70% aqueous ethanol *vs.* 60% aqueous acetone? The UCLA duo reasoned that the S_N1 solvolysis rate constant for t-butyl chloride should go up when solvent polarity increases. So they adopted 80% ethanol as a "standard" medium and defined an ionizing power parameter "Y" by the equation

$$Y = \log k - \log k_0$$

Here k and k_0 are the kinetically determined rate constants for t-butyl chloride at 25°C in the chosen solvent and in the 80% ethanol "standard," respectively. Defined thus, Y is zero for the "standard" and ranges from about $+4$ to -3 for the most common ionizing media. But solvents unable to sustain S_N1 processes can't be tagged with Y-values, so the acetone *vs.* DMF dilemma still remained.

A decade later, Edward Kosower, then at the University of Wisconsin, dived in and came up with a treasure—"Z-values."[40] He measured UV-VIS spectra of N-alkylpyridinium iodides (e.g., **45**) in numerous solvents and discovered that the position (λ_{CT}) for the longest wavelength band, namely, the charge-transfer absorption (**45** → **46**), was remarkably sensitive to the nature of the medium. And for ionizing milieux like aqueous methanol, aqueous ethanol, and aqueous acetone, these λ_{CT} positions correlated linearly with the Winstein–Grunwald Y-values. From then on, the physical–organic clan began to enjoy more leisure time, because recording spectra was easier

45   **46**

than running kinetics. And besides, these charge-transfer bands showed up in a wide range of solvents, unlike the limited scope for Y-numbers. Ultimately, Professor Kosower adopted the specific pyridinium iodide **45** as his workhorse—the ethyl group helped solubility, and the CO_2CH_3 kept λ_{CT} nicely clear of other absorptions by the pyridinium ion. And he urged chemists to use the transition energy of the charge-transfer band in **45** (expressed as kcal/mol to give convenient numbers) as an empirical measure of solvent polarity. Dr. Kosower called such numbers Z-values. For example, in methanol **45** has λ_{CT} 342.1 nm, and so Z (kcal/mol) = 83.6. For acetone, λ_{CT} 435.4 nm; $Z = 65.7$. The Z-number gets bigger as polarity increases, and Professor Kosower's writings (and those of others[41]) tell you why.[40]

The Grunwald–Winstein "Y" publication and Kosower's "Z" papers became citation classics.[42,43] But how did these men of letters land at the end of the alphabet? Professor Grunwald happened on Y while writing his Ph.D. thesis just because he wasn't using that symbol for anything else.[44] And Kosower, who eschewed the temptation to dub his numbers K-values, settled on Z-values simply because they came on the scene after Y-values.[43,45] We can't fault this logic. But, if chemist Kosower uncovers yet another measure of solvent polarity, what would he then call it?

In 1972, Peter Gund at Princeton University also contributed to the chemical alphabet with "Y-delocalization" to symbolize the outline of the π-system in structures such as trimethylenemethane (**47**) and the carbonate ion (**48**).[46] Furthermore, he supported the view that a Y-system with six π-electrons gains stability from "Y-aromaticity";[47] an example is the dianion from addition of two electrons to **47**. A year

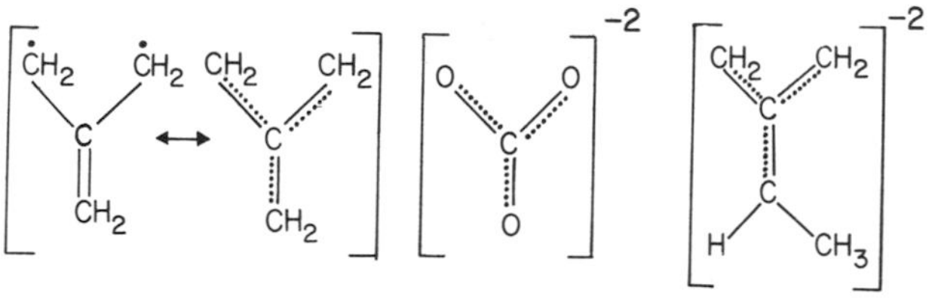

47   **48**   **49**

later, Joseph Klein's group at Hebrew University, Jerusalem, prepared this dianion.[48] In 1981, Nancy Mills and her students provided experimental evidence for the formation of the monomethyl homolog, **49**,[49] but Professor Mills has questioned the importance of Y-aromaticity to the stability of such dianions.[50] Klein has reviewed this topic,[51] and further support for Y-aromaticity has appeared.[52]

In our alphabet, W and Y live close to each other. They also visit back and forth, at least when these letters define shape in NH_2OH, or in isoelectronic species like the $^-CH_2OH$ carbanion. Such genre adopt two minimum-energy conformations termed W and Y.[53] In hydroxylamine, a Newman projection of W (seen from the nitrogen end) shows the OH hydrogen on the internal bisector of the HNH angle (**50**), whereas Y shows OH on the external bisector (**51**). Our molecular postal service speeds these

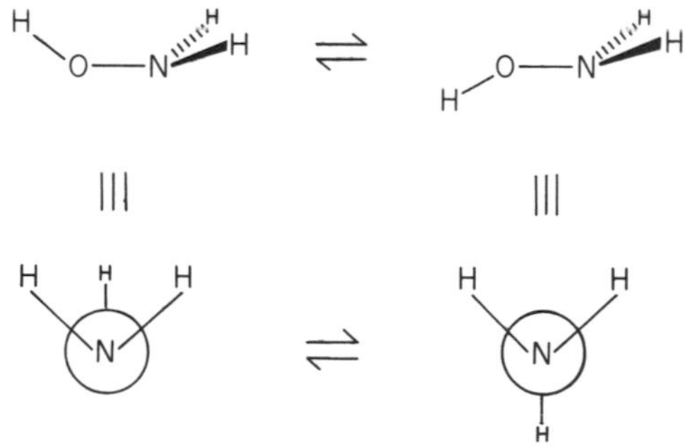

50   **51**

letters back and forth, either by torsional rotation around the O–N bond or by pyramidal inversion at nitrogen. For substituted analogs, the preferred path depends on who is travelling, but usually the atoms don't dally at other addresses en route. Raban and Kost have announced the best ways to make these trips for different passengers as well as in other neighborhoods (e.g., sulfenamides, $R_1SNR_2R_3$).[54]

In biology, Y's don't last long. According to IUPAC-IUB nomenclature rules, the symbol Y was to be used for pyrimidine nucleosides of unknown structure.[55] Sometimes, researchers applied terms such as "Y"-bases for unidentified purine derivatives as well.[56] In any case, the French are happy to see Y in print. For it is a small town in France with the shortest name in Europe. And Y should they mind the publicity?

Students would be delighted to see on their report cards structure **52**, created by Richard Eisenberg and Clifford Kubiak at the University of Rochester.[57] A front view resembles the letter "A," so they refer to such structures as "A-frame." It may have started a trend of sorts, because in 1983 "B-frame" hit the consumer market. But in this case the term stands for "a polyhedrol borane cluster to which one or more metal centers can be added as additional cluster vertices." If B-frame matrices pique your interest, then take a tour of those fascinating complexes courtesy of their founders, a cluster of inorganic specialists at the University of Leeds.[58]

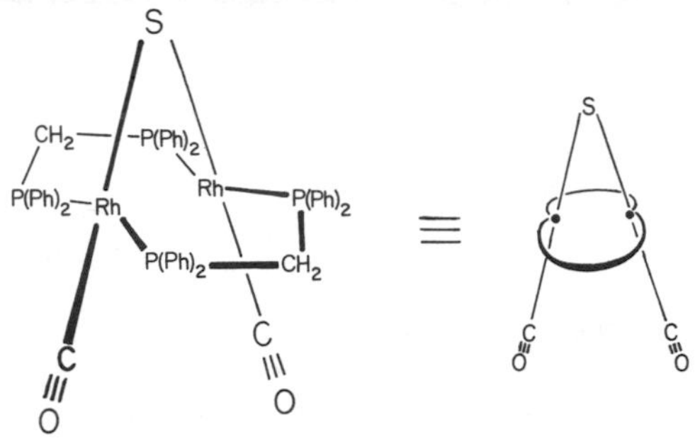

52

Acronyms (from Greek *akros* for "tip" or "outmost" and *onyma* for "name") are terms usually made up from the first letters or syllables of other words. Acronyms are handy in science, so let's sample a few.[59] At the Shell Development Company laboratory in the 1960s, Schrauzer, Bastian, and Fosselius were exploring alkene polymerization with binuclear catalysts such as $Zn[Co(CO)_4]_2$.[60] In an initial trial with norbornadiene (**53**) and too much catalyst, their mixture reacted violently and propelled the product upward. Dr. Bastian dutifully scraped the solid off the ceiling and

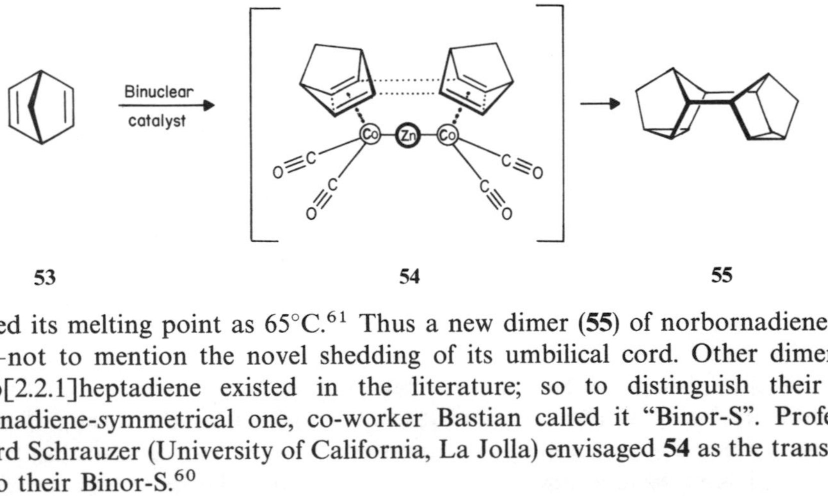

53   **54**   **55**

recorded its melting point as 65°C.[61] Thus a new dimer (**55**) of norbornadiene was born—not to mention the novel shedding of its umbilical cord. Other dimers of bicyclo[2.2.1]heptadiene existed in the literature; so to distinguish their *bis-norbornadiene-symmetrical* one, co-worker Bastian called it "Binor-S". Professor Gerhard Schrauzer (University of California, La Jolla) envisaged **54** as the transition state to their Binor-S.[60]

Richard Lawton and Thomas Holmes (University of Michigan) have nicknamed the reagent **56** so that you will remember what it does. They call it "Cyssor I," derived from

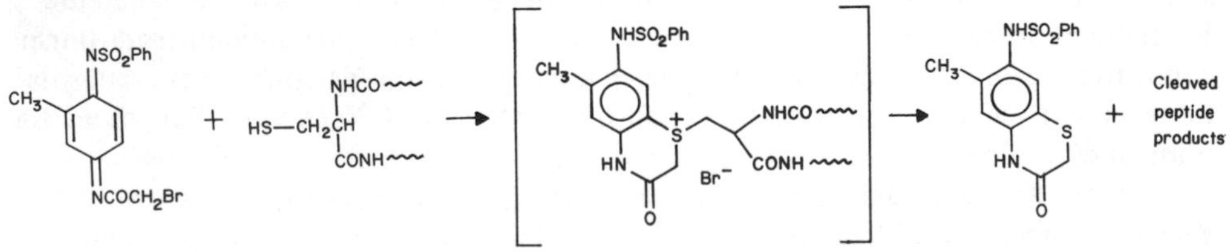

56   **57**   **58**   **59**

*cy*steine specific *s*cission by *o*rganic *r*eagent.[62] It reacts with sulfhydryl groups in polypeptides (**57**) and can result in chain cleavage at a cysteine site. They think sulfonium salt **58** is involved; and the sulfur ends up in **59**. Cyssor I became available commercially,[63] and when reagent II comes along you will have a pair of Cyssors to cut proteins.

The acronym "elplacarnet" gives no clue for its *raison d'être*. But read R. Bruce King's proclamation about sp²-carbon, perhapto ligands in transition-metal complexes to appreciate why he coined the word.[64] Elplacarnet stands for *el*emental *pla*nar *car*bon *net*works.[65] According to King, professor at the University of Georgia, only about 13 flat, unsaturated carbon networks exist with dimensions sufficiently compact to allow total bonding to a single transition metal. Five of these "elplacarnets" are **60–64**. In contrast, skeletons with larger diameter (such as **65–69**) either bind only partially to a single metal or they require two or more metals for complete bonding. By

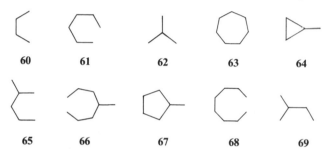

arranging planar networks according to their two-dimensional breadth, Dr. King grew an "elplacarnet tree" to help us understand why some polyene complexes involve one metal, whereas others need two.[64]

When writing about accomplishments by a team of investigators, you ordinarily save space by mentioning only the senior author or, perhaps, first author plus *et al*. The remaining colleagues must rest content with epitaphs in the bibliography cemeteries. But, you can use acronyms—not necessarily pronounceable ones—to allot all coauthors their just dues. For example, when Laidig and Schaefer discussed a paper by Collins, Dill, Jemmis, Apeloig, Schleyer, Seeger, and Pople, they referred to it throughout as CDJASSP.[66] It's short; it's unambiguous; and it doesn't slight anyone. By the same token, the 49-author opus from Harvard we mentioned on p. 67 would become WLNKWABBCCCFFGGHHHHHHHIJKKKKLLMMMOPRRSSTTT-UUUVVWWW. And if space is really tight, how about alphabetizing it to $AB_2C_3F_2G_2H_6IJK_5L_3M_3NOPR_2S_2T_3U_3V_2W_5$? Or regrouping it to $AIJNOP(BFGRSV)_2(CLMTU)_3(KW)_5H_6$? We started to go through the same exercise for the 142-author physics paper cited on p. 67 but decided that "Arnison *et al*." is not so bad after all.

Few of us would disagree that a pert acronym can expedite communication—if not for everyone at least for those "in the know." But besides that, concocting abbreviations is sometimes fun and can add spice and amusement to a sober science. The field of magnetic resonance spectroscopy abounds in delightful examples, and its practitioners tailor their terminology with seeming abandon. As a sampling, consider the following acronyms: FOCSY, NOESY, COSY, SECSY, INEPT, INADEQUATE. One (unnamed) colleague confessed he didn't know what they stood for—except, perhaps, the story of his life. But, to aficionados these titillating terms represent specialized NMR techniques for probing molecular structure.[67] The first four acronyms showed up in publications from the Zurich ETH laboratories of Kurt Wüthrich and Richard Ernst;

and the last two are by courtesy of Ray Freeman's group at Oxford University.*
Editors and authors don't always see eye to eye on the matter of trivial names and
abbreviations. For a time, Professor Wüthrich was at odds with the *European Journal
of Biochemistry* for its restrictive policy on nicknames. Ultimately, however, the
Editorial Board conceded that COSY, NOESY, and SECSY are indeed handier than
the full technical descriptions.[74]

At Oxford, Professor Freeman and company shopped for better broadband
decoupling sequences with the help of computer simulation and came up with one,
symbolized as

$$R = 90° \, (X) \, 180° \, (-X) \, 270° \, (X) = 1\bar{2}3.$$

The shorthand notation at the end indicates the number of 90° pulses in the
sequence, and the bar denotes a phase inversion. In NMR jargon their complete cycle
became

$$R R \bar{R} \bar{R} = 1\bar{2}3 \; 1\bar{2}3 \; \bar{1}2\bar{3} \; \bar{1}2\bar{3}$$

In step with the times, the Freeman troupers advertised this progression as WALTZ-
4; they also gave lessons in WALTZ-8 and WALTZ-16. The word stands for
*W*ideband, *A*lternating-phase, *L*ow-power *T*echnique for *Z*ero residual splitting; but
their journal article states that the inventor of the acronym wished to remain
anonymous.[75]

At the same studio, Shaka and Freeman had fun with composite pulses—a cluster of
two or more conventional ones designed to give mutual cancellation of errors. The
Oxford pair observed that composite pulses with "dual compensation greatly extend
the permissible 'working-space' available to the operator." And to convey this idea,
they offered the code name GROPE (*G*eneralized compensation for *R*esonance *O*ffset
and *P*ulselength *E*rrors). Their specific sequence was GROPE-16.[76]

Practitioners of two-dimensional "J" spectroscopy often see spurious blips. These
responses are usually weak replicas of the main signals and have come to be known as
"phantom" and "ghost" responses. Wizards Geoffrey Bodenhausen, Ray Freeman, and
David L. Turner held a seance in Oxford's hallowed halls and conjured up a phase-
cycling sequence to suppress "phantoms" and "ghosts." This blithe trio dubbed their
sequence "exorcycle."[77]

The acronyms DANTE (Delays Alternating with Nutations for Tailored Excitation)
and INFERNO (Irradiation of Narrow-Frequency Envelopes by Repeated Nutation)
have an interesting literary tie.[78] In DANTE spectroscopy, the nuclear spin trajectories
follow a fairly complicated path in the rotating reference frame. Starting at the North
Pole, the tip of the magnetization vector moves along a small arc down a meridian then
executes a full revolution (nutation) along a circle of latitude. Then, another small step
down the same meridian of longitude, followed by another complete latitudinal trip.
This itinerary continues until the vector tip reaches the equator.

Ray Freeman's team agonized over a simple way to title this unusual trajectory. His
student Gareth Morris came forth to point out that, except for a reversal in vertical
direction, this was the path prescribed by Dante for souls in Purgatory.[73c] Dante's
Purgatory consists of ten circular ledges, one above the other, around a mountain.

---

*FOCSY: foldover-corrected spectroscopy.[68] NOESY: two-dimensional nuclear Overhauser enhance-
ment spectroscopy.[69] COSY: correlated spectroscopy.[70] SECSY: two-dimensional spin-echo-correlated
spectroscopy.[71] INEPT: insensitive nuclei enhanced by polarization transfer.[72] INADEQUATE: incredible
natural-abundance double-quantum-transfer experiment.[73]

Travellers progress toward Heaven by circumnavigating each ledge before moving up to the next. Every level corresponds to a human frailty or shortcoming, with more interesting sins near the top. (If you forget which direction to go, just remember: souls aim for Heaven; spectroscopy heads the other way.) This literary analogy extends yet further. Dante's Inferno has the form of a cone whose apex is at the earth's center. The trajectory of magnetization during an INFERNO experiment lies on the surface of just such a cone.[78] It is oft said that scientists should not forsake the classics. At Oxford, they surely don't.

John Waugh at MIT frequently lamented the dryness and pomposity that pervades some publications by scientists.[79] So when he, Alexander Pines, and Michael Gibby developed a new NMR method to study solids, they mischieviously titled it *Proton-Enhanced Nuclear Induction Spectroscopy*.[80] Its *acronym* was not used in their paper, so the title aroused no editorial suspicion. But the scientific community soon got the point. Professor Pines (University of California at Berkeley) professes that the acronym is but an *anagram* of his surname and fits rather closely its pronunciation in Hebrew.[79] Wa-Hu-Ha and HOHAHA!* Whatever the rationale, no one can argue that acronyms are SIMPLE and have their SPOTS in science; and NMR folk are certainly DEFT at them.

However, we don't want to leave you with the notion that spectroscopists hold a monopoly on salaried amusement. Other disciplines also get in on the act. For a starter, you might look up what computer chemists do with their EROS and SECS.† W. Todd Wipke at University of California, Santa Cruz, settled for SECS after a systematic title search by computer. For this act he had written a program that generated all possible acronyms from a list of key words, such as organic, chemical, synthesis, simulation, etc.[90] Recognizing a good theme, other researchers later came up with their EROS.[88]

Neologism by computers is not new in chemistry and is not restricted to acronyms. For example, in the 1950s the Pfizer Pharmaceutical Co. programmed an instrument to churn out thousands of titles that might be suitable for new drugs.[91] Input criteria included being easy to pronounce and spell; easy to remember; easy to transliterate into foreign languages; having a "medical sound," etc. When Pfizer developed a new product, the decision makers needed only to scan their lexicon of "drugless names" for a wide selection.

Nowadays, however, you don't need the resources of a giant corporation or of a marketing consultant for suggestions. To come up with names that communicate, differentiate, are memorable, and that may connote an important quality or characteristic all you require is a personal computer and a software package called *NAMER by SALINON*.[92] This program was designed to help businesses or individuals originate terms for new products, services, or whatever. The computer relies on conditional probabilities, adaptive learning techniques, cryptographic algorithms, and linguistic data bases to generate names; but the user need only touch a few keys. According to what meaning or image you want to convey, the program can create names from Latin, Greek, or English roots and can assemble completely new words

---

*Wa-Hu-Ha: a pulse sequence used to obtain high-resolution NMR spectra of solids.[81] The method was originated at Massachusetts Institute of Technology by J.S. Waugh, L.M. Huber, and U. Haeberlen.[82] The acronym sprung from the first two letters in the surnames of the three authors.[83] HOHAHA: homonuclear Hartmann–Hahn coherence transfer process.[84] SIMPLE: secondary isotope multiplets of partially labeled entities.[85] SPOTS: spin-polarization torsional spectroscopy.[86] DEFT: driven-equilibrium Fourier transform spectroscopy.[87]

†EROS: Elaboration of Reactions for Organic Synthesis.[88] SECS: Simulation and Evaluation of Chemical Synthesis.[89]

that sound English. It even triggers a siren if a chosen name has a profane or vulgar connotation in any of five foreign languages—English, Spanish, French, Italian, and German. Besides all its seriousness, *NAMER by SALINON* also sounds like fun.

Have you ever hit upon a word that would sound good as a chemical name? Eiji Ōsawa at Japan's Hokkaido University likes "banzaine," because it reflects his nationalistic pride. Someday, Professor Ōsawa hopes to construct a molecule worthy of this title.[93]

And what about BMC? Britons might identify these letters with their venerable British Motor Company. But the French will grin and change the subject. This chemical story goes back to the mid-1950s at the University of Barcelona and the research of Manuel Ballester, graduate student Carlos Molinet, and postdoctorate Juan Castañer. They worked with chlorinating mixtures containing S_2Cl_2, $AlCl_3$, and SO_2Cl_2. From earlier experiments by O. Silberrad in England, such concoctions were known to convert benzene to hexachlorobenzene.[94] The Spaniards modified this mixture and dramatically increased its power for nuclear chlorination.[95] For example, their reagent handily converts benzotrichloride (**70**) to perchlorotoluene (**72**). No mean feat, considering that all previous attempts to produce this hydrogenless oddity never got past seven chlorines (**71**).[96] (Evidently, the halogens impose such congestion that one ortho site stayed impenetrable. Overly vigorous treatment did not help; it merely lopped off the benzylic carbon to afford hexachlorobenzene.) The Ballester–Molinet–Castañer reagent, like a shoehorn, slips in the final chlorine (**71 → 72**) despite the tight squeeze.[95]

70   **71**   **72**

After this synthetic breakthrough, interest in perchlorinated aromatic molecules burgeoned. Among other things, they serve as precursors for an array of remarkably stable free radicals.[97] The chlorinating mixture that made all this possible is called BMC after its three Spanish inventors. That abbreviation appears in Professor Ballester's papers[98] and in citations to their work.[99] The acronym seemed fine until Dr. Ballester extolled the virtues of BMC in a lecture at the CNRS in Grenoble, France. Giggles and whispers swelled through the audience when he related their "success with BMC" and how "BMC worked wonders" in the laboratory. Later, his host André Rassat discreetly explained: In France, BMC was a household term for "Bordel Militaire de Campagne," a mobile legion of female prostitutes that serviced French armed forces during military operations in Algeria and elsewhere.[100,101]

Dr. Ballester did not visit The Netherlands on that trip, and it was just as well. Otherwise, funsters Hans Wynberg, Hepke Hogeveen, Jan Engberts, and Richard Kellogg at the University of Groningen might have taken him to task for "suggestive" terminology. This bevy of chemists signed a missive decrying obscenity in chemical journals. According to an account in *Chemical and Engineering News*[102] Professor Wynberg wrote:

> The recent ruling by the Supreme Court restricting obscenity in books, magazines, and movies requires that we reexamine our own journals for lewd contents. The recent chemical literature provides many examples of words and

concepts whose double meaning and thinly veiled overtones are an affront to all clean chemists. What must a layman think of "coupling constants," "tickling techniques," or "increased overlap"? The bounds of propriety are surely exceeded when heterocyclic chemists discuss homoenolization.

After some further examples, Wynberg and his cosigners closed with a plea to "cleanse our journals and our minds."

The plea did not go unheeded. University of Tennessee chemists Ronald Magid and Fred Schell, along with graduate student Jay E. Parker III, immediately took up the crusade by forming a coalition titled "Chemists Rise Up Against Pornography" (CRAP) to cull the literature for salacious material.[103] They sent Wynberg about 40 more phrases that "innocent students see in our scientific journals." Their list included

> backside attack
> orbital penetration
> aromatic sex-tet
> insertion reaction
> puckered conformation
> degenerate rearrangement
> well-greased male and female joints

Coalition CRAP even uncovered an objectionable word in a paper coauthored by Dr. Wynberg![104] And they implied that he deliberately tried to conceal the word by spreading it to the 2nd, 3rd, 22nd, 23rd, and 25th letters in the second paragraph. "Surely no impartial observer would attribute this to coincidence" they asserted. Look up the publication to see if you agree.

REFERENCES AND NOTES

1. (a) Barnett, W.E.; Needham, L.L.; Powell, R.W. *Tetrahedron* **1972**, *28*, 419–424, (b) Lambert, J.B.; McConnell, J.A.; Schultz, W.J., Jr. *J. Am. Chem. Soc.* **1986**, *108*, 2482–2484.
2. Bailey, D.; Bailey, K.C. *An Etymological Dictionary of Chemistry and Mineralogy* (Edward Arnold & Co., London, England, 1929).
3. Fieser, L.F.; Fieser, M. *J. Am. Chem. Soc.* **1933**, *55*, 3010–3018.
4. Boekelheide, V.; Langeland, W.E. *J. Am. Chem. Soc.* **1951**, *73*, 2432–2435.
5. Boekelheide, V.; Vick, G.K. *J. Am. Chem. Soc.* **1956**, *78*, 653–658.
6. Barton, J.W.; Walker, R.B. *Tetrahedron Lett.* **1978**, 1005–1008.
7. For a discussion of sesterterpenes see Hanson, J.R., in *Chemistry of Terpenes and Terpenoids*, Newman, A.A. (Ed.), (Academic Press, New York, 1972), pp. 200–206.
8. Egan, H.; Godly, E.W. *Chem. Brit.* **1980**, *16*, 16–25.
9. (a) Gutman, I. *J. Chem. Educ.* **1985**, *62*, 674, (b) Ball, D.W. *J. Chem. Educ.* **1985**, *62*, 787–788.
10. *International Newsletter on Chemical Education*, No. 11, September, 1979.
11. Rawls, R. *Chem. Eng. News* **1982**, Oct. 11, pp. 27–28.
12. (a) Jeannin, Y.; Holden, N.E. *Nature (London)* **1985**, *313*, 744, (b) Loening, K.L. *Chem. Eng. News* **1985**, Aug. 19, p. 3.
13. (a) *Chem. Brit.* **1983**, *19*, 698, (b) Seaborg, G.T. *J. Chem. Educ.* **1985**, *62*, 463–467.
14. Reetz, M.T. *Angew. Chem. Int. Ed. Engl.* **1972**, *11*, 129–130.
15. Reetz, M.T., private communication, March, 1981.
16. Barton, D.H.R.; Prabhakar, S. *J. Chem. Soc., Perkin Trans. 1* **1974**, 781–792.
17. Grob, C.A.; Winstein, S. *Helv. Chim. Acta* **1952**, *35*, 782–802.
18. Mauthner, J. *Monatsh. Chem.* **1906**, *27*, 421.
19. Miller, W.W.; Richards, J.H. *J. Am. Chem. Soc.* **1969**, *91*, 1498–1507.
20. Mastryukova, T.A.; Aldzheva, I.M.; Leont'eva, I.V.; Petrovski, P.V.; Fedin, E.I.; Kabachnik, M.I. *Tetrahedron Lett.* **1980**, *21*, 2931–2934.

21. Kornblum, N.; Smiley, R.A.; Blackwood, R.K.; Iffland, D.C. *J. Am. Chem. Soc.* **1955**, *77*, 6269–6280.
22. Kornblum, N., private communication, May, 1983.
23. (a) Reutov, O.A.; Beletskaya, I.P.; Kurtz, A.L. *Ambident Anions* (Plenum Press, New York, 1983), (b) Olah, G.A.; Gupta, B.G.B.; Garcia–Luna, A.; Narang, S.C. *J. Org. Chem.* **1983**, *48*, 1760–1762, and references cited therein.
24. According to a historical account [Brewster, J.H. *Aldrichimica Acta* **1987**, *20*, No. 1, 3–8], publication number 1000 is Brown, H.C.; Kim, K.–W.; Cole, T.E.; Singaram, B. *J. Am. Chem. Soc.* **1986**, *108*, 6761–6764.
25. Brown, H.C.; Bartholomay, H., Jr.; Taylor, M.D. *J. Am. Chem. Soc.* **1944**, *66*, 435–442.
26. (a) Brown, H.C.; Gerstein, M. *J. Am. Chem. Soc.* **1950**, *72*, 2926–2933, (b) Brown, H.C.; Fletcher, R.S.; Johannesen, R.B. *J. Am. Chem. Soc.* **1951**, *73*, 212–221.
27. Brown, H.C. *Record Chem. Progr. (Kresge–Hooker Sci. Lib.)* **1953**, *14*, 83–97.
28. Geibel, J.; Cannon, J.; Campbell, D.; Traylor, T.G. *J. Am. Chem. Soc.* **1978**, *100*, 3575–3585.
29. Bates, R.B.; Carnighan, R.H.; Staples, C.E. *J. Am. Chem. Soc.* **1963**, *85*, 3031–3032.
30. Bates, R.A., private communication, September, 1975.
31. Nickon, A.; Werstiuk, N.H. *J. Am. Chem. Soc.* **1967**, *89*, 3914–3915.
32. Nickon, A.; Weglein, R.C.; Mathew, C.T. *Can. J. Chem.* **1981**, *59*, 302–313.
33. Davis, D.D.; Chambers, R.L.; Johnson, H.T. *J. Organometal. Chem.* **1970**, *25*, C13–C16.
34. Davis, D.D., private communication, March, 1986.
35. (a) Davis, D.D.; Black, R.H. *J. Organometal. Chem.* **1974**, *82*, C30–C34, (b) David D.D.; Johnson, H.T. *J. Am. Chem. Soc.* **1974**, *96*, 7576–7577.
36. Shiner, V.J., Jr.; Ensinger, M.W.; Kriz, G.S. *J. Am. Chem. Soc.* **1986**, *108*, 842–844.
37. Park, H.; King, P.F.; Paquette, L.A. *J. Am. Chem. Soc.* **1979**, *101*, 4773–4774.
38. (a) *IUPAC Nomenclature of Organic Chemistry*, Sections A and B, 2nd ed. (Butterworths, London, 1966), (b) Eckroth, D.R. *J. Org. Chem.* **1967**, *32*, 3362–3365.
39. Grunwald, E.; Winstein, S. *J. Am. Chem. Soc.* **1948**, *70*, 846–854.
40. Kosower, E.M. *J. Am. Chem. Soc.* **1958**, *80*, 3253–3260, 3261–3267, 3267–3270.
41. Dimroth, K.; Reichardt, C.; Siepmann, T.; Bohlmann, F. *Liebigs Ann. Chem.* **1963**, *661*, 1–37.
42. Grunwald, E. *Current Contents* **1984**, *24*, No. 43, p. 18.
43. Kosower, E.M. *Current Contents* **1980**, *20*, No. 29, p. 11.
44. Grunwald, E., Brandeis University, private communication, June, 1984.
45. Kosower, E.M., private communication, April, 1984.
46. Gund, P. *J. Chem. Educ.* **1972**, *49*, 100–103.
47. Finnegan, R.A. *Ann. N.Y. Acad. Sci.* **1969**, *159*, 242–266.
48. Klein, J.; Medlik, A. *J. Chem. Soc., Chem. Commun.* **1973**, 275–276.
49. Mills, N.S.; Shapiro, J.; Hollingsworth, M. *J. Am. Chem. Soc.* **1981**, *103*, 1263–1264.
50. Mills, N.S. *J. Am. Chem. Soc.* **1982**, *104*, 5689–5693.
51. Klein, J. *Tetrahedron* **1983**, *39*, 2733–2759.
52. (a) Rajca, A.; Tolbert, L.M. *J. Am. Chem. Soc.* **1985**, *107*, 698–699, (b) Agranat, I.; Skancke, A. *J. Am. Chem. Soc.* **1985**, *107*, 867–871.
53. Wolfe, S.; Tel, L.M.; Csizmadia, I.G. *Can. J. Chem.* **1973**, *51*, 2423–2432.
54. Raban, M.; Kost, D. *Tetrahedron* **1984**, *40*, 3345–3381.
55. IUPAC–IUB Commission on Biochemical Nomenclature (CBN). Abbreviations and Symbols for Nucleic Acids, Polynucleotides, and their Constituents. *J. Mol. Biol.* **1971**, *55*, 299–310.
56. Frihart, C.R.; Feinberg, A.M.; Nakanishi, K. *J. Org. Chem.* **1978**, *43*, 1644–1649.
57. (a) Kubiak, C.P.; Eisenberg, R. *J. Am. Chem. Soc.* **1977**, *99*, 6129–6131, (b) *Ibid.* **1980**, *102*, 3637–3639.
58. (a) Bould, J.; Crook, J.E.; Greenwood, N.N.; Kennedy, J.D.; McDonald, W.S. *J. Chem. Soc., Chem. Commun.* **1983**, 949–950, (b) Bould, J.; Crook, J.E.; Greenwood, N.N.; Kennedy, J.D. *J. Chem. Soc., Chem. Commun.* **1983**, 951–952.
59. For an alphabetic list of hundreds of acronyms used in synthetic chemistry to describe reagents, solvents, functional groups, protective groups, etc., see Daub, G.H.; Leon, A.A.; Silverman, I.R.; Daub, G.W.; Walker, S.B. *Aldrichimica Acta* **1984**, *17*, 13–23 (a publication of the Aldrich Chemical Co.). A list of acronyms that abound in catalysis studies is available (Haggin, J. *Chem. Eng. News*, 1987, March 2, pp. 9–15).
60. Schrauzer, G.N.; Bastian, B.N.; Fosselius, G.A. *J. Am. Chem. Soc.* **1966**, *88*, 4890–4894.

61. Schrauzer, G.N., private communication, January, 1981.
62. Holmes, T.J., Jr.; Lawton, R.G. *J. Am. Chem. Soc.* **1977**, *99*, 1984–1986.
63. Aldrich Chemical Co., Inc., Milwaukee, WI.
64. King, R.B. *Israel J. Chem.* **1976/77**, *15*, 181–188.
65. King, R.B., private communication, March, 1978.
66. Laidig, W.D.; Schaefer, H.F., III. *J. Am. Chem. Soc.* **1978**, *100*, 5972–5973.
67. Benn, R.; Günther, H. *Angew. Chem. Int. Ed. Engl.* **1983**, *22*, 350–380.
68. Nagayama, K.; Kumar, A.; Wüthrich, K.; Ernst, R.R. *J. Magn. Reson.* **1980**, *40*, 321–334.
69. Kumar, A.; Wagner, G. Ernst, R.R.; Wüthrich, K. *J. Am. Chem. Soc.* **1981**, *103*, 3654–3658.
70. Wüthrich, K. *Biochem. Soc. Symp.* **1981**, *46*, 17–37.
71. Nagayama, K.; Wüthrich, K.; Ernst, R.R. *Biochem. Biophys. Res. Commun.* **1979**, *90*, 305–311.
72. Morris, G.A.; Freeman, R. *J. Am. Chem. Soc.* **1979**, *101*, 760–762.
73. (a) Sorensen, O.W.; Freeman, R.; Frenkiel, T.; Mareci, T.H.; Schuck, R. *J. Magn. Reson.* **1982**, *46*, 180–184, (b) The term first appeared in an informal newsletter circulated among NMR spectroscopists, (c) Freeman, R., private communication, May, 1983.
74. Wüthrich, K., private communication, May, 1983.
75. Shaka, A.J.; Keeler, J.; Frenkiel, T.; Freeman, R. *J. Magn. Reson.* **1983**, *52*, 335–338.
76. Shaka, A.J.; Freeman, R. *J. Magn. Reson.* **1983**, *55*, 487–493.
77. Bodenhausen, G.; Freeman, R.; Turner, D.L. *J. Magn. Reson.* **1977**, *27*, 511–514.
78. Morris, G.A.; Freeman, R. *J. Magn. Reson.* **1978**, *29*, 433–462.
79. Pines, A., private communication, May, 1976.
80. Pines, A.; Gibby, M.G.; Waugh, J.S. *J. Chem. Phys.* **1972**, *56*, 1776–1777.
81. *Chem. Brit.* **1980**, *16*, 383–384.
82. Waugh, J.S.; Huber, L.M.; Haeberlen, U. *Phys. Rev. Letters* **1968**, *20*, 180–182.
83. We thank Professor R.K. Harris, University of East Anglia, England, for enlightening us about WaHuHa in private correspondence, January, 1984.
84. Davis, D.G.; Bax, A. *J. Am. Chem. Soc.* **1985**, *107*, 7197–7198.
85. Christofides, J.C.; Davies, D.B. *J. Am. Chem. Soc.* **1983**, *105*, 5099–5105.
86. (a) Hallsworth, R.S.; Nicoll, D.W.; Peternelj, J.; Pintar, M.M. *Phys. Rev. Lett.* **1977**, *39*, 1493–1496, (b) Nicoll, D.W.; Halsworth, R.S.; Peternelj, J.; Pintar, M.M. *J. Magn. Reson.* **1979**, *34*, 421–424.
87. Becker, E.D.; Ferretti, J.A.; Farrar, T.C. *J. Am. Chem. Soc.* **1969**, *91*, 7784–7785.
88. Brandt, J.; Friedrich, J.; Gasteiger, J.; Jochum, C.; Schubert, W.; Ugi, I. Computer Programs for the Deductive Solution of Chemical Problems on the Basis of a Mathematical Model of Chemistry. In *Computer-Assisted Organic Synthesis* Wipke. W.T.; Howe, W.J. (Eds.), (American Chemical Society, Washington, DC, 1977), Symposium Series No. 61, chapter 2.
89. Wipke, W.T.; Braun, H.; Smith, G.; Choplin, F.; Sieber, W. Simulation and Evaluation of Chemical Synthesis: Strategy and Planning. In *Computer-Assisted Organic Synthesis*, Wipke, W.T.; Howe, W.J. (Eds.), (American Chemical Society, Washington, DC, 1977), Symposium Series No. 61, chapter 5.
90. Wipke, W.T., private communication, June, 1983.
91. *Chem. Eng. News* **1956**, *34*, p. 774.
92. A trademark of The Salinon Corporation, 7430 Greenville Avenue, Dallas, TX, 75231. We are grateful to Michael L. Carr of that organization for promotional information about their product.
93. Ōsawa, E., private communication, November, 1975.
94. Silberrad, O. *J. Chem. Soc.* **1922**, *121*, 1015–1022.
95. Ballester, M.; Molinet, C.; Castañer, J. *J. Am. Chem. Soc.* **1960**, *82*, 4254–4258.
96. Ballester, M.; Molinet, C. *Chem. Ind. (London)* **1954**, 1290.
97. (a) Ballester, M.; Riera, J.; Castañer, J.; Badía, C.; Monsó, J.M. *J. Am. Chem. Soc.* **1971**, *93*, 2215–2225, (b) Ballester, M.; Castañer, J.; Pujadas, J. *Tetrahedron Lett.* **1971**, 1699–1702, (c) Ballester, M.; Castañer, J.; Riera, J.; Ibáñez, A.; Pujadas, J. *J. Org. Chem.* **1982**, *47*, 259–264, (d) Ballester, M.; Riera, J.; Castañer, J.; Rodríguez, A.; Rovira, C.; Veciena, J. *J. Org. Chem.* **1982**, *47*, 4498–4505, (e) Ballester, M.; Riera, J.; Castañer, J.; Rovira, C.; Veciana, J.; Onrubia, C. *J. Org. Chem.* **1983**, *48*, 3716–3720, (f) Ballester, M. *Acc. Chem. Res.* **1985**, *18*, 380–387.
98. (a) Ballester, M. *Bull. Soc. Chim. Fr.* **1966**, 7–15, (b) Ballester, M.; Castañer, J.; Riera, J. *An.*

Quim. **1977**, *73*, 546–556, (c) Ballester, M.; Castañer, J.; Riera, J.; Parés, J. *An. Quím.* **1980**, *76C*, 157–170.

99. Fieser, L.F.; Fieser, M. *Reagents for Organic Synthesis* (John Wiley & Sons, New York, 1967), p. 1131.
100. Ballester, M., personal discussion, November, 1982.
101. Rassat, A., private communication, May, 1984.
102. Reese, K.M. *Chem. Eng. News* **1973**, Oct. 8, p. 60.
103. Schell, F.M., private communications, January, 1976 and March, 1984.
104. Wieringa, J.H.; Strating, J.; Wynberg, H. *Tetrahedron Lett.* **1970**, 4579–4582.

Chapter 15

MOLECULES HAVE HOME TOWNS, TOO!

What do these have in common: "Tennessee Ernie" Ford (country and western singer); the "Georgia Peach" (baseball player Ty Cobb); "Minnesota Fats" (pool player); and Seattle Slew (racehorse)? Each was named (or nicknamed) in honor of a home town or state in America. Many chemical compounds in nature have also acquired names that reflect their source; that is, the plant or animal in which they are found. Let us look at but a small sampling.

A great many of us have taken a penicillin drug to cure an ailment. This class of antibiotics has the general structure $\mathbf{1}$,[1] and a common one is penicillin G, in which R $= C_6H_5CH_2$. These antibiotics are isolated from molds of the *Penicillium* family, so called because the molds form chains of spores arranged in a brush-like head (Latin *penicillum* means "a small brush"). Alexander Fleming* observed the antibacterial

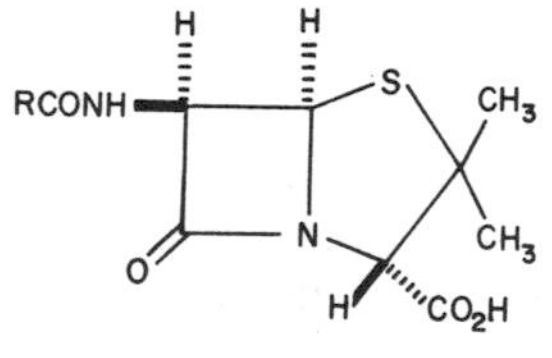

1

action of *Penicillium notatum* molds in his laboratory at St. Mary's Hospital, London, in 1928 and published his landmark paper in 1929.[2] He has received well-deserved acclaim for this most important work, including knighthood and a Nobel Prize. Furthermore, the fiftieth anniversary of his achievement was heralded philatelically by the British Commonwealth nation of Mauritius in 1978 (see **2**); and Hungary honored Sir Alexander on a stamp in 1981, the centenary of his birthdate (**3**).[3] A fascinating, little-known fact is that a 21-year-old French medical student, Ernest Duchesne, had noticed (and described in a dissertation at the Army Medical Academy in Lyons) the bactericidal effect of *Penicillium glaucum* molds back in 1896! The discovery went unnoticed, and his thesis gathered dust until a librarian accidentally came across it a

---

*Nobel Prize in physiology/medicine, 1945, shared by Sir Alexander Fleming, Howard W. Florey, and Ernst B. Chain.

half-century later.[4] What a pity the likes of *Dissertation Abstracts* didn't exist in those days. *C'est la vie!*

In fact, John Sanderson (1870), Joseph Lister (1871), William Roberts (1874), and John Tyndall (1876) recorded that *Penicillium* molds prevent bacterial growth even before Dr. Duchesne did. But these pre-Fleming scientists may not have been seeing the effect of a penicillin, since the molds also secrete other bacteriostatic agents.[5] Incidentally, Duschesne became an army doctor and died of illness in 1912 at the age of 38. However, his contribution to bacteriology was not forgotten entirely. In 1974 the principality of Monaco issued a stamp (**4**) to commemorate the centennial of his birth.[3]

3

2

4

The saga of penicillin, from original discovery to commercial development as a miracle drug, unfolded many memorable events—including some that may be apocryphal. One of these appeared in the British press[6] and has been recounted elsewhere.[7]

The story goes that Alexander Fleming, as a farm lad in Scotland, had saved from drowning in a lake a London youth who was there on holiday. The city teenager eventually became a noted politician, and during an important trip to the Middle East he became gravely ill with pneumonia. A preparation of penicillin was flown to his bedside, and within hours the amazing drug had performed its life-saving mission. For the second time Alexander Fleming had saved the life of Winston Churchill, the youngster who had gone swimming in that lake so many years ago!

According to one subsequent biography, Sir Alexander and Lady Fleming had a good laugh over that story, especially as he would have been just seven at the time of the "drowning" incident.[6] Furthermore, it seems that the medication sped to save Churchill's life had not been penicillin but a sulfapyridine, known then by the code name M&B 693. The letters stood for May & Baker, manufacturer of the drug, but Churchill jokingly called it Moran and Bedford after the names of his two doctors.[8a] Be that as it may, Alexander Fleming and Winston Churchill did know each other personally, and the Prime Minister had used penicillin ointments to curb eye infections.[8b]

Now let us switch scene to the University of Virginia where cheerleaders may have stimulated a consortium of chemists to call **5** "megaphone."[9]* At first, the name raised editorial eyebrows at the *Journal of Organic Chemistry*; but it was accepted for publication.[11] After all, this cytotoxic neolignan is a ketone and comes from the roots of

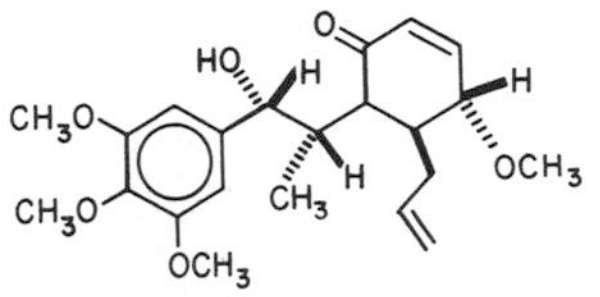

5 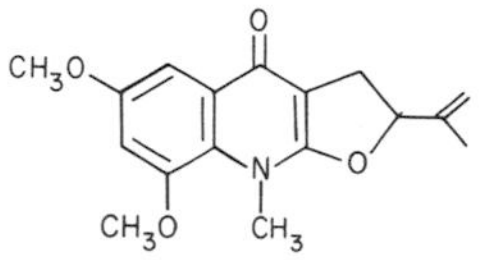   6

the plant *Aniba megaphylla* Mez.[9] George Büchi's band at MIT developed a synthesis that is truly sound.[12] Another acoustic coup bellowed from Northern Ireland's New University of Ulster, where Gaston and Grundon announced the first synthesis of "ptelefolone" (**6**).[13] We suppose you could then dial Professor Grundon direct to learn that this ketonic alkaloid comes from the hop tree, *Ptelea trifoliata* (L.).[14]

All of us are aware that some chemical reactions can explode. But did you know that even simple mixing of reagents can produce a snap, crackle, or pop? When Jan Reedijk, then at Delft University, and co-workers Joop van Ooijen and Eric van Tooren combined zinc chloride with pyrazine, a 1:1 complex precipitated, emitting a rather strong cracking sound.[15] They reproduced their "*krakende Kristallen*" for colleagues within earshot, and thereby confirmed that Dutch chemists get a bang out of their work.[16] You can join the experience by tuning in to an extensive collection of audible reactions compiled in 1981.[17]

But if you prefer not to listen to such things, try your hand doing what peptide people are very good at. They quietly develop protective groups for amino acids and get their kicks by penning noisy-sounding names. For example, L-histidine safeguarded on the ring nitrogen by a benzyloxymethyl unit and on the side-chain nitrogen by a tertiary-butoxycarbonyl group (**7**) becomes "Boc-His(πBom)-OH."[18] One way or another, it seems that chemists will be heard.

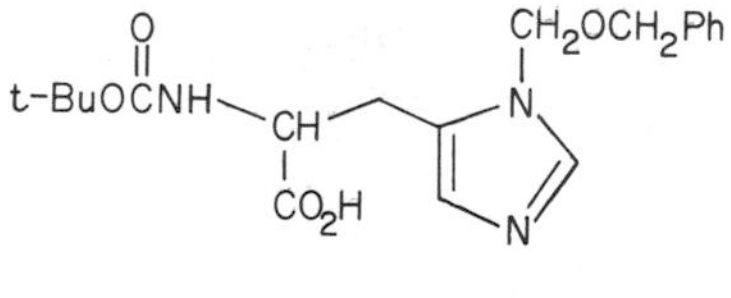

7

At the University of California, Berkeley, K. Peter Vollhardt and Timothy Weidman found a noiseless way to draw a crowd. They synthesized the thermally labile bisruthenium compound **8** and showed that it isomerized cleanly to **9** at room temperature in tetrahydrofuran. The same rearrangement took place sharply when they heated **8** in the crystalline state to 208°C.[19] Professor Vollhardt informed us that the sudden and drastic change in geometry that accompanies the solid-phase **8** → **9** convolution causes the crystals to leap inches into the air.[20] After landing on all fours, the Berkeley jumping beans need only be dissolved in tetrahydrofuran and exposed to

---

*In accord with SI numerology, 1 megaphone = 10^{12} microphones.[10]

the California sun. The light converts yellow **9** back to colorless **8**, ready to perform again.[19]

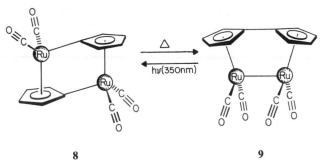

In 1962, a Czechoslovakian trio isolated an alkene from rhizomes of the plant *Petasites albus* (L.) Gaertn and pertly named it "albene."[21] Ten years later, Frantisek Sorm and co-workers assigned structure **10** to albene.[22] In 1978, Kreiser and Janitschke at the Technical University of Braunschweig synthesized **10** but found that

their compound differed in properties from natural albene.[23] They proved that albene has in fact the stereoisomeric structure **11**; so their synthetic product, **10**, became "isoalbene." The same pair of workers clinched the matter with the first total synthesis of (−)-albene in 1979.[24] Routes to the racemic olefin soon followed,[25] as did assignment of absolute configuration to the natural product.[26]

A research squad at Belgium's Université Libre de Bruxelles gave the title "propyleine" to **12**, even though there is not a propyl group in sight.[27] This onomastic oddity becomes less puzzling when you consider that alkaloid **12** was extracted from ladybird beetles (commonly called ladybugs) of the species *Propylaea quatuordecimpunctata* (L.). Evidently, one must "*cherchez la femme*" even in beetle land.

Richard Mueller and his student, Mark Thompson, at Yale University discovered that the structural situation is more complex. On the basis of total synthesis and other evidence, they believe the substance isolated from the ladybug is not **12** but actually a 25:75 equilibrium mixture of **12** and **14** ("isopropyleine").[28] These two go back and forth easily, presumably by facile protonation to a nitrogen-stabilized carbocation (**13**).

Frank Stermitz's team at Colorado State University obtained an alkaloid from the Puerto Rican tree *Zanthoxylum punctatum* and proved that its formula is **15**.[29] This tree is sometimes called "alfiler"; it has sharp spines on its branches, and *alfiler* means "pin" in Spanish. Hence, Dr. Stermitz's group named **15** "alfileramine."[29] The tree is also known in Puerto Rico as the "toothache tree." It seems that chewing its bark numbs the mouth and alleviates toothaches. We wonder if the structure proof of **15** was painless.

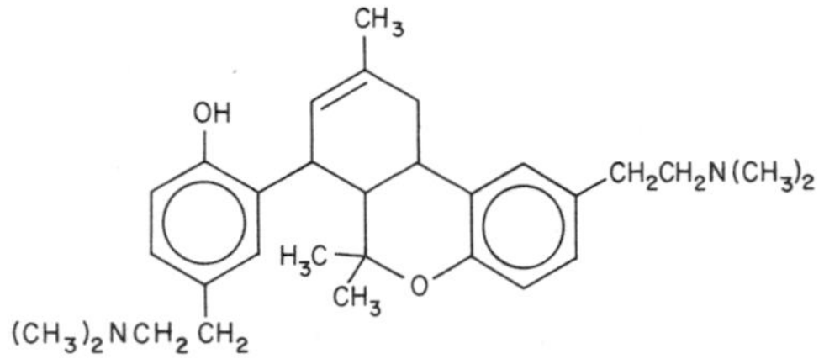

15

The dipolar ion **16** is known as "betaine." You might think it was named for a fraternity or sorority. Not so. The first four letters have nothing to do with the Greek letter beta, but represent the Latin word for "beet." Consequently, the word is pronounced "beeta-een"; and the compound is found in sugar beets, or *Beta vulgaris*.[30]

Incidentally, "betalains" are colored plant alkaloids with carbohydrate residues.[31] Red beets owe their pleasing hue to a group of red–violet betalains known as "betacyanins" (e.g., "beet blue," **17**).[32] Note the betaine-like dipolar moiety in **17**. Betacyanines can differ in the nature of the sugar part or in the stereochemistry of a carboxyl unit.[32]

By the way, "betylates" and "norbetylates" are sulfonic esters of the type $ROSO_2(CH_2)_nN(CH_3)_3^+X^-$ and $ROSO_2(CH_2)_nNH(CH_3)_2^+X^-$, respectively, which can be prepared from ROH.[33,34] At the University of Western Ontario, King and his subjects demonstrated that a wide spectrum of nucleophiles (Z) attack these sulfonates readily at R; so [n]betylates and [n]norbetylates allow S_N2 conversion of primary and secondary ROH to RZ.[35] The by-product is a betaine (e.g., $^-O_3S(CH_2)_n\overset{+}{N}(CH_3)_3$),

$(CH_3)_3\overset{+}{N}CH_2CO_2^-$

16

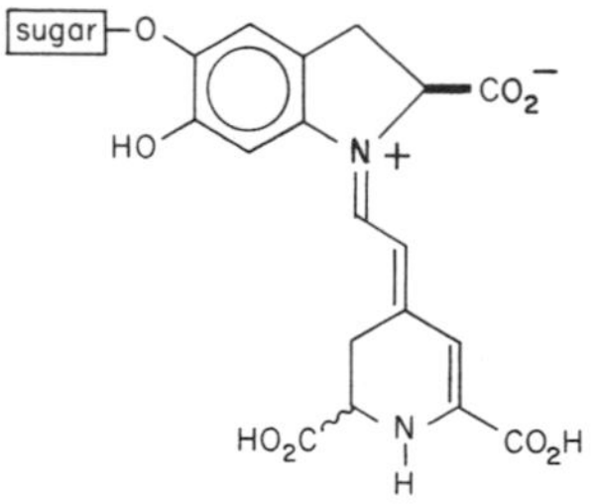

17

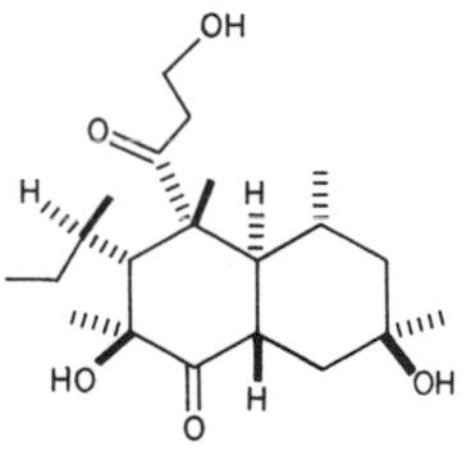

$(CH_3)_2\overset{+}{\underset{H}{N}}-CH_2-CO_2^-$

18

19

which prompted the term "betylate."[33] Try out Professor King's methodology to see if you support the slogan, "You can't beat a betylate."[36] And, while beating you numb with nomenclature, let us point out that "betaenones" (e.g., betaenone B, **18**) are phytotoxins from *Phoma betae* (Fr.), the fungus responsible for leaf spot disease in sugar beets.[37] One small reminder regarding casual terminology of dipolar ions. When nonadjacent* plus and minus charges result from a *proton* transfer (e.g., **19**), call it a zwitterion. Otherwise its a betaine—no matter how you say it.[39]

The pronunciation of chemical words can perplex students and even professionals. Of course, we can expect distinctive vocalizations from different languages. But inconsistencies flourish even within a monolingual country. For example, in the United States organic chemists do not pronounce chlorides as chlorids, or sulfides as sulfids. Yet for amides they utter amid; and ylide comes across as ylid.[40] Perhaps the German spellings of these words simply exerted a greater influence in some cases than in others.

John Galt and Jack Stocker assured the community that an ACS subcommittee, in cooperation with IUPAC, was studying this issue and would publish a rational guide to pronunciation.[41] Among its other recommendations, this body would propose that root soundings be retained in contracted expressions. Thus "chemotherapy" (coined by Paul Ehrlich in 1909)[42] stands for "chemical therapy" and should be sounded with a short "e" (as in "chemical") and not as "keemo-therapy," a frequent variant. (Maybe "hemoglobin" with a long "e" helped popularize that latter usage.) By the way, Dr. Galt agreed with us unofficially on how to pronounce betaine.[43]

But we wonder if the pronunciation logic for betaine should be extended to some other amines. For example, the alkaloid "cocaine" occurs in the leaves of several South American shrubs of the genus *Erythroxylon* (e.g., *E. coca*), and its name was formed from "coca + ine." Despite its trisyllabic beginning however, the word cocaine most commonly is rendered in two syllables, as "ko-kane." So much for consistency.

In the naming process, even the ocean bottom sometimes gets on stage. A class of corals known as "gorgonians" was named after the three extremely unattractive Gorgon sisters of Greek mythology.[44] The cortex (outer covering) of these corals resembles a petrified plant; and, according to legend, anyone upon whom the Gorgon sisters looked was turned to stone (i.e., *literally* petrified).[45] Gorgonians provide a rich source of interesting organic molecules.

Leon Ciereszko's researchers at the University of Oklahoma isolated from *Pseudodoptegorgia americana* a sesquiterpene they christened "β-gorgonene."[46] Dick van der Helm's team at the same institution proved (by X-ray inspection of the silver nitrate adduct) it has structure **20**.[47] The corals are also talented enough to produce steroids containing a cyclopropane ring in the side chain. For example, Carl Djerassi's devotees at Stanford established that "gorgosterol" is represented by **21**.[48]

20   **21**

---

*With *adjacent* charges, you enter the domain of "ylides," etc.[38]

Chemists at Osaka University in 1973 cleverly combined the source, taste, and functionality of **22** when they concocted the euonym "hypacrone."[49] It seems that **22** comes from the fern *Hypolepsis punctata*, has an *ac*rid (bitter) taste, and, of course, it is a di*one*. Another Japanese contingent (in joint research from Fukui and Hokkaido Universities) synthesized the molecule in 1981.[50] Their elegant work will surely not leave a bitter taste in your mouth.

We now meet quinine (**23**), an important antimalarial drug. It was also named after its source—the bark of the cinchona tree (*quina* is Spanish spelling for the Peruvian word *kina* for "bark"). Furthermore, the tree itself was named after a *person*. In 1639, Countess Anna del Chinchon, wife of the Spanish Viceroy of Peru, contracted the

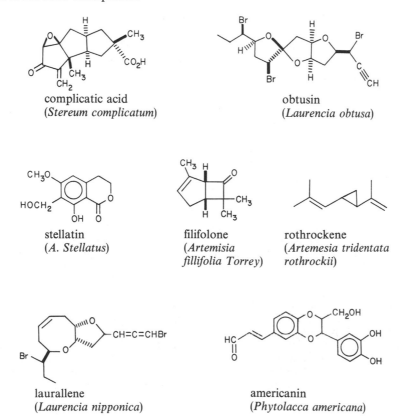

22   **23**

deadly disease, malaria. Jesuit missionaries had learned from natives that the bark of certain trees indigenous to Peru could reduce the fever and, perhaps, cure the patient. It worked, and the countess's surname (misspelled during translation) gave rise to "cinchona" for this family of trees.[51]

Finally, we list without comment a few more natural products with engaging names derived from their birthplaces.[52-60]

complicatic acid
(*Stereum complicatum*)

obtusin
(*Laurencia obtusa*)

stellatin
(*A. Stellatus*)

filifolone
(*Artemisia
fillifolia Torrey*)

rothrockene
(*Artemesia tridentata
rothrockii*)

laurallene
(*Laurencia nipponica*)

americanin
(*Phytolacca americana*)

catherine
(*Catharanthus roseus*)

reflexine
(*Rauwolfia reflexa*)

But what about these birthplaces? In 1735, Swedish naturalist Carl Linnaeus decreed that every animal and plant should have a generic and a specific name, preferably in language stemming from Latin or Greek. Since then zoologists, botanists, and their ilk have been playing the name game, not infrequently with tongue-in-cheek. Science journalist Richard Conniff described some delightful cases from entomology.[61] For example, a Britisher, George Kirklady, sought legitimate names for stink bugs, squash bugs, and seed bugs. So he created a Greek-sounding ending "-chisme" (pronounced "kiss me") and proceeded to name these genera *Polychisme, Peggichisme*, and *Dolichisme*. (Do you suppose residents of Kissimmee, Florida, know of these winsome wenches?) Another scientist found a new species of wasp in the genus *Lalapa*, so he called this species *lusa* and thus immortalized *Lalapa lusa*. Harrison G. Dyar of the Smithsonian's National Museum of History was initially pleased that someone had named a new genus of moths after him. His enthusiasm waned, however, when he realized the bestower chose *Dyaria*. Arnold Menke of the U.S. Department of Agriculture's Systematic Entomology Laboratory opened a package from a colleague containing wasp specimens from Australia. Menke uttered the phrase that at once named the genus and began the title of his next publication: "*Aha*, a new genus of Australian Sphecidae." If his colleague sends him another new find, Menke planned to dub it *Ohno* and to title his paper: "*Ohno*, another new genus of Australian Sphecidae." It's nice to know that onomastic frivolity thrives elsewhere besides chemistry.

REFERENCES AND NOTES

1. Lowe, G. In *Comprehensive Organic Chemistry*, Barton, D.; Ollis, W.D. (Eds.), (Pergamon Press, Oxford, 1979), vol. 5, pp. 289–320.
2. Fleming, A. *Brit. J. Exptl. Path.* **1929**, *10*, 226.
3. (a) Gratton, R. *Philatelia Chimica* **1985**, 7, 30–33, (b) Gratton, R. private communication, August, 1985. We are grateful to Dr. Richard Gratton (Rolland Inc., Mont-Rolland, Quebec) for lending us the Mauritius, Hungary, and Monaco stamps for photography and also for copies of philatelic articles he wrote on biochemistry and chemistry in *Philatélie Québec* and in *La Philatélie au Québec*.
4. *Scient. Amer.* **1978**, *239*, Nov., pp. 90–91.
5. Abraham, E.P. *Scient. Amer.* **1981**, *244*, June, pp. 76–86.
6. Rowland, J. *The Penicillin Man: The Story of Sir Alexander Fleming* (Lutterworth Press, London, 1957), pp. 129–130.
7. Recounted by Barbara Hyde in *The Numismatist* **1985**, *98*, 913–915 (official publication of the American Numismatic Association). The father of author Hyde (married name) had been on the chemistry faculty at the University of Nebraska, and he had assisted her in gathering information for the article. Hyde, B., private communication, August, 1985.
8. (a) Slinn, J. *A History of May & Baker, 1834–1984* (Hobson Ltd., Cambridge, England, 1984). We are grateful to Professor W. David Ollis for copies of relevant pages from Judy Slinn's

book, (b) Moran, C.M.W. *Churchill; Taken from the Diaries of Lord Moran* (Houghton Mifflin Co., Boston, MA, 1966).

9. Kupchan, S.M.; Stevens, K.L.; Rohlfing, E.A.; Sickles, B.R.; Sneden, A.T.; Miller, R.W.; Bryan, R.F. *J. Org. Chem.* **1978**, *43*, 586–590.

10. *Chem. Eng. News* **1983**, Oct. 31, p. 48.

11. Greene, F.D. Editor of *J. Org. Chem.*; private communication, July, 1977.

12. Büchi, G.; Chu, P.–S. *J. Am. Chem. Soc.* **1981**, *103*, 2718–2721. For another synthesis see Zoretic, P.A.; Bhakta, C.; Khan, R.H. *Tetrahedron Lett.* **1983**, *24*, 1125–1128.

13. Gaston, J.L.; Grundon, M.F. *Tetrahedron Lett.* **1978**, 2629–2632.

14. Reisch, J.; Sendrei, K.; Pápay, V.; Novák, I.; Minker, E. *Tetrahedron Lett.* **1970**, 3365–3368.

15. van Ooigen, J.A.C.; van Tooren, E.; Reedijk, J. *J. Am. Chem. Soc.* **1978**, *100*, 5569–5570.

16. Reedijk, J., private communication, August, 1982.

17. Betteridge, D.; Joslin, M.T.; Lilley, T. *Anal. Chem.* **1981**, *53*, 1064–1073.

18. Brown, T.; Jones, J.H.; Richards, J.D. *J. Chem. Soc., Perkin Trans. 1* **1982**, 1553–1561.

19. Vollhardt, K.P.C.; Weidman, T.W. *J. Am. Chem. Soc.* **1983**, *105*, 1676–1677.

20. Vollhardt, K.P.C., private communication, April, 1984.

21. Hochmannová, J.; Novotný, L.; Herout, V. *Coll. Czech. Chem. Commun.* **1962**, *27*, 2711–2713.

22. Vokác, K.; Samek, Z.; Herout, V.; Sorm, F. *Tetrahedron Lett.* **1972**, 1665–1668.

23. Kreiser, W.; Janitschke, L. *Tetrahedron Lett.* **1978**, 601–604.

24. Kreiser, W.; Janitschke, L. *Chem Ber.* **1979**, *112*, 408–422.

25. (a) Baldwin, J.E.; Barden, T.C. *J. Org. Chem.* **1981**, *46*, 2442–2445, (b) Trost, B.M.; Ranaut, P. *J. Am. Chem. Soc.* **1982**, *104*, 6668–6672.

26. Baldwin, J.E.; Barden, T.C. *J. Org. Chem.* **1983**, *48*, 625–626.

27. Tursch, B.; Daloze, D.; Hootele, C. *Chimia* **1972**, *26*, 74–75.

28. Mueller, R.H.; Thompson, M.E. *Tetrahedron Lett.* **1980**, *21*, 1097–1100.

29. Caolo, M.A.; Stermitz, F.R. *Tetrahedron* **1979**, *35*, 1487–1492.

30. Büchi, G., private communication, August, 1979.

31. Ayer, W.A., private communication, August, 1979.

32. Büchi, G.; Fliri, H.; Shapiro, R. *J. Org. Chem.* **1978**, *43*, 4765–4769.

33. King, J.F.; Loosmore, S.M.; Lock, J.D.; Aslam, M. *J. Am. Chem. Soc.* **1978**, *100*, 1637–1639.

34. King, J.F.; Lee, T.M.–L. *Can. J. Chem.* **1981**, *59*, 356–361, 362–372.

35. King, J.F.; Loosmore, S.M.; Aslam, M.; Lock, J.D.; McGarrity, M.J. *J. Am. Chem. Soc.* **1982**, *104*, 7108–7122.

36. King, J.F., private communication, April, 1984.

37. Ichihara, A.; Oikawa, H.; Hayashi, K.; Sakamura, S.; Furusaki, A.; Matsumoto, T. *J. Am. Chem. Soc.* **1983**, *105*, 2907–2908.

38. March, J. *Advanced Organic Chemistry*, 3rd ed. (John Wiley & Sons, New York, 1985), p. 35–37.

39. For suggested classification and nomenclature of heterocyclic mesomeric betaines, see Ollis, W.D.; Stanforth, S.P.; Ramsden, C.A. *Tetrahedron* **1985**, *41*, 2239–2329.

40. See a letter by S. Toby in *Chem. Eng. News* **1985**, Sept. 2, p. 4.

41. (a) See individual letters by J.F. Gall and J.H. Stocker in *Chem. Eng. News* **1985**, May 13, p. 36.

42. Koob, R.P. *Chem. Eng. News* **1985**, March 4, p. 4.

43. Gall, J.F., discussion by phone July, 1985.

44. Ciereszko, L., private communication, September, 1975.

45. Woodcock, P.G. *Dictionary of Mythology* (Philosophical Library, New York, 1953).

46. Ciereszko, L.; Sifford, D.H.; Weinheimer, A.J. *Ann. N.Y. Acad. Sci.* **1960**, *90*, 917–919.

47. Hossam, M.B.; van der Helm, D. *J. Am. Chem. Soc.* **1968**, *90*, 6607–6611.

48. Ling, N.C.; Hale, R.L.; Djerassi, C. *J. Am. Chem. Soc.* **1970**, *92*, 5281–5282.

49. Hayashi, Y.; Nishizawa, M.; Sakan, T. *Chem. Lett.* **1973**, 63–66.

50. Sakan, F.; Minanu, Y.; Shirahama, H.; Matsumoto, T. *Bull. Chem. Soc. Jpn.* **1981**, *54*, 2235.

51. Sainsbury, M. *Chem. Brit.* **1979**, *15*, 127–130.

52. Mellows, G.; Mantle, P.G.; Feline, T.C.; Williams, D.J. *Phytochemistry* **1973**, *12*, 2717–2720.

53. Chatterjee, A.; Chakrabarty, M.; Ghosh, A.K.; Hagaman, E.W.; Wenkert, E. *Tetrahedron Lett.* **1978**, 3879–3882.

54. Simpson, T.J. *J. Chem. Soc., Chem. Commun.* **1978**, 627–628.

55. Woo, W.S.; Kang, S.S.; Wagner, H.; Chari, V.M. *Tetrahedron Lett.* **1978**, 3239–3242.

56. Fukuzawa, A.; Kurosawa, E. *Tetrahedron Lett.* **1979**, 2797–2800.
57. Howard, B.M.; Fenical, W.; Arnold, E.V.; Clardy, J. *Tetrahedron Lett.* **1979**, 2841–2844.
58. Bates, R.B.; Onore, M.J.; Paknikar, S.K.; Steelink, C.; Blanchard, E.P. *Chem. Commun.* **1967**, 1037–1038.
59. Guilhem, J.; Ducruix, A.; Riche, C.; Pascard, C. *Acta Cryst.* **1976**, *B32*, 936–938.
60. Epstein, W.W.; Gaudioso, L.A. *J. Org. Chem.* **1982**, *47*, 175–176.
61. Conniff, R. *Science 82*, **1982**, June, pp. 66–67. (A former publication of the American Association for the Advancement of Science.)

Chapter 16

EVERYTHING EQUALS THE SUM OF ITS PARTS

At one time or another, all of us have taken something apart to repair or to see how it works. On reassembly, we may have had a piece or two left over, and so the item didn't function properly. Surely, a mechanical object is equal to the sum of its parts.* Molecules are no different; their constituent parts determine how they behave. Usually, functional groups govern chemical properties, and hydrophilic and hydrophobic regions regulate solubility. Realizing all this, chemists have adopted names to describe the components of molecules.

The term "triquinacene," coined for **1** at Harvard by R.B. Woodward and Tadamichi Fukunaga, who first prepared it, describes its makeup nicely.[2] The molecule consists of three (tri-) five-membered (-quin-) alkene (-ene) rings.[3] We met triquinacene in chapter 2; it can be had by heat treatment[4] of crown-like diademane, **2**, and can be

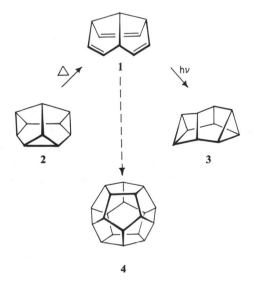

---

*A similar credo applies to research professors, according to renowned chemist and Nobel laureate Leopold Ruzicka. He held that "No professor is worth more than the sum of his co-workers."[1]

irradiated[5] to give hat-like barettane, **3**. Workers in this area need to use their heads. Several additional syntheses of **1** are on record since the original one.[4,6-10]

Professor Woodward perceived that **1** could, hypothetically, dimerize to dodeca-hedrane (**4**).[2] Alas, such a marriage was neither blessed in heaven nor consummated in the flask.[11,12] Also, some investigators have wondered whether or not the double bonds in **1** interact with one another. A molecular model suggests that hominal p-orbitals stand close and parallel enough for communion. And, with its six π-electrons, the triene might even be homoaromatic.[13] Again, alas; the π-merger is, at best, very weak.[2,8,9,14-16] The same holds true in "hexaquinacene" (**5**)—a homolog, synthesized in Paquette's laboratory, that logically extends the triquinacene concept.[17] Dr. Paquette's builders looked upon hexaquinacene as a firm foundation on which to erect dodecahedrane (**4**); four additional methine units would do the trick.[18]

A brief aside: James Cook and explorers at University of Wisconsin–Milwaukee donated the general term "polyquinenes" for fused cyclopentanoids loaded with unsaturation.[19] In joint efforts with other teams, they constructed several prototypes,[20] including **6**. Be sure not to relate such compounds to quinene[21] (a long-known transformation product of quinine) or to quinarenes (p. 220). The saturated counterpart of **6** can be classed as a polyquinane[22] and titled officially as tetracyclo[6.6.0.0^{1,5}.0^{8,12}]tetradecane. Collaborator Ulrich Weiss at the National Institutes of Health found this a mouthful, so he called it "parlerane" to save time (and taxpayers' money) in long-distance telephone discussions with Professor Cook. The name was strictly in-house and not for publication. Why parlerane? When this polyquinane is drawn as in **7**, its silhouette reminded the NIH scientist of symbol **8**, the logo used to mark works by Meister Peter von Gmünd (known in Prague, Czechoslovakia simply as Peter Parler), a leading architect and sculptor of the fourteenth century. Among other achievements, Parler had been principal architect of St. Vitus Cathedral in Prague, beloved native city of Dr. Weiss.[23] Now back to triquinacene.

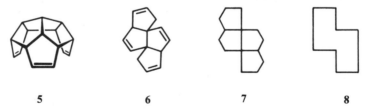

5	6	7	8

Dr. Fukunaga did not forsake this first-born after completing his studies at Harvard. In fact, at the Du Pont Central Research Laboratory he mobilized colleagues Howard Simmons, John Wendoloski, and Michael Gordon. Together they wiped out the central CH of triquinacene (see **9**) along with the three H's at the starred positions. For such tasks their employer offered the full resources of the E.I. du Pont de Nemours & Co, but in this case all it took was an eraser. The conceptualization left three nude carbons, each with a spoke-like sigma electron and an orthogonal p-electron (**10**). But, at close quarters "two is company—three is a crowd"; so the Du Pont squad thought the molecule might fare better if one electron from the middle hopped out to join the π-network. The outcome would be a 10 π-electron perimeter with the three starred atoms tethered to a core of two electrons (**11**). Such an entity could represent a new type of delocalized system, termed a "trefoil aromatic." Its character is a $[4n + 2]$ annulene and, at the hub, a three-center two-dot bond ("trefoil bond") to take the place of an atom.[24] It's just possible that Du Pont trefoils could provide "better things for better living," but not everyone may agree.[25]

During the 1960s, University of Michigan's Richard Lawton (who whetted us with nautical nomenclature, p. 75) pondered the probable geometry of **12**, which was then unknown. Naturally, the five benzene rings would like to occupy one plane to maximize neighborly melding of p-lobes. But, if each corner of the central pentagon is a normal 108°, the adjoining angles in the hexagons would have to strain to 126°. On the other hand, a saucer shape (exaggerated in **13**) could alleviate the discomfort. This intriguing geometric dilemma prompted Professor Lawton and his student, Wayne Barth, to synthesize **12**.[26,27]

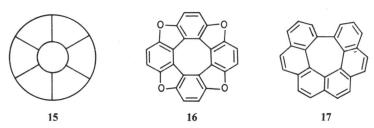

12   13   14

Seeking a name, Dr. Lawton considered "corolene," derived from "corolla" (English for "flower-like petals").[28] (Drawing **13** does look like a flower with benzene petals.) But he settled on "corannulene," from the Latin *cor* ("heart" or "within" and *annulus* ("ring"). This term emphasizes the presence of one circle within another and, as a bonus, subtlely honored Dr. Lawton's wife, Ann. (The word annulene had been introduced earlier by Franz Sondheimer (chapter 8) for rings with alternating double and single bonds.) As it turned out, X-ray crystallography of **13** confirmed a warped shape.[27] Lawton and Barth also appreciated that the dipolar resonance structure facing you in **14** has both perimeters formally aromatic; the outer one hosts 14 π-electrons, the inner one, 6. Molecular orbital calculations by Gerald Gleicher of Oregon State University suggested that **14** contributes substantially, nonplanarity notwithstanding.[29] Other theoretical treatments of corannulenes are available.[30]

Corannulene became a member of a family dubbed "circulenes" in 1975 by Hans Wynberg's group at the University of Groningen.[31,32] In $[m]$-circulene, m aromatic rings are arranged in a circle (**15**); thus, corannulene is $[5]$-circulene. As might be expected, $[6]$-circulene is flat as a pancake.[33] Its formula happens to fill the reverse side of

15   16   17

an Egyptian five pound coin struck in 1984 by the Cairo mint, but there is no chemical connotation. Rather, the honeycomb network served as a symbol to commemorate Egypt's Diamond Jubilee of Cooperation.[34] 7-Circulene, synthesized in 1983 in Japan, is saddle shaped.[35] Molecular models indicate that circulenes with $m > 7$ are corrugated; the skeleton looks like an unending wave when viewed from the side.[31] Wynberg's laboratory delivered "heterocirculenes" containing sulfur;[31] and Högberg and Erdtman provided several oxa-analogs of the type **16**.[36] An Australian squad captained by James Reiss donated a truncated [7]-circulene (**17**).[37]

The circulenes comprise but one branch of the family tree made up of ortho-fused benzenes. Walter Jenny's team in Basel, Switzerland, started to climb another limb in 1965. They coined "coronaphenes" (accent on the second syllable, derived from Latin *corona* for "crown") for molecules with aromatic rings lined *non*circularly around an empty inner space.[38] Examples would be the triangular [9]-coronaphene, **18**, and the rhomboid [10]-coronaphene, **19**.[39] This terminology does not accommodate different types of annellations. Thus "[10]-coronaphene" could just as well mean **19** or the racetrack isomer, **20**.

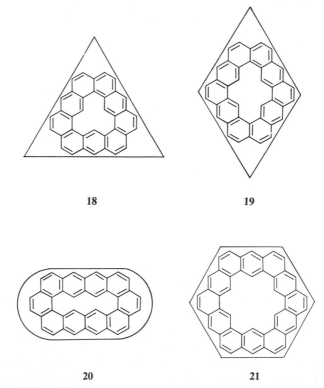

18   19

20   21

At the Max-Planck-Institute in Heidelberg, Heinz Staab's school appreciated the limitations of coronaphene terminology[40] and proposed for such polycycles an alternative name, "cycloarenes." Their definition: "polycyclic aromatic compound in which, by a combination of angular and linear annellations of benzene units, fully annellated macrocyclic systems are present enclosing a cavity into which carbon–hydrogen bonds point."[41] Kekulene (**21**), which we met in chapter 10, happened to be the first bottled representative of this family; and **19** was the second.[42] Incidentally, cycloarenes have masses that are multiples of 50 (viz., 450 for **18**, 500 for **19** and **20**, and 600 for **21**). The Heidelberg ringsters came up with a simple lettering system to distinguish isomers. Their notation tags **19** as cyclo[d.e.d.e.e.d.e.d.e.e]decakisbenzene and **20** as cyclo[d.d.e.e.e.d.d.e.e.e]decakisbenzene. But we suspect kekulene (**21**) will not be overjoyed to learn that its alias is cyclo[d.e.d.e.d.e.d.e.d.e.d.e]dodecakisbenzene.

Our ortho-fused brood proliferates yet in other directions. These progeny are more deeply rooted in history, so we present them only briefly and refer you to Eric Clar's excellent compendium for details.[43] In "acenes" the rings travel out linearly (**22**); in "phenes" they take a 60° turn (**23**); and in "starphenes" they branch as in **24**.[44] Tetracene (**25**) and pentaphene (**26**) exemplify the first two types; and we have already gazed at starphene (p. 94).

Some ortho-condensed structures, like "zethrene" (**27**) and "triangulene" (**28**), are not members of these clans. Earlier we befriended zethrene (p. 111), in which two 120° turns produce a Z shape. The other outcast, "triangulene," has long intrigued chemists and theoreticians. In 1941, Professor Clar pointed out that this unsaturated ring system has no Kekulé form and should be unstable.[45] And Longuet–Higgins predicted from theory that the ground state of such an "alternant hydrocarbon" (i.e., all atoms in a plane; all carbons sp^2; no odd-membered rings) should exhibit paramagnetism and behave as a radical.[46] Abortive attempts to prepare and isolate triangulene underscored its instability.[47] However, its dianion is accessible and has been characterized spectroscopically.[48]

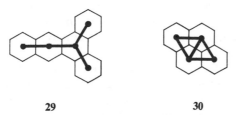

By the way, you will occasionally meet "catacondensed" (or "cataannellated") and "pericondensed" in literature about polycyclic aromatics.[43] These terms, whose usage is not always uniform, arise from Greek *kata* (cata) and *peri*, meaning "down" and "around," respectively. For catacondensed polybenzenoids (e.g., **25**, **26**), no carbon belongs to more than two rings. In contrast, pericondensed systems (e.g., **27**, **28**) contain one or more atoms common to three rings. You can also call upon dual graph theory to distinguish cata and peri. Simply join the centers of all neighboring hexagons to see whether your resulting graph is "tree" (e.g., **29**) or "nontree" (e.g., **30**), respectively.[49]

How about a loop of rings with a filled inner space (in contrast to the emptiness in circulenes)? If benzene units completely surround a smaller system with an established name, the prefix "circo-" or "circum-" comes in handy. For example, 31 is known as "circumnaphthalene."[43] It also goes by the name "ovalene" because of its oval shape.[43] But, "ellipticine," penned for natural product 32 by Goodwin, Smith, and Horning at the National Institutes of Health,[50] has naught to do with molecular contour; the name reflects its plant source, *Ochrosia elliptica*.[51]

31   32

What about ortho-annealed olefinic cycles that are not benzenoid? When two such loops are identical, we attach the traditional "-alene" ending of naphthalene to a prefix that denotes ring size. For example, 33, 34 and 35 are "pentalene,"[52] "heptalene,"[53] and "octalene,"[54] respectively. Octalene first saw the light of day in Emanuel Vogel's laboratory at the University of Cologne.[54] Collaboration with Jean Oth and his team from the University of Zürich established that the double bonds stay fixed as in 35 and not as in 36.[55] The ^{13}C NMR spectrum at room temperature has 7 signals (14 at −150°C because of conformational considerations[55]); this spectral result fits the symmetry expected for 35 but not for 36.

33   34   35   36

37   38   39

By the way, don't confuse pentalene with "pentalenene" (37), a natural sesquiterpene with a triquinane torso.[56] Leo Paquette and Gary Annis at Ohio State University showed chemists how to make it;[57] and Brown University's David Cane and Ann Marie Tillman found out how a cell-free extract of *Streptomyces* likes to get it.[58] Finally, "trialenes" are *not* derivatives of 38 but explosive mixtures of TNT (trinitrotoluene), hexogen (1, 3, 5-trinitro-1, 3, 5-triazinane, 39), and aluminum powder used in World War II bombs and torpedoes.[59] (Actually, with all its strain, 38 might well burst also.)

Helium, neon, argon, krypton, xenon, and radon are called noble gases since, like some noblemen, they remain aloof from other elements. Their atoms deign to fraternize only with a select clientele; and they completely refuse to bond one another. In some respects, benzenoid rings mimic these snobbish inorganic elements. The molecular bonding orbitals of an aromatic ménage are satiated, and the structure gains stability

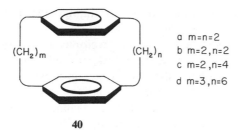

a m=n=2
b m=2, n=2
c m=2, n=4
d m=3, n=6

40

by resonance. So, such networks have little inclination to bother others. Donald Cram and his students at the University of California at Los Angeles decided, in the late 1940s, to force two of these aloof benzene units to meet face-to-face. They set out to synthesize compounds of type **40**. While their work was steaming along, C.J. Brown and A.C. Farthing of Imperial Chemical Industries, Manchester, England, isolated (in 1949) a small amount of **40a** from the polymerization of p-xylene.[60] The Cram crowd reported its own directed syntheses of **40a–d**, in 1951.[61]

The British team dubbed **40a** "di-p-xylylene,"[60] whereas the Americans introduced, for this breed of compounds, the expression "paracyclophanes" (para + cycle + phenyl, made euphonious[62]). Since 1951, the terminology grew to include metacyclophanes (e.g., compound **41**). Analogs with more than two "bridges" between rings cannot be described simply with para- and meta- prefixes. So, nowadays, numerals in *brackets* denote the number of methylenes in the bridges, whereas digits in *parentheses* identify positions of attachment to the aromatic moieties. For example, **42** is [2, 2, 2](1, 2, 4)(1, 3, 5)cyclophane.[63] Cyclophanes also include analogs like **43** and **44**, in which a single benzenoid skeleton is part of a larger cycle.[64]

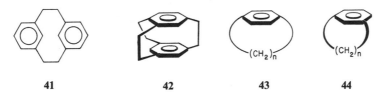

41   **42**   **43**   **44**

Incidentally, "cyclophanes" strictly implies analogs with benzene rings. For nonbenzene cases, Smith[65] suggested replacing cyclo- by a descriptive prefix as in pyridinophane, thiophenophane, naphthalenophane, and purinophane.[66] Vögtle recommended the general class name "phanes" to embrace all such bridged aromatic structures.[67]

Confrontation of two benzene faces affects the molecule in interesting ways. Short bridges, as in **40a**, act like strong springs that force the plates to buckle. X-ray crystallography showed that each hexagon has two 11° bends (see **45**),[60,61] which contort the p-overlap in each circuit. Furthermore, nearness of the rings allows π-orbitals of one to perturb those of the other and plays havoc with the ultraviolet absorption. Whereas **40c** and **40d** exhibit normal spectra (i.e., similar to that of the model compound **46**), the pressed decks in analogs **40a** and **40b** change the UV absorption drastically.[61]

Let's climb out on some boughs of the cyclophane tree that have grown since those pioneering days. These include the mixed paracyclotropolonaphane **47**[68] as well as an ansa-type tropolonaphane **48**.[69] Many metacyclophanes have been reported.[64] Synthesis of **49**[70] shows that in a monobenzene as few as five carbons can span two meta positions. And, with some agony, the same chain length can even connect two para carbons; [5]paracyclophane (**50**) shows up on irradiation of its Dewar benzene analog at −60°C. This strained phane stays around if it keeps its cool.[71] [2,2]Metacyclophane (**41**), which was first synthesized in 1899,[72] has the *anti*, or

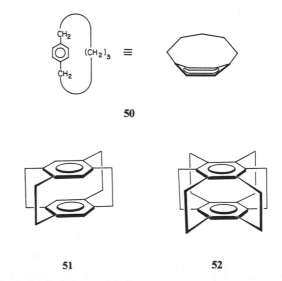

47   48   49

"ladder," conformation.[73] The *syn*, or "butterfly," structure[74] of **41** has been netted, but unless kept cold it reshapes handily to the ladder.[75]

Multi*layered* cyclophanes have been fashioned (e.g. "chochins," p. 101) in which alkane bridges stack more than two benzene units above one another.[76] Multi*bridged* cyclophanes (such as **42** and **51**) remind us of spiders on a mirror.[77,78] At the University

50

51   **52**

of Oregon, Virgil Boekelheide took this concept all the way by synthesizing **52** in 1979.[79,80] This structure had brewed in Professor Boekelheide's mind in 1974–1975, during a sabbatical leave at the University of Karlsruhe. He discussed the molecule with Henning Hopf (then at Karlsruhe), who immediately exclaimed: "Aha; superphane!"[81] The rest is history; "superphane" it became.[79]

Dr. Boekelheide's researchers found that superphane does some unearthly things.[80] For example, carbenoid assault on one of the unsaturated sites, by action of ethyl diazoacetate/Cu(II), produced **54**. (Only the reacting ring is shown.) Reduction to the primary alcohol and treatment with BF_3 gave (after a hydride shift) tropylium-like ion **56**. A trace of water reconverted this ion to superphane! The Boekelheide crew formulated this reincarnation as proceeding via alcohol **55**, valence tautomerization to

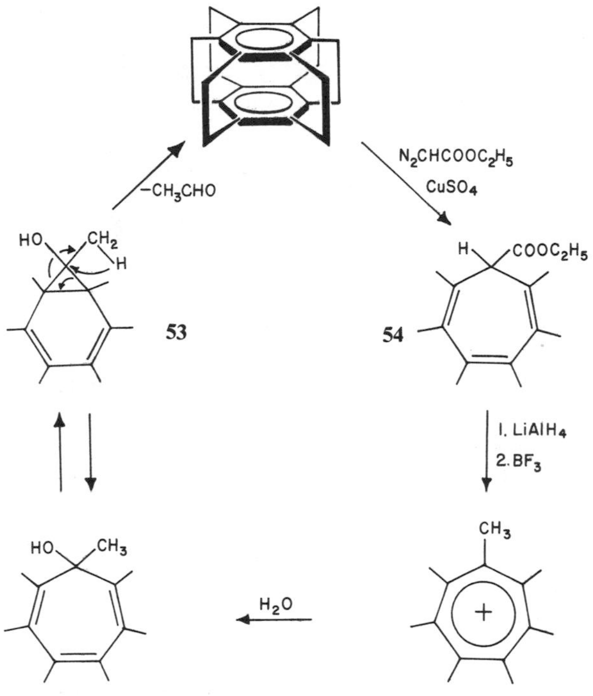

53

54

N$_2$CHCOOC$_2$H$_5$
CuSO$_4$

−CH$_3$CHO

HO CH$_2$ H

H COOC$_2$H$_5$

1. LiAlH$_4$
2. BF$_3$

HO CH$_3$

H$_2$O

CH$_3$
(+)

55

56

cyclopropyl alcohol **53**, and then bond reorganization. Sometimes we progress in chemistry by ending up where we start.

In 1985 the foundry at Oregon forged yet another intriguing cyclophane, namely **58**. It was a product from gas-phase pyrolysis of the benzodicyclobutene **57**. The three benzene rings in this highly symmetrical prototype are held rigid and closely face to face. An end-view silhouette reminded Boekelheide and co-workers of the Greek capital letter delta (Δ), so they christened this hydrocarbon "deltaphane."[82] The compound is pale yellow, so inner π clouds may interact with one another. Deltaphane

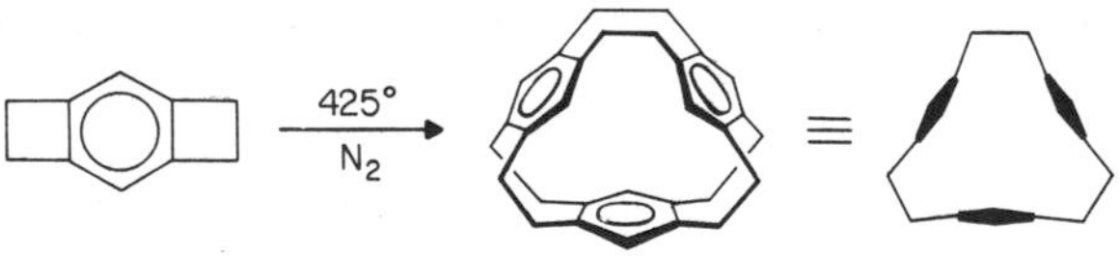

425°
N$_2$

≡

57

58

and silver triflate formed a 1:1 complex, in which the silver ion positions itself *outside* the cavity entrance. The Oregon researchers suggested that the next higher homolog of **58** (i.e., four peripheral benzene rings) might be cavernous enough to admit a cation or even a slender, rod-like molecule capable of charge-transfer complexation to the arene inner lining (see π-cryptation, p. 239).

At the Technical University in Braunschweig, Germany, Henning Hopf and Ph.D. candidate Manfred Psiorz took a synthetic stroll through clover and came up with **61**. This beautiful hydrocarbon melts above 360°C, and they made it in five steps from [2, 2]paracyclophane (**59**) via the presumed transient cyclophyne **60**.[83] Student Psiorz knew from his biology class that plants of the genus *Trifolium* have three leaflets, so he coined "trifoliaphane" for **61**. Incidentally, Professor Hopf liked to challenge his students to devise syntheses for unusual alkanes and phanes, and in lighter moments he

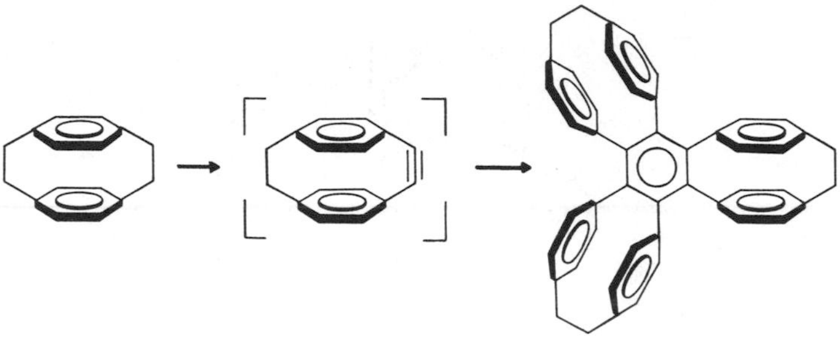

59   60   61

even proferred hypothetical targets such as insane, cellophane, and the ubiquitous acetylene, asinyne.[84]

Phanes and more phanes! At Harvard, William Roberts and Gil Shoham construc-ted the dithiatriquinacenophane **62**.[85] They established the structure by X-ray analysis. Teddy Traylor's tacticians at the University of California, San Diego, prepared **63**, in

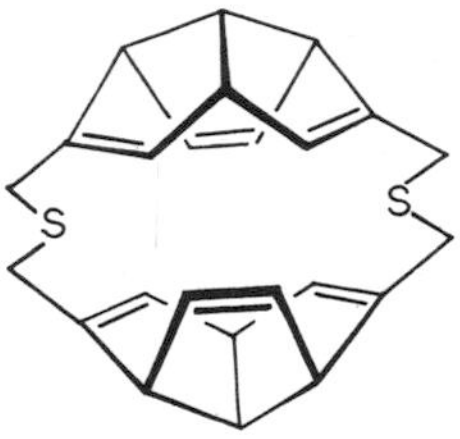

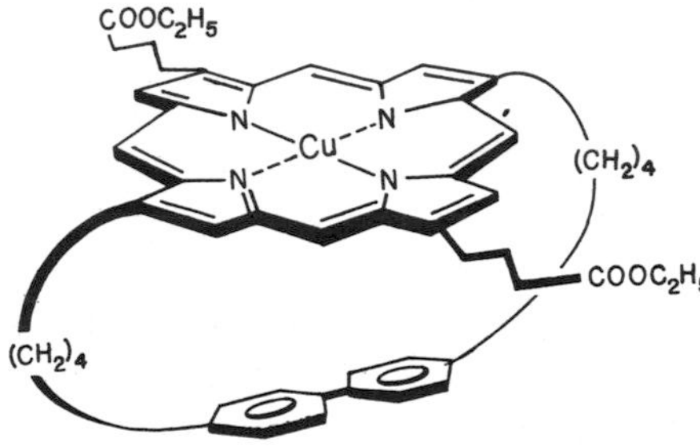

62   63

which one of the aromatic arrays is a porphyrin.[86] Philip Eaton's students contributed perhydro $[0,0](1,4)$-cyclophane (**64**),[87] which looks like biplanene (p. 7) with wings clipped. Lucio Randaccio's X-ray team at the University of Trieste elucidated precise geometries of some heterophanes containing nitrogen, oxygen, and sulfur.[88] Fritz Vögtle at the University of Bonn chained the bridgeheads of triptycene to produce "triptycenophanes" **65**, which are also paddlanes (chapter 3).[89] In Tokyo, Masao Hisatome's troupe at the Science University has forged gigantic ferrocenoporphyrino-phanes.[90] And at the University of Bielefeld, Hans–Fr. Grützmacher's ward may have delighted new parents by delivering "cyclophane-annelated cyclophanes." The West German chemists christened these progeny "twin-cyclophanes" (in German, "Zwillings-cyclophane") because of the siamese-like fusion of two subunits.[91] James

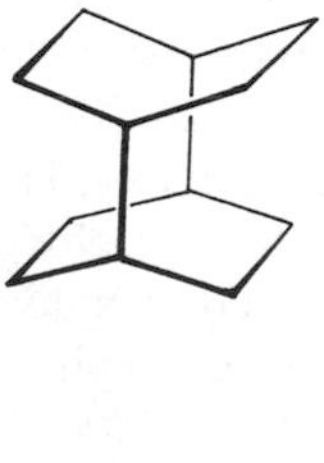

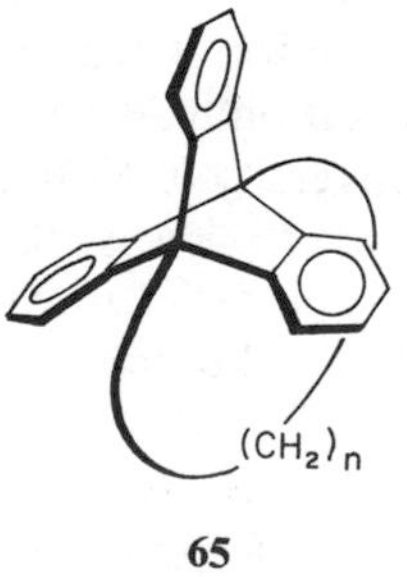

64   65

Collman's crew at Stanford brought two whole porphyrin moieties face to face.[92] And Heinz Staab's strategists at Heidelberg's Max-Planck-Institute opposed a benzenoid ring with a flavin in their "flavinocyclophane" **66**.[93]

Gourmets can select phane sandwiches with different fillings, courtesy of organometallic chefs Ulrich Zenneck at the University of Marburg (**67**)[94] and Koji Yamakawa at the University of Tokyo (**68**).[95] And to order the superphane of the sandwich set you should try the Tokyo outlet, where they obligingly carried the idea of **68** to its ultimate by synthesizing **69** in 1982.[96] Here, the iron is neatly and completely wrapped; so its creators referred to **69** as a "package compound," presumably ready to go. But, examine it carefully at home and you will discover that **69** is not what you think—one bridge is shorter than the other four. No problem. The Tokyo affiliates guarantee 100% satisfaction. So, in 1986, you could exchange **69** for its completely symmetrical counterpart. Not surprisingly, managers Hisatome, Watanabe, Yamakawa, and Iitaka advertized this superphane of the ferrocene crowd as "superferrocenophane."[97] For more about the captivating world of cyclophanes, read the authoritative monographs edited by Keehn and Rosenfeld.[98]

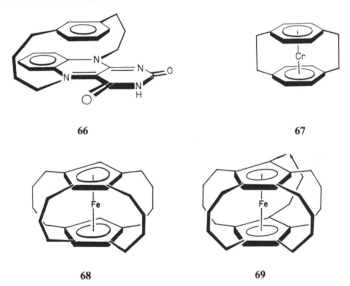

| 66 | 67 |

| 68 | 69 |

About a decade before his work on janusene (p. 141), Stanley Cristol (with student Robert Snell) was the first to synthesize a ring system like the one in **70** and **71**. Simple irradiation of **72** to **73** did the job.[99,100] Dr. Cristol sought for parent skeleton **70** a title that emphasized its four rings. (The number of rings in any polycycle equals the *minimum* number of cuts that would open it to a noncyclic structure.[101] For example, in **70** we may sever the four bonds of the four-membered ring.) He first suggested "nortetracyclene,"[99] because **71** had been dubbed "tetracyclene" in 1941 by Maria Lipp, who tried, unsuccessfully, to prepare it.[102] (See p. 147 for a discussion of the prefix "nor-.") But, to avoid muddle with tetracycline antibiotics and with tetracyclone

| 70 | 71 |

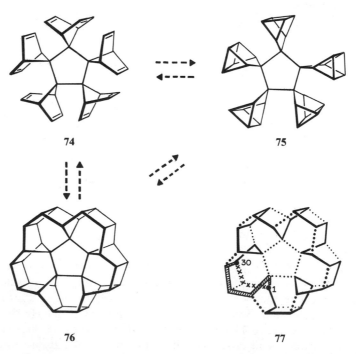

72   73

(*tetra*phenyl*cyclo*pentadien*one*, p. 219) Professor Cristol later adopted "quadricy-clene"[100] (modified to quadricyclane by some investigators).[103]

At the University of Freiburg, Horst Prinzbach's personnel conceived ways to pit the intracyclic norbornadiene → quadricyclene change against an intercyclic counterpart. For example, norbornadienylidene units surrounding a ring, as in the pentanorbor-nadienylidene **74**, might isomerize in stages to the fivefold quadricyclanylidene **75**. Of special interest here is whether alternative, transannular [2 + 2] closures could transform these analogs to still other isomers, such as **76**.[104] Professor Prinzbach and coHorsts refer to such interrelated polycycles as "spirocoganes,"[105] and, with simpler model compounds, they sought ways to control the intracyclic *vs.* intercyclic competition.

If you like the challenge of naming complex structures à la IUPAC rules,[106] try your hand at monster **76**.

74   75

76   77

A missive from Dr. Prinzbach cited its official handle as heneicosa-cyclo$[28.4.1.0^{2,5}.0^{3,33}.0^{4,8}.0^{6,10}.0^{7,14}.0^{7,34}.0^{9,12}.0^{11,15}.0^{13,17}.0^{14,21}.0^{16,19}.0^{18,22}.0^{20,24}$ $.0^{21,28}.0^{23,26}.0^{25,29}.0^{27,31}.0^{28,34}.0^{32,35}]$pentatriacontane.[105] We'll take his word for it. But, before you renounce allegiance to IUPAC derive some comfort in portrayal **77**. Two bold dots identify the bridgeheads he chose. And the three strands (viz., C_{28}, C_4, C_1) linking them appear, respectively, as a solid line, a hatched line, and a string of x's. The rest is up to you.

For nomenclature masochists, let us dwell a bit on naming complex polycyclics.

Consider, for example, the polyhedron $C_{60}H_{60}$ (**78**). This exquisite figure features 12 regular pentagons and 20 regular hexagons. One glimpse at its global shape tells why, in 1983, Josep Castells and Felix Serratosa, at the University of Barcelona, dubbed it "footballane" or, alternatively, "soccerane."[107] The 1982 World Cup matches in Spain were still fresh in their minds. Stylized versions of **78** show up everywhere as a symbol of the sport. For example, **79** is even cast on a coin issued by Haiti to commemorate the 1978 World Cup in Argentina.

78   **79**

Spheroid **78** was one of the hollow polyhedrals drawn by Leonardo da Vinci back in 1509; thus it has waited a long time to be baptized by IUPAC. The Barcelona chemists maintain that naming it is not hopeless if you heed advice published by David Eckroth.[106b]

In 1967, Professor Eckroth steered readers to so-called "Schlegel diagrams" for drawing polycycles in planar projection. To create a Schlegel projection, rest the polycycle on one of its faces. Next, view it from below, with the perspective greatly

5 FACES = PENTACYCLO   **II FACES = UNDECACYCLO**

80   **81**

exaggerated so that all the other faces appear to lie within the bottom one. For Schlegel snaps of cubane and dodecahedrane, see **80 and 81**, respectively.[108] Soccerane, a semiregular polygon, has two Schlegels, according to whether you peer through a pentagonal window (**82**) or a hexagonal one (**83**).[109] (The number of faces in your window—excluding the frame itself—corresponds to the total number of rings in the molecule. Soccerane has 31 of them, hence it will be hentriacontacyclo-.)

Then, find a continuous circuit (called a Hamiltonian line) that passes once—and only once—through as many ring atoms as possible and that returns to its starting point. The Hamiltonian path is delineated by bold print in **80** and **81**,[108] and by shaded regions in **82** and **83**.[109] A given Schlegel projection can have more than one Hamiltonian. For example, **84** depicts an alternative for the hexagonal window. But, according to hierarchial numbering rules, IUPAC chooses the Hamiltonian that allows you to bifurcate the circuit as symmetrically as possible and to keep numerical superscripts as small as possible. On this basis, **83** wins over **84**. Professor Serratosa assured that either **82** or **83** generates the following official name for soccerane: hentriacontacyclo[$29.29.0.0^{2,14}.0^{3,29}.0^{4,27}.0^{5,13}.0^{6,25}.0^{7,12}.0^{8,23}.0^{9,21}.0^{10,18}.0^{11,16}.0^{15,60}.$ $0^{17,58}.0^{19,56}.0^{20,54}.0^{22,52}.0^{24,50}.0^{26,49}.0^{28,47}.0^{30,45}.0^{32,44}.0^{33,59}.0^{34,57}.0^{35,43}.0^{36,55}.$ $0^{37,42}.0^{38,53}.0^{39,51}.0^{40,48}.0^{41,46}$]hexacontane.[110] But if you disagree, don't call us—call him.

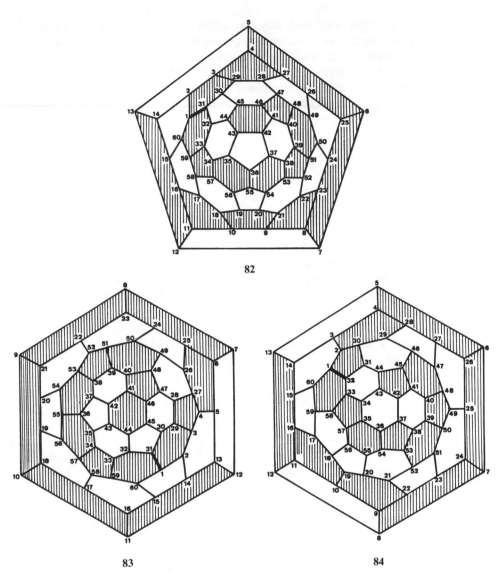

82

83   84

But why worry about the official name of soccerane? Would anyone be brash enough to try to synthesize it? The answer could be yes—thanks to chemical physicists and physical chemists at Rice University (Houston, Texas). Here's some background.[111]

Professor Harry Kroto (University of Sussex, U.K.) had entertained views on how long-chain polyalkynes arise in interstellar dust and outer shells of stars; and to test the ideas he collaborated at Rice with Professors Richard Smalley and Robert Curl and graduate students Jim Heath and Sean O'Brien. Using sophisticated equipment at Houston, they vaporized graphite with a pulsed laser beam at extremely high temperatures, a proven technique for producing strings of 20–30 carbons as well as larger aggregates (up to 190 carbons). By time-of-flight mass spectrometry, they observed what seemed a particularly prominent cluster, identified as C_{60}. Puzzled over its apparent remarkable stability, they were inspired to postulate a ball-like, rather than a chain-like, shape. This inspiration came from images of geodesic domes, architect Buckminster Fuller's famed frameworks of hexagons and pentagons connected so as to distribute stress equally. The Rice team proposed for their C_{60} cluster the shape of a truncated icosahedron (i.e., **78** devoid of hydrogens). Its cavity (diameter *ca.* 7 Å) should be large enough to entrap an atom, such as lanthanum.[112]

This unique globular structure, with an outer as well as an inner lining of 60 p-orbitals, had been on people's minds before. In fact, several theoretical treatments of its unusual π-network had preceded (and also paralleled) the experimental break-through.[113] The structure has 12,500 Kekulé forms (e.g., **85**); and Hückel calculations

85

86

predicted a large π-delocalization energy (0.55β per carbon) despite the surface curvature.[113f,g,h] If the C_{60} cluster is indeed **85**, then those forces that governed its fiery birth obey the rules of molecular syntax. In effect, they sentence the 60 carbons to a cage having sizeable π-overlap, all σ valences satisfied, and no dangling particles.[113g]

During his theoretical considerations of **85**, Anthony Haymet (University of California, Berkeley) referred to it as "footballene," a reasonable description of this highly symmetrical C_{60} isomer. His term made its rounds and was in common (often humorous) use at Berkeley before the experimental works at Houston appeared.[114]

Professor Haymet also pointed out that other structures for C_{60} are possible. One of them (delocalization energy 0.54β but possibly destablized by angle strain) he called "graphitene." This hypothetical C_{60} isomer consists of two overlayed kekulenes (chapter 10) appropriately linked to create pentagonal rings.[113f,g,115]

In the original announcement from Rice University, the C_{60} cluster **85** was named "buckminsterfullerene" in honor of the late architect.[116] (That mouthful slowed communication, so the term sometimes degenerated to "buckyball" in the labora-tory[111] and to "BF" in print.[113h]) The authors themselves lamented over the length of their original coined name and alluded to some possible shorter ones ("ballene," "spherene," "soccerene," "carbosoccer").[116a] We can speculate why they did not adopt either soccerene or footballene, precedent[107] notwithstanding. After all, the research was an Anglo-American effort; also, the experiments took place in Texas, where soccer and football are not exactly bedfellows. Presumably, Berkeley's Dr. Haymet also faced this dilemma in terminology, because his publications pointedly identify soccerene as a United States equivalent of footballene.[113f,115]

Professor Kroto thought that "fullerenes" would be an appropriate name for the family of hollow caged polyenes (C_n) made up of such pentagons and hexagons. And he proposed that "magic" clusters with $n = 70, 60, 50, 36$, and 28 should have enhanced stability.[116b]

Incidentally, Dr. Haymet's calculations led him to predict another good candidate for a stable spheroidal cluster, namely C_{120}. Its structure (**86**) should be that of a truncated icosidodecahedron with 0.49β delocalization energy and less dihedral strain than **85**. The ten-membered portals could allow metal ions in and out. He labelled this hypothetical compound "archimedene"; like the Archimedean solid **78**, its faces are regular polygons but not all the same (contrast Platonic solids, chapter 6).[115] Also, look back at p. 130 and you will see that archimedene is a spherically tesselated *anti-kekulene*.

Most residents of North America know football as a game played not with a round projectile but with an egg-shaped one. So, for their sake let us remold the spheroidal polyhedron **78** by removing hexagons while retaining the 12 pentagons; and let us not fret over slight distortions of bonds and angles. We find that the 16-hedron **87** (4 hexagons), the 15-hedron **88** (3 hexagons), and the 14-hedron **89** (2 hexagons) progressively flatten toward an oblate football.[117] These squashed "fulleranes" also stand as enticing—albeit awesome—synthetic goals.

| 87 | 88 | 89 |

But if you are content just to name such structures, look to P. Anantha Reddy for a helpful hint. At Bangalore's Indian Institute of Science, Dr. Reddy worried about counting up rings correctly in saturated polycycles. So he came up with a simple equation for cross-checking.[118] If a = number of ring *atoms*, b = number of ring *bonds*, and c = number of *cycles*, then

$$c = 1 - a + b$$

In cubane, for example, $a = 8$ and $b = 12$; so the molecule is pentacyclic ($c = 5$). It's good to have a Reddy rule for peace of mind.

In 1978, Richard Ranieri and Gary Calton of W.R. Grace Company isolated and characterized the antitumor agent **90**. This ketone has four rings and they graced it "quadrone."[119] Collaboration by Samuel Danishefsky and Kenward Vaughan of Yale University and Robert Gadwood and Kazuo Tsuzuki of the University of Pittsburgh resulted in the first total synthesis of quadrone.[120] Since then, other routes to it have more than quadrupled.[121]

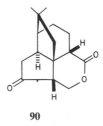

90

In 1965, Alex Nickon and his protégés at the Johns Hopkins University prepared the C_9 polycycles **91**, **92**, and **93**, all of which contain the bicyclo[2, 2, 1]heptane (i.e., norbornane) skeleton (**94**).[122-124] They christened **91** "brexane," as it has the extra

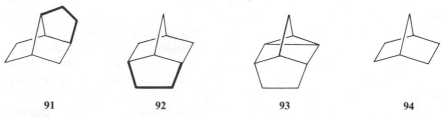

| 91 | 92 | 93 | 94 |

*bri*dge attached *exo* to the norbornane system. Hydrocarbon **92** became "brendane," since its extra *bridge* is attached *endo*. Involvement by a real-life Brenda would have made a better story—but facts are facts.

Brexane has intriguing symmetry. The figure is chiral and in it we can find two norbornyl units.[124,125] A substituent (Z) at C2 is simultaneously exo to one norbornyl unit and endo to the other (see **95a** and **95b**). If you interchange Z and H at C2, the result is neither an epimer nor an enantiomer but, instead, a structure superposable on the

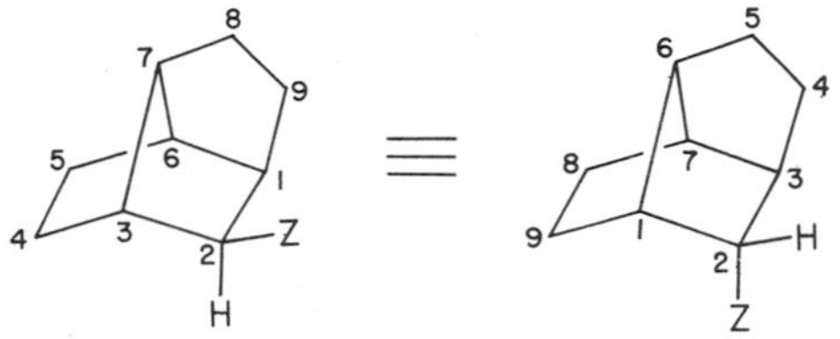

95a   **95b**

original. The ionization behavior of brexyl derivatives holds considerable interest in connection with the classical ion *vs.* nonclassical ion interpretation of norbornyl brosylate solvolyses (chapter 3).[124,126]

A key feature in **93** is its four rings, but names beginning with tetracyclo- and quadricyclo- were already in place for other structures discussed in this chapter. As "delta" comes fourth in the Greek alphabet, the Johns Hopkins team dubbed **93** "deltacyclane."[124a] In chemical shorthand the symbol for the Greek capital letter delta, namely Δ, commonly means "heat"[127] and also represents a cyclopropane ring. How nice that one of the rings in deltacyclane is three-membered.

While courting deltacyclane, we musn't slight its unsaturated sister, "deltacyclene" (**96**), first prepared by Lawrence Cannell of the Shell Development Corporation and then, differently, by Thomas Katz's contingent at Columbia University.[128] When undergraduate James Carnahan achieved the synthesis of **96**, he asked Professor Katz

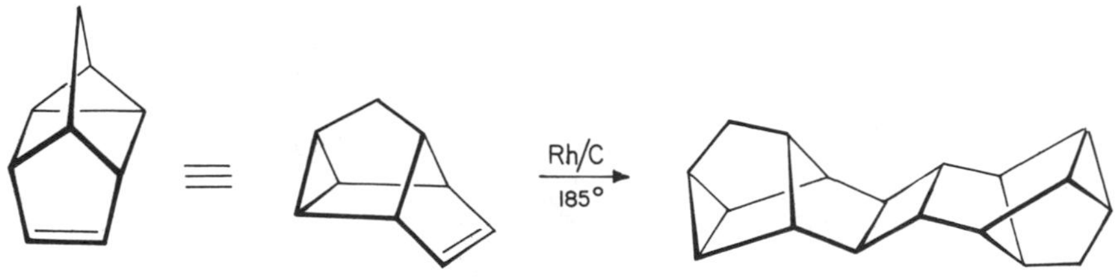

96   **97**

to suggest an in-house title for it. "George" was the first name that came to Dr. Katz; and since trivial titles are often arbitrary, he saw no reason why a more esoteric one was needed. Katz and co-workers had just previously discovered that rhodium catalyzes cycloadditions, so he asked Jim to heat George with rhodium. By George, it produced **97**. And that is just what they called this dimer: "Bi-George"![129,130]

Dr. Katz frequently used informal designations in his laboratory. On another occasion, a student produced a gas chromatogram containing three peaks; these were dubbed "Tom," "Dick," and "Harry." The Columbia researcher said it's easy to remember which one elutes from the column first.[129]

Let us bid *adieu* to deltacyclene, but with a parting caution: Don't confuse it with the similar-sounding "decacyclene" (**100**). This comely structure has been around since 1903 when Karl Dziewoński at the Academy of Science in Krakow made it by dehydrogenating acenaphthene (**98**) with sulfur at 290°C.[131] His short name "decacyclene" reminded the populace that this yellow hydrocarbon has ten rings. More recently, it has been shown to arise from several reactions that may involve transient

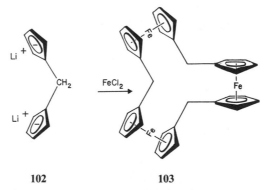

acenaphthyne (**99**).[132,133] Pyrolysis of the bis(tosylhydrazone) salt **101** is but one of these reactions.[133,134] But, if you need decacyclene in a hurry, Aldrich might still sell it.[135]

The most famous chemical sandwich, ferrocene (chapter 8), can be made from sodium cyclopentadienide and ferrous chloride.[136] Hence, Thomas Katz's team reasoned that the dilithium derivative of bis(cyclopentadienyl) methane (**102**) and ferrous chloride might give a dimer sandwich, or even a trimer (**103**) and higher oligomers, which they termed "ferrocenophanes" based on nomenclature proposed earlier by Smith.[65] And, by George, it proved so. They isolated and characterized the dimer, trimer, tetramer, and pentamer.[137]

The accepted designation for **103** was [1, 1, 1]ferrocenophane.[65,138] (The digits tell us how many carbons are in each bridge between rings.) Dr. Katz felt such coding would become cumbersome for large oligomers, so he suggested either [1^n]ferrocenophane—where n is the degree of polymerization—or the use of Latin prefixes. Thus, the hexamer becomes [1^6] or sexiferrocenophane (*sexi-* is Latin for "six").[132] This prefix is a titillating switch from the more common "hexa-," but we doubt that the molecule qualifies for a [1^6] centerfold. Along similar lines, the pentamer becomes [1^5]ferrocenophane or quinqueferrocenophane; (the Latin prefix *quinque** means five and is pronounced almost like "kinky.") Even though Columbia

---

*For example, the prefix "quinque-" appears in "quinquereme," an ancient galley with five rows of oars on each side.[139] Kinki University, Japan, did not inherit its name from this Latin source but from the province of Kinki.

University is not far from Greenwich Village, Dr. Katz thought it best to avoid kinky nomenclature, so he suggested yet another notation: [ЦНТ]-ferrocenophene.[137] This proposal led to a friendly exchange with a referee, who averred that the tally marks should be drawn thus: [ТНĻ].[129] Perhaps Professor Katz traveled between New York and California, whereas the referee spent summers in Alaska and winters in Florida. The Columbia chemist stood up for his own version by citing an encyclopedia.[137]

For some reason, this topic reminds us that scientists at Scripps Institution of Oceanography, La Jolla, California, worked with substances extracted from mollusks called "nudibranchs."[140] But rumors that the institution changed its name to "Stripps" are false.

In 1935, W. Dilthey and co-workers at the University of Bonn introduced the name "tetracyclone" for *tetra*phenyl*cyc*lopentadien*one* (104).[141] The molecule is best known as an aggressive diene for Diels–Alder events. For example, it mates with phenyl vinyl ketone to give 105.[142] Alkynes can serve as dienophiles; and some adducts* of 104 expel carbon monoxide to become benzenoid (106).[141] It would be nice if a real cyclone then blew away this harmful gas.

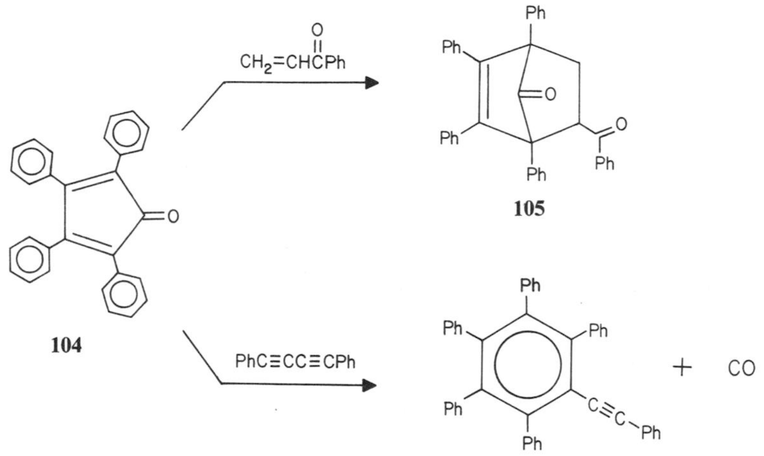

During the 1960s, a contingent at Tohoku University that included Kazuko Takahashi and Tetsuo Nozoe began to investigate polycyclic molecules with a quinoid ring in the center (e.g., 107).[144,145] These workers were gripped by the thought that 107 is a vinylog (i.e., an ethenylog) of pentaheptafulvalene (chapter 20); thus, a dipolar

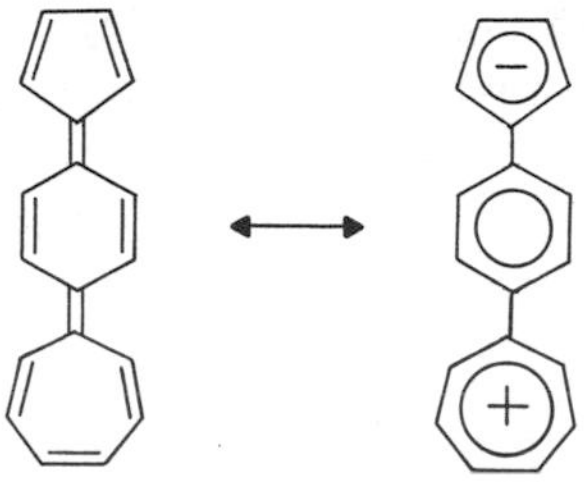

---

*"Adduct" is a common contraction for "addition product." However, "educt" is not shorthand for elimination product. Educt, in its usual chemistry context, means isolated starting material. Occasionally, in journal articles, it simply stands for starting compound (as distinct from product).[143]

resonance structure, **108**, is possible with all three rings aromatic. Professor Nozoe titled this family of compounds "quinarenes," because they contain a *quin*oid ring within a potentially aromatic (*arene*) system.[146] Numbers indicate the sizes and sequence of the cycles. Thus, **107** is [5.6.7]quinarene.[145,146b]

Dr. Takahashi and collaborators showed that the dipolar contributor has much to say about the ground state of certain quinarenes. For example, the [3.6.5.] analog **109** is stable when dry; but water rapidly converted it to the hydroxycyclopropene derivative, **111**.[147] The regioselectivity of this hydration suggests the three-membered ring had positive character (**110**). Similarly, the seven-membered cycle in [5.6.7] quinarene (**112**) seems electron-poor, since methanolic sodium bicarbonate whipped it over to the salt **114**.[144] The visible spectrum of **112** provided further evidence; the longest-wavelength absorption moved to shorter wavelength (hypsochromic shift) when the solvent polarity was increased.[148]

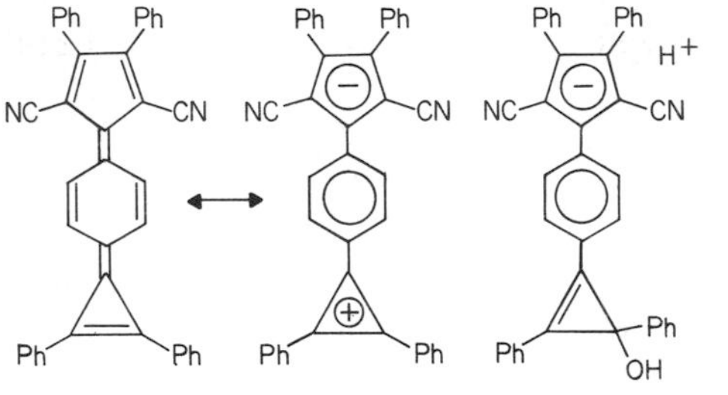

                109   **110**   **111**

Presumably, polar media preferentially stabilize the ground state (polarized as in **113**). Consequently, more energy (i.e., shorter wavelength) is needed for excitation.[148,149]

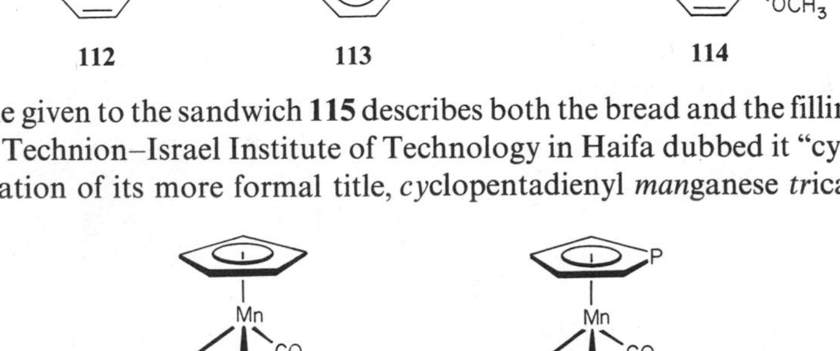

        112   **113**   **114**

The name given to the sandwich **115** describes both the bread and the filling. Michael Cais of the Technion–Israel Institute of Technology in Haifa dubbed it "cymantrene," an abbreviation of its more formal title, *cy*clopentadienyl *man*ganese *tri*carbonyl.[150]

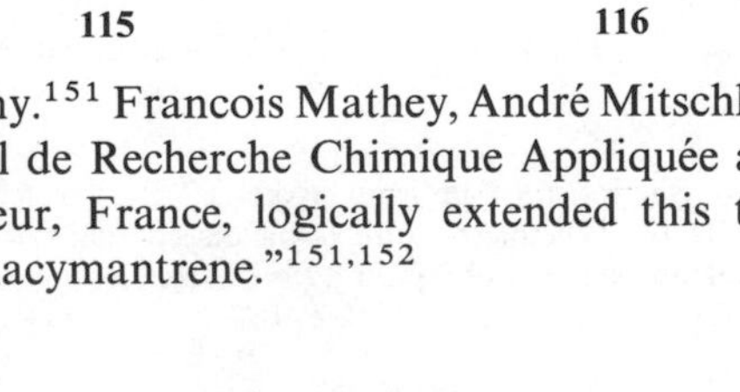

            115   **116**

The "-ene" is for euphony.[151] Francois Mathey, André Mitschler, and Raymond Weiss of the Institut National de Recherche Chimique Appliquée and the Institut Le Bel, Université Louis Pasteur, France, logically extended this terminology when they addressed **116** "phosphacymantrene."[151,152]

When the British firm, Cambrian Chemicals Ltd., marketed a versatile reagent, *to*sylmethyl *iso*cyanide (**117**), the ads called it "Tosmic."[153] This perky term is a

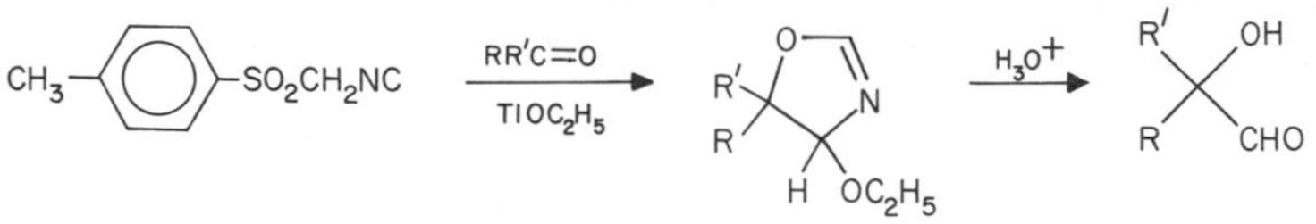

$$CH_3-\langle\bigcirc\rangle-SO_2CH_2NC \xrightarrow[\text{TlOC}_2\text{H}_5]{\text{RR'C=O}}$$

117 **118** **119**

euphonious shorthand for the constituent parts.[154] One of its many talents is to transform a ketone to an ethoxyoxazoline (**118**), which can be hydrolyzed to an α-hydroxyaldehyde (**119**).[155] The reagent first came to light in 1971 through the work of Leusen and Strating.[156] You could buy it from Aldrich Chemical Co. as "TosMIC"[157] and read about its marvelous deeds in a 1980 review.[158]

In 1975, Philip Stotter and stalwarts at the University of Texas, San Antonio, converted enolates of type **120** to *me*thoxymethyl ethers (**121**) by treatment with chloromethyl methyl ether. They came up with a touching term for the desired products: MOMyl ethers.[159] Professor Stotter felt that chances for success in research might be greater with motherhood on their side.[160] (Because the chloromethyl reagent is carcinogenic, maternal protection is further welcomed.)

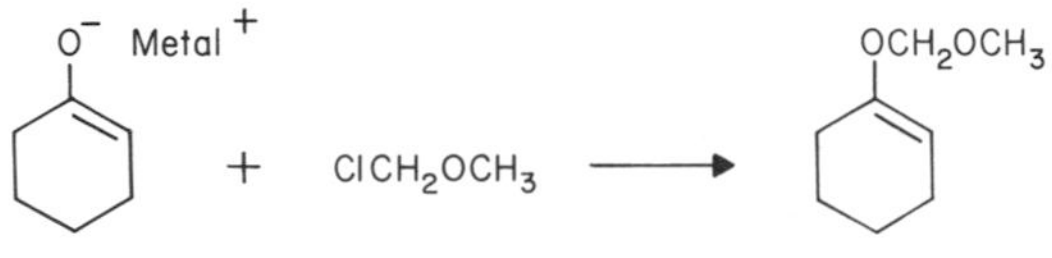

120 + $ClCH_2OCH_3$ → **121**

120 **121**

$$ROH \quad + \quad ClCH_2OCH_2CH_2OCH_3 \longrightarrow RO\!-\!CH_2OCH_2CH_2OCH_3$$

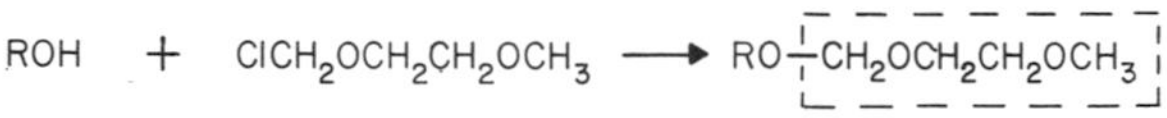

122 **123**

What Dr. Stotter did for motherhood, Elias Corey did for *mem*bership. He and his corps at Harvard came up with the abbreviation MEM for the β-*me*thoxy*e*thoxy*m*ethyl group (see **123**).[161] The MEM unit, introduced by means of its chloride (**122**), safeguards an OH as in **123**. A mild Lewis acid (e.g., $ZnBr_2$) disMEMbers the protective group after it has done its duty.[161] If you deal with OH, NH, SH, COOH, or C=O and are looking for protection, Theodora Greene wrote a book that's comforting and saves you oodles of time. It summarizes what you need, what to expect, and the likelihood of success.[162]

Herbert Brown came up with the catchy contractions "diglyme" and "triglyme" for the solvents *di*ethylene *glyc*ol *m*ethyl *e*ther [$CH_3O(CH_2CH_2O)_2CH_3$] and *tri*methylene *glyc*ol *m*ethyl *e*ther [$CH_3O(CH_2CH_2O)_3CH_3$], respectively.[163] His research school also popularized "disiamyl" (for *dis*econdary-*isoamyl*) and "thexyl" (for *tert-hexyl*). These short forms come in handy when you decide to pelt an olefin with disiamylborane, $[CH_3CH(CH_3)CH(CH_3)]_2BH$,[164a] or with thexylborane, $CH_3CH(CH_3)C(CH_3)_2BH_2$.[164b]

"Atrane" is not something we board at a railroad station but an expression that M.G. Voronkov, Director of the Institute of Organic Chemistry, Siberian Division of the Academy of Sciences, Irkutsk, Russia, contrived in 1965 for a structure of type **124**.[165]

The word appeared in an article in English first in 1966.[166] The M represents an atom of valence n (3, 4 or 5); and R can be hydrogen, halogen, or an organic group. When M is pentavalent, R can also stand for a doubly bonded oxygen. For example, if M is silicon (125), the compound is a "silatrane"; this derives from "*aminotriethoxysilane*."[167] Dr.

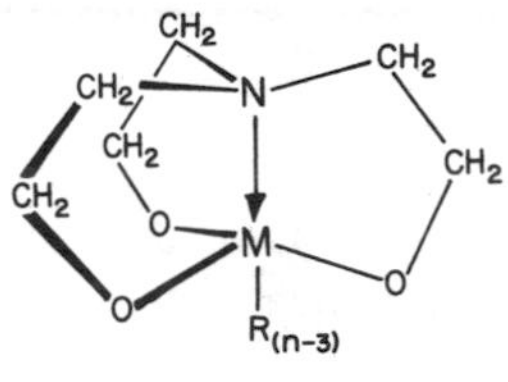

124 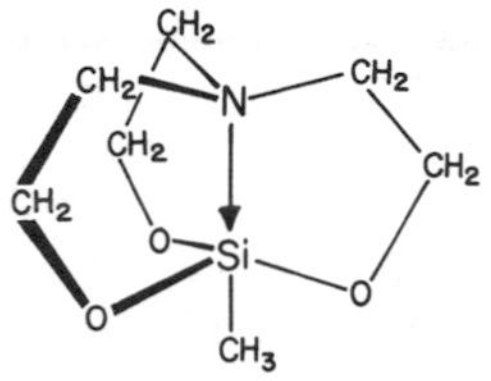   125

Voronkov moved "sil" to the front but otherwise kept everything on the same track. John Verkade and co-workers at Iowa State University studied "phosphatranes" (M = phosphorus). They learned that the N → P coordination can be strong or weak, according to the nature of R. In the latter circumstance, the bonds to phosphorus are tetrahedral and those to nitrogen are nearly planar—an overall geometry similar to that in manxine (p. 123).[168,169]

"Tipyl" (abbreviated "tip") is yet another pert term that describes the parts of a structure. It stands for 2, 4, 6-*triisopropylphenyl* (126) and was penned by Reynold Fuson's group in 1946[170] and independently by Arnold Miller in 1971,[171–173] all at the University of Illinois. This aryl moiety is of interest because it can stabilize certain enols (e.g., 127) with respect to rearrangement to a carbonyl form (e.g., 128).[170,172] (For most monoaldehydes and monoketones the tautomers interconvert readily, and the carbonyl isomer is favored by far.[174,175]) Enol 127 persists probably because tipyl

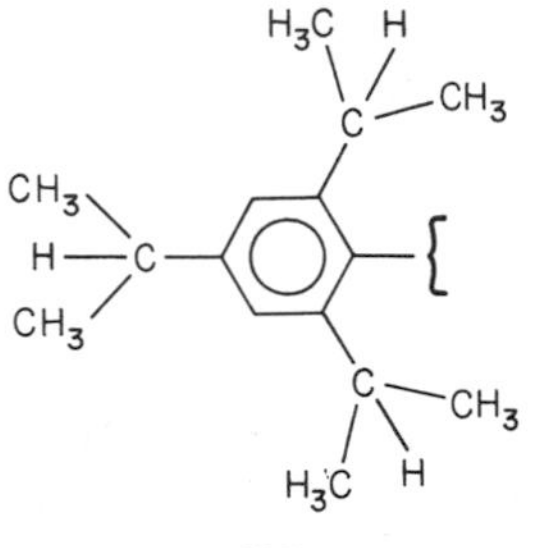

126

shields the carbon to be protonated;[170] Dr. Miller has discussed this relationship between bulkiness and enol longevity.[172] The Illinois chemists deserve thanks for their tip on how to preserve enols.

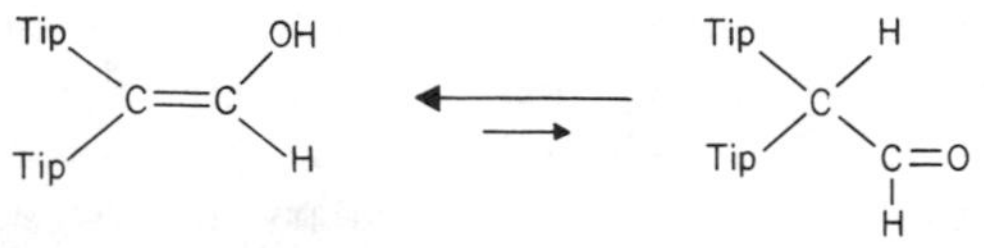

127   128

The "-sylate" family of *sulfonyl* esters and salts is well established in organic chemistry. Its patriarchs include tosylate, brosylate, and mesylate. In 1976, Robert Bergman and his student, Chaim Sukenik, at the California Institute of Technology added "amsylate" for the p-trimethyl*ammonium* benzenesulfonates.[176] At first they had considered including the "tr" to give "tramsylates," but decided to stay off the rails.[177]

In 1979, Paul Peterson's party at the University of South Carolina enlarged the tribe by pursuing "damsylates."[178] The name might suggest despair or visions of a lovely damsel. But in the flask you'll have the p-*dimethylamino*benzene*sulfonyl* derivative.

For convenience we've gathered the "sylates" for a family portrait. The sitting includes kin you may not have met. But Professor W. Cornard Fernelius of Kent State

amsylate	p-trimethylammonium benzene sulfonate	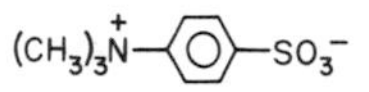 $(CH_3)_3\overset{+}{N}-\langle\bigcirc\rangle-SO_3^-$
besylate[179]	benzenesulfonate	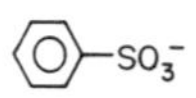 $\langle\bigcirc\rangle-SO_3^-$
brosylate	p-bromobenzenesulfonate	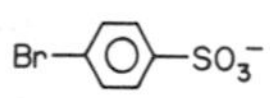 $Br-\langle\bigcirc\rangle-SO_3^-$
camsylate[179]	camphor-10-sulfonate	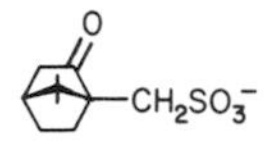 (camphor structure)$-CH_2SO_3^-$
closylate[179]	p-chlorobenzenesulfonate	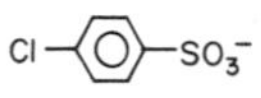 $Cl-\langle\bigcirc\rangle-SO_3^-$
damsylate	p-dimethylaminobenzenesulfonate	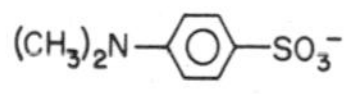 $(CH_3)_2N-\langle\bigcirc\rangle-SO_3^-$
edisylate[179]	1, 2-ethanedisulfonate	$RO_3SCH_2CH_2SO_3^-$
esylate[179]	ethanesulfonate	$CH_3CH_2SO_3^-$
isethionate[179]	2-hydroxyethanesulfonate	$HO-CH_2CH_2SO_3^-$
mesylate	methanesulfonate	$CH_3SO_3^-$
napsylate[179]	2-naphthalenesulfonate	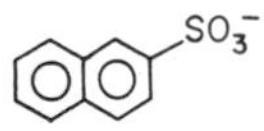 (naphthalene)$-SO_3^-$
nosylate[182]	p-nitrobenzenesulfonate	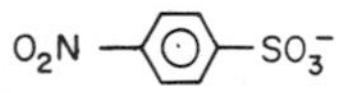 $O_2N-\langle\bigcirc\rangle-SO_3^-$
tosylate	p-toluenesulfonate	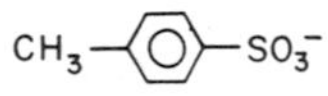 $CH_3-\langle\bigcirc\rangle-SO_3^-$
tresylate[183]	2, 2, 2-trifluoroethanesulfonate	$F_3CCH_2SO_3^-$
triflate	trifluoromethanesulfonate	$F_3CSO_3^-$
trimylate[184]	2, 4, 6-trimethylbenzenesulfonate	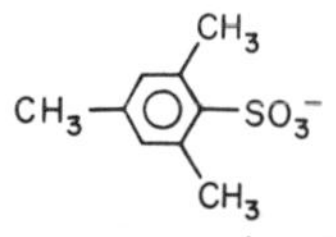 $CH_3-\langle\bigcirc\rangle-SO_3^-$ with CH_3 groups
trisylate[185]	2, 4, 6-triisopropylbenzenesulfonate	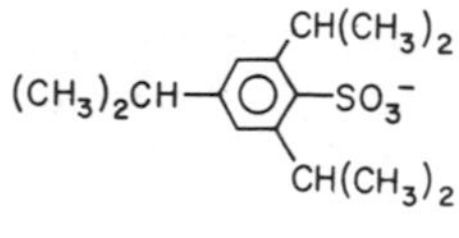 $(CH_3)_2CH-\langle\bigcirc\rangle-SO_3^-$ with $CH(CH_3)_2$ groups

University pointed out that some of these distant relatives show up in nonproprietary names for drugs and related pharmaceuticals, as recommended by various national (e.g., U.S. Adopted Names Council) and international (e.g., World Health Organization) agencies.[179] (The WHO prefers the suffix to be spelled "silate"). We warn you, however, that some "sylate" children have had an identity crisis. So, researchers occasionally adopt different nicknames for the same relative, or pick the same name for a brood of different lineage. For example, esters of camphor-10-sulfonic acid are "camsylates" but have also been dubbed "casylates."[180] And snappy "trisyl"—in use earlier as a synonym for 2, 4, 6-triisopropylbenzenesulfonyl—has also been bestowed on $[(CH_3)_3Si]_3C$ and abbreviated "Tsi."[181] The world outside chemistry knowns how to avoid such weighty problems; there the younger offspring would simply become "Trisyl Jr."

REFERENCES AND NOTES

1. Taken from an inverview with R.B. Woodward and A. Eschenmoser, published in *Nachr. Chem. Techn.* **1972**, *20*, 147–150.
2. Woodward, R.B.; Fukunaga, T.; Kelly, R.C. *J. Am. Chem. Soc.* **1964**, *86*, 3162–3164.
3. Fukunaga, T., private communication, July, 1977.
4. de Meijere, A.; Kaufmann, D.; Schallner, O. *Angew. Chem. Int. Ed. Engl.* **1971**, *10*, 417–418.
5. Bosse, D.; de Meijere, A. *Chem. Ber.* **1978**, *111*, 2223–2242.
6. Wyvratt, M.J.; Paquette, L.A. *Tetrahedron Lett.* **1974**, 2433–2436.
7. Mercier, C.; Soucy, P.; Rosen, W.; Deslongchamps, P. *Synth. Commun.* **1973**, *3*, 161–164.
8. Jacobson, I.T. *Acta Chem. Scand.* **1967**, *21*, 2235–2246.
9. Prinzbach, H.; Stusche, D. *Helv. Chim. Acta* **1971**, *54*, 755–759.
10. (a) Lannoye, G.; Honkan, V.; Weiss, U.; Bertz, S.; Cook, J.M. *Abstracts of Papers, 16th Annual Meeting, Great Lakes American Chemical Society Region, Illinois State University, Normal, IL, June 7–9, 1982*, (American Chemical Society, Washington, DC, 1982), No. 201, (b) Bertz, S.H.; Lannoye, G.; Cook, J.M. *Tetrahedron Lett.* **1985**, *26*, 4695–4698.
11. Jefford, C.W. *J. Chem. Educ.* **1976**, *53*, 477–482.
12. Repic, O., Ph.D. Thesis, Harvard University, October, 1976.
13. March, J. *Advanced Organic Chemistry*, 3rd ed. (John Wiley & Sons, New York, 1985), pp. 63–64.
14. Bischof, D.; Bosse, R.; Gleiter, M.J.; Kukla, A.; de Meijere, A.; Paquette, L.A. *Chem. Ber.* **1975**, *108*, 1218–1223.
15. Bünzli, J.C.; Frost, D.C.; Weiler, L. *Tetrahedron Lett.* **1973**, 1159–1162.
16. Stevens, E.O.; Kramer, J.D.; Paquette, L.A. *J. Org. Chem.* **1976**, *41*, 2266–2269.
17. Christoph, G.G.; Muthard, J.L.; Paquette, L.A.; Böhm, M.C.; Gleiter, R. *J. Am. Chem. Soc.* **1978**, *100*, 7782–7784.
18. Osborn, M.E.; Kuroda, S.; Muthard, J.L.; Kramer, J.D.; Engel, P.; Paquette, L.A. *J. Org. Chem.* **1981**, *46*, 3379–3388.
19. (a) Venkatachalam, M.; Jawdosiuk, M.; Deshpande, M.; Cook, J.M. *Tetrahedron Lett.* **1985**, *26*, 2275–2278, (b) Cook, J.M., private communication, July, 1985.
20. (a) Deshpande, M.N.; Jawdosiuk, M.; Kubiak, G.; Venkatachalam, M.; Weiss, U.; Cook, J.M. *J. Am. Chem. Soc.* **1985**, *107*, 4786–4788, (b) Venkatachalam, M.; Kubiak, G.; Cook, J.M.; Weiss, U. *Tetrahedron Lett.* **1985**, *26*, 4863–4866, (c) Venkatachalam, M.; Deshpande, M.N.; Jawdosiuk, M.; Kubiak, G.; Wehrli, S.; Cook, J.M.; Weiss, U. *Tetrahedron* **1986**, *42*, 1597–1605.
21. Turner, R.B.; Woodward, R.B. In *The Alkaloids*, Manske, R.H.F.; Holmes, H.L. (Eds.), (Academic Press, New York, 1953), vol. III, chapter 16.
22. For a review of polyquinanes, see Paquette, L.A. *Top. Curr. Chem.* **1979**, *79*, 41–165.
23. Weiss, U., private communication, August, 1985.
24. Fukunaga, T.; Simmons, H.E.; Wendoloski, J.J.; Gordon, M.D. *J. Am. Chem. Soc.* **1983**, *105*, 2729–2734.
25. Alder, R.W.; Petts, J.C.; Clark, T. *Tetrahedron Lett.* **1985**, *26*, 1585–1588.
26. Barth, W.E.; Lawton, R.G. *J. Am. Chem. Soc.* **1966**, *88*, 380–381.
27. Barth, W.E.; Lawton, R.G. *J. Am. Chem. Soc.* **1971**, *93*, 1730–1745.

28. Lawton, R.G., private communications, June, 1977 and March, 1984.
29. Gleicher, G.J. *Tetrahedron* **1967**, *23*, 4257–4263.
30. Randić, M.; Trinajstić, N. *J. Am. Chem. Soc.* **1984**, *106*, 4428–4434.
31. Dopper, J.H.; Wynberg, H. *J. Org. Chem.* **1975**, *40*, 1957–1966.
32. Wynberg, H., private communication, March, 1978.
33. Robertson, J.M.; White, J.G. *J. Chem. Soc.* **1945**, 607–617.
34. *Proof Collectors Corner*, **1985**, May/June issue, p. 79 (a newsletter published by the World Proof Numismatic Association, Pittsburgh, PA, 15201).
35. Yamamoto, K.; Harada, T.; Nakazaki, M.; Naka, T.; Kai, Y.; Harada, S.; Kasai, N. *J. Am. Chem. Soc.* **1983**, *105*, 7171–7172.
36. (a) Erdtman, H.; Högberg, H.-E. *Chem. Commun.* **1968**, 773–774; *Tetrahedron Lett.* **1970**, 3389–3392, (b) Högberg, H.-E. *Acta Chem. Scand.* **1972**, *26*, 2752–2758; *ibid.* **1973**, *27*, 2591–2596.
37. Jessup, P.J.; Reiss, J.A. *Aust. J. Chem.* **1976**, *29*, 173–178.
38. Jenny, W.; Peter, R. *Angew. Chem. Int. Ed. Engl.* **1965**, *4*, 979–980.
39. Peter, R.; Jenny, W. *Helv. Chim. Acta* **1966**, *49*, 2123–2135.
40. Vögtle, F.; Staab, H.A. *Chem. Ber.* **1968**, *101*, 2709–2716.
41. Staab, H.A.; Diederich, F. *Chem. Ber.* **1983**, *116*, 3487–3503.
42. (a) Funhoff, D.J.H.; Staab, H.A. *Angew. Chem. Int. Ed. Engl.* **1986**, *25*, 742–744. We thank Professor Staab for a prepublication copy of the manuscript, (b) For comments on earlier attempts by Peter and Jenny, see Staab, H.A.; Diederich, F.; Čaplar, V. *Liebigs Ann. Chem.* **1983**, 2262–2273.
43. Clar, E. *Polycyclic Hydrocarbons* (Academic Press, New York, 1964), vol. 1, pp. 3–11.
44. Clar, E.; Mullen, A. *Tetrahedron* **1968**, *24*, 6719–6724.
45. Clar, E. *Aromatische Kohlenwasserstoffe* (Springer-Verlag, Berlin, 1941), p. 311; 2nd ed., 1952, pp. 93, 461.
46. Longuet–Higgins, H.C. *J. Chem. Phys.* **1950**, *18*, 265–274.
47. Clar, E.; Stewart, D.G. *J. Am. Chem. Soc.* **1953**, *75*, 2667–2672.
48. Hara, O.; Tanaka, K.; Yamamoto, K.; Nakazawa, T.; Murata, I. *Tetrahedron Lett.* **1977**, 2435–2436.
49. (a) Ohkami, N.; Motoyama, A.; Yamaguchi, T.; Hosoya, H.; Gutman, I. *Tetrahedron* **1981**, *37*, 1113–1112, (b) For leading references and development of a formula periodic table for benzenoid polycyclic aromatic hydrocarbons see Dias, J.R. *Acc. Chem. Res.* **1985**, *18*, 241–248.
50. Goodwin, S.; Smith, A.F.; Horning, E.C. *J. Am. Chem. Soc.* **1959**, *81*, 1903–1908.
51. Sainsbury, M., private communication, September, 1975.
52. Dauben, H.J., Jr.; Bertelli, D.J. *J. Am. Chem. Soc.* **1961**, 83, 4659–4660.
53. Hafner, K.; Suda, M. *Angew. Chem. Int. Ed. Engl.* **1976**, *15*, 314–315.
54. Vogel, E.; Runzheimer, H.V.; Hogrefe, F.; Baasner, B.; Lex, J. *Angew. Chem. Int. Ed. Engl.* **1977**, *16*, 871–872.
55. Oth, J.F.M.; Müllen, K.; Runzheimer, H.V.; Mues, P.; Vogel, E. *Angew. Chem. Int. Ed. Engl.* **1977**, *16*, 872–874.
56. Seto, H.; Yonehara, H. *J. Antibiot.* **1980**, *33*, 92–93.
57. (a) Annis, G.D.; Paquette, L.A. *J. Am. Chem. Soc.* **1982**, *104*, 4505–4506, (b) Paquette, L.A.; Annis, G.D. *J. Am. Chem. Soc.* **1983**, *105*, 7358–7363, (c) For later syntheses and leading references, see Mehta, G.; Rao, K.S. *J. Chem. Soc., Chem. Commun.* **1985**, 1464–1466.
58. Cane, D.; Tillman, A.M. *J. Am. Chem. Soc.* **1983**, *105*, 122–124.
59. Meyer, R. *Explosives* (Verlag Chemie International, New York, 1977), p. 297.
60. Brown, C.J.; Farthing, A.C. *Nature (London)* **1949**, *164*, 915–916.
61. Cram, D.J.; Steinberg, H. *J. Am. Chem. Soc.* **1951**, *73*, 5691–5704.
62. Cram, D.J., private communication, July, 1977.
63. Nakazaki, M.; Yamamoto, K.; Yasuhiro, M. *J. Org. Chem.* **1978**, *43*, 1041–1044.
64. Griffin, R.W., Jr. *Chem. Rev.* **1963**, *63*, 45–54.
65. Smith, B.H. *Bridged Aromatic Compounds* (Academic Press, New York, 1964).
66. Akahori, K.; Hama, F.; Sakata, Y.; Soichi, M. *Tetrahedron Lett.* **1984**, *25*, 2379–2382.
67. Vögtle, F. *Tetrahedron Lett.* **1969**, 3193–3196.
68. Kato, N.; Fukazawa, Y.; Ito, S. *Tetrahedron Lett.* **1979**, 1113–1116.
69. Saito, H.; Fujise, Y.; Ito, S. *Tetradedron Lett.* **1983**, *24*, 3879–3882.
70. Prelog, V.; Wiesner, K. *Helv. Chim. Acta* **1947**, *30*, 1465–1471.

71. (a) Jenneskens, L.W.; de Kanter, F.J.J.; Kraakman, P.A.; Turkenburg, L.A.M.; Koolhaas, W.E.; de Wolf, W.H.; Bickelhaupt, F.; Tobe, Y.; Kakiuchi, K.; Odaira, Y. *J. Am. Chem. Soc.* **1985**, *107*, 3716–3717, (b) Tobe, Y.; Ueda, K.; Kakiuchi, K.; Odaira, Y. *Angew. Chem. Int. Ed. Engl.* **1986**, *25*, 369–371 (they *isolated* [6] paracycloph-3-ene).
72. Pellegrin, M.M. *Rec. Trav. Chim. Pays-Bas* **1899**, *18*, 457–465.
73. Brown, C.J. *J. Chem. Soc.* **1953**, 3278–3285.
74. Gordon, A.J.; Gallagher, J.P. *Tetrahedron Lett.* **1970**, 2541–2544.
75. Mitchell, R.H.; Vinod, T.K.; Bushnell, G.W. *J. Am. Chem. Soc.* **1985**, *107*, 3340–3341.
76. Nakazaki, M.; Yamamoto, K.; Tanaka, S.; Kametani, H. *J. Org. Chem.* **1977**, *42*, 287–291.
77. Gray, R.; Boekelheide, V. *Angew. Chem. Int. Ed. Engl.* **1975**, *14*, 107–108.
78. For reviews see (a) Vögtle, F.; Neumann, P. *Top Curr. Chem.* **1974**, *48*, 67–129, (b) Vögtle, F.; Hohner, G. *Top Curr. Chem.* **1978**, *74*, 1–29.
79. Sekine, Y.; Brown, M.; Boekelheide, V. *J. Am. Chem. Soc.* **1979**, *101*, 3126–3127.
80. Sekine, Y.; Boekelheide, V. *J. Am. Chem. Soc.* **1981**, *103*, 1777–1785.
81. Boekelheide, V., private communication, June, 1979.
82. Kang, H.C.; Hanson, A.W.; Eaton, B.; Boekelheide, V. *J. Am. Chem. Soc.* **1985**, *107*, 1979–1985.
83. Psiorz, M.; Hopf, H. *Angew. Chem. Int. Ed. Engl.* **1982**, *21*, 623–624.
84. Hopf, H., private communication, November, 1982.
85. Roberts, W.P.; Shoham, G. *Tetrahedron Lett.* **1981**, 4895–4898.
86. Dieckmann, H.; Chang, C.K.; Traylor, T.G. *J. Am. Chem. Soc.* **1971**, *93*, 4068–4070.
87. Eaton, P.E.; Chakraborty, U.R. *J. Am. Chem. Soc.* **1978**, *100*, 3634–3635.
88. Pahor, N.B.; Calligaris, M.; Randaccio, L. *J. Chem. Soc., Perkin Trans. 2* **1978**, 38–42; *ibid.*, **1978**, 42–45.
89. Vögtle, F.; Mew, P.K.T. *Angew. Chem. Int. Ed. Engl.* **1978**, *17*, 60–62.
90. Hisatome, M.; Takano, S.; Yamakawa, K. *Tetrahedron Lett.* **1985**, *26*, 2347–2350.
91. Grützmacher, H.–Fr.; Husemann, W. *Tetrahedron Lett.* **1985**, *26*, 2431–2434.
92. Collman, J.P.; Chong, A.O.; Jameson, G.B.; Oakley, R.T.; Rose, E.; Smittou, E.R.; Ibers, J.A. *J. Am. Chem. Soc.* **1981**, *103*, 516–533.
93. Zipplies, M.F.; Krieger, C.; Staab, H.A. *Tetrahedron Lett.* **1983**, *24*, 1925–1928.
94. Elschenbroich, C.; Möckel, R.; Zenneck, U. *Angew. Chem. Int. Ed. Engl.* **1978**, *17*, 531–532.
95. Hisatome, M.; Kawiziri, Y.; Yamakawa, K.; Iitaka, Y. *Tetrahedron Lett.* **1979**, 1777–1780.
96. Hisatome, M.; Kawaziri, Y.; Yamakawa, K. *Tetrahedron Lett.* **1982**, *23*, 1713–1716.
97. Hisatome, M.; Watanabe, J.; Yamakawa, K.; Iitaka, Y. *J. Am. Chem. Soc.* **1986**, *108*, 1333–1334.
98. Keehn, P.M.; Rosenfeld, S.M., Eds. *Cyclophanes* (Academic Press, New York, 1983), vols. 1 and 2.
99. Cristol, S.J.; Snell, R.L. *J. Am. Chem. Soc.* **1954**, *76*, 5000.
100. Cristol, S.J.; Snell, R.L. *J. Am. Chem. Soc.* **1958**, *80*, 1950–1952.
101. Frèrejacque, M. *Bull. Soc. Chim. Fr.* **1939**, [5], *6*, 1008–1011.
102. Lipp, M. *Chem. Ber.* **1941**, *74B*, 1–6.
103. Cristol, S.J. *Tetrahedron* **1986**, *42*, 1617–1619.
104. Prinzbach, H.; Weidmann, K.; Trah, S.; Knothe, K. *Tetrahedron Lett.* **1981**, *22*, 2541–2544.
105. Prinzbach, H., private communication, March, 1983.
106. (a) *IUPAC Nomenclature of Organic Chemistry, Sections A and B*, 2nd ed. (Butterworths, London, 1966), (b) Eckroth, D.R. *J. Org. Chem.* **1967**, *32*, 3362–3365.
107. Castells, J.; Serratosa, F. *J. Chem. Educ.* **1983**, *60*, 941. (See also reference 109.)
108. Moyano, A.; Serratosa, F.; Camps, P.; Drudis, J.P. *J. Chem. Educ.* **1982**, *59*, 126–127.
109. Castells, J.; Serratosa, F. *J. Chem. Educ.* **1986**, *83*, 630. We thank Professor Serratosa for a copy of his diagrams prior to publication.
110. Serratosa, F., private communications, December, 1983 and January, 1984.
111. (a) Baum, R.M. *Chem. Eng. News* **1985**, Dec. 23, pp. 20–22, (b) *Chem. Brit.* **1986**, *22*, p. 11, (c) *Sci. News* **1985**, *128*, 325 and 396, (d) Baum, R.M. *Chem. Eng. News* **1986**, June 16, pp. 16–17, (e) Kroto, H.W. *Proc. R. Inst. Gr. Brit. London*, **1986**, *58*, 45–72.
112. (a) Heath, J.R.; O'Brien, S.C.; Zhang, Q.; Liu, Y.; Curl, R.F.; Kroto, H.W.; Tittel, F.K.; Smalley, R.E. *J. Am. Chem. Soc.* **1985**, *107*, 7779–7780, (b) For views that challenged the postulated spheroidal structure of C_{60}La, see Cox, D.M.; Trevor, D.J.; Reichmann, K.C.; Kaldor, A. *J. Am. Chem. Soc.* **1986**, *108*, 2457–2458.
113. (a) Jones, D.E.H. *New Scientist* **1966**, Nov. 3, under a pseudonym in his "Daedalus" column

on "Ariadne's" page, (b) Yoshida, Z.; Ōsawa, E. *Aromaticity* (in Japanese), (Kagakudojin, Kyoto, 1971) pp. 174–178. An English translation by Professor Ōsawa was made available to us by Professor H. Kroto, (c) Bochvar, D.A.; Gal'pern, G.E. *Dokl. Akad. Nauk SSSR*, **1973**, *209*, 610–612; *Dokl. Chem. (Engl. Transl.)*, **1973**, *209*, 239–241, (d) Davidson, R.A. *Theoret. Chim. Acta (Berlin)* **1981**, *58*, 193–231, (e) Stankevich, I.V.; Nikerov, M.V.; Bochvar, D.A. *Russ. Chem. Rev. (Engl. Transl.)* **1984**, *53*, 640–654, (f) Haymet, A.D.J. *J. Am. Chem. Soc.* **1986**, *108*, 319–321, (g) Klein, D.J.; Schmalz, T.G.; Hite, G.E.; Seitz, W.A. *J. Am. Chem. Soc.* **1986**, *108*, 1301–1302, (h) Newton, M.D.; Stanton, R.E. *J. Am. Chem. Soc.* **1986**, *108*, 2469–2470.

114. Haymet, A.D.J., private communication, February, 1986.

115. Haymet, A.D.J. *Chem. Phys. Lett.* **1985**, *122*, 421–424.

116. (a) Kroto, H.W.; Heath, J.R.; O'Brien, S.C.; Curl, R.F.; Smalley, R.E. *Nature (London)* **1985**, *318*, 162–163, (b) Kroto, H., private communications, June, 1987.

117. Williams, R. *The Geometrical Foundation of Natural Structure* (Dover Publications, New York, 1979).

118. (a) Reddy, P.A. *Abstracts of Papers, 190th National Meeting of the American Chemical Society, Chicago, IL* (American Chemical Society: Washington, DC, 1985), ORGN 298. (b) Reddy, P.A. *J. Chem. Educ.* **1987**, *64*, 400, (c) Reddy, P.A., private communications, January, 1986 and January, 1987.

119. Ranieri, R.L.; Calton, G.J. *Tetrahedron Lett.* **1978**, 499–502.

120. Danishefsky, S.; Vaughan, K.; Gadwood, R.C.; Tsuzuki, K. *J. Am. Chem. Soc.* **1980**, *102*, 4262–4263; **1981**, *103*, 4136–4141.

121. (a) Bornack, W.K.; Bhagwat, S.S.; Ponton, J.; Helquist, P. *J. Am. Chem. Soc.* **1981**, *103*, 4647–4648, (b) Burke, S.D.; Murtiashaw, C.W.; Saunders, J.O.; Dike, M.S. *J. Am. Chem. Soc.* **1982**, *104*, 872–874, (c) Takeda, K.; Shimono, Y.; Yoshii, E. *J. Am. Chem. Soc.* **1983**, *105*, 563–568, (d) Smith, A.B., III; Konopelski, J.P.; Wexler, B.A.; Tu, C.–Y. *Abstracts of Papers, 187th National Meeting of the American Chemical Society. St. Louis MO* (American Chemical Society, Washington, DC, 1984), ORGN 9, (e) Burke, S.D.; Murtiashaw, C.W.; Saunders, J.O.; Oplinger, J.A.; Dike, M.S. *J. Am. Chem. Soc.* **1984**, *106*, 4558–4566, (f) Piers, E.; Moss, N. *Tetrahedron Lett.* **1985**, *26*, 2735–2738.

122. Nickon, A.; Kwasnik, H.; Swartz, T.; Williams, R.O.; DiGiorgio, J.B. *Abstracts, 149th National Meeting of the American Chemical Society, Detroit, MI* (American Chemical Society, Washington, DC, 1965), p. 2P.

123. Nickon, A.; Kwasnik, H.; Swartz, T.; Williams, R.O.; DiGiorgio, J.B. *J. Am. Chem. Soc.* **1965**, *87*, 1613–1615.

124. (a) Nickon, A.; Kwasnik, H.; Swartz, T.; Williams, R.O.; DiGiorgio, J.B. *J. Am. Chem. Soc.* **1965**, *87*, 1615–1616, (b) Nickon, A.; Weglein, R.C.; Mathew, C.T. *Can J. Chem.* **1981**, *59*, 302–313.

125. Nakazaki, M.; Naemura, K.; Kadowaki, H. *J. Org. Chem.* **1976**, *41*, 3725–3730.

126. (a) Nickon, A.; Weglein, R.C. *Tetrahedron Lett.* **1986**, *27*, 2675–2678, (b) Nickon, A.; Swartz, T.D.; Sainsbury, D.M.; Toth, B.R. *J. Org. Chem.* **1986**, *51*, 3736–3738.

127. The triangle as a symbol for heat likely descended from early alchemists, who used this geometric figure to portray the element Fire.

128. (a) Cannell, L.G. *Tetrahedron Lett.* **1966**, 5967–5972, (b) Katz, T.J., Carnahan, J.C., Jr.; Boecke, R. *J. Org. Chem.* **1967**, *32*, 1301–1304, (c) Mrowca, J.J.; Katz, T.J. *J. Am. Chem. Soc.* **1967**, *89*, 4012–4017.

129. Katz, T.J., private communication, January, 1976.

130. *Chemistry* **1967**, July–August, p. 37.

131. Dziewoński, K. *Chem. Ber.* **1903**, *36*, 962–971.

132. Chapman, O.L.; Gano, J.; West, P.R.; Regitz, M.; Maas, G. *J. Am. Chem. Soc.* **1981**, *103*, 7033–7036.

133. Nakayama, J.; Segiri, T.; Ohya, R.; Hoshino, M. *J. Chem. Soc., Chem. Commun.* **1980**, 791–792.

134. Nakayama, J.; Ohshima, E.; Ishii, A.; Hoshino, M. *J. Org. Chem.* **1983**, *48*, 60–65.

135. Aldrich Chemical Company, Inc., Milwaukee, WI.

136. Wilkinson, G.; Cotton, F.A.; Birmingham, J.M. *J. Inorg. Nucl. Chem.* **1956**, *2*, 95–113.

137. Katz, T.J.; Acton, N.; Martin, G. *J. Am. Chem. Soc.* **1969**, *91*, 2804–2805.

138. Watts, W.E. *Organometal. Chem. Rev.* **1967**, *2*, 231–254.

139. Foley, V.; Soedel, W. *Scien. Amer.* **1981**, *244*, April, pp. 148–163.

140. Hochlowski, J.E.; Faulkner, D.J. *Tetrahedron Lett.* **1981**, *22*, 271–274.
141. Dilthey, W.; Schommer, W.; Höschen, W.; Dierichs, H. *Chem. Ber.* **1935**, *68B*, 1159–1162.
142. Allen, C.H.F.; Bell, A.C.; Bell, A.; Van Allan, J. *J. Am. Chem. Soc.* **1940**, *62*, 656–664.
143. Examples: (a) Frenking, G.; Schmidt, J. *Tetrahedron* **1984**, *40*, 2123–2132, (b) Grimme, W.; Mauer, W.; Sarter, C. *Angew. Chem. Int. Ed. Engl.* **1985**, *24*, 331–332, (c) Gleiter, R.; Sander, W. *Angew Chem. Int. Ed. Engl.* **1985**, *24*, 566–568.
144. Takahashi, K.; Oikawa, I.; Takase, K. *Chem. Lett.* **1974**, 1215–1218.
145. Takahashi, K.; Takenaka, S.; Kikuchi, Y.; Takase, K.; Nozoe, T. *Bull. Chem. Soc. Jpn.* **1974**, *47*, 2272–2276.
146. (a) Nozoe, T. Lecture presented at the 19th Annual Meeting of the Chemical Society of Japan, Tokyo, April 1, 1966 (unpublished). (b) Takahashi, K., private communication, October, 1979.
147. Takahashi, K.; Ishikawa, F.; Takase, K. *Tetrahedron Lett.* **1976**, 4655–4658.
148. Takahashi, K.; Takase, K. *Tetrahedron Lett.* **1975**, 245–248.
149. For leads to other quinarenes and to quinarenones see (a) Takahashi, K.; Ohnishi, K.; Takase, K. *Tetrahedron Lett.* **1984**, *25*, 73–76, (b) Takahashi, K.; Nozoe, T.; Takase, K.; Kudo, T. *Tetrahedron Lett.* **1984**, *25*, 77–80.
150. Cais, M. Lecture at XIXth International Congress of Pure and Applied Chemistry, London, July 10–17, 1963 (unpublished). See footnote 4 in Maoz, N.; Mandelbaum, A.; Cais, M. *J. Chem. Soc. A* **1968**, 3086–3095.
151. Mathey, F., private communication, February, 1981.
152. Mathey, F.; Mitschler, A.; Weiss, R. *J. Am. Chem. Soc.* **1978**, *100*, 5748–5755.
153. "Tosmic," Technical Brochure, Cambrian Chemicals Ltd., Croydon, England; see *Chem. Brit.* **1976**, *12*, No. 10, Advertisement, p. 8.
154. Tucker, G.J., Cambrian Chemicals Ltd., Croydon, England, private communication, July, 1977.
155. Oldenziel, O.H.; van Leusen, A.M. *Tetrahedron Lett.* **1974**, 163–166; *ibid.*, 167–170.
156. van Leusen, A.M.; Strating, J.Q. *Rep. Sulfur Chem.* **1970**, *5*, 67.
157. Aldrich Chemical Co., Advertisement in *J. Chem. Soc., Chem. Commun.* **1984**, No. 1 (January).
158. van Leusen, A.M. *Lect. Heterocyclic Chem.* **1980**, *5*, S111.
159. Stotter, P.L.; Edwards, C.L.; Lade, R.E. *Abstracts, Southeast–Southwest Regional Meeting of the American Chemical Society, Memphis, TN, October, 1975* (American Chemical Society, Washington, DC, 1975), pp. 380 and 381.
160. Stotter, P.L., private communication, February, 1976.
161. Corey, E.J.; Gras, J.-L.; Ulrich, P. *Tetrahedron Lett.* **1976**, 809–812.
162. Greene, T.W. *Protective Groups in Organic Synthesis* (John Wiley & Sons, New York, 1981).
163. Brown, H.C., private communication, August, 1977.
164. (a) Brown, H.C.; Zweifel, G. *J. Am. Chem. Soc.* **1961**, *83*, 1241–1246, (b) Brown, H.C.; Moerikofer, A.W. *J. Am. Chem. Soc.* **1962**, *84*, 1478–1484.
165. Voronkov, M.G.; Zelcans, G.I. *Khim. Geterotsikl. Soedin.* **1965**, 51–57.
166. Voronkov, M.G. *Pure Appl. Chem.* **1966**, *13*, 35–59.
167. Voronkov, M.G., private communication, July, 1981.
168. Verkade, J.G., private communication, February, 1981.
169. Verkade, J.G.; Milbrath, D.S. *J. Am. Chem. Soc.* **1977**, *99*, 6607–6613.
170. Fuson, R.C.; Chadwick, D.H.; Ward, M.L. *J. Am. Chem. Soc.* **1946**, *68*, 389–393.
171. Miller, A.R., private communication, February, 1977.
172. Miller, A.R. *J. Org. Chem.* **1976**, *41*, 3599–3602.
173. Miller, A.R.; Curtin, D.Y. *J. Am. Chem. Soc.* **1976**, *98*, 1860–1865.
174. March, J. *Advanced Organic Chemistry*, 3rd ed. (John Wiley & Sons, New York, 1985), p. 66.
175. Chiang, Y.; Kresge, A.J.; Tang, Y.S. *J. Am. Chem. Soc.* **1984**, *106*, 460–462.
176. Sukenik, C.N.; Bergman, R.G. *J. Am. Chem. Soc.* **1976**, *98*, 6613–6623.
177. Sukenik, C.N., private communication, May, 1981.
178. Peterson, P.E.; Vidrine, D.W. *J. Org. Chem.* **1979**, *44*, 891–893.
179. (a) Fernelius, W.C. *J. Chem. Educ.* **1982**, *59*, 572–573, (b) Fernelius, W.C., private communication, August, 1982, (c) United States Adopted Names, **1974**, Supplement *10*, Appendix 1, 93–96. We thank Professor Fernelius for a copy of this material.
180. McManus, S.P.; Roberts, F.E.; Lam, D.H.; Hovanes, B. *J. Org. Chem.* **1982**, *47*, 4386–4388.
181. Eaborn, C.; Safa, K.D. *J. Organometal. Chem.* **1980**, *204*, 169–179.

182. Monitz, M.; Whiting, M.C. *J. Chem. Soc., Perkin Trans. 2* **1982**, 613–616. These workers used "nosylic acid" for *p*-nitrobenzenesulfonic acid.
183. Shiner, V.J., Jr.; Fisher, R.D. *J. Am. Chem. Soc.* **1971**, *93*, 2553–2554.
184. Bertz, S.H.; Dabbagh, G. *J. Org. Chem.* **1983**, *48*, 116–119. "Trimsyl" rather than "trimyl" might have maintained a closer family tie.
185. Chamberlin, A.R.; Stemke, J.E.; Bond, F.T. *J. Org. Chem.* **1978**, *43*, 147–154. These researchers and others[184] used "trisyl" for 2, 4, 6-triisoproplbenzenesulfonyl. Actually, "tipsyl" might be better as it builds logically upon tipyl, and abbreviation of triisopropylphenyl (see references 170–173.)

Chapter 17

I WANT TO BE ALONE!

Buffs of vintage movies may have heard about the immortal plea "I want to be alone," by screen star Greta Garbo. She did not relish fuss and attention. Organic chemistry has its Greta Garbos also: functional groups that couch behind chemically inert "shields" and show little or no desire to confront reagents. We have already met keuthynes with their Garboid triple bonds (p. 156), let us look at some others.

In 1967, James Marshall (then at Northwestern University) and Alex Nickon (Johns Hopkins University) dined together at a chemical conference in Natick, Massachusetts. In the course of conversation, it came out that each had independently thought of a way to allow a carbon–carbon double bond to "be alone." It could be sheltered from marauders if it were part of a *trans*-bicycloalkene structure **1**, in which two alkyl chains, one above and the other below the olefinic plane, block the double bond. In a spirit of friendly competition they agreed not to say any more to each other about their ideas. For a time, little was done on this problem by either party; each thought the other would surely pursue these interesting systems and therefore kept the project on the back burner.

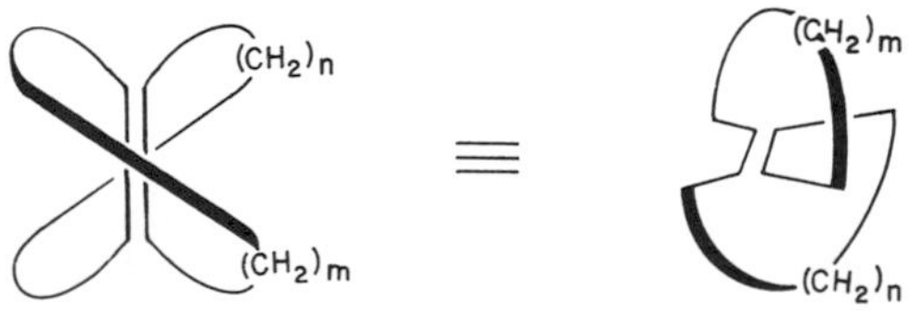

1

Finally, in 1977, Professor Marshall broke the ice.[1] A molecular model of **1** suggested that m and n should be eight or larger to allow each chain to loop to the opposite side without excessive strain. While his team labored to synthesize such alkenes (initially dubbed "loopanenes" by postdoctoral associate Mary Delton), mentor Marshall came up with the name "$[m, n]$betweenanenes." After all, *between* two alk*ane* chains the structures contain an alk*ene*.[2] His collaborators, Mary Delton, Alfonse Runquist, and Morris Llewellyn, reached their first betweenanene (the [10, 10]), after 16 unambiguous steps.[1]

Virtually simultaneously, a research group headed by Masao Nakazaki of Osaka University reported a short synthesis of [10, 8]betweenanene from the known triene **2**.[3] Selective hydrogenation of its disubstituted double bonds left the *cis*-bicycloalkene **3**.

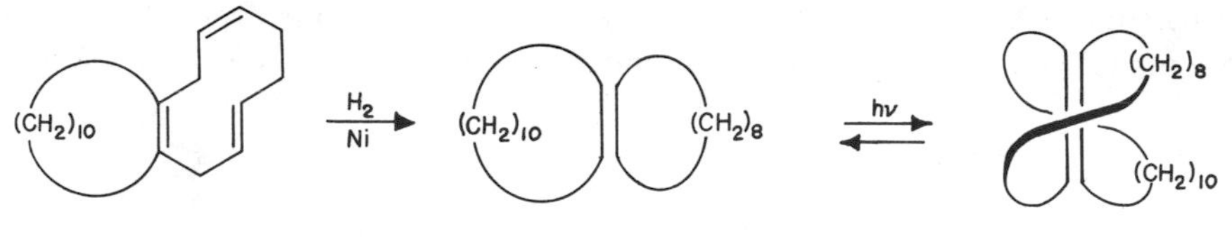

Then, irradiation in the presence of xylene as a photosensitizer afforded a mixture of **3** and **4**, difficult to separate. To isolate **4**, the Osaka team capitalized on the expectation that its double bond "wants to be alone." When exposed to dichlorocarbene, *cis* isomer **3** gave an adduct (**5**), which was easily removed; but **4** came through unscathed. The secluded π-unit in **4** was also untouched by hydrogenation under conditions that reduced **3**.[3]

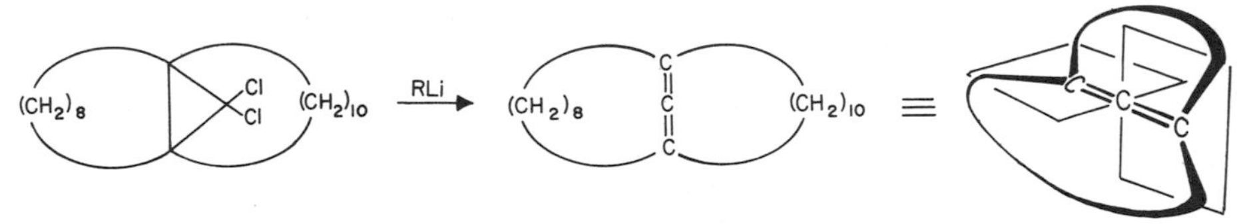

In 1981, Professor Nakazaki and his cohorts introduced the caconym* "screwene" for such doubly bridged perverts.[5a] For example, **4** became [8][10]screw[1]ene; and the "betweenallene"[5b] **6**, which the Nakazaki team prepared from **5**, is [8][10]screw[2]ene. This terminology is not popular, but it does have virtue; the digit in the third bracket tells how many cumulated double bonds there are.

In 1980, Johns Hopkins University chemists added [11, 11]betweenanene (**7**) to the growing collection of "domed" olefins by their three-step synthesis from commercially available cyclododecanone as shown.[6] And also at JHU, 1984 ushered in the first member having a heteroatom (sulfur) attached directly to the hidden double bond.[7]

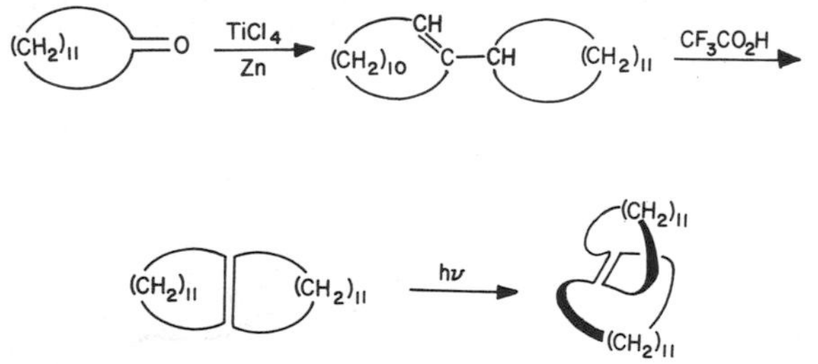

The most imperative chore for metal–porphyrin complexes such as hemoglobin is to transport molecular oxygen. Therefore, these complexes (**8**) must bind the gas and then release it when needed; the process **8** ⇌ **9** is called reversible oxygenation.[8] However, it is best if **9** does *not* react with another molecule of **8**, because that route leads irrevocably to the "μ-oxo dimer," **10**, which *cannot* carry dioxygen.[8] (All six of the

---

*Caconym: a name that is undesirable for linguistic reasons. Austin M. Patterson, noted authority on chemical nomenclature, thought that the most uncouth word in the English language is "cheilognath-ouranoschisis."[4] It also has an unpleasant meaning: "hare lip combined with cleft palate."

bonding sites to M are filled.) The process $8 \rightarrow 10$ amounts to irreversible oxidation. Like some teenagers who want to assert independence, 9 should stay clear of its own progenitor, 8.

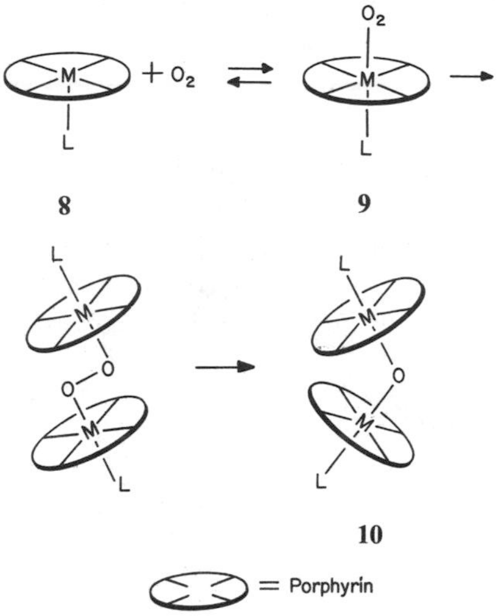

8   9

10

= Porphyrin

L = Ligand (e.g. an imidazole ring)
M = Metal (usually in the +2 oxidation state)

Experimentalists have found ways to slow down or prevent irreversible oxidation. For example, the research groups of Teddy Traylor (University of California at San Diego),[9] Fred Basolo (Northwestern University),[10] and Jack Baldwin (then at MIT)[11] did it by keeping things cold (-45 to $-78°C$). And, low concentrations ($< 10^{-4}$ M) also helped by lessening the chances of collision between 9 and 8.[9,11] A more specialized approach involves bulky groups on the side of 8 opposite the ligand. The idea is to allow the (relatively small) oxygen molecule to sneak in but to prevent the large, sterically hindered 8 and 9 from combining. This tactic is exactly what James Collman and co-workers at Stanford had in mind when they constructed their "picket fence" porphyrin, (p. 65, X=NHCOC(CH$_3$)$_3$); four bulky amide groups, pointing in the same direction, did the trick.

Baldwin's team also obtained oxygenation without oxidation with a "capped" porphyrin, 11. Their specific compound was 12,[12,13] whose structure surrendered to X-

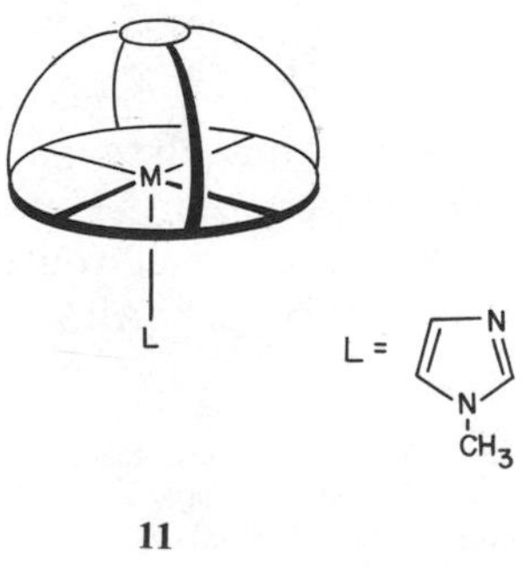

L =

11 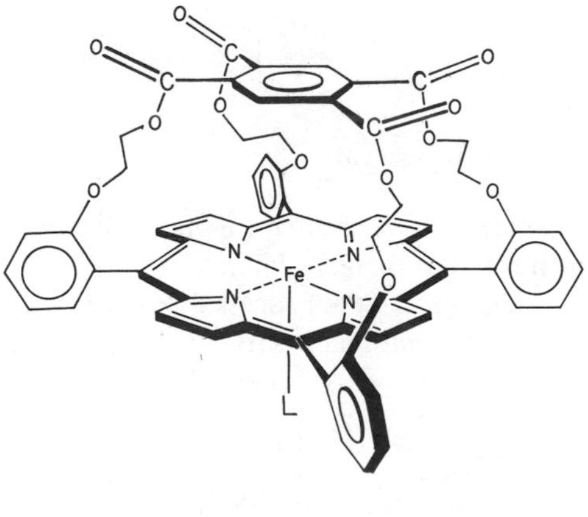   12

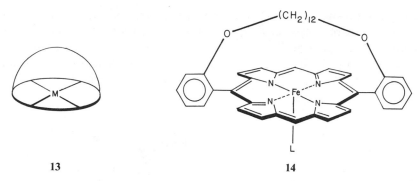

13   **14**

ray study.[14] Blockage by just one chain, as in a "strapped" porphyrin (**13**) was not enough; Professor Baldwin and co-workers found that **14** oxidized irreversibly.[8,15]

But strapping both decks can inhibit the loathsome dimerization, according to Michel Momenteau and cohorts at the University of Orsay.[16] In 1979, they used alkyl chains to join pairs of phenolic oxygens in tetrakis (*o*-hydroxyphenyl)porphyrin **15** and produced "basket handle" porphyrins—the desired one **16**, and two isomers **17** and **18**. (Meanwhile, on the other side of the English Channel, Allan Battersby's bunch at Cambridge University had also fashioned double-bridged porphyrins[17] analogous to **16**, as well as single-bridged ones.[18] These Britons used amide links to fasten their handles.) The Orsay researchers observed that the Fe(II) complex of basket-handle **16** simultaneously binds 1-methylimidazole and dioxygen while resisting the undesired μ-oxo coupling. Dr. Momenteau's team also constructed handles with pyridine or

15   **16**   **17**   **18**

imidazole units. In such porphyrins, the "hanging base" serves as an internal, fifth ligand for a central iron; and the derived complexes (**19** and **20**) grab molecular oxygen faster than do nonhanging counterparts.[19]

19   **20**

Eric Rose and co-workers, at Paris's Université Pierre and Marie Curie, "handled" porphyrins with two loops fixed to the same pair of sites (**21**) and called them "gyroscope-like" porphyrins.[20] The two great circles here bring to mind the armillary rings discussed on p. 45. In gyroscope porphyrins the two handles are mechanically linked and, presumably, so would be their conformational motions. For example, in an

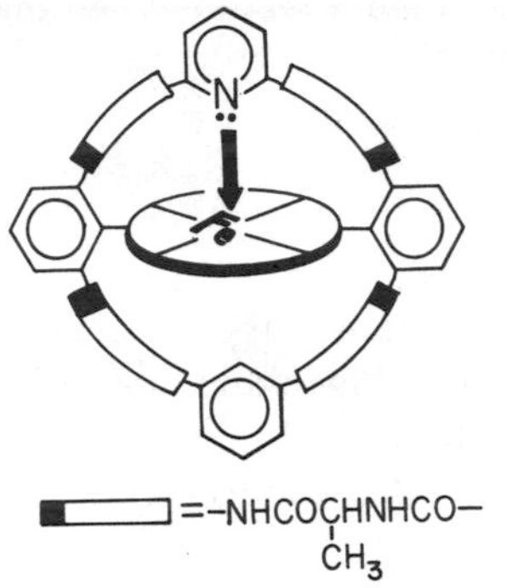

$\blacksquare = -NHCOCHNHCO-$
$\overset{|}{C}H_3$

21   **22**

iron complex like **22**, movement of the lower hydrophobic handle could be restricted by a ligated hanging base. Trust chemists to come to grips with a problem by putting handles on it!

Chi Kwong Chang of Michigan State University prepared a "crowned" porphyrin (**23**) in which a crown ether ring (chapter 2) covers one face of the π-rich macrocycle.[21] An oxygen molecule can slip in and out from under the crown; and the irreversible oxidation intrudes only very slowly.

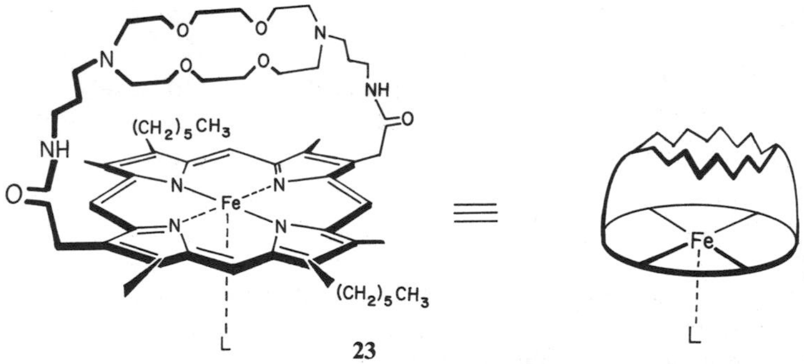

23

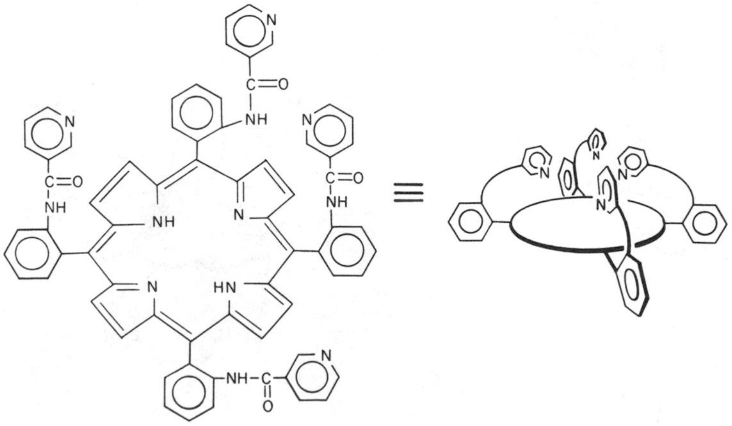

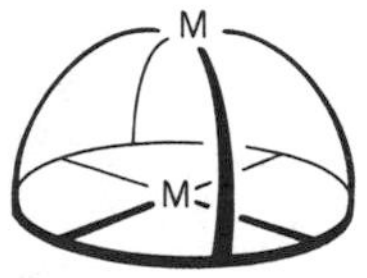

24   **25**

David Buckingham and his students at Australian National University converted one of Collman's picket fences (p. 65, $X = NH_2$) into a cap.[22] They accomplished this sartorial magic with the derived tetramide **24**. When treated with a metal salt, **24** binds one cation to its porphyrin posts (as usual), while another metal gathers the four pyridines to fashion the "cap" (**25**).

Teddy Traylor's team recruited yet another interesting member, **26**, to this clan of porphyrins. (The large hoop stands for the porphyrin moiety.) The array of rings atop the figure reminded the California researchers of the series of roofs on a pagoda; so they registered it as "pagoda" porphyrin.[23]

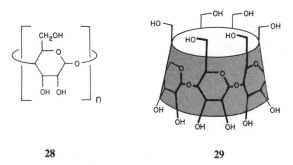

26   **27**

At the University of Freiburg, Horst Prinzbach likewise had a keen eye for oriental style. In 1983 he, co-worker Wolf-Dieter Fessner, and collaborator Grety Rihs (Ciba–Geigy Company, Basel) synthesized the 11 ring, compact polycycle **27** and christened it "pagodane."[24] Its layered ledges reminded them of a pagoda-style temple, especially if you disregard the reflection symmetry.[25] The European architects assembled this $C_{20}H_{20}$ polyquinane with the hope of eventually transforming it to dodecahedrane. The remodelling contract went to Paul Schleyer and his student Wolfgang Roth at the University of Erlangen–Nürnberg.[26] In a collaborative coup, the Erlangen and Freiburg conglomerate amazingly succeeded in isomerizing pagodane to dodecahedrane![27]

Capping also became chic for cyclodextrins (**28**), which are cyclic oligomers of glucose.[28] When $n = 6$, 7, or 8, they are called α-, β-, and γ-cyclodextrin, respectively;[29] and such molecules adopt overall shapes resembling truncated cones (**29**).[29,30] Primary

28   **29**

OH groups dangle near the narrow end, and secondary OH's guard the wider entrance.

The 6–10 Å (600–1000 pm) cavity allows these oligomers to form inclusion complexes with a variety of guest molecules.[29]

In 1982, Iwao Tabushi of Kyoto University wanted a cyclodextrin with two of the primary OH's replaced. So he and his colleagues deftly added a cap as in **30**, then doffed it by nucleophilic displacement, first with azide ion (**31**) then with p-tert-butylphenylthiolate ion (**32**). They delightfully referred to the sequence as "flamingo-type capping."[31] The flamingo, a graceful bird, stands on both its spindly legs (cf. **33**) but can raise one (cf. **34**) and then the other and fly away (**35**).[32] Dr. Tabushi's ornithological analogy was inspired by the logo (see **35**) of Japan Air Lines (JAL) as well as by a former graduate student, Lung–Chi Yuan, who was fond of pink clothes (à la flamingos).[32] If you yearn to make cyclodextrins, a review by Croft and Bartsch glides you nicely through the literature.[33]

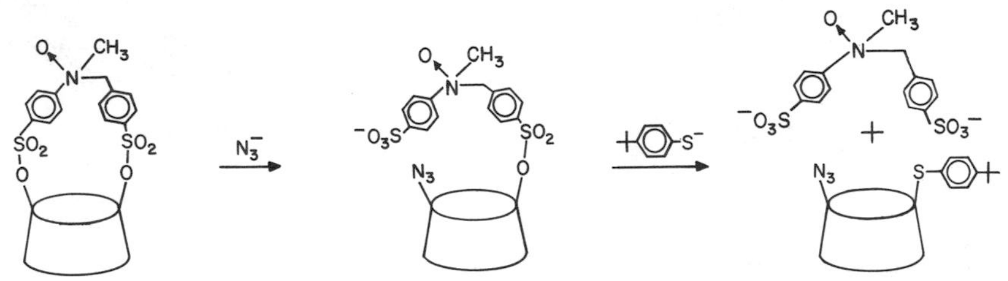

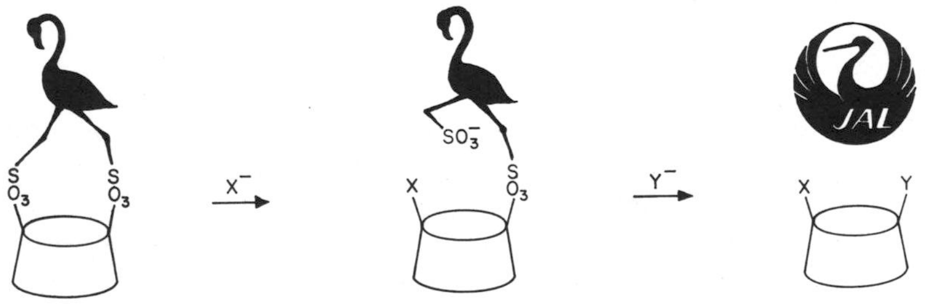

The Kyoto researchers were also into heavy construction. For example, they assembled the dimer-type porphyrin **36** and presented evidence that its binding to metal cations (M) in the presence of dipyridylmethane involves a gable-like conformation (**37**). Professor Tabushi and co-worker Tomikazu Sasaki recommended such "gable porphyrins" and their complexes as models to mimic some of the unique electron transport properties of cytochrome c_3.[34]

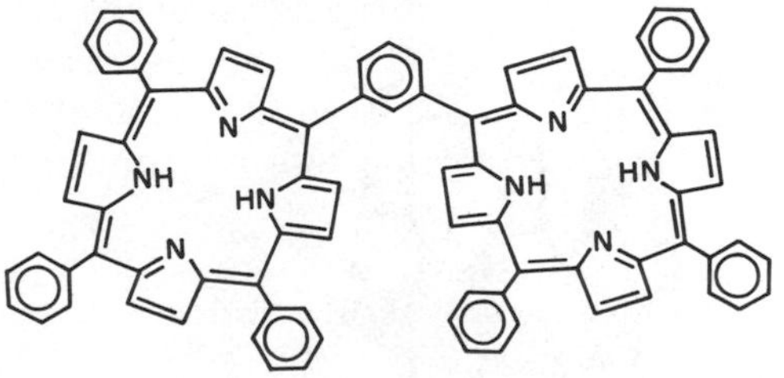

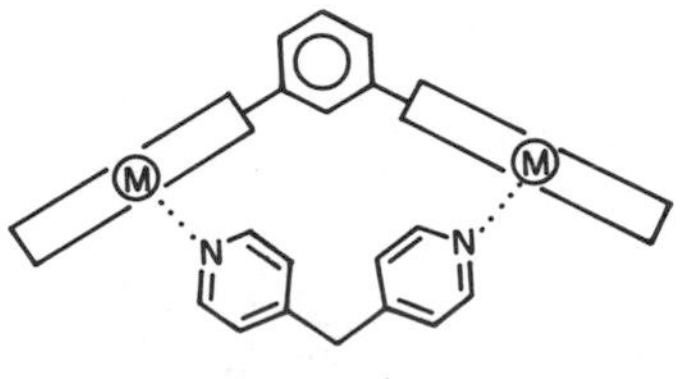

Ions often want to be alone. For example, crown ligands (chapter 2) seclude metal cations quite effectively by ringing them with oxygens. Jean–Marie Lehn, of Louis

Pasteur University in Strasbourg, propelled this idea into three dimensions with compounds such as **38**.[35,36] Dr. Lehn's researchers found that various ions can be

38   **39**

captured within such a structure by the heteroatoms (**39**).[35] They coined "cryptate," from the Greek *kryptos* ("hidden") and from the Latin *crypta* ("cavity") to describe these complexes.[37] And they chose "cryptand"* for the free host molecule **38**;[35,37] note that this word ends in the same three letters as ligand. The numbers of heteroatoms in the three chains (excluding the bridgehead nitrogens) are placed in brackets; thus, **38** became a [2, 2, 2]cryptand.[36]

Cryptates are generally more stable than the corresponding crown complexes. For example, K^+ and Ba^{++} bind 10^5 times tighter to **38** than they do to the crown counterpart, **40**.[43] Once inside a cryptand, ions can be very much isolated from the medium. Professor Lehn's students showed that two protons within the [1, 1, 1]-cryptate **41** exchange with solvent only very slowly, even in a strongly basic aqueous

40   **41**

medium.[44] A research trio led by Alexander Popov of Michigan State University found that the ^7Li NMR chemical shift in $^7Li^+$-[2, 1, 1]cryptate is the same in polar and nonpolar solvents.[45] Evidently, in its crypt, lithium is oblivious to exterior changes. Francois Peter and Maurice Gross of Louis Pasteur University, Strasbourg, reported that ions imprisoned in cryptates are harder to reduce than ions on the outside.[46]

Howard Simmons, Chung Ho Park, and colleagues at the Du Pont Company's Central Research Laboratory realized that cryptands such as **42** can, in theory, exist as three topological isomers (**a, b**, and **c**), because the lone pair on each nitrogen can turn inward (toward the cavity) or outward.[47–49] For this type of isomerism, they offer a choice of the terms "in–out"[47] or "homeomorphic"[50] (from Greek *homoios*, meaning "same," and *morphē*, meaning "form"); thus, **a, b**, and **c** are the in–in, in–out, and out–out forms, respectively. In general, homeomorphic isomers could interconvert by

---

*This term even went commercial. The Parish Chemical Company of Orem, Utah, referred to its line of cryptands as "Kryptofix."[38] But crypts are not always what they seem, at least in the chemical literature. For instance, "cryptal" and "cryptone" have nothing to do with Professor Lehn's cavities or with his complexes. In fact, both names stood for a natural product isolated from various eucalyptus oils in the 1920s. For years the compound masqueraded as an aldehyde and so was called "cryptal."[39] But in 1931, its structure was proved to be that of 4-isopropyl-2-cyclohexenone.[40] So Cooke and Macbeth in 1938 at the University of Adelaide appropriately rechristened it "cryptone,"[41] a sound to which this ketone still answers.[42]

conformational contortion that passes one chain through the ring of the other two. With bridgehead nitrogens an alternative path exists, via pyramidal inversion at each N.

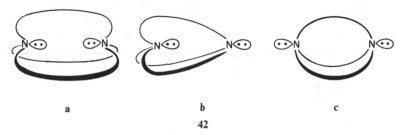

a   b   c

42

Raymond Weiss's group at Louis Pasteur University reported that **38** prefers in–in; but the **38**—$(BH_3)_2$ complex is out–out.[51] Evidently, boron acts as a suction on unshared electron pairs.

At the University of Bristol, Professor Roger Alder and his students put together several bicyclic diamines of type **42**, as well as mononitrogen counterparts.[52] Through deals made by Alder, these bicyclics earned labels according to the number of atoms in each bridge. For example, **41** was simply a [5.5.5]diamine. Their Bristol slang for a [4.4.4]monoamine was "hiddenamine," because the "in" lone pair is hidden from the outside world. And, naturally, an unsaturated analog became "hiddenenamine."[53]

Cute names have also been hatched for some first cousins of cryptates and cryptands. Drs. Park and Simmons have dubbed in–in dihydrochlorides such as **43**, in which one chloride ion is encapsulated, "katapinates," from the Greek *katapinos* (to "swallow" or "engulf").[49b, 54]

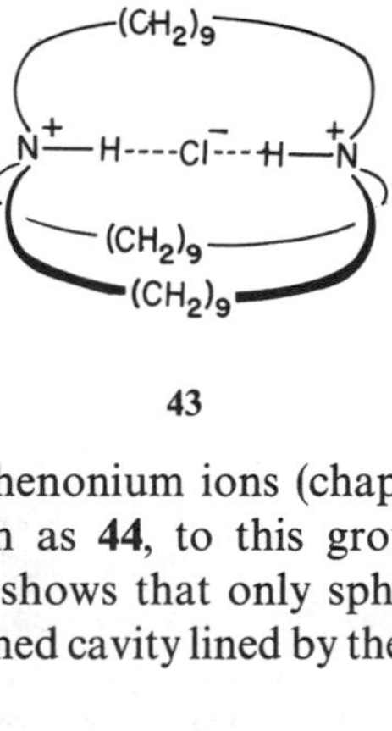

$$N^+\!\!-\!\!H\cdots Cl^-\cdots H\!\!-\!\!N^+$$

43

Donald Cram, who vended phenonium ions (chapter 12) and cyclophanes (chapter 16), added "spherands," such as **44**, to this growing assortment of interesting ligands.[55] A space-filling model shows that only spheroidal guests (e.g., monatomic cations) fit snugly into the preformed cavity lined by the oxygens (**45**). The captive ion in

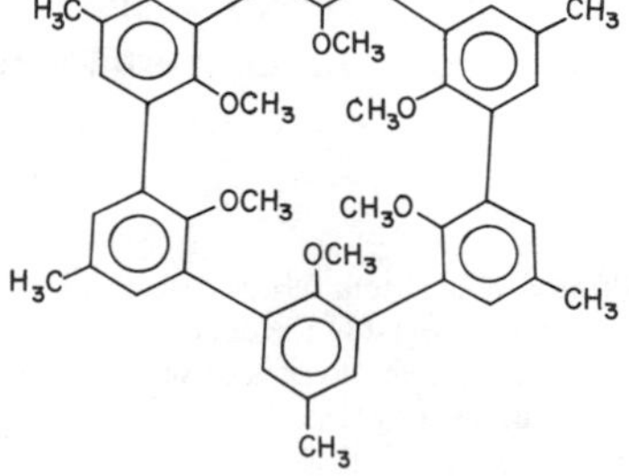

44 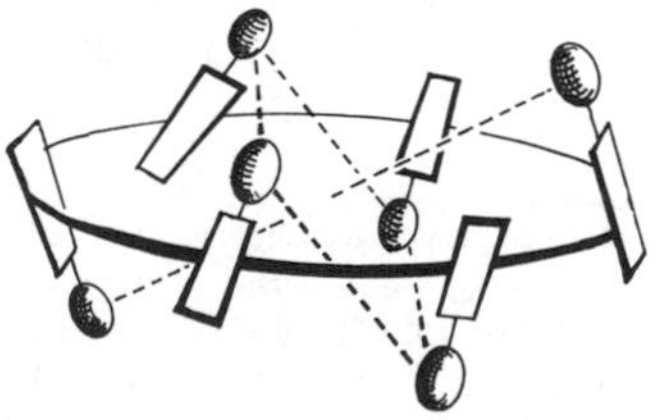   45

such "metallospherium" salts[55] hides behind six benzene rings; it truly "wants to be alone." Professor Cram's constituency also designed and constructed partial spheres[56] and called such captors "hemispherands."[55]

Chemists in Australia prepared cobalt compound **46** and named it a "sepulchrate," from the Latin *sepulcrum* ("grave" or "tomb").[57] Fritz Vögtle and Edwin Weber at the University of Bonn gave us "coronand" (from the Latin *corona*, meaning "crown") for monocyclic ligands that form crown-shaped complexes (see 18-crown-6 and other crown ethers in chapter 2).[58] The Bonn duo also recommended "podand" (from the Greek *podos*, genetive for "foot") for ligands (e.g., **47**) in which the donor atoms are not part of a ring.[59,60] Octopus and hexapus molecules (chapter 1), with six such feet or arms, are hexapodands.

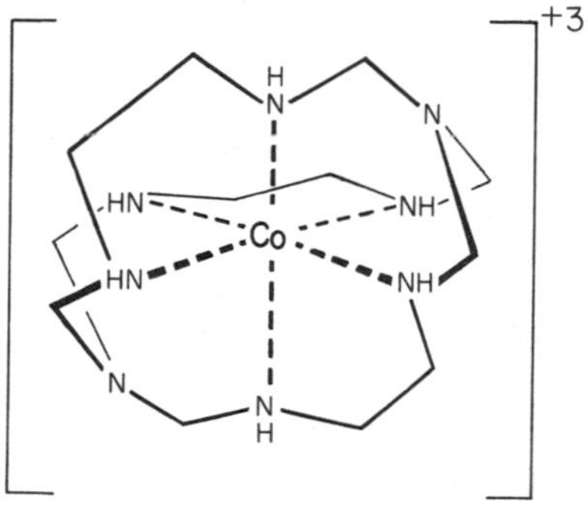

46 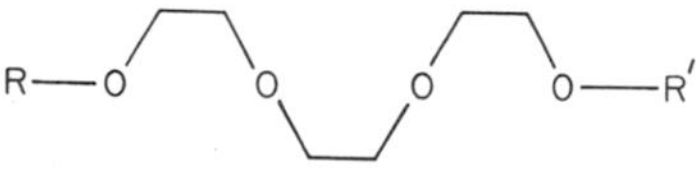   47

James Damewood (University of Delaware) had tunnel vision when he dreamed up crown-like cryptands that might bind acetylcholine and other sausage-shaped polar goodies of biological interest. At an intercollegiate conference in 1986, his students reported progress on the synthesis of their "tunnelanes."[61]

Scientists in Grenoble, France, have even induced paracyclophanes to play the "complex" game.[62] Specifically, they noted that paracyclophane **48** looks like a shallow cylinder (see **49**) whose inner wall of π-electrons could tether a metal ion. Indeed, this molecule forms a stable, crystalline, 1:1 aggregate with silver triflate; and the stability constant in methanol solution is *ca.* 100-fold greater than usual for arene–Ag$^+$ complexes. A perceived prism-like juxtaposition of the three benzene rings (see **50**) led the Grenoble investigators to dub this type of host a "π-prismand" and the phenomenon, "π-cryptation." X-ray studies elsewhere suggested that the Ag$^+$ in these complexes has not penetrated the phane cavity but hovers near the entrance.[63] In any case, π-prismands and π-cryptation are good terms to keep in mind if you want to corner the silver market.

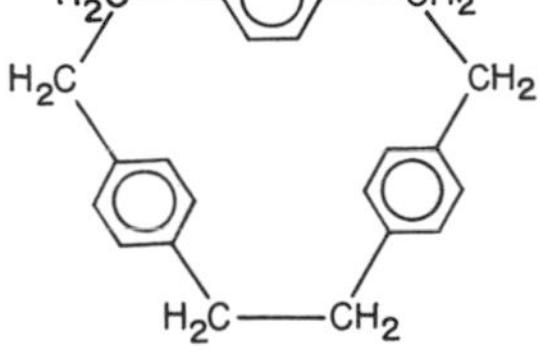

48    49    50

André Collet and colleagues at Collège de France in Paris used three strands to join two saucer-shaped fragments and ended up with the likes of **51**. The hollow can accommodate small molecules; and optically active hosts can serve as chiral NMR shift reagents. In fact, HCFClBr was partially resolved through this type of enantioselective complexation (**52**). Hosts like **51** embody structural characteristics of cryptands and of cyclophanes, so Collet *et al.* aptly named this family "cryptophanes."[64]

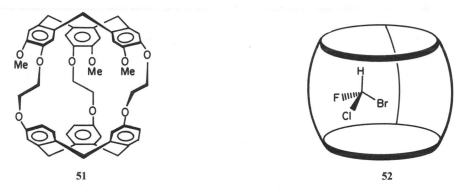

51   **52**

Complexation intrigues chemists, because they crave to understand and to mimic the remarkable specificity of enzyme–substrate interactions, of antibody–antigen phenomena, and of ionophore–metal ion selectivity.[65] But if you delve into the "complex" literature, be ready to falter over parlance that is not always uniform or well defined. You will meet broad categories such as "cavitand"[66] and "speleand" (from Greek *spelaion* meaning "cave")[67] for *hosts* with more-or-less rigid cavities. Picturesque *ligand descriptors* include "hinged," "belted,"[68] and "soccer."[69] Derived *complexes* have been designated as "cascade,"[70] "perching," "nesting," "strapped," "capsular," and "closed jaw."[68] *Host–guest recognition* can involve "central," "lateral,"[71] "circular," and "trigonal" discrimination.[72] And don't wince when you realize "chorand"[73] means virtually the same as coronand,[58] or when you see katapinate[49b] spelled catapinate.[69]

In 1980, Edwin Weber and Fritz Vögtle began to systematize the nomenclature in this field;[74] and in 1983 Weber and Josel proposed a comprehensive classification and terminology for host–guest-type compounds.[75,76] Their recommendations take account of different types of host–guest interactions as well as topology of the aggregate and the number of components. The authors retain old favorites like clathrate (see later), as well as youngsters like cryptate and cavitate. But they also proffered some new ones, such as "tubulate" (Latin *tubus* means "tube") for channel-type topology; and "aediculate" (from latin *aedicula*) for pocket- or niche-type topology. So, if you are an ion or a small molecule looking for privacy, a cryptato-clathrate or a tubulato-cavitate might be just right.[75] But, a coronato complex could be more interesting, especially since "cyclosexipyridines" became available as hosts. George Newkome and graduate pupil H.–W. Lee at Louisiana State University synthesized the parent sexibody **53**;[77] and John Toner at Eastman Kodak Company

53   **54**

assembled partially clothed analogs and isolated them as 1:1 NaOAc coronato complexes (e.g., **54**).[78] Having met sexiferrocenophanes previously (p. 218), we can now see virtue in "sexipyridines."

Thomas Bell and Albert Firestone at State University of New York (Stony Brook) made **53** even sexier. How? By adding *six* more *six*-membered rings (to get **55**, R = *n*-butyl). Such compounds have the potential to bind metal ions (**56**), and their semirigid toroidal shapes earned these hosts the class name "torands."[79] Note that **55** is in fact an elaborated kekulene-like framework (chapter 10).

METAL   ION

55   **56**

The venerable term "clathrate" (from Latin *clathratus*, meaning "enclosed or protected by cross bars of a grating") was coined in 1948 by Oxford University crystallographer H.M. Powell.[80] Dr. Powell recognized that a small molecule could be completely enclosed within a cavity formed by one or more molecules of another compound. The encapsulated entity cannot escape without overcoming the strong forces that created the cage. For example, water can crystallize in cage-like networks that entrap guest molecules. Such "clathrate hydrates" also have parallels in silicate chemistry. And in 1982, University of Kiel mineralogists Hermann Gies, Friedrich Liebau, and Horst Gerke coined "clathrasils" for clathrates with SiO_2 host lattices. One of their synthetic silicates became "dodecasil-3C." This name identifies the type of lattice (*dodeca*hedron); the host compound (*sil*ica); the number of layers in the unit cell (3); and the idealized crystal system (*c*ubic).[81]

Another famous "sil" is well known to those concerned with zeolite structures. Zeolites are framework aluminosilicates with characteristic cavities, channels, and supercages; they enjoy wide application as sorbents and catalysts, especially in the petroleum industry.[82] At Mobil Oil Company, George Kokotailo and colleagues worked out the structures of specific ones called ZSM-5 and ZSM-11. These two are high in silica and comprise oxygen-containing five-membered rings fused as in **57**. Dr. Kokotailo coined for this repeating structural unit the name "pentasil" (penta for five; sil for silicon). The zeolites in question have their networks of pentasils linked up differently, and one possible combination is shown (**58**).

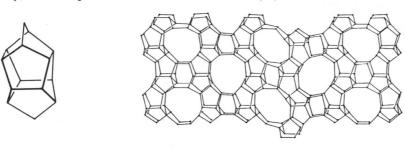

57   **58**

The "ZSM" call letters were first adopted by Julius Ciric and stood for "Zeolite Socony Mobil" (the company was at that time called Socony Mobil). Those letters continued to be used even for new zeolites because the name of the vice president in charge of research in those days was Seymour Meisel, and no one dared to alter the lettering.[83]

At Columbia University zeolites have found use as storage containers, of sorts. It began when Nicholas Turro and Charles Doubleday, Jr., and their associates discovered that photic α-cleavage of large-ring 2-phenylcycloalkanones produced paracyclophanes by a specific mode of biradical collapse. Appreciating that such a transformation alters the overall size and shape of a molecule, they irradiated (for example) 2-phenylcyclododecanone that had been first adsorbed into the supercage interior of a zeolite by normal deposition from pentane solvent. After the photic event, the corpulent paracyclophane proved too big to slip out, and so remained trapped. The port city team called this irreversible encapsulation their "ship in the bottle" strategy and pictoralized it as **59 → 60 → 61**.[84]

59   60   61

For precursors too large to get into the bottle initially, the same workers devised a "reptation" approach, which involved irradiation of a ketone adsorbed onto the zeolite's outer surface. The biradical intermediate was sufficiently pliant to slither inside. There the radical termini joined up and doomed the paracyclophane forever to remain in its house of birth. Thanks to our colleagues at Columbia, parents can now explain to children how ships get into bottles with narrow necks.

REFERENCES AND NOTES

1. (a) Marshall, J.A.; Llewellyn, M. *J. Am. Chem. Soc.* **1977**, *99*, 3508–3510, (b) Marshall, J.A. *Acc. Chem. Res.* **1980**, *13*, 213–218.
2. Marshall, J.A., private communication, August, 1978.
3. Nakazaki, M.; Yamamoto, K.; Yanagi, J. *J. Chem. Soc., Chem. Commun.* **1977**, 346–347.
4. Patterson, A.M. *Chem. Eng. News* **1952**, *30*, 5086.
5. (a) Nakazaki, M.; Yamamoto, K.; Maeda, M. *Chemistry Lett.* **1981**, 1035–1036, (b) Marshall, J.A.; Rothenberger, S.D. *Tetrahedron Lett.* **1986**, *27*, 4845–4848.
6. Nickon, A.; Zurer, P.S. *Tetrahedron Lett.* **1980**, *21*, 3527–3530.
7. Nickon, A.; Rodriguez, A.; Shirhatti, V.; Ganguly, R. *Tetrahedron Lett.* **1984**, *25*, 3555–3558.
8. Baldwin, J.E.; Klose, T.; Peters, M. *J. Chem. Soc., Chem., Commun.* **1976**, 881–883.
9. Chang, C.K.; Traylor, T.G. *J. Am. Chem. Soc.* **1973**, *95*, 5810–5811.
10. Anderson, D.L.; Weschler, C.J.; Basolo, F. *J. Am. Chem. Soc.* **1974**, *96*, 5599–5600.
11. Almog, J.; Baldwin, J.E.; Dyer, R.L.; Huff, J.; Wilkerson, C.J. *J. Am. Chem. Soc.* **1974**, *96*, 5600–5601.
12. Almog, J.; Baldwin, J.E.; Dyer, R.L.; Peters, M. *J. Am. Chem. Soc.* **1975**, *97*, 226–227.
13. Almog, J.; Baldwin, J..E.; Huff, J. *J. Am. Chem. Soc.* **1975**, *97*, 227–228.
14. Jameson, G.B.; Ibers, J.A. *J. Am. Chem. Soc.* **1980**, *102*, 2823–2831.
15. Baldwin, J.E.; Perlmutter, P. *Top Curr. Chem.* **1984**, *121*, 181–220.

16. (a) Momenteau, M.; Loock, B.; Mispelter, J.; Bisagni, E. *Nouv. J. Chim.* **1979**, *3*, 77–79, (b) Momenteau, M.; Mispelter, J.; Loock, B.; Bisagni, E. *J. Chem. Soc., Perkin Trans. 1* **1983**, 189–196.
17. Battersby, A.R.; Hartley, S.G.; Turnbull, M.D. *Tetrahedron Lett.* **1978**, 3169–3172.
18. Battersby, A.R.; Buckley, D.G.; Hartley, S.G.; Turnbull, M.D. *J. Chem. Soc., Chem. Commun.* **1976**, 879–881.
19. (a) Momenteau, M.; Lavalette, D. *J. Chem. Soc., Chem. Commun.* **1982**, 341–343, (b) Momenteau, M.; Loock, B.; Lavalette, D.; Tetreau, C.; Mispelter, J. *J. Chem. Soc., Chem. Commun.* **1983**, 962–964, (c) Momenteau, M.; Mispelter, J.; Loock, B.; Lhoste, J.-M. *J. Chem. Soc., Perkin Trans 1* **1985**, 61–70, 221–231.
20. Boitrel, B.; Lecas, A.; Renko, Z.; Rose, E. *J. Chem. Soc., Chem. Commun* **1985**, 1820–1821.
21. Chang, C.K. *J. Am. Chem. Soc.* **1977**, *99*, 2819–2822.
22. Buckingham, D.A.; Gunter, M.J.; Mander, L.N. *J. Am. Chem. Soc.* **1978**, *100*, 2899–2901.
23. Traylor, T.G.; Campbell, D.; Tsuchiya, S. *J. Am. Chem. Soc.* **1979**, *101*, 4748–4749.
24. Fessner, W.-D.; Prinzbach, H.; Rihs, G. *Tetrahedron Lett.* **1983**, *24*, 5857–5860.
25. Prinzbach, H., private communication, February, 1984.
26. Schleyer, P.v.R., private communications, March, 1984 and June, 1986.
27. Roth, W. Doctoral dissertation, University of Erlangen–Nürnberg, 1984. For citations to these and related preliminary results, see (a) Carceller, E.; García, M.L.; Moyano, A.; Pericas, M.A.; Serratosa, F. *Tetrahedron* **1986**, *42*, 1831–1839, (b) Prakash, G.K.S.; Krishnamurthy, V.V.; Herges, R.; Bau, R.; Yuan, H.; Olah, G.A.; Fessner, W.-D.; Prinzbach, H. *J. Am. Chem. Soc.* **1986**, *108*, 836–838.
28. Tabushi, I.; Shimokawa, K.; Shimizu, N.; Shirakata, H.; Fujita, K. *J. Am. Chem. Soc.* **1976**, *98*, 7855–7856.
29. Cramer, F.; Saenger, W.; Spatz, H.-Ch. *J. Am. Chem. Soc.* **1967**, *89*, 14–20.
30. Breslow, R.; Overman, L.E. *J. Am. Chem. soc.* **1970**, *92*, 1075–1077.
31. Tabushi, I.; Nabeshima, T.; Kitaguchi, H.; Yamamura, K. *J. Am. Chem. Soc.* **1982**, *104*, 2017–2019.
32. Tabushi, I., private communication, May, 1982.
33. Croft, A.P.; Bartsch, R.A. *Tetrahedron* **1983**, *39*, 1417–1474.
34. Tabushi, I.; Sasaki, T. *Tetrahedron Lett.* **1982**, *23*, 1913–1916.
35. Dietrich, B.; Lehn, J.-M.; Sauvage, J.P. *Tetrahedron Lett.* **1969**, 2889–2892.
36. Lehn, J.-M. *Accts. Chem. Res.* **1978**, *11*, 49–57.
37. Lehn, J.-M., private communication, March, 1976.
38. Advertisement, *J. Org. Chem.* **1979**, *44*, p. 2A.
39. Penfold, A.R. *J. Chem. Soc., Transactions* **1922**, *121*, 266–269, and references cited there.
40. Cahn, R.S.; Penfold, A.R.; Simonsen, J.L. *J. Chem. Soc.* **1931**, 1366–1369.
41. Cooke, R.G.; Macbeth, A.K. *J. Chem. Soc.* **1938**, 1408–1413.
42. Thomas, A.F. The Synthesis of Monoterpenes. In *The Total Synthesis of Natural Products*, ApSimon, J. (Ed.), (John Wiley & Sons, New York, 1983), vol. 2, pp. 1–195. See other chapters in this series for additional natural substances with "crypt-" names.
43. Lehn, J.-M.; Sauvage, J.P. *J. Am. Chem. Soc.* **1975**, *97*, 6700–6707.
44. Cheney, J.; Lehn, J.-M. *J. Chem. Soc., Chem. Commun.* **1972**, 487–489.
45. Cahen, Y.M.; Dye, J.L.; Popov, A.I. *J. Phys. Chem.* **1975**, *79*, 1289–1291.
46. (a) Peter, F.; Gross, M.J. *Electroanal. Chem.* **1974**, *53*, 307–315, (b) *Ibid.* **1975**, *61*, 245–248.
47. Simmons, H.E.; Park, C.H.; Uyeda, R.T.; Habibi, M.F. *Trans N.Y. Acad. Sci.* **1970**, *32*, 521–534.
48. Simmons, H.E.; Park, C.H. *J. Am. Chem. Soc.* **1968**, *90*, 2428–2429.
49. (a) Park, C.H.; Simmons, H.E. *J. Am. Chem. Soc.* **1968**, *90*, 2429–2431, (b) *Ibid.* **1968**, *90*, 2431–2432.
50. Park, C.H.; Simmons, H.E. *J. Am. Chem. Soc.* **1972**, *94*, 7184–7186.
51. Metz, B.; Moras, D.; Weiss, R. *J. Chem. Soc., Perkin Trans. 2* **1976**, 423–429.
52. Alder, R.W. *Acc. Chem. Res.* **1983**, *16*, 321–327.
53. Alder, R.W., private communication, February, 1986.
54. Simmons, H.E., phone discussion, June, 1978.
55. Cram, D.J.; Kaneda, T.; Helgeson, R.C.; Lein, G.M. *J. Am. Chem. Soc.* **1979**, *101*, 6752–6754.
56. Koenig, K.E.; Lein, G.M.; Stuckler, P.; Kaneda, T.; Cram, D.J. *J. Am. Chem. Soc.* **1979**, *101*, 3553–3566.
57. Creaser, I.I.; Harrowfield, J. MacB.; Herlt, A.J.; Sargeson, A.M.; Springborg, J.; Gene, R.J.; Snow, M.R. *J. Am. Chem. Soc.* **1977**, *99*, 3181–3182.

58. Vögtle, F.; Weber, E. *Angew. Chem. Int. Ed. Engl.* **1974**, *13*, 814–816.
59. Vögtle, F.; Weber, E. *Angew. Chem. Int. Ed. Engl.* **1979**, *18*, 753–776.
60. Vögtle, F., private communication, June, 1981.
61. (a) Hadad, C.M.; Seidl, E.T.; Damewood, J.R., Jr. Paper delivered at the annual conference of Intercollegiate Student Chemists, West Chester University, PA, April 12, 1986 (unpublished), (b) Damewood, J.R., Jr., personal discussion, April 12, 1986.
62. Pierre, J.-L.; Baret, P.; Chautemps, P.; Armand, M. *J. Am. Chem. Soc.* **1981**, *103*, 2986–2988.
63. (a) Cohen–Addad, C.; Baret, P.; Chautemps, P.; Pierre, J.-L. *Acta Crystallogr., Sect. C* **1983**, *39*, 1346–1349, (b) Kang, H.C.; Hanson, A.W.; Eaton, B.; Boekelheide, V. *J. Am. Chem. Soc.* **1985**, *107*, 1979–1985.
64. Canceill, J.; Lacombe, L.; Collet, A. *J. Am. Chem. Soc.* **1985**, *107*, 6993–6996.
65. (a) Host–Guest Complex Chemistry I and II. In *Topics in Current Chemistry*, Vögtle, F. (Ed.), (Springer-Verlag, Berlin, vol. 98, 1981), vol. 101, **1982**, (b) *Host–Guest Complex Chemistry/Macrocycles*, Vögtle, F.; Weber, E. (Eds.), (Springer-Verlag, New York, 1985).
66. Moran, J.R.; Karbach, S.; Cram, D.J. *J. Am. Chem. Soc.* **1982**, *104*, 5826–5828.
67. Canceill, J.; Collet, A.; Gabard, J.; Kotzyba–Hibert, F.; Lehn, J.-M. *Helv. Chim. Acta* **1982**, *65*, 1894–1897.
68. Cram, D.J.; Trueblood, K.N. *Top. Curr. Chem.* **1981**, *98*, 43–106.
69. Vögtle, F.; Sieger, H.; Müller, W.M. *Top. Curr. Chem.* **1981**, *98*, 107–161.
70. (a) Lehn, J.-M. *Pure Appl. Chem.* **1977**, *49*, 857–870, (b) Lehn, J.-M.; Simon, J. *Helv. Chim. Acta* **1977**, *60*, 141–151.
71. Lehn, J.-M. *Pure Appl. Chem.* **1978**, *50*, 871–892.
72. Weber, E.; Vögtle, F. *Top. Curr. Chem.* **1981**, *98*, 1–41.
73. Cram, D.J.; Dicker, I.B.; Lein, G.M.; Knobler, C.B.; Trueblood, K.N. *J. Am. Chem. Soc.* **1982**, *104*, 6827–6828.
74. Weber, E.; Vögtle, F. *Inorg. Chim. Acta* **1980**, *45*, L65–L67.
75. Weber, E.; Josel, H.-P. *J. Incl. Phenom.* **1983**, *1*, 79–85.
76. Weber, E., private communications, March and September, 1983.
77. Newkome, G.R.; Lee, H.-W. *J. Am. Chem. Soc.* **1983**, *105*, 5956–5957.
78. Toner, J.L. *Tetrahedron Lett.* **1983**, *27*, 2707–2710.
79. Bell. T.W.; Firestone, A. *Abstracts of Papers, 192nd National Meeting of the American Chemical Society, Anaheim, CA.* (American Chemical Society, Washington, DC, 1986), ORGN 23.
80. Powell, H.M. *J. Chem. Soc.* **1948**, 61–73.
81. Gies, H.; Liebau, F.; Gerke, H. *Angew. Chem. Int. Ed. Engl.* **1982**, *21*, 206–207.
82. (a) Fyfe, C.A.; Kokotailo, G.T.; Kennedy, G.J.; DeSchutter, C. *J. Chem. Soc., Chem. Commun.* **1985**, 306–308, (b) Kokotailo, G.T. In *Zeolites: Science and Technology*, Ribeiro, F.R.; Rodriguez, A.E.; Rollmann, L.D.; Naccache, C. (Eds.), (Martinus Nijhoff Publishers, The Hague, 1984).
83. Kokotailo, G.T., private communication, January, 1986.
84. Lei, X.; Doubleday, C.E., Jr.; Zimmt, M.B.; Turro, N.J. *J. Am. Chem. Soc.* **1986**, *108*, 2444–2445.

Chapter 18

THAT'S THE WAY IT GOES!

Up to now, we have focused mainly on names of substances. But chemists also preoccupy themselves with reactions and with step-by-step descriptions of their mechanisms. Some transformations have catchy titles; for example, we have already wheeled out the bicycle reaction (chapter 5). Let us continue this exercise by taking some walks.

In 1978, Robert Hutchins's research hustlers at Drexel University converted α, β-unsaturated aldehydes and ketones to tosylhydrazones and reduced these to alkenes in which the olefinic link had moved. As an example, **1** gave **2**, even though the π-bonds are then unconjugated. The investigators called this migration an "alkene walk."[1]

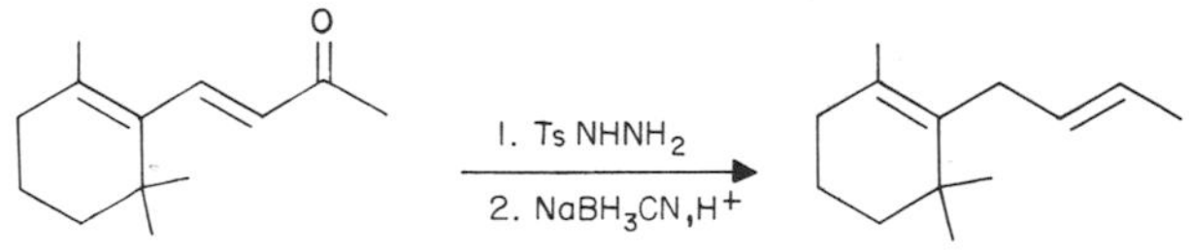

Sulfurs need exercise, too, so Colin Day's team at Oxford University interconverted the cyanothiophenes **3** and **4** by irradiation.[2] They could trap no thiophene and therefore proposed two intramolecular electrocyclic steps with an internal "sulfur walk" inbetween.*

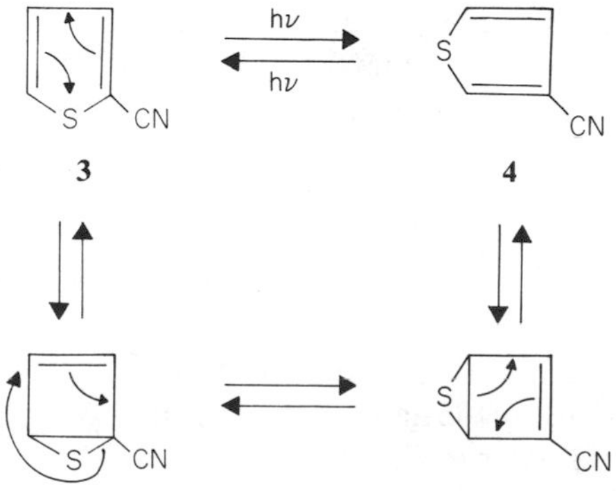

---

*A review by Ronald Childs of McMaster University, Ontario, labeled such "walks" around rings as "circumambulatory rearrangements."[3]

Frank–Gerrit Klarner's squad at the University of Bochum, Germany, noticed the "walk" of a $C(CH_3)CO_2CH_3$ moiety when they transformed **5** to **6** at $0°C$.[4] This [1, 3]sigmatropic stroll inverts the configuration at the "walking" carbon, as decreed by the Woodward–Hoffmann canons.

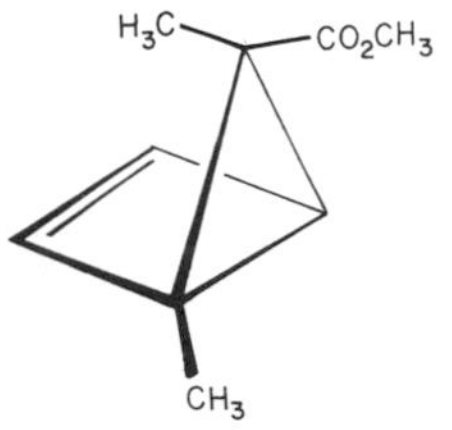

| **5** | **6** |

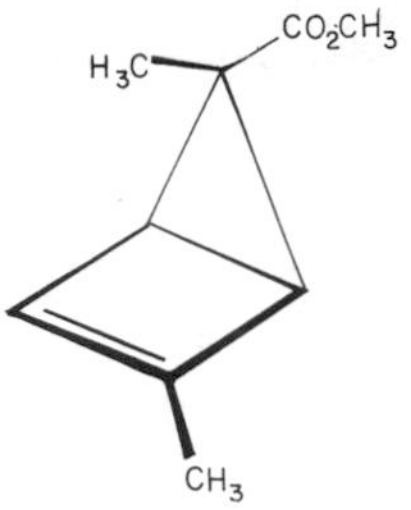

UCLA's Donald Cram enjoyed walking, so, in 1964, he switched from "facing" benzene rings (chapter 16) to conducting tours. His research students generated carbanions by abstracting acidic hydrogens (or deuterium).[5] They discovered that optically active 2-deuterio-2-phenylbutanenitrile (**7**) with tri-*n*-propylamine in tetrahydrofuran/*t*-butyl alcohol lost rotatory activity twenty times faster than it

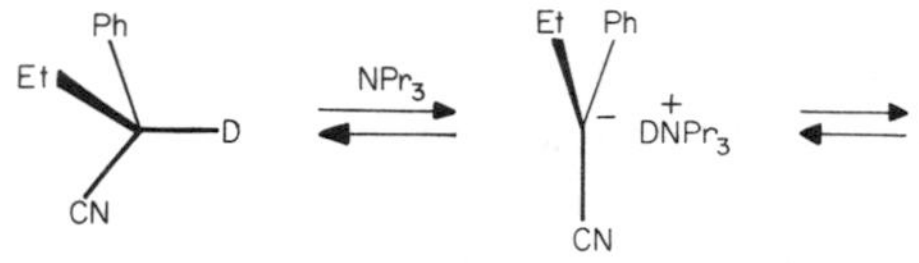

| **7** | **8** |

| **9** | **10** | **11** |

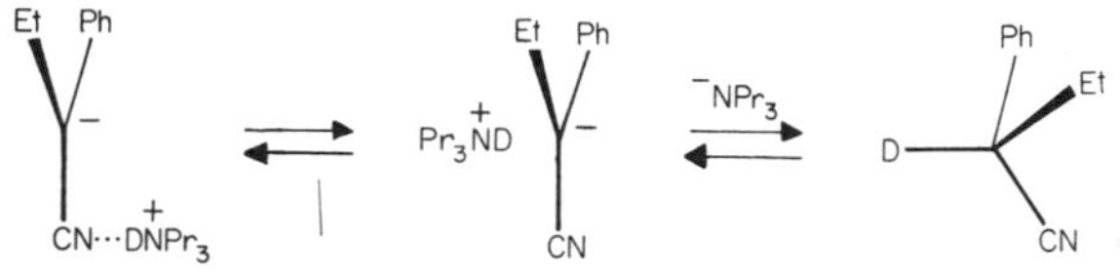

released its deuterium to the protic solvent.[6] Evidently, the D can travel to the other side of the asymmetric carbon without becoming detached from the rest of the molecule. At the suggestion of Harvard's Paul Bartlett, Dr. Cram called this phenomenon "isoracemization."[6] One possible mechanism involves an ion pair (**8**). The tripropylammonium partner then journeys past the cyano nitrogen, to which it forms a transient hydrogen bond, **9**. Once on the opposite side (**10**), the cation deposits its deuterium passenger to give **11**, the enantiomer of **7**. Professor Cram felt that tripropylamine escorted deuterium from one side of carbon to the other. When he proposed the expression "conducted tour mechanism" to his students they were horrified. He asked them to say it to themselves repeatedly for a week. Even after that period they still thought little of the name, but it began to grow on them.[7] The outcome? It appeared in print[6] and has been used since.[8]

The Wagner–Meerwein rearrangement, in which a group moves to a positive carbon (**12 → 13**)[9] has engendered interesting terminology. Sometimes, the migrating R seems to assist the parting of X, and this aspect raises the question of timing in the overall

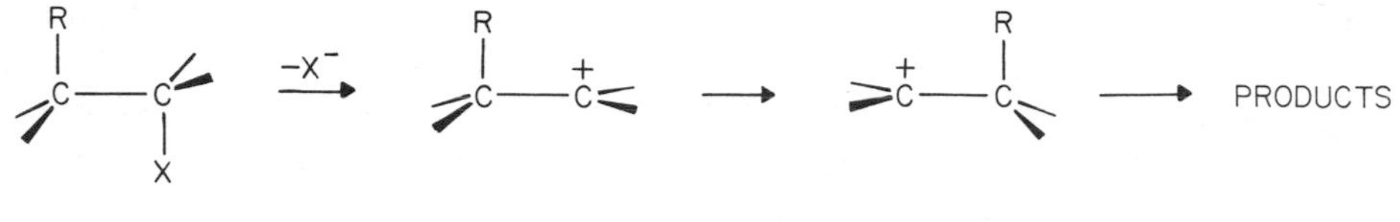

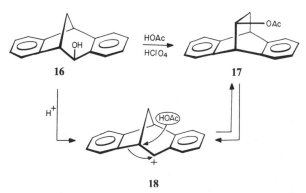

12   **13**   PRODUCTS

process. Specifically, Saul Winstein (of UCLA) believed that R and its electrons begin to move as X departs, when exo-2-norbornyl brosylate undergoes acetolysis (see the "windshield wiper" controversy, chapter 3). Sir Christopher Ingold, at University College, London, invoked similar participation of sigma electrons to explain why exo-2-norbornyl chloride (**14**) solvolyzes 70,000 times more rapidly than does pinacolyl chloride (**15**).[9] The Winstein school labeled this kinetic lift by a neighboring group "anchimeric assistance," based on the Greek *anchi-* ("neighboring") and *meros*

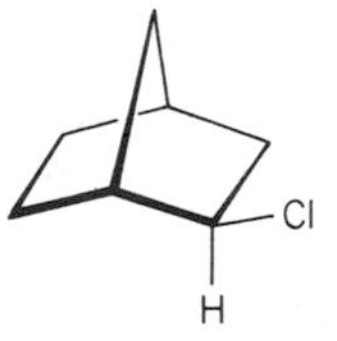

$(CH_3)_3 CCHCH_3$
$\quad\quad\quad |$
$\quad\quad\quad Cl$

14 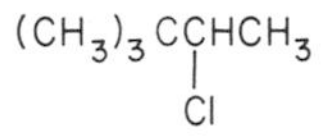   **15**

("parts").[10] Ingold's circle chose the term "synartetic acceleration."[9] (If you need a noun, use "synartesis";[11] but don't confuse it with synarthrosis, the medical term for a joint that permits no motion between its parts.) Outside of chemistry, "synartetic" usually applies to a line of poetry; it means that the meter flows smoothly and is not divided into disparate units. The word probably comes from the Greek *synartetein* ("to join together"). Thus, the heterolysis and migration stages of the Wagner–Meerwein performance need not be separate but flow together smoothly; rearrangement begins before heterolysis ends.[9] No wonder chemists often wax poetic about their science.

In 1965, the research contingent of Stanley Cristol reported that **16** rearranged to **17** in acetic acid containing perchloric acid.[12] Their data suggested that the nucleophile

16   HOAc/HClO₄ →   **17**

H⁺   HOAc   **18**

attacks transient cation **18** at a carbon next to the positive center while a bond shifts to that center (see arrows in **18**). With the help of Mrs. Joy King of the Classics Department at the University of Colorado, Professor Cristol coined "geitonodesmic" for this mechanism. This word derives from the Greek *geiton* ("neighbor") and *desmos* ("bond").[12] The concept is somewhat like Sir Christopher's "synartetic"; but now rearrangement and *capture* of the cation are joined together, rather than rearrange-

ment and *formation* of the cation. In fact, Dr. Cristol also considered the ananym*
"citetranys," which is "synartetic" spelled backwards, but decided against heading in
that direction.[12,13] Subsequent studies by the Colorado investigators cast some doubt
on the validity of the geitonodesmic path,[14] but we can't knock its name.

Use of "desmos" for "bond" may have spawned other words with this root. In 1970,
John Pople's people at Carnegie Mellon University recommended "isodesmic" (*isos* is
Greek for "equal") to describe cases where reactants and products have equal numbers
of bonds of a given formal type.[15,16] For example, the accompanying stoichiometric
equation is isodesmic; on each side it contains one CC single bond, one CC double
bond, one CO double bond, and twelve CH single bonds.[15] Isodesmic analysis can help
you estimate strain energies in some molecules.[17]

$$CH_3-CH=C=O \;+\; 2\;CH_4 \;\longrightarrow\; CH_3-CH_3 \;+\; CH_2=CH_2 \;+\; H_2C=O$$

In 1975, Philip George (University of Pennsylvania), Mendel Trachtman (Weizmann Institute), Charles Bock, and Alistair Brett (both at Philadelphia College of
Textiles and Science) proposed "homodesmotic" (*homos* is Greek for "same") to
categorize some (but not all) isodesmic reactions.[18,19] Homodesmotic pertains only to
hydrocarbons; reactants and products must contain equal numbers of carbons with: (a)
a given hybridization (sp, sp^2 or sp^3); and (b) a given number of attached hydrogens. An
example is shown below.[18] Formally all carbons are sp^2, and each side of the balanced

$$3\;CH_2=CH-CH=CH_2 \;\longrightarrow\; \langle\!\bigcirc\!\rangle \;+\; 3\;CH_2=CH_2$$

equation has six carbons with one attached hydrogen and six with two. This example
completes our bunch of names based on "bond."

Now sit back and think about a cyclohexane chair. A substituent (such as methyl)
can be attached by an equatorial (**19**) or an axial† (**20**) bond; these conformational
isomers interconvert by inversion of the ring from one chair to the other. An alkyl
group ordinarily likes to be equatorial, so the equilibrium favors **19**.[22] In 1965, Ernest
Eliel (then at the University of Notre Dame), Norman Allinger (then at Wayne State
University), Stephen Angyal (University of New South Wales), and George Morrison

19   **20**

(University of Leeds) coined the term "biased"[23] or "biassed"[24] (either spelling is
acceptable) for a situation where the steady state favors one conformer virtually
exclusively (e.g., **19**, R = tBu). A year later, Marc Anteunis, Dirk Tavernier, and Frans
Borremans at the University of Ghent, Belgium, introduced the adjective "anan-

---

*Ananym: a pseudonym consisting of the real name written backwards.
†In the fledgling days of conformational analysis, axial bonds were called "polar" bonds because they
projected upward or downward from the ring, pointing, as it were, toward the North and South Poles of the
earth.[20] The name was changed in 1953 to avoid possible confusion with the more common meaning of
polar in chemistry.[21]

comeric" for such an extreme equilibrium.[25] This word has more international appeal, as it derives from Greek, and Professor Eliel has since adopted it.[26] It comes from the noun *ananche* ("fate" or "necessity") and the related verb "ananchein" ("to force by some natural law").[27] In this case, the natural law involves hominal repulsion between a bulky R and axial hydrogens shown in **20**.

Electrophilic aromatic substitutions normally bring about replacements ortho, meta, or para to something already on the ring. But, for some reactions these terms aren't enough to describe all possible sites of substitution. When Charles Perrin (University of California at San Diego) and his co-workers nitrated p-bromoanisole (**21**) in acetic anhydride, 31% of the product was *p*-nitroanisole (**22**).[28] Nitration had occurred at the site of the halogen itself. Professor Perrin was familiar with the term *ipso facto* ("by the fact itself"), so he knew that *ipso* is Latin for "itself". He thus came up with

$$CH_3O{-}\langle\bigcirc\rangle{-}Br \longrightarrow CH_3O{-}\langle\bigcirc\rangle{-}NO_2 + CH_3O{-}\langle\bigcirc\rangle{-}Br$$

$$\qquad\qquad\qquad\qquad\qquad\qquad\qquad\qquad\qquad\qquad\qquad\qquad Br$$

$$\textbf{21}\qquad\qquad\qquad\qquad\qquad\textbf{22}$$

"ipso position" for the ring carbon bearing a substituent and "ipso substitution" for replacement of that substituent.[28,29] To top it off, he referred to a substituent's tendency to be replaced as its "ipso factor."[28] Did we hear a groan from Dr. Perrin's high school Latin teacher?

Roger Hahn's team at Syracuse University[30] and Philip Myhre's at Harvey Mudd College[31] have also observed electrophilic ipso attack; and James Traynham of Louisiana State University pointed out that free radical substitution can be ipso, also.[32]

Ordinarily, nucleophilic aromatic substitutions are ipso events, but many examples exist in which the nucleophile takes a seat other than the one vacated by the incumbent.[33] Historically, the first of these was probably the nineteenth century von Richter reaction. For example, p-chloronitrobenzene (**23**) reacts with potassium cyanide in aqueous ethanol to give m-chlorobenzoic acid (**24**) rather than the para isomer.[34] Surprisingly, m-chlorobenzonitrile is not an intermediate; Joseph Bunnett (then at the University of North Carolina) and his students worked out a mechanism.[35]

$$Cl{-}\langle\bigcirc\rangle{-}NO_2 \xrightarrow[\text{C}_2\text{H}_5\text{OH}]{\text{KCN,H}_2\text{O}} Cl{-}\langle\bigcirc\rangle$$

$$\qquad\qquad\qquad\qquad\qquad\qquad\qquad\qquad\qquad CO_2H$$

$$\textbf{23}\qquad\qquad\qquad\qquad\qquad\textbf{24}$$

Professor Bunnett and Roland Zahler (then at Reed College, Portland, Oregon) coined "cine-substitution" for instances where the substituent's site differs in reactant and product.[33] "Cine" derives from the Greek *kinein* ("to move"); the same root is found in "cinema" ("moving picture") and "kinetic" ("of or resulting from motion"). In other words, it's as though the site occupied by the substituent moves. Drs. Bunnett and Zahler did not intend "cine" to mean that the new and old sites had to be adjacent (ortho) to each other.[33,36] Most of their cases involved ortho sites, but two meta types had been reported.[37] Some doubt has been expressed about the validity of the meta examples.[38]

In any case, others thought "cine" implied only adjacent sites;[39-42] so, in 1977, Guiseppe Guanti of the University of Genoa adopted the descriptor "tele-substitution" for nucleophilic aromatic substitutions in which the positions are nonadjacent.[39] (The Greek prefix *tēle-* means "far away.") Actually, Dr. Guanti did not dream up the term tele-substitution; Josef Arens of the University of Utrecht, The Netherlands, used it a decade earlier to specify reactions in which only the terminal atoms of a system change hybridization.[43] A case in point is the conversion of 3-bromo-1-butyne (**25**) to 1-(phenyltellurio)-1, 2-butadiene (**26**).[44] (In fact, Dr. Arens's terminology also included "tele-eliminations," "tele-isomerizations," and "tele-carbene formation."[45])

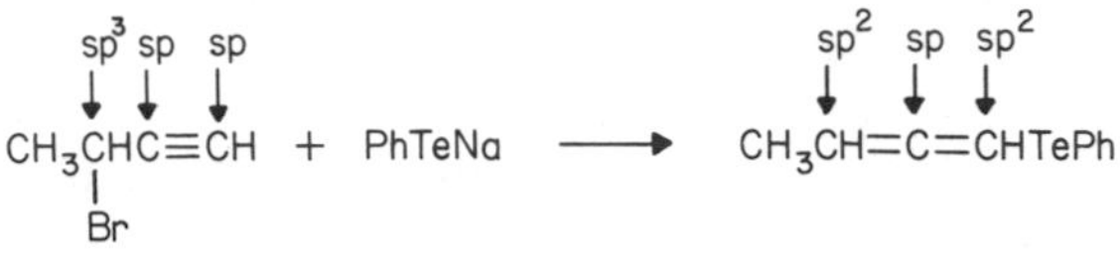

25   **26**

Hendrik van der Plas's personnel at Agricultural University, Wageningen, The

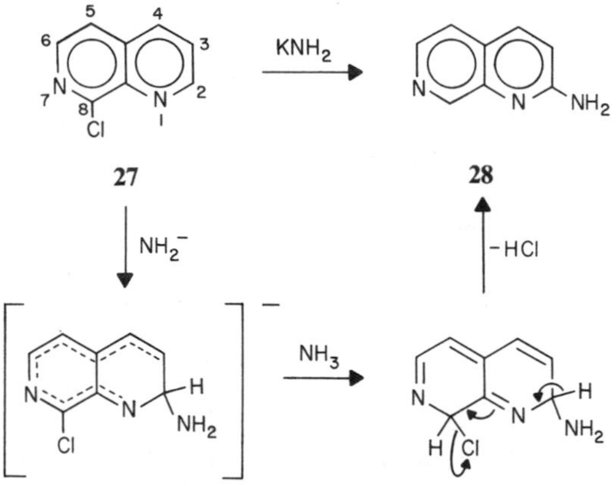

27   **28**

29   **30**

Netherlands, provided an example of aromatic tele-substitution; 8-chloro-1, 7-naphthyridine (**27**) and potassium amide give some 2-amino product **28**.[46] (That's "tele-" enough to compete with telephones and telegraphs). The Dutch chemists reasoned that amide ion assaults **27** at C2. The resulting delocalized species (**29**) picks up a proton at C8 to give **30**, which undergoes 1,4-dehydrohalogenation. This last step could be classed as a tele-elimination. Another case is the dehalogenation of 1, 4-dibromo-2-butyne (**31**).[43,47]

$$Br-CH_2-C\equiv C-CH_2-Br \;+\; Zn \;\longrightarrow\; CH_2{=}C{=}C{=}CH_2 + ZnBr_2$$

31

Nucleophilic aromatic replacements ordinarily involve a leaving group on the *ring* (e.g., the chlorine in **27**). But what if the incoming nucleophile also carries its own departing group? In 1978, Mieczyslaw Makosza's laboratory at Technical University, Warsaw, reported that nitrobenzene and phenyl chloromethyl sulfone (**32**) convert to phenyl *p*-nitrobenzyl sulfone (**33**) in the presence of KOH. Their suggested mechanism is shown. At first glance it appears that a hydride ion (a very poor nucleofuge) has been ousted. But actually the hydride merely migrated while the chloride departed. The researchers proffered "vicarious substitution" for this type of phenomenon.[48] "Vicari-

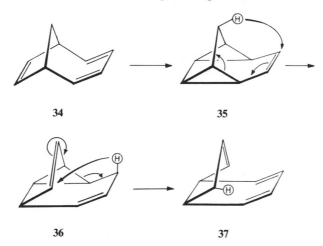

32   33

ous" means "performed by one person in place of another"; in Dr. Makosza's case the chloride departed in place of the hydride. Hydrogen was most appreciative, for "parting is such sweet sorrow."[49]

In 1971, Jerome Berson and his stalwarts at the University of Wisconsin and then at Yale University decided to compete with basketball players in rebounding. Professor Berson used "rebound" to describe a process in which two successive migrations occur in opposite directions.[50] Heating **34** gave **37**; and deuterium labeling showed that some of the product arises via **35** and **36**. The game began with an intramolecular Diels–

34   35

36   37

Alder blitz (**34** → **35**); then the rebounding hydrogen hopped into action with two consecutive 1,5-shifts (**35** → **36** → **37**). Another example sprang from Leo Paquette's research bench, which is a mother lode of fascinating molecules (e.g., see chapters 1, 4, 6). It seems that the heat of the game transformed **38** into **39**, evidently via two [1,5]sigmatropic shifts.[51] Stereochemical details are not shown.

38   39

Roald Hoffmann, caterer to picnic tables (p. 102), joined up with Robert Woodward to enrich our science with the vocabulary of orbital symmetry.[52] "Conrotatory," "disrotatory," "suprafacial," "antarafacial," "sigmatropic," and "cheletropic," from the Greek *chēlē* ("claw") and *tropos* ("turning"), have become household words in organic chemistry. When they pondered the names "conrotatory" and "disrotatory," sages Woodward and Hoffmann also considered "domino" and "anti domino," suggested by Harvard's Professor William Doering in seminars.[53] His terms indeed suit the orbital motions (see **40** and **41** respectively). A chemist of Greek ancestry wrote Drs. Woodward and Hoffmann that, instead of "con-" and "disrotatory" (Latin), they could

40	**41**

have chosen "sys-strophic" and "dia-strophic" (*sys, dia,* and *strophē* are Greek for "together," "apart," and "turning or twisting," respectively). However, these suggestions came too late. Besides, they bring to mind "catastrophic," so Professor Hoffmann remained happy with the original choices.[53] Messrs. W & H had some difficulty coming up with a good expression to denote the "other side of" a molecule. The most common sources, Greek and Latin, were barren; but Sanskrit came to the rescue. *Antara* is Sanskrit for "the other"; thus "antarafacial" was born.[53]

William Klyne and David Kirk took a liking to the prefixes "con" and "dis" and, in 1983, proposed the handy words "consignate" and "dissignate" to improve discussions about optically active compounds.[54] Since the pioneering work of Djerassi's school in the late 1950s, chemists have appreciated that substituents could perturb the chiroptical properties of a parent chromophore (e.g., the amplitude and sign of a Cotton effect in optical rotatory dispersion or the dichroic absorption, $\Delta\varepsilon$, in circular dichroism). Whether the substituent contributes in a plus or minus sense to the value of a chiral property depends upon its spatial relationship to the absorbing chromophore, as was adumbrated in convenient rules (e.g., octant rule).[55] Sometimes, however, the experimental outcome proved contrary to expectations and led to awkward expressions like "anti-octant," "reverse-octant," "anti-reverse-octant," and so forth. K and K simplified life by recommending "consignate" when a perturbation obeys the rule, and "dissignate" when it does not. For example, in a conformationally fixed cyclohexanone, an α-axial alkyl or Br is virtually always consignate, whereas F or OAc is usually dissignate.[54] These words remain in common use and so were not just "signs" of the times.[56]

The suffix "-facial" has given rise to the term "cofacial" for certain organic and inorganic compounds. Virgil Goedken's gang at Florida State University used

42	**43**

"cofacial" for structures in which two **42** moieties are held "face to face" by a Ru—Ru bond, as depicted in **43**.[57] But chemists at Kyoto,[58] Stanford,[59] Michigan State,[60] and Rockefeller Universities[61] had adopted "cofacial" (as well as "face to face," and "*strati*-") for porphyrin cyclophanes in which two flat polycycles occupy parallel planes. (Strati- comes from Latin *stratum* for "covering.") The Rockefeller researchers synthesized *strati*-bisporphyrins and recommend the prefix to denote any fixed "overlapping of planar rings."[61]

The cofacial sections need not both be porphyrins. For example, in **44** a quinone ring provides one face.[62] Such molecules were intriguing for model studies of electron transfer in photosynthesis.[63] We already knew porphyrins to be a beleaguered lot.

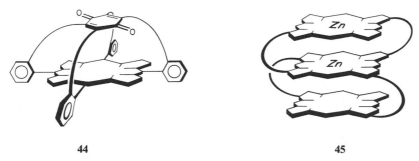

| 44 | 45 |

They have been bridged, capped, crowned, and strapped (chapter 17), and fenced, pocketed, and picket-pocketed (chapter 5). And then, to enhance their beauty: the cofacial. We are all for it, especially when elegant structures like **44** result. But if you were still unfulfilled esthetically, chemists at Argonne National Laboratory extended the cosmetology with their "doubly cofacial" porphyrin trimers **45**.[64] And to reflect on all these good works, we should look into the "open-well effect"[65] of certain metalloporphyrins to the accompaniment of tetrapyrrole ligands with "accordion" structures. This latter type of macrocycle can orchestrate a pair of metal ions to nestle, side-by-side, within its bosom.[66]

Polarity is a concept often drilled into chemistry students. In a carbonyl group the fractional positive charge on carbon invites nucleophiles; but electrophiles had better look elsewhere. In 1965 at Harvard, Elias Corey and Dieter Seebach began to change all that.[67] They converted (for example) acetaldehyde to the 2-methyl-1, 3-dithiane **46** and then removed a proton with alkyllithium to give **47**.[68] Suddenly, the carbon that was *positive* in the aldehyde has become *negative*. In 1969, Professor Seebach (then at the University of Karlsruhe) coined "umpolung" to describe this type of ionic transvestism.[69] That noun derives from the German verb *umpolen*, which means "to interchange the positive and negative pole."[70]

As expected, **47** likes to couple with electron deficient centers. For example, it joins up with the isopropyl carbon of 2-iodopropane to produce **48**. Similarly, the carbonyl carbon of benzophenone and of ethyl cyclohexanecarboxylate succumb, leading to **49** and **50**. As we know (chapter 13), mercuric ions love sulfur; hence, $HgCl_2$ cleaves these elaborated thioketals to carbonyl compounds, and in this way umpolung opens up new routes to structures like **51**, **52**, and **53**.[71] Chemists have ramified this principle extensively.[72] The moral of all this? Reversing things doesn't always mean going backwards.

During the mid-1960s, Cyril Grob's go-getters at the University of Basel found that γ-haloamines (**54**) in aqueous or alcoholic solution gave products from fragmentation (see equation) as well as from conventional substitutions and eliminations.[73] In fact, solvolysis of *N, N*-dimethyl-3-bromo-1-adamantylamine (**55**) proceeded only by fragmentation, as in **56**.[74] (The term fragmentation is stretched a bit here, since the pieces stay connected through other bonds.) The bromine in **55** comes off 520 times faster than in the carbon analog **57**. Evidently, the nitrogen gets into the act as shown and accelerates ionization via the fragmentation path. We shall see in chapter 19 that the stereochemical relationships in **55** play an important role. This participation by nitrogen reminded Dr. Grob of "anchimeric acceleration" (p. 247), but he wanted to

46 47 48 49 50 51 52 53

54

55 56 57

distinguish his effect with a unique name. He first considered "schizomeric," from the Greek words *schizein* ("to break or cleave") and *meros* ("part").[75] Because "schizo-" has unpleasant overtones, he turned to the poikilonym* "frangomeric"; *frangere* is Latin for "break" or "cleave." Professor Grob realized that purists object to such Latin–Greek mixing but defended his etymological hybrid with precedent. He pointed out that "automobile" comes from Greek *autos* ("self") and Latin *mobilis* ("movable"). Strictly, the vehicle should be an ipsomobile. At this point, let's drive on to another topic.

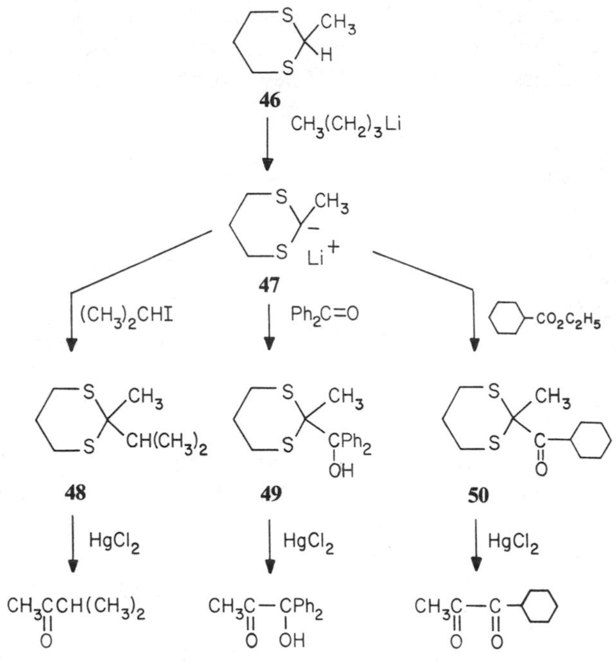

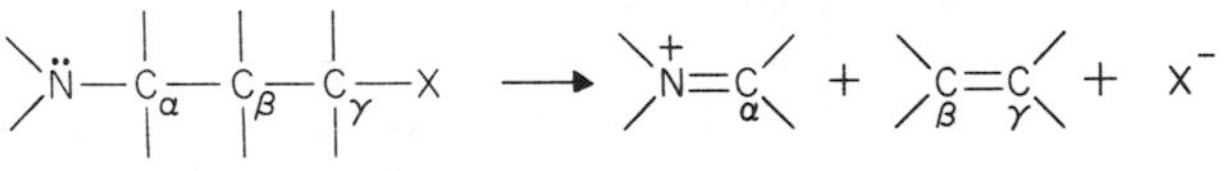

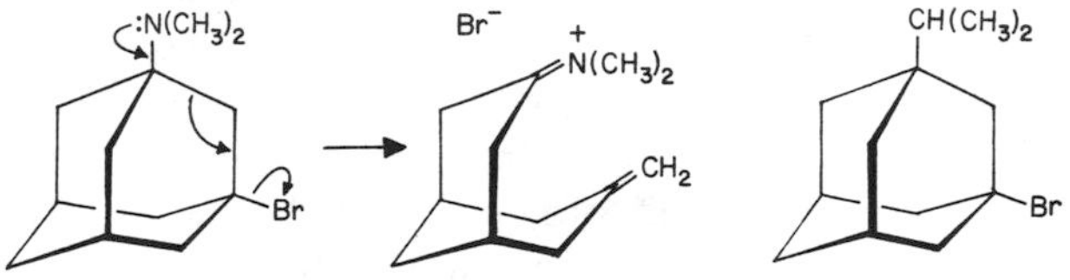

In 1971, Dutch chemist Hendrik van der Plas, who figured in telechemistry (p. 250), uncovered a new mechanism for nucleophilic aromatic substitution. His coresearchers had shown that 6-bromo-4-phenylpyrimidine (**58**) became the 6-amino derivative **60** by

---

*Poikilonym: a name derived from different systems of nomenclature.

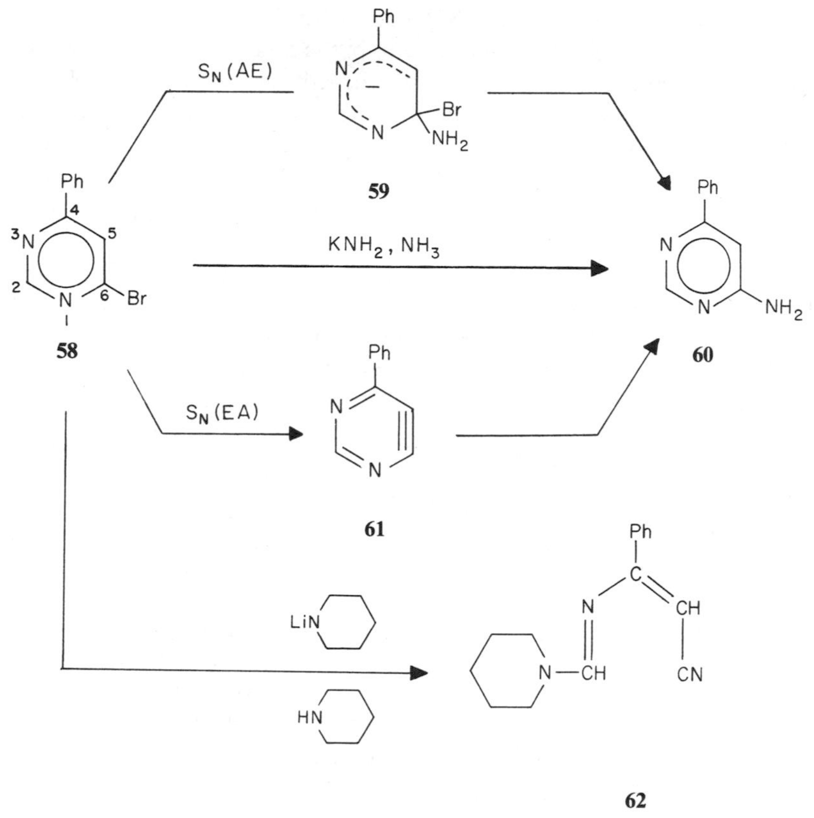

59

58

S$_N$(AE)

KNH$_2$, NH$_3$

60

S$_N$(EA)

61

LiN⟨piperidine⟩

HN⟨piperidine⟩

62

action of KNH$_2$ in ammonia.[76] At the time, two well-established paths could explain this result. First: an S$_N$(AE) mechanism, which involves *N*ucleophilic *S*ubstitution via *A*ddition–*E*limination. The symbolism originated in Thomas Kauffmann's laboratory at the Technische Hochschule Darmstadt;[77] and this route would involve intermediate **59**. Second: an S$_N$(EA) mechanism, via **61**, essentially akin to John Roberts's famous benzyne path (chapter 20).[78] However, when Professor van der Plas found that lithium piperidide pummels C2 and opens **58** to **62**,[79] he wondered whether the KNH$_2$ reaction also prefers attack at C2 and ring scission.

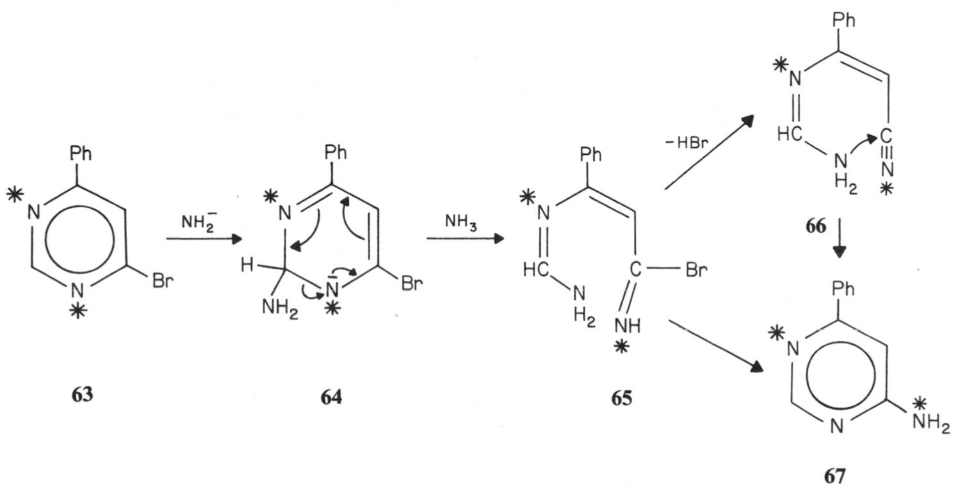

63   **64**   **65**

NH$_2^-$ NH$_3$ −HBr

66

67

To find out, his research group used **58** labeled with ^{15}N (**63**). The S$_N$(AE) and S$_N$(EA) scenarios require that all of the isotope remains in the ring. In contrast, ring opening (**63** →**64**→**65**) and subsequent reclosure to **67** (directly or via **66**) leaves half the ^{15}N outside the ring (**67**). By now you've probably guessed correctly that almost 50% of the marker wound up on the NH$_2$,[80] Dr. van der Plas dubbed his new mechanism S$_N$(ANRORC), from *N*ucleophilic *S*ubstitution via *A*ddition of the *N*ucleophile, *R*ing *O*pening, and *R*ing *C*losure.[81] We hope you like the name, even though it is an anagram of RANCOR.

Another time-saving abbreviation for a ring closure came from Gary Posner and company at The Johns Hopkins University. In 1981 they applied two Michael additions in tandem (*Mi*chael *Mi*chael), followed by a *r*ing *c*losure, to build a steroid skeleton. The three steps are carried out in one flask without isolation of intermediates; they dubbed their approach the "MIMIRC" reaction sequence.[82] Actually, a year earlier Little and Dawson at the University of California, Santa Barbara, had coined MIRC (*M*ichael-*I*nduced *R*ing *C*losure) for conjugate addition to an α, β-enone system followed by a cyclization in the derived enolate.[83] Strictly, then, Posner's "I's" in MIMIRC don't stand for the same word as the "I" in MIRC. If this bothers you, Johns Hopkins graduate student Martin Hulce had a happy suggestion: call Professor Posner's *s*equential *M*ichael *r*ing *c*losure "SMIRC."[84]

Duke University's Ned Porter, Andrew McPhail, and graduate student Nicholas Roe longed to form rings sequentially—but with the help of radicals and molecular oxygen. They induced hydroperoxide **68** to submit to two consecutive cyclizations when dosed with tert-butyl hydroperoxy radicals and air. After reduction, the product contained several stereoisomers of bisperoxide **69**. They baptized the event "serial cyclization" and suggested it could play a role in autoxidation of natural polyenes.[85]

68   **69**

In 1970, Manfred Schlosser, then at the University of Heidelberg, came up with a catchy acronym. He and his students had ramified Wittig methodology as represented by the scheme shown. A lithio β-oxido phosphorous ylide is a key intermediate and gave the E olefin with high stereoselectivity. Overall, the sequence fastens an electrophile (e.g., deuterium) to the α-carbon and converts an erstwhile carbonyl (e.g., of benzaldehyde) to a trisubstituted alkene. Thus, Dr. Schlosser views it as α-*s*ubstitution plus *c*arbonyl *o*lefination via β-*o*xido *p*hosphorous *y*lides, or "SCOOPY."[86] Scoopy may sound like a distant relative of "large scoupene" (p. 15). But actually, Professor

phosphorous ylide   phosphorous betaine   β-oxido phosphorous ylide

α - substituted betaine   substituted olefin

Schlosser's choice was influenced by Snoopy, the dog in the American comic strip "Peanuts," which was popular in Europe during the 1960s.[87]

Free radicals get especially well stabilized when simultaneously regaled by electron-donating and electron-accepting groups. For example, in radical **70** the morpholinyl nitrogen "donates" (represented by resonance form **71**), while the cyano group "accepts" (**72**).[88] This effect has been honored by three terms, one more than the President of the United States may have.

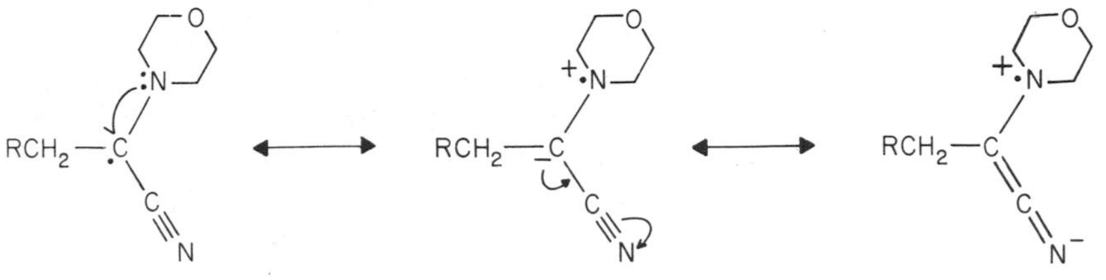

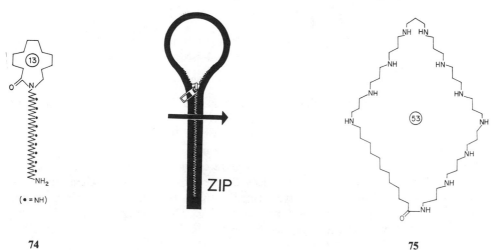

70   **71**   **72**

Alexander Balaban of The Polytechnic, Bucharest, started the fun in 1971 with the phrase "push–pull" effect.[89,90] (Others, including John Roberts of the California Institute of Technology and Ronald Breslow of Columbia University, had used push–pull earlier to describe stabilization of cyclobutadienes by electron donors and acceptors[91,92].) Two years later, Alan Katritzky's coterie (then at the University of East Anglia, England) called the effect "merostabilization" because of a similarity to the resonance stabilization in merocyanine dyes (**73**).[93] Then, in 1978, Heinz Viehe's team at the University of Louvain, Belgium, introduced "capto-dative," from the Latin *captus* ("taking") and *datus* ("giving").[88] A nice bit of give and take involving three research groups.

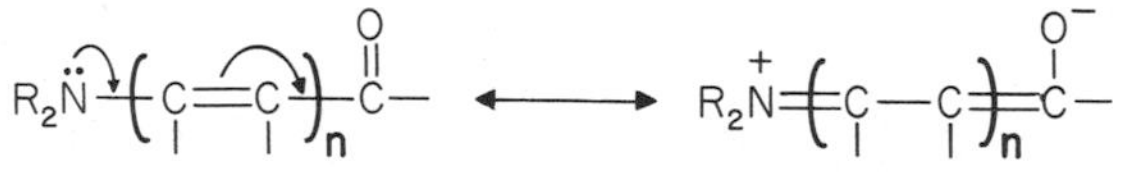

73

In 1978, Manfred Hesse of the University of Zürich and his co-workers synthesized **74**, which consists of a 13-membered cycle and a 40-membered polyamine chain.[94]

ZIP

74   **75**

Treatment with the potent base potassium 3-amino-propylamide (KAPA) brought about a net internal transamidation, and (presto chango) all 53 skeletal atoms are now in the ring (75)! This transformation reminded the Zürich team of the unzipping of a sweater; a hole just large enough for the neck is suddenly much larger.[95] Hence, they adopted the nifty term, "zip reaction."[94] The reaction, with accompanying zipper, adorned the cover of the March, 1978 issue of *Angewandte Chemie*. Since then, Dr. Hesse's house has fashioned a zipper solely of carbon atoms. [96]

Incidentally, Charles Brown's group, then at Cornell University, first fabricated KAPA (the compound and the name) in 1975.[97] This base (sometimes called PAPA[98] rather than KAPA) adroitly isomerizes internal acetylenes to 1-alkynes. The coiner of KAPA has a famous papa, namely Nobelist Herbert C. Brown. And KAPA reminds us of another illustrious organic chemist, Sir John Warcup Cornforth, known to friends and colleagues by the nickname "kappa." When Professor Cornforth was an undergraduate in chemistry at Sydney University, he wrote Greek symbols to identify his beakers and vials. The Greek alphabet has no letters that correspond to his initials, so JWC used $\iota\omega\kappa$ as the nearest approximation. As a result, a chum started calling him "Kappa," and the nickname stuck forever.[99] Kappa has been deaf since boyhood, yet he won the Nobel Prize in chemistry in 1975!

Some chemical tailors work with snaps instead of zippers. The simultaneous loss of two atoms or groups from a saturated molecule (76 → 77) reminded Sidney Benson of the University of Southern California of the unfastening of a snap, so he termed such processes "snap-out reactions".[100] As a specific example, propane ejects H_2 on vacuum ultraviolet photolysis.[101]

76   **77**

Herbert C. Brown's battalion convinced the world that organoborons are not organomorons. His prodigious crew at Purdue "stitched" together olefin units with boron thread, then "riveted" them lastingly with a carbon atom. One example involved dainty stitching of all-*trans* cyclododecatriene (78) with BH_3 to give the organoboron 79 (largely one stereoisomer). Then, at the foundry, a carbon monoxide "rivet" was

78   **79**   **80**   **81**

pounded in with heat, pressure, and ethylene glycol. Finally, annealing the intermediate dioxaborole 80 in tepid alkaline hydrogen peroxide produced tertiary alcohol 81 ("tercyclanol").[102] It's amazing what you can do with B's in chemistry.

Our brotherhood often refers to the departing unit in a substitution reaction as a "leaving" group. In a 1960 paper in *Angewandte Chemie*, Drs. Jean Mathieu, André Allais, and Jacques Valls of the Services de Recherches Roussel–Uclaf in Paris introduced a pair of more distinctive terms.[103] In nucleophilic substitution, we get "nucleofugal departure"; *fugitivus* is Latin for "fleeing." The leaving moiety departs *with* its bonding electron pair and is termed a "nucleofugal group" or a "nucleofuge."[104] Conversely, something that exits *without* its bonding electron pair is an "electrofugal

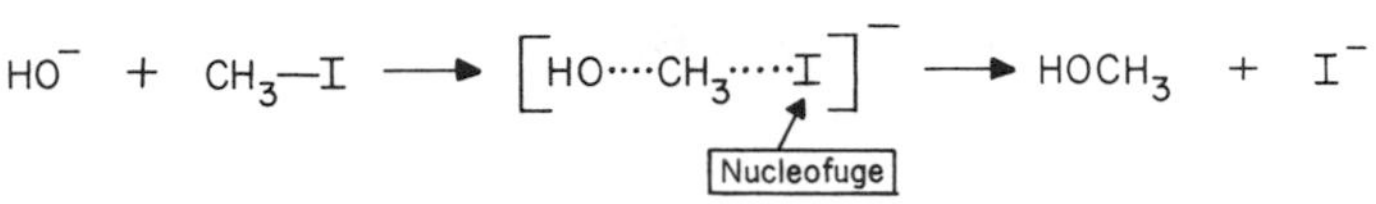

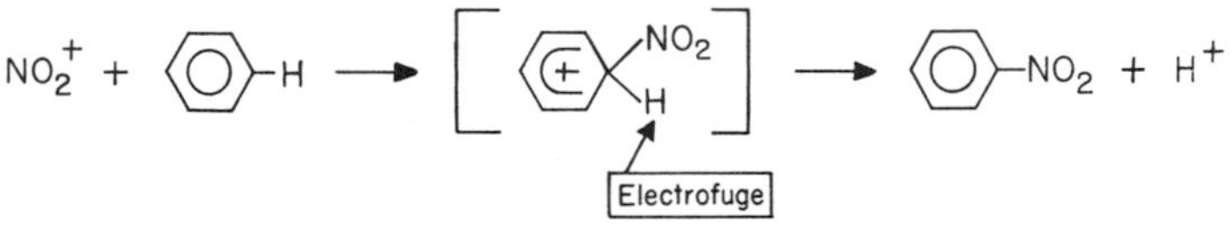

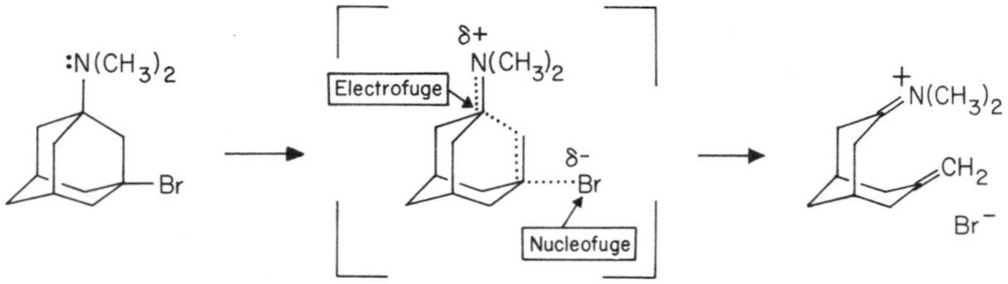

$$HO^- + CH_3{-}I \longrightarrow \left[HO\cdots CH_3\cdots I \right]^{-} \longrightarrow HOCH_3 + I^-$$

group" or "electrofuge."[103] Both can show up in the same reaction; in his fragment-ations, Cyril Grob referred to X and $\overset{+}{N}{=}C$ as nucleofugal and electrofugal groups, respectively.[73] Illustrations are given in the accompanying equations.

Very sensibly, Charles Stirling of University College of North Wales referred to the ability of a group to serve as a nucleofuge as its "nucleofugality."[104] (Would reluctance to be a nucleofuge be termed nucleofrugality?) As we know, triflate and nonaflate abound in nucleofugality (chapter 11); they aren't worried about the high cost of leaving. Why should they? That would be an exercise in nucleofutility.

In 1953, Kurt Alder and Marianne Schumacher at the University of Cologne transformed a common chemical suffix, -ene, into a descriptive name of a reaction. They

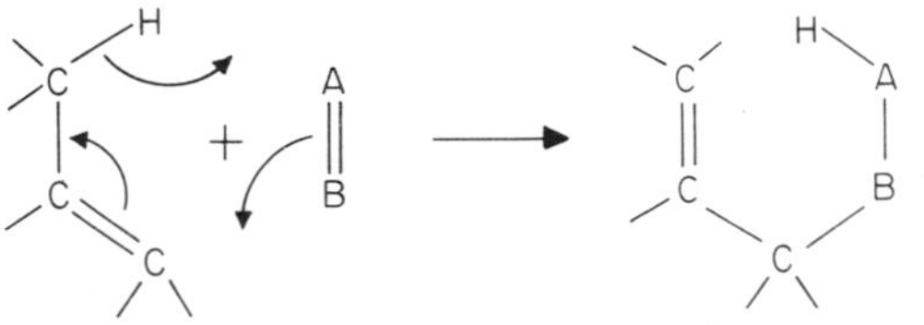

82

conjured up "ene synthesis" for the "indirect substitutive" addition of an olefin having an allylic hydrogen (the "ene") to a recipient with a double or triple bond (the

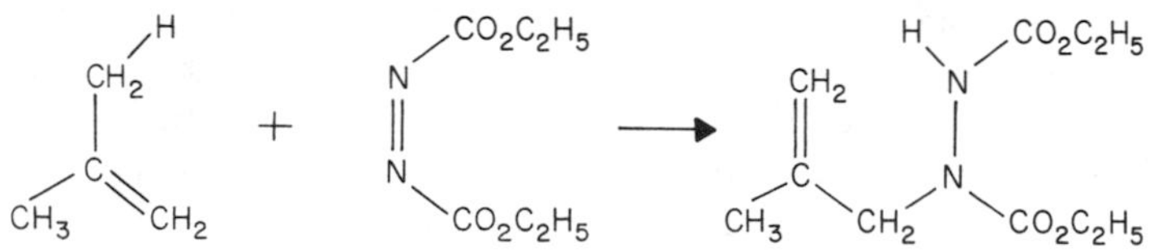

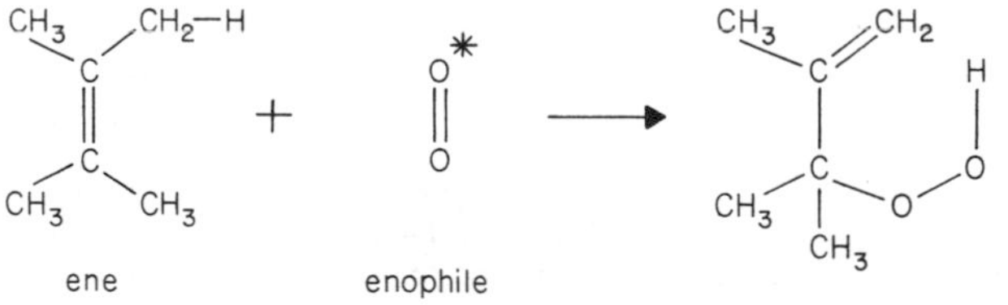

ene   enophile

"enophile"),* as in **82**.[105] A 1969 review by Hans Hoffmann[106] used the dionym "ene reaction," and other researchers subsequently followed suit.[107,108] For example, Warren Thaler and Boris Franzus promoted diethyl azodicarboxylate as an enophile.[109] And, excited molecular oxygen (i.e., singlet oxygen) behaves similarly.[110] The adjoining equations provide examples. In 1981, Barry Snider's research squad, then at Princeton University, called upon an ene operation to prepare methyl (2R, 4S)-2-bromo-4-methyl-5-hexenoate (**83**).[108] The transformation was remarkably stereo-

83

specific; only 5% of another stereoisomer showed up. Günter Kresze and co-workers at the University of Munich found exceptional enophilic reactivity in compounds featuring an N=S bond. For example, N-sulfinyl-p-toluenesulfonamide (**84**) and allylbenzene produced **85** at room temperature. Thus, **84** earned the title "superenophile,"[111] and the process became a "superene reaction."[112] Enes are in. Can ane, ol, one, al, and ate reactions be far behind?

84   **85**

Coined names often make papers easier to read and talks more fun to hear. But, for the serious tasks of information storage and retrieval (such as compiling, listing, and indexing *a lá* "*Chemical Abstracts*") we rely on accurate, systematic nomenclature. To this end, the International Union of Pure and Applied Chemistry (IUPAC) devises schemes for naming *transformations* to go along with its system for naming *compounds*.

Joseph Bunnett has written up the recommendations of the IUPAC's Commission on Physical Organic Chemistry.[113] For example, these specify that the name of a *substitution* process should consist of (1) the name of the entering group, (2) a hyphen, (3) the syllable "de", (4) another hyphen, (5) the name of the leaving group, and (6) the suffix "ation." (IUPAC is slightly flexible and allows omission of hyphens and minor spelling changes for euphony.) As one illustration, the change $PhN_2^+ + I^- \rightarrow PhI + N_2$ is "iodo-de-diazoniation." Accordingly, Heinrich Zollinger (ETH, Zurich) wrote "dediazoniation" in the title of one of his papers.[114] As a further illustration, the change $CH_3CN + H_2O \rightarrow CH_3CO_2H$ becomes "hydroxy, oxo-de-nitrilo-tersubstitution," but we think chemistry teachers will not rush to break this news to their students.

Some inorganic mechanisms also have fetching names. For example, Albrecht Rabenau's researchers at the Max-Planck-Institut für Festkörperforschung (solid-state research) in Stuttgart felt that conduction of protons in solids such as $HUO_2AsO_4 \cdot 4H_2O$ occurs (at least in part) by a "vehicle mechanism."[115] The proton travels through the crystal on a vehicle (in this case, a water molecule); the carrier

---

*By the way, enophiles turn up everywhere. Not surprising, when you realize the word also denotes "lover of wine" (from "oeno," a combining form of Greek *oinos* meaning "wine," and *philos* meaning "loving").

(H_3O^+) moves in one direction while the unladen wagon (H_2O) goes the opposite way. Let us hope the travelers signal whenever they decide to turn.

At Texas A&M University, F. Albert Cotton's cadre rounded up two perky terms to depict fluxional motion in $(\eta^6$-cyclooctatriene) (hexacarbonyl) diiron (Fe—Fe).[116] Proton NMR had disclosed[117] that this compound at about $-100°C$ adopts the same skew structure (**86**) as in the crystal state.[118] (In our drawings, large spheres denote Fe; small ones, CO.) But at room temperature, this molecule in solution attained hydrogen symmetry corresponding to rapid enantiomerzation, **86** ⇌ **88**.[117] The Texas titans surmised that such averaging might take place via symmetrical intermediates like **87** or **89**.[116] The **86** → **87** → **88** path looked jerky—like a twitch; so the Cotton people called it a "twitching process." Note that each iron moves but little and stays at its end of the yard. In contrast, the trip through symmetrical state **89** equalizes the two metal atoms, and they would lose their original identities. Such motion requires the Fe's to glide about; so Cotton *et al.* called it the "gliding process."

But they did not simply sit on their surmises. By thorough ^{13}C NMR analysis, they confirmed the occurrence of enantiomerization at $-65°C$ to $8°C$, and also discovered

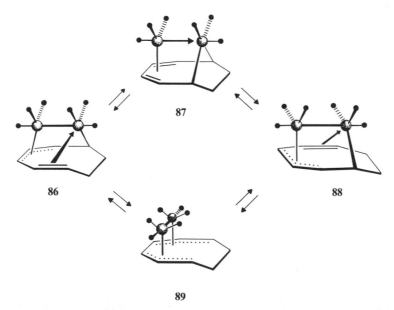

that simultaneous with this change the three CO's on the allyl-bound iron became shuffled. (At higher temperatures, e.g. 8–110°C, the trio of CO's on the other Fe couldn't contain their excitement and also scrambled among themselves.) After pondering the two fluxional itineraries and the body of spectroscopic evidence in hand, the Cotton crew ultimately picked the twitch and let the glide slide.[116]

REFERENCES AND NOTES

1. Hutchins, R.O.; Natale, N.R. *J. Org. Chem.* **1978**, *43*, 2299–2301.
2. Baltrop, J.A.; Day, A.C.; Irving, E. *J. Chem. Soc., Chem. Commun.* **1979**, 881–883.
3. Childs, R.F. *Tetrahedron* **1982**, *38*, 567–608.
4. (a) Klärner, F.-G.; Adamsky, F. *Angew. Chem. Int. Ed. Engl.* **1979**, *18*, 674–675, (b) Klärner, F.-G. Walk Rearrangements in [*n*.1.0] Bicyclic Compounds. In *Topics in Stereochemistry*, Eliel, E.L.; Wilen, S.H.; Allinger, N.L. (Eds.), (Wiley–Interscience, New York, 1984), vol. 15, pp. 1–42.

5. Cram, D.J. *Fundamentals of Carbanion Chemistry* (Academic Press, New York, 1965), pp. 98–103.
6. Cram, D.J.; Gosser, L. *J. Am. Chem. Soc.* **1964**, *86*, 2950–2952.
7. Cram, D.J., private communication, July, 1977.
8. March, J. *Advanced Organic Chemistry*, 3rd ed. (John Wiley & Sons, New York, 1985), pp. 519, 525.
9. Brown, F.; Hughes, E.D.; Ingold, C.K.; Smith, J.F. *Nature* (*London*) **1951**, *168*, 65–67.
10. Winstein, S.; Lindegren, C.R.; Marshall, H.; Ingraham, L.L. *J. Am. Chem. Soc.* **1953**, *75*, 147–155.
11. Cooper, C.N.; Jenner, P.J.; Perry, N.B.; Russell–King, J.; Storesund, H.J.; Whiting, M.C. *J. Chem. Soc., Perkins Trans. 2* **1982**, 605–611.
12. Cristol, S.J.; Parungo, F.P.; Plorde, D.E.; Schwarzenbach, K. *J. Am. Chem. Soc.* **1965**, *87*, 2879–2886.
13. Cristol, S.J., private communication, January, 1976.
14. Cristol, S.J.; Kochansky, M.C. *J. Org. Chem.* **1975**, *40*, 2171–2179.
15. Hehre, W.J.; Ditchfield, R.; Radom, L.; Pople, J.A. *J. Am. Chem. Soc.* **1970**, *92*, 4796–4801.
16. Hehre, W.J., private communication, September, 1975.
17. Fuchs, R. *J. Chem. Educ.* **1984**, *61*, 133–136.
18. George, P.; Trachtman, M.; Bock, C.W.; Brett, A.M. *Theoret. Chim. Acta* (*Berl.*) **1975**, *38*, 121–129.
19. (a) George, P.; Trachtman, M.; Bock, C.W.; Brett, A.M. *Tetrahedron* **1976**, *32*, 317–323, (b) *J. Chem. Soc., Perkin Trans 2* **1976**, 1222–1227.
20. Beckett, C.W.; Pitzer, K.S.; Spitzer, R. *J. Am. Chem. Soc.* **1947**, *69*, 2488–2495.
21. Barton, D.H.R.; Hassel, O.; Pitzer, K.S.; Prelog, V. *Nature* (*London*) **1953**, *172*, 1096–1097; *Science* **1954**, *119*, 49.
22. Eliel, E.L. *J. Chem. Educ.* **1960**, *37*, 126–133.
23. Eliel, E.L.; Allinger, N.L.; Angyal, S.J.; Morrison, G.A. *Conformational Analysis*, (Wiley–Interscience, New York, 1965), p. 71.
24. Eliel, E.L. *Angew. Chem. Int. Ed. Engl.* **1965**, *4*, 761–774.
25. Anteunis, M.; Tavernier, D.; Borremans, F. *Bull. Soc. Chim. Belg.* **1966**, *75*, 396–412.
26. Eliel, E.L., private communication, October, 1975.
27. Tavernier, D., private communication, August, 1977.
28. Perrin, C.C.; Skinner, G.A. *J. Am. Chem. Soc.* **1971**, *93*, 3389–3394.
29. Perrin, C.C., private communication, August, 1977.
30. Hahn, R.C.; Stock, D.L. *J. Am. Chem. Soc.* **1974**, *96*, 4335–4337.
31. (a) Myhre, P.C. *J. Am. Chem. Soc.* **1972**, *94*, 7921–7923, (b) Barnes, C.E.; Myhre, P.C. *J. Am. Chem. Soc.* **1978**, *100*, 973–975.
32. Traynham, J.G. *Tetrahedron Lett.* **1976**, 2213–2216.
33. Bunnett, J.F.; Zahler, R.E. *Chem. Revs.* **1951**, *49*, 273–412.
34. von Richter, V. *Chem. Ber.* **1875**, *8*, 1418–1425.
35. Bunnett, J.F.; Rauhut, M.M. *J. Org. Chem.* **1956**, *21*, 944–948.
36. Bunnett, J.F., private communication, October, 1977.
37. (a) Hodgson, H.H.; Leigh, E. *J. Chem. Soc.* **1938**, 1031–1034, (b) *Ibid.* **1939**, 1094–1096.
38. Reference 33, p. 388, footnote 7a.
39. Guanti, G.; Thea, S.; Novi, M.; Dell'Erba, C. *Tetrahedron Lett.* **1977**, 1429–1430.
40. Guanti, G., private communication, November, 1977.
41. van der Plas, H.C., private communication, August, 1977.
42. Self, D.P.; West, D.E. *J. Chem. Soc., Chem. Commun.* **1980**, 281–282.
43. Arens, J.F. *Bull. Soc. Chim. Fr.* **1968**, 3037–3044.
44. Pourcelot, G. *Compt. Rend.* **1965**, *260*, 2847–2850.
45. Arens, J.F. *Chem. Weekbl.* **1967**, *63*, 513–525.
46. van der Plas, H.C.; Wozniak, M.; van Veldhuizen, A. *Tetrahedron Lett.* **1976**, 2087–2090.
47. Schubert, W.M.; Lanka, W.A. *J. Am. Chem. Soc.* **1952**, *74*, 569.
48. Golinsky, J.; Makosza, M. *Tetrahedron Lett.* **1978**, 3495–3498.
49. Shakespeare, W., *Romeo and Juliet. The Complete Works of Shakespeare*, Ribner, I.; Kittredge, G.L. (Eds.) (John Wiley & Sons, New York, 1971), pp. 963–1005.
50. Berson, J.A.; Boettcher, R.R.; Vollmer, J.J. *J. Am. Chem. Soc.* **1971**, *93*, 1540–1541.
51. Detty, M.R.; Paquette, L.A. *J. Chem. Soc., Chem. Commun.* **1978**, 365–366.
52. Woodward, R.B.; Hoffmann, R. *The Conservation of Orbital Symmetry* (Academic Press, New York, 1970).

53. Hoffmann, R., private communication, April, 1978.
54. Klyne, W.; Kirk, D.N. *Tetrahedron Lett.* **1973**, 1483–1486.
55. Moffitt, W.; Woodward, R.B.; Moscowitz, A.; Klyne, W.; Djerassi, C. *J. Am. Chem. Soc.* **1961**, *83*, 4013–4018.
56. For a review see Kirk, D.N. *Tetrahedron* **1986**, *42*, 778–818.
57. Warren, L.F.; Goedken, V.L. *J. Chem. Soc., Chem. Commun.* **1978**, 909–910.
58. Ogoshi, H.; Sugimoto, H.; Yoshida, Z. *Tetrahedron Lett.* **1977**, 169–172.
59. Collman, J.P.; Elliot, C.M.; Halbert, T.R.; Tovrog, B.S. *Proc. Natl. Acad. Sci.* **1977**, *74*, 18–22.
60. (a) Chang, C.K.; Kuo, M.-S.; Wang, C.-B. *J. Heterocycl. Chem.* **1977**, *14*, 943–945, (b) Chang, C.K.; Abdalmuhdi, I. *J. Org. Chem.* **1983**, *48*, 5388–5390.
61. Kagan, N.E.; Mauzerall, D.; Merrifield, R.B. *J. Am. Chem. Soc.* **1977**, *99*, 5484–5486.
62. Lindsey, J.S.; Mauzerall, D.C. *J. Am. Chem. Soc.* **1982**, *104*, 4498–4500.
63. Ganesh, K.N.; Sanders, J.K.M.; Waterton, J.C. *J. Chem. Soc., Perkin Trans. 1* **1982**, 1617–1624.
64. (a) Wasielewski, M.R.; Niemczyk, M.P.; Svec, W.A. *Tetrahedron Lett.* **1982**, *23*, 3215–3218, (b) For other stacked trimeric porphyrins, see Dubowchik, G.M.; Hamilton, A.D. *J. Chem. Soc., Chem. Commun.* **1986**, 665–666.
65. Bortolini, O.; Meunier, B. *J. Chem. Soc., Perkin Trans. 2* **1984**, 1967–1970.
66. Acholla, F.V.; Mertes, K.B. *Tetrahedron Lett.* **1984**, *25*, 3269–3270.
67. Corey, E.J.; Seebach, D. *Angew Chem. Int. Ed. Engl.* **1965**, *4*, 1075–1077; *Ibid.* **1965**, *4*, 1077–1078.
68. Seebach, D.; Corey, E.J. *J. Org. Chem.* **1975**, *40*, 231–237.
69. Seebach, D. *Angew. Chem. Int. Ed. Engl.* **1969**, *8*, 639–649.
70. Seebach, D., private communication, August, 1977.
71. Seebach, D.; Jones, N.R.; Corey, E.J. *J. Org. Chem.* **1968**, *33*, 300–305.
72. Seebach, D. *Angew. Chem. Int. Ed. Engl.* **1979**, *18*, 239–258.
73. Grob, C.A. *Angew. Chem. Int. Ed. Engl.* **1969**, *8*, 535–546.
74. Grob, C.A.; Schwarz, W. *Helv. Chim. Acta* **1964**, *47*, 1870–1878.
75. Grob, C.A., private communication, August, 1977.
76. van der Plas, H.C.; Smit, P.; Koudijs, A. *Tetrahedron Lett.* **1968**, 9–13.
77. Kauffmann, T.; Risberg, A.; Schulz, J.; Weber, R. *Tetrahedron Lett.* **1964**, 3563–3568.
78. Roberts, J.D.; Semenow, D.A.; Simmons, H.E.; Carlsmith, L.A. *J. Am. Chem. Soc.* **1956**, *78*, 601–611.
79. van der Plas, H.C.; Koudijs, A. *Rec. Trav. Chim. Pays–Bas* **1970**, *89*, 129–132.
80. (a) de Valk, J.; van der Plas, H.C. *Rec. Trav. Chim. Pays–Bas* **1971**, *90*, 1239–1245, (b) *Ibid.* **1972**, *91*, 1414–1422.
81. van der Plas, H.C. *Acc. Chem. Res.* **1978**, *11*, 462–468.
82. Posner, G.H.; Mallamo, J.P.; Black, A.Y. *Tetrahedron* **1981**, *37*, 3921–3926.
83. Little, R.D.; Dawson, J.R. *Tetrahedron Lett.* **1980**, *21*, 2609–2612.
84. Posner, G.H., personal discussions, March, 1983.
85. (a) Porter, N.A.; Roe, A.N.; McPhail, A.T. *J. Am. Chem. Soc.* **1980**, *102*, 7574–7576, (b) Roe, A.N.; McPhail, A.T.; Porter, N.A. *J. Am. Chem. Soc.* **1983**, *105*, 1199–1203.
86. (a) Schlosser, M. In *Topics in Stereochemistry*, Eliel, E.L.; Allinger, N.L. (Eds.) (John Wiley & Sons, New York, vol. 5, pp. 1–30, (b) Schlosser, M.; Christmann, K.F. *Synthesis* **1969**, 38–39, (c) Schlosser, M.; Christmann, K.F.; Piskala, A.; Coffinet, D. *Synthesis* **1971**, 29–31, (d) Schlosser, M.; Tuong, H.B.; Respondek, J.; Schaub, B. *Chimia* **1983**, *37*, 10–11.
87. Schlosser, M., private communication, October, 1978.
88. Stella, L.; Janousek, Z.; Merényi, R.; Viehe, H.G. *Angew. Chem. Int. Ed. Engl.* **1978**, *17*, 691–692.
89. Balaban, A.T. *Rev. Roumaine Chim.* **1971**, *16*, 725.
90. Balaban, A.T. Caproiu, M.T.; Nagoita, N.; Baican, R. *Tetrahedron* **1977**, *33*, 2249–2253.
91. Manatt, S.L.; Roberts, J.D. *J. Org. Chem.* **1959**, *24*, 1336–1338.
92. Breslow, R.; Kivelevich, D.; Mitchell, M.J.; Fabian, W.; Wendel, K. *J. Am. Chem. Soc.* **1965**, *87*, 5132–5139.
93. Baldock, R.W.; Hudson, P.; Katritzky, A.R.; Soti, F. *Heterocycles* **1973**, *1*, 67–71.
94. Kramer, U.; Guggisberg, A.; Hesse, M.; Schmid, H. *Angew. Chem. Int. Ed. Engl.* **1978**, *17*, 200–202.
95. Hesse, M., private communication, March, 1981.
96. Nakashita, Y.; Hesse, M.; *Angew. Chem. Int. Ed. Engl.* **1981**, *20*, 1021.

97. Brown, C.A.; Yamashita, A. *J. Am. Chem. Soc.* **1975**, *97*, 891–892.
98. Lindhoudt, J.C.; van Mourik, G.L.; Pabon, H.J.J. *Tetrahedron Lett.* **1976**, 2565–2568.
99. Cornforth, J.W., private communication, May, 1983.
100. Benson, S.W. *Adv. Photochem.* **1964**, *2*, 1–23.
101. Okabe, H.; McNesby, J.R. *J. Chem. Phys.* **1961**, *34*, 668–669; *Ibid.* **1962**, *36*, 601–604; *Ibid.* **1962**, *37*, 1340–1346.
102. Brown, H.C. *Organic Syntheses Via Boranes* (John Wiley & Sons, New York, 1975), pp. 131, 159. We thank Professor Thomas Bryson (University of South Carolina) for helpful correspondence on this topic.
103. Mathieu, J.; Allais, A.; Valls, J. *Angew. Chem.* **1960**, *72*, 71–74. Drs. Mathieu and Allais actually coined the terms three years earlier in their book *Principes de Synthese Organique* (Masson and Co., Paris, 1957). We thank Professor Mathieu for alerting us to this book during correspondence in March, 1984.
104. Stirling, C.J.M. *Acc. Chem. Res.* **1979**, *12*, 198–203.
105. Alder, K.; Schumacher, M. *Fortschr. Chem. Org. Naturst.* **1953**, *X*, 1–118.
106. Hoffmann, H.M.R. *Angew. Chem. Int. Ed. Engl.* **1969**, *8*, 556–577.
107. Metzger, J.; Knöll, P. *Angew. Chem. Int. Ed. Engl.* **1979**, *18*, 70–71.
108. Snider, B.B.; Duncia, J.V. *J. Org. Chem.* **1981**, *46*, 3223–3226.
109. Thaler, W.A.; Franzus, B. *J. Org. Chem.* **1964**, *29*, 2226–2235.
110. (a) Denny, R.W.; Nickon, A. *Org. React.* **1973**, *20*, 133–336, (b) *Singlet Oxygen*, Wasserman, H.H.; Murray, R.W. (Eds.) (Academic Press, New York, 1979).
111. Bussas, R.; Kresze, G. *Liebigs Ann. Chem.* **1980**, 629–649.
112. Munsterer, H.; Kresze, G.; Brechbiel, M.; Kwart, H. *J. Org. Chem.* **1982**, *47*, 2679–2681.
113. Bunnett, J.F. *Pure Appl. Chem.* **1981**, *53*, 305–321.
114. Szele, I.; Zollinger, H. *J. Am. Chem. Soc.* **1978**, *100*, 2811–2815.
115. Kreuer, K.–D.; Rabenau, A.; Weppner, W. *Angew. Chem. Int. Ed. Engl.* **1982**, *21*, 208–209.
116. Cotton, F.A.; Hunter, D.L.; Lahuerta, P. *J. Am. Chem. Soc.* **1975**, *97*, 1046–1050.
117. Cotton, F.A.; Marks, T.J. *J. Organometal. Chem.* **1969**, *19*, 237–240.
118. Cotton, F.A.; Edwards, W.T. *J. Am. Chem. Soc.* **1969**, *91*, 843–847.

Chapter 19

IT'S THE PRINCIPLE OF THE THING

Some students approach organic chemistry with qualm because of warnings about the mountain of material to learn. In time they come to realize that certain underlying principles, when mastered, make life a lot simpler. In this chapter we look at a few principles with catchy come-ons; and we will meet some enticing chemical effects, methods, and phenomena as well.

Thousands of reactions in living systems are promoted by enzymes, nature's incredibly efficient catalysts. In 1975, William Jencks of Brandeis University wrote a review summarizing our knowledge of how enzymes go about their business.[1] Most scientists agree that catalyst and substrate first merge to form a complex. In this combined state the enzyme positions the substrate and distorts it in just the right way to favor a subsequent reaction. For maximum efficiency, the catalyst and substrate(s) should fit together well, just like pieces in a jigsaw puzzle. Hence, enzymes are snobbishly specific; a given one usually promotes a single reaction (or a small number of closely related ones), but that is all. Not surprisingly, any interference with these complex molecules can wreak havoc. We take advantage of that; sulfa drugs[2] and penicillins[3] can make us well by gumming up enzymes of invading bacteria. In Dr. Jencks' description, an enzyme lures a substrate "into a site in which it undergoes an extraordinary transformation of form and structure."[1] He turned to Greek mythology (Homer's "Odyssey") for a suitable term for this phenomenon and came up with the "Circe effect." Circe (which he pronounced "sir-see") was a temptress who lured men to her palace on the island of Aeaea, between Italy and Sicily, and then with her potions turned them into wolves, lions, or swine. What a cruel enzyme!

Researchers are not at a loss for other words to describe an enzyme's holding of reactive moieties near one another;[4] these include "propinquity" (Koshland, 1972),[5] "orbital steering" (Koshland, 1971),[6] "togetherness" (Jencks, 1972)[7] and "*Freezing at Reactive Centers of Enzymes* (FARCE)" (Mildvan, with tongue in cheek, 1972).[8] The first one may have engendered Robert B. Woodward's maxim: "We all know that enforced propinquity often leads on to greater intimacy...."[9] As a step toward curbing this proliferation of terms, Professor Jencks graciously withdrew his own "togetherness."[4,10]

Life often gets complicated, so chemists are pleased when someone solves a problem or explains a phenomenon simply and easily. A common plea from listeners at meetings

265

is "KISS" (Keep It Simple, Stupid). The principle is not modern at all, however. It was expressed in the fourteenth century by William of Ockham* (sometimes spelled "Occam"), an English theologian and philosopher, and member of the Franciscan friars. His expressions were much more elegant than "KISS." He wrote: "Plurality is not to be assumed without necessity"; and "What can be done with fewer assumptions is done in vain with more."[11] This principle has become known as "Ockham's Razor."

Chemists often invoke the principle when they feel that the simplest explanation of their observations is the preferred one. For example, in 1977 at the University of Utah, Evan Allred and ensemble (who lit up our lives with lampane, p. 36) used it to justify their mechanism for the reaction $1 \rightarrow 2$.[12] The same year at a Gordon conference, Dennis Tanner of the University of Alberta delivered a paper titled "Application of Ockham's Razor to the Mechanism of Free Radical Bromination."[13]

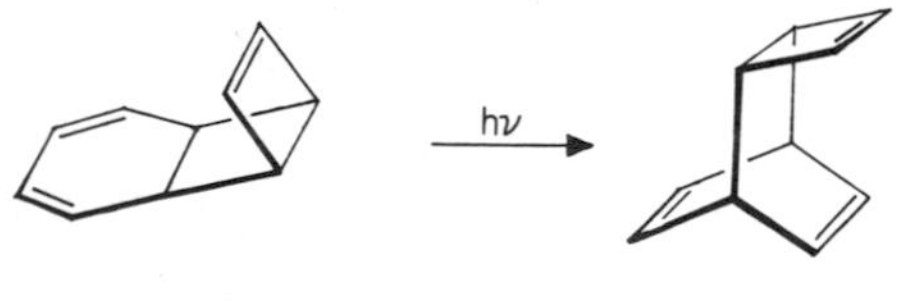

1   2

However, as Paul Gassman of the University of Minnesota pointed out, Ockham's Razor is not always sharp.[14,15] In 1972, his investigators (then at the Ohio State University) converted **3** to **4**. They considered a simple mechanism (breakage of bonds a and b) and a more complex one (breakage of b and c, followed by hydrogen migration).

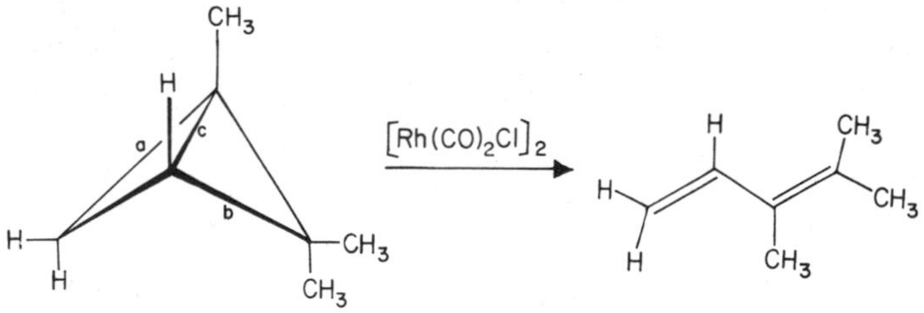

3   4

Deuterium labeling at the bridgehead (**5**) solved this quandary: The product was **7**. William of Ockham would have preferred the simpler scheme (**5** → **6**), but Paul of Ohio decreed otherwise.

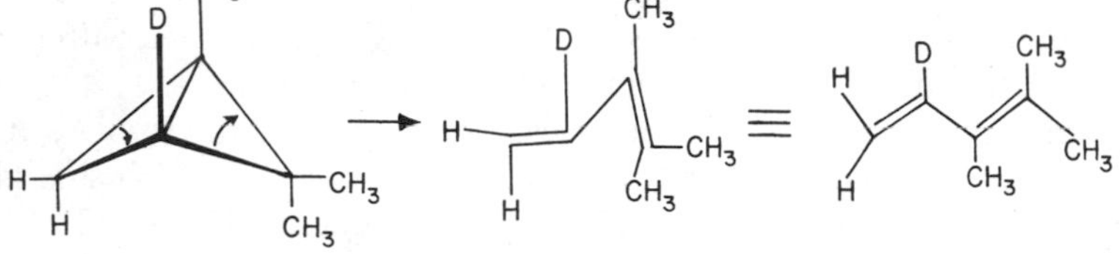

5   6

*William of Ockham was born in the 1280s in Ockham, near London. He entered the Franciscan order, studied at Oxford University, and received the degree of master of theology around 1319. His written views on philosophical and theological matters thrust him into an ongoing controversy with Pope John XXII. As a result, Ockham fled to Germany, whereupon the pope excommunicated him. He died in 1349 of the Black Plague.[11]

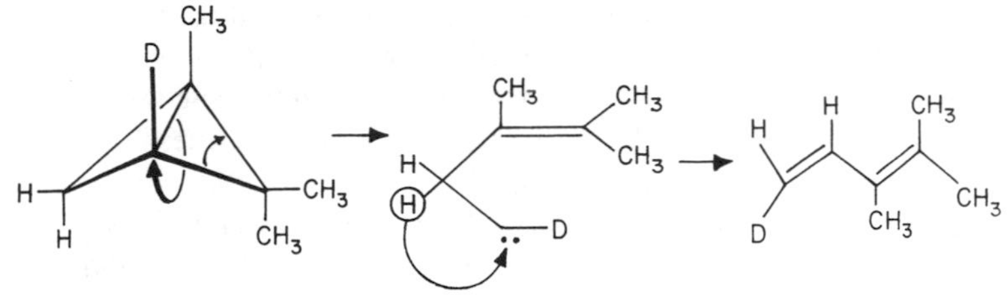

In a review on carbenoids, David Wulfman and Bruce Poling of the University of Missouri at Rolla warned readers that Ockham's Razor (like its steel namesake) must be wielded cautiously.[16] It is well established that photolysis and thermolysis of diazomethane generate methylene; this knowledge allows researchers to write simple carbene mechanisms for other reactions of diazo compounds. But chemists, invoking Ockham's Razor indiscriminately, have tended to assume that all diazo compounds behave similarly even when they might not.[16] The message from Drs. Wulfman and Poling was clear: Don't shave too closely.

We have talked about chair ⇆ chair interconversions in cyclohexanes and learned that the equilibrium can weigh strongly to one side (p. 248). How are such equilibrium constants (and hence standard free-energy changes) determined experimentally? Many analysts use low-temperature NMR spectroscopy. With sufficient cooling the interconversion slows, so that each chair form gives its own characteristic signals, with peak areas in proportion to its population. Early workers[17] relied on ^1H NMR, but ^{13}C NMR has furnished even better results.[18] Let us look at a specific example.

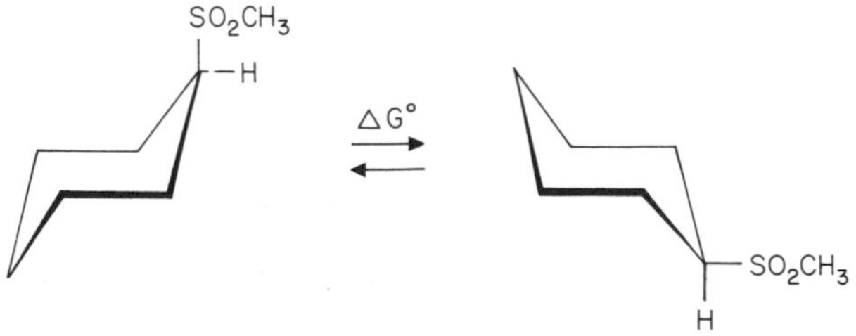

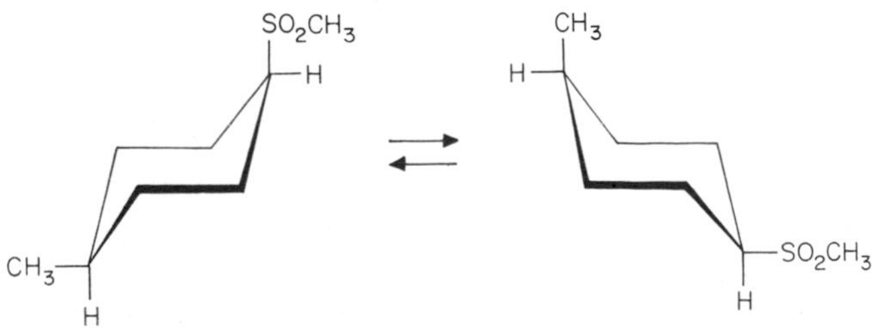

Ernest Eliel's researchers at the University of North Carolina sought the difference in standard free energy ($\Delta G°$) between the two chair isomers of methyl cyclohexyl sulfone (8a ⇆ 8b). This difference is called the A value[19] of the CH_3SO_2 group. The ^{13}C NMR spectrum, taken at $-90°C$, revealed a highly unbalanced equilibrium. It was not possible to obtain an accurate A for CH_3SO_2 because too little of the axial form (8a) was there to measure reliably. Undaunted, the North Carolina chemists resorted to an earlier strategem of theirs.[17] They measured K_{equil} for the analog with a *cis*-CH$_3$ at C4 (**9**). As each chair now had one axial *and* one equatorial substituent, the steady state was less one-sided and could be determined more accurately. The methyl acted as a conformational restoring group by offsetting the dictatorial methylsulfonyl; so Dr. Eliel's team called this the "counterpoise method."[17,18] A counterpoise (Latin *contra*, meaning "against" and Middle English *pois*, meaning "weight") is a weight that balances another one. The counterpoising alkyl, in this case CH$_3$, should be one whose own *A* value is already in hand. After that, only a bit of arithmetic was needed to compute *A* for the original equilibrium **8a** $\leftrightarrows$ **8b**.[18] Evidently, things go better in research if you keep your poise.

For reactions to take place, electrons in the transition state need to drift into the right regions of space. Elias Corey and Richard Sneen alerted the community to this fact of life and loosely called this requirement the stereoelectronic factor.[20] Stereo, from the Greek, *stereos* ("solid"), implies position in three-dimensional space. The S$_N$2 process serves as a simple example. In the nucleophile (Y), the electron pair that forms the new

10

bond prefers to attack *opposite* the leaving group (X) and to pass through transition state **10**. When this stereoelectronic imperative cannot be met, as in **11** where rear approach is blocked, the S$_N$2 path normally does not prevail.[20]

11

Stereoelectronic factors play essential roles in frangomeric cleavage of amines; for a speedy split the lone-pair orbital on nitrogen should be able to get antiperiplanar* to the C$_\alpha$—C$_\beta$ bond, which also should be antiperiplanar to the leaving group.[23] The amine on p. 254 meets these requirements.

In 1978, Robert Fraser and Philippe Champagne at the University of Ottawa reported a dramatic example of apparent stereoelectronic control.[24] In CH$_3$OD—NaOCH$_3$ the proton labeled H$_f$ (f = fast) in twistan-4-one (**12**; see p. 291 for the twist on this name) exchanged 290 times swifter than the one labeled H$_s$ (s = slow). Note that the C—H$_f$ bond stands nearly erect with the p-orbital on the carbonyl carbon, whereas the C—H$_s$ bond is about 60° out of line. Thus, if we begin to remove H$_f^+$, the electrons left behind flow nicely into the π-system. Not so for H$_s$. Just as with the wheels of an ipsomobile (p. 254), the better the alignment the smoother the trip. But if you prefer a

---

*For definitions of antiperiplanar (and the related stereochemical descriptors synperiplanar, anticlinal, and synclinal), see Klyne and Prelog.[21] These terms have no connection with Periplanone-A and -B, sex pheromones of the American cockroach (*Periplanata americana*).[22]

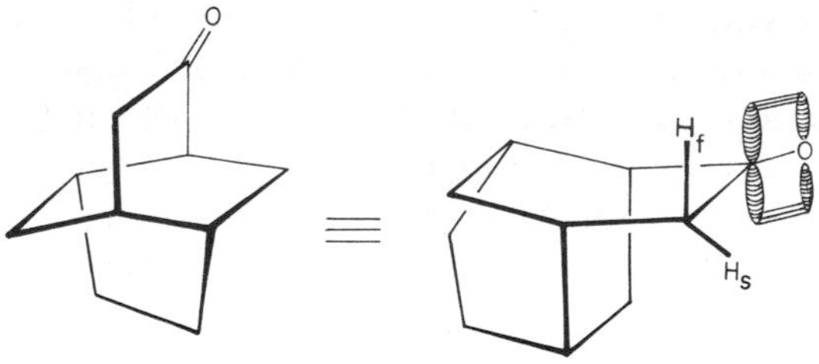

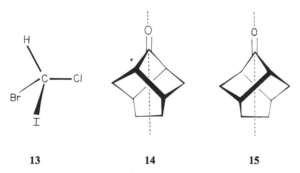

12

change of scenery for the journey, chauffeurs Fraser and Champagne can also get you there just as quickly by a different route—and with less worry about alignment.[25] Their compatriot Pierre Deslongchamps at Quebec's University of Sherbrooke championed stereoelectronic control and did much to consolidate contemporary thought on this topic.[26]

Another staunch advocate, Anthony Kirby (Cambridge University), sometimes referred to the theory of stereoelectronic control as the Antiperiplanar Lone-Pair Hypothesis (ALPH). In a report on enzymatic hydrolysis of acetals, Professor Kirby noted that English poet Samuel Taylor Coleridge (1772–1834) prophetically may have presaged the importance of orbital alignment as well as the difficulties of probing interactions at enzymic active sites.[27] Coleridge's poem "Kubla Khan" contains these lines:

> Where Alph, the sacred river, ran
> Through caverns measureless to man

Most chiral molecules (e.g., **13**) are also asymmetric; that is, they have no point, axis, or plane of symmetry. Therefore, ketone **14**, prepared by Ronald Sauers's team at Rutgers University, holds particular interest.[28] It is not superposable on its mirror image (**15**) and is thus chiral. However, it *does* possess a C_2 axis of symmetry (shown as a dotted line). Rotate the structure around this axis by 180° and you get what you started with.

13   **14**   **15**

Masao Nakazaki and his band at Osaka University coined "gyrochiral" for a figure that is chiral but not asymmetric, such as **14**.[29] "Gyro" comes from Latin *gyrare* ("to turn or twirl"); it refers to what we do to **14** to make it look like itself. But "chiral" derives from the Greek *cheir* ("hand"), the chemistry teacher's ever-present example of a dissymmetric object. Thus, Dr. Nakazaki's "gyrochiral" is another Latin–Greek hybrid (see p. 254).

The symmetry sages at Osaka dazzled the world with a host of fascinating gyrochiral structures. Consider a cage-like family Professor Nakazaki called "triblattanes" (from German *Blatt* for "leaf").[30] These locked polycycles consist of a twisted bicyclo[2.2.2]octane core (16), with one, two, or three additional diagonal bridges. Accordingly, triblattanes can be tri-, tetra-, or pentacyclic (see 17, 18, and 19). Nakazaki's nomenclature proclaims them [m]-, [m,n]-, and [m,n,p]triblattanes, respectively; the letters stand for the number of CH_2's in each extra span, and zero would mean a direct connection. Many members are chiral (i.e., dissymetric) despite having high symmetry, and the Osaka laboratory has churned out numerous optically active prototypes.[30]

16

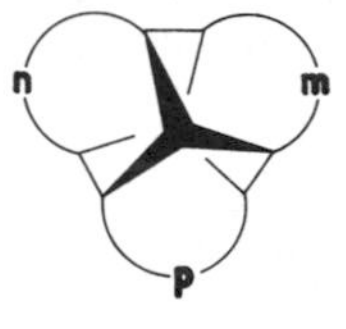

17   18   19

     Meanwhile, in 1978, Hans–Dieter Martin (then at the University of Würzburg) and co-worker Peter Pföhler transformed the bishomocubanedione 20 to the appealing pentacycle 21. This diene has C_2 symmetry and can be viewed architecturally as a benzene dimer from a four-point landing of one ring atop another.[31] The skeleton in 21 could pass for a basket with handles, but "basketane" was already taken. So, Martin and Pföhler instead envisaged a cube with two handles and blended Latin *ansa* (for "handle") and New Greek *sari* (for "cube") to concoct the mellifluous "ansarane" (a contraction of ansa-sarane).[32] On this basis, 21 became C_2-ansaradiene. (The two handles can be elsewhere on a cube, so don't neglect the symmetry symbol). When Osaka's Dr. Nakazaki leafed through that autumn 1978 publication from Würzburg, he spotted familiar foliage: C_2-ansaradiene is in fact an unsaturated [2.2.0]triblattane (see perspective 22).[30] Chemists certainly love to twist molecules, but not always for the same reasons.

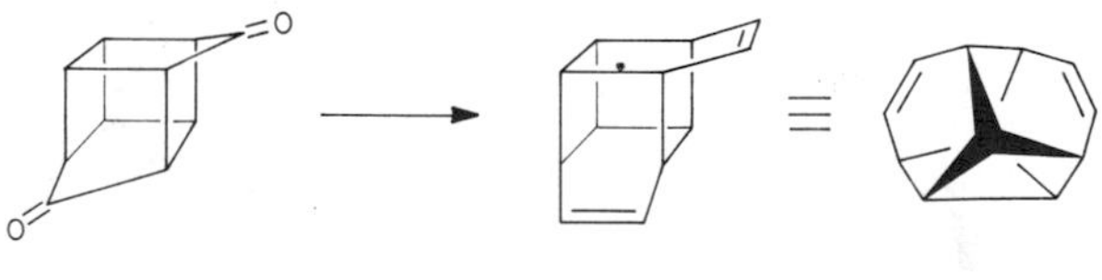

20   21   22

     Wilhelm Maier at the University of California, Berkeley, wondered how to deform an olefinic link to the limit and yet sustain it in a stable (or at least experimentally observable) molecule. So he designed the unique, tetracyclic alkene 23.[33] Several different theoretical computations by Dr. Maier and co-worker David Jeffrey

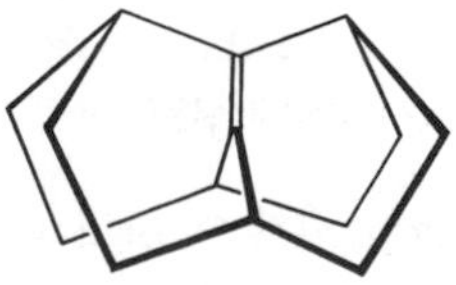

23

convinced them that the double bond in **23** would be distorted $\sim 90°$. Yet, it should remain a ground state because of stabilization by the suprastructure. In a preliminary announcement, the Berkeley researchers termed this achiral figure "orthogonene" and recommended it as a worthy synthetic target.[34]

Incidentally, the words "chiral" and "chirality" were first introduced by University of Glasgow physicist William Thompson (Lord Kelvin)* in a lecture at The Johns Hopkins University on October 17, 1884.[36] In 1966, Cahn, Ingold, and Prelog popularized the terms through their classic paper on specification of molecular chirality.[37] Chiral (and achiral) have precise meanings when applied to geometric figures or models. But for real systems (such as molecules, solvents, or reagents) the definitions are not sharp and can depend on the conditions of experimental measurements.[38]

In 1983, Stephen Hanessian, synthetic chemist at the University of Montreal, took pieces from "chiral" and "synthon"[39] to create the smooth portmanteau word "chiron."[40] It means an "optically active synthon," and strategist Hanessian made it up to steer our thinking about asymmetric synthesis of natural products. But the name was old hat to students of Greek mythology.[41]

Chiron (Greek *Cheiron*) was born a Centaur (i.e., half man, half horse); and most Centaurs were known to be wild and lawless. But Chiron was raised by the Greek god Apollo and his sister Artemis, and they educated him in the arts of healing, music, prophecy, and hunting. So Chiron grew up learned, very wise, and kind. He tutored such eminent Greek heroes as Achilles, Hercules, Jason, and Asclepius (who became the god of medicine). Legend continues that Chiron was accidentally wounded by an arrow of Hercules, chose mortality, and was transformed into the constellation Sagittarius (the "Archer") by Zeus, the supreme deity of the Greeks.[42]

Closer to our time, at the University of Zurich, André Dreiding and Karl Wirth had adopted the word "chiron" for another purpose. It stood for the smallest conceivable, chemically feasible, subarrangement of elements that can exist in a left- and right-handed form; to wit, a nonplanar cluster of four different atoms. You can find their erudite communication in a 1980 issue of the mathematical chemistry journal *Match*.[43]

Chemists often depict energy changes by potential-energy curves, which contain maxima (transition states) and minima (intermediates and products). An energy profile may also include an extended flat region; Roald Hoffmann's contingent at Cornell University called such a flatland a "twixtyl."[44] "Twixt" is a poetic form of "betwixt," which comes from Old English, and means "between." The phrase "betwixt and between" is tautologous and simply means an intermediate position, neither completely one nor the other. Thus, a twixtyl is neither a maximum nor a minimum, but something inbetween. (Note: twixtyl sounds adjectival but is a noun). A footnote in the Cornell publication asserts that the new term won the approval of an *ad hoc* committee in Ithaca, New York.[44] That city is where Dr. Hoffmann's research group hangs out, so the committee was undoubtedly anancomeric (p. 248) and could have used some counterpoise (p. 268).

Carbon can find itself betwixt and between during E2 eliminations. University of California, Santa Cruz researchers directed by Joseph Bunnett probed E2 transition

---

*The Kelvin scale (absolute temperature) was named in honor of Thompson. This scale serves scientists in many ways and even saved the day for one author of a submitted manuscript. The paper stated that a particular experiment was carried out in a cold room where the ambient temperature was 4–7°C. A referee complained that this amounted to a temperature fluctuation of about 45%, which was unacceptable. In reponse, the author changed 4–7°C to 277–280 K and pointed out to the editor that the temperature variation was now only 1%.[35]

states and drew attention to instances where C—X cleavage ran well ahead of C—H cleavage (24) and also to cases where the reverse prevailed (25). In 1967, when Dr. Bunnett was a visiting professor at the University of Canterbury, Christ Church, New

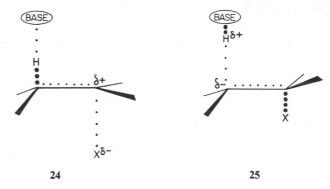

24   25

Zealand, he fretted over how best to describe E2 states that fell just short of involving true carbocations or carbanions.[45] His host, Professor John Vaughan, hopped over to the Classics Department and returned with the Latin prefix *paene-*, which means "almost." Bunnett happily took it on and in 1969 used "paenecarbonium" and "paenecarbanion" in a published review of variable transition states.[46] The former term later evolved to "paenecarbenium" in accord with nomenclature suggestions by Olah (see p. 40).[47] Incidentally, the same prefix is recognizable in words like penultimate (almost last) and peninsula (almost an island). Professor Bunnett's "paenecarbenium" and "paenecarbanion" aptly describe two extreme E2 transition states; but in retrospect he regrets that their distinction comes at the end of the words rather than at their beginning (as, for example, in terms like electrophilic and nucleophilic). In our paene-humble opinion, however, last does not mean least.

We have already encountered exo and endo terminology in the norbornane skeleton (26). An exo bond projects outside the solid angle of intersection of planes abcd and

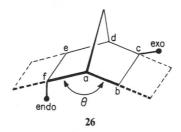

26

afed; the designation comes from Greek *exon* ("without; on the outside"). Similarly, "endo" derives from the Greek *endon* ("within"). These terms form part of the names brexane and brendane (p. 216).

Richard Dickerson and Irving Geis of the California Institute of Technology pointed out that exo and endo are also used for enzymes.[48] An "exocatenase" catalyzes cleavage of a terminal amino acid from a peptide or protein chain. (Recall that *catena* is Latin for "chain"). In other words, it promotes hydrolysis at the end of the line. An "endocatenase" operates at an inner link. We see now why an enzyme that chops a chunk from the *end* of a chain is *not* an endopeptidase.

If double or triple bonds are conjugated, the p-orbitals on contiguous atoms want to merge. For 1,3-butadiene (27), representation 28 shows such orbital interaction. Howard Simmons and Tadamichi Fukunaga of the Du Pont Company built on this

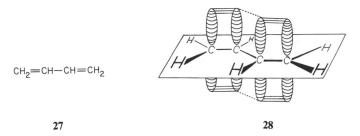

CH_2=CH—CH=CH_2

27   **28**

concept when they coined the word "spiroconjugation" for p-overlap between nonadjacent carbons in molecules of type **29**.[49] The sketch **30** shows only those p-lobes partaking in spiroconjugation. As these orbitals are not next to each other and are disposed orthogonally, they overlap less zealously than in **28**.

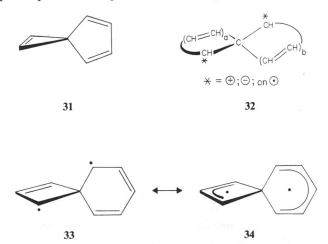

(CH=CH)$_m$

C

(CH=CH)$_n$

29   **30**

While they were at it, Drs. Simmons and Fukunaga introduced us to "[m, n]spirenes"—molecules that show spiroconjugation. Thus, **31** is [1, 2]spirene.[49] The prefix "spir" implies that the plane turns, like a spiral, as we glide mentally from one

31   **32**

ring to the other. Roald Hoffmann's team at Cornell had the same concept in mind when they dubbed species of type **32**, "spirarenes."[50] Thus, diradical **33**, which can also be written as **34**, is [3, 5]spirarene. (Here, the digits specify the number of interacting p-orbitals in each ring.) We met spiro structures earlier in helixanes (chapter 6).

Around 1970, Cornell's Melvin Goldstein sat down with colleague Roald Hoffmann for loop-drawing sessions. They pondered ways to connect the ends of polyene segments by means of saturated "insulators" and yet retain topographies that permit cyclic flow of the π-electrons. For convenience, the Cornell duo called a polyene segment with conjugated p-orbitals a "ribbon."[51] We depict it here by a thick,

33   **34**

unbroken partial loop; and we use a thin loop for the insulating chain and a dashed line for p overlap across space.

In a single ribbon (**35**), the π-electrons can travel cyclically only via a conventional "pericyclic" topology in which the termini interact with each other. (*Peri* is a Greek combining form meaning "all around.") Below the drawing we show a suitable hydrocarbon representative, arbitrarily chosen to be a delocalized anion.

Two ribbons joined end to end also can produce a net pericyclic array (**36**). But, if each tip in one ribbon connects to *two* termini in the other ribbon (**37**) rather than just to one, Goldstein and Hoffmann call the π-topology "spirocyclic," because it involves

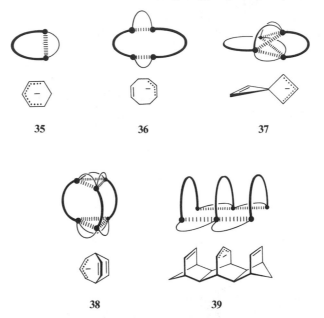

spiroconjugation.[52] With three ribbons, the topological options increase. For example, besides peri and spiro, a three-ribbon network can also be "longicyclic" (*longi* in Latin means "long") and "laticyclic" (*lati* is Latin for "wide; broad"). In longicyclic arrays every terminus has a connectivity of two (**38**), whereas in laticyclic only the middle ribbon is so blessed (**39**). The exercise gets even more fascinating with four or more ribbons; but by now you have the idea. The Cornell professors assured that all this is not just fun and games with molecular doodles. Among other things, they presaged what types of π-rich bicyclic ions would be stabilized (i.e., "bicycloaromatic") and destabilized ("antibicycloaromatic").[51] Subsequent experiments with various polycyclic carbanions proved these forecasts to be "right on."[53]

Molecules that prey on one another in solution can sometimes coexist peacefully when tethered to a polymer backbone. In 1977, Abraham Patchornik and co-workers at the Weizmann Institute of Science in Rehovot found that trityllithium (**40**) and *o*-nitrophenyl benzoate (**41**) individually attached to insoluble polymer chains ((P)) are content to stay that way until a soluble reactant (e.g., benzyl cyanide) passes by.[54] Biblically inspired, the Weizmann scientists termed these processes "wolf and lamb" reactions; according to Isaiah 11:6, "...the wolf shall dwell with the lamb...." This approach makes it possible to produce **42** (and other acylated carbon acids) in excellent yield.[54] Things do not fare nearly so well when the entire menagerie is in solution, since trityllithium is a "wolf" and *o*-nitrophenyl benzoate is a "lamb," and the inevitable happens.

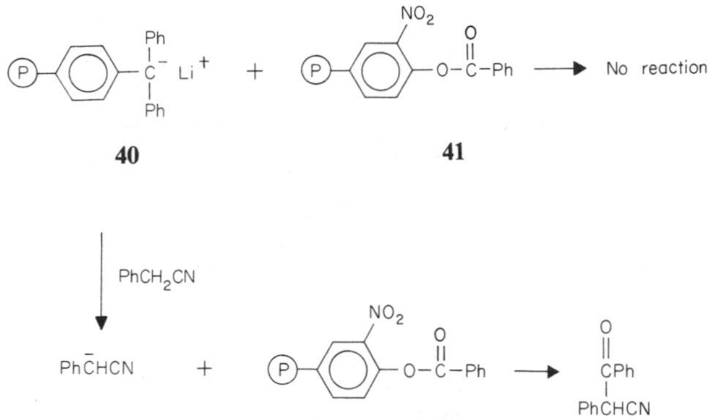

40   **41**

42

By the way, when Professor Patchornik and colleagues developed new methodology for peptide synthesis, they liked to make up names with Jewish or Israeli flavor.[55] For instance, in 1976 Amit, Ben-Efraim, and Patchornik observed that irradiation of N-acyl-5-bromo-7-nitroindolines (e.g., **43**) predisposes the acyl function to cleavage by nucleophiles like HOH, ROH, RNH_2, and so on.[56] Among other things, the Weizmann Institute chemists employed this finding to couple two peptide segments (A and B in **44**) free of racemization.[57] It seemed only natural when the Israeli innovators, at

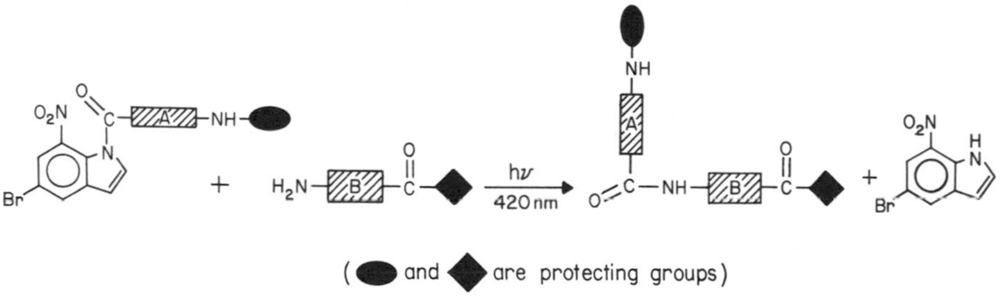

(⬤ and ◆ are protecting groups)

43   **44**   **45**

symposia[55] and in publications,[57] referred to 5-*bromo*-7-*nitro*indoline (**45**) as "Bni." But this shorthand is more than just an acronym. In Hebrew, *Bni* means "my son." The laboratories of Abraham wanted to remind the world that *they* begat this useful reagent.[55]

Do you remember the delightful character "Yente" in the musical production, *Fiddler on the Roof*? Yente was a shadchan (Yiddish spelling is *shadchen*)—a matchmaker or go-between; someone who brought couples together for marriage. According to Professor Patchornik, a documented use of a shadchan in the Bible involved Abraham's eldest servant, who brought together Rebekah and Abraham's son Isaac for marriage (Genesis 24). Abraham dispatched his servant to a distant town to find a suitable girl, to break her family ties, to induce her to return with him, and to release her upon meeting with Isaac.

At the Weizmann Institute, Patchornik *et al.* developed a chemical shadchan—a soluble carrier molecule (S in **46**) that can displace an amino acid unit (see A in **46**) from a heterogeneous[58] home on one polymer and transport it to a different polymer.[55,59] There, the agent deposits its protégé at the marrying end (B) of an anchored peptide segment and, task completed, flows back to escort yet another eligible mate to the amide altar. Not all chemical shadchans confine their talents to peptide weddings.

Some have served the Israelis in matters of acylation, phosphorylation, deuteriation, and bromination.[55]

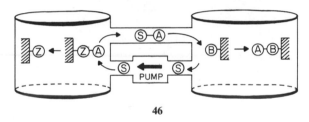

46

Organic chemistry texts usually teach catalytic hydrogenation of alkenes in an early chapter. Students accept the concept easily and, because hydrogenation normally involves *syn* addition,[60] they begin to appreciate stereochemistry. The intimate details of hydrogen uptake promoted by a metal are not thoroughly understood; heterogeneous processes resist scrutiny more than do homogeneous ones. Evidently, alkene and hydrogen molecules sit on the catalyst surface and somehow become activated; the hydrogen atoms (not necessarily both from the same H_2 molecule) then hitch on (one at a time) to the alkene from the catalyst side.[61]

If the two faces of a double bond have different surroundings, as in **47**, another stereochemical question emerges.[62] Which surface will sit on the catalyst? If (as with most of us) the bottom side of **47** sits, we get a product with a *trans* ring junction, **48**. We expect such an outcome if R sterically blocks catalyst approach from the top. On the

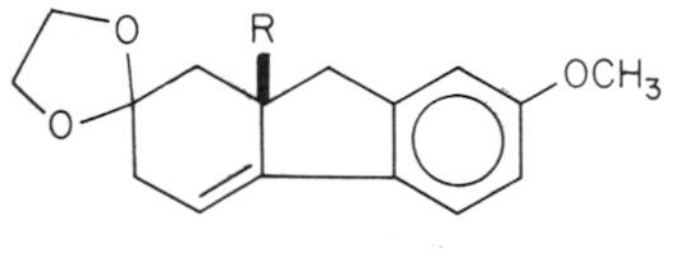

47

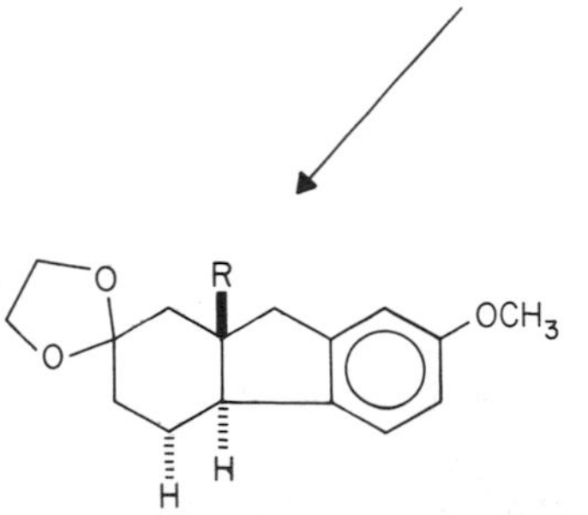

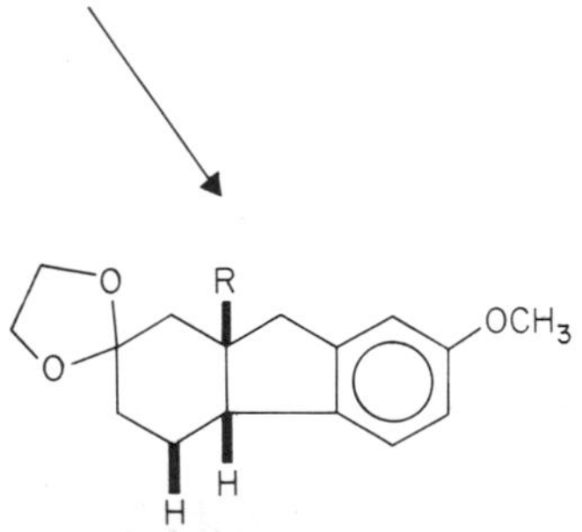

48   49

other hand if, for some reason, R is drawn to the catalyst, the top side would "sit," and *cis* isomer **49** would form. In 1973, Hugh Thompson and his team at Rutgers University suggested "haptophilicity" to describe the ability of a group to attract the catalyst surface.[62] He derived the word from the Greek *hapto* ("I attach or fasten myself to") and *philio* ("I love"). It thus joins other terms of affection such as electrophilic, nucleophilic, enophilic, and apicophilic.*

In a footnote, the Rutgers publication explained the genesis of "haptophilic" by noting that "the word is not found in classical Greek, nor does it occur among the

---

*"Electrophilic" and "nucleophilic" were introduced by Sir Christopher Ingold.[63] "Apicophilicity" is the desire of a substituent to occupy an apical site (see p. 135) when attached to a pentavalent atom such as phosphorus.[64]

writers of the Renaissance, nor anywhere else." Professor Thompson admitted he lifted this curious disclaimer verbatim from one used by Hilaire Belloc in a book "*The Bad Child's Book of Beasts*."[65] (In that illustrated album of witty verse, author Belloc invented the word "tuptophilist," from Greek stems meaning "I strike" and "I love.") The Rutgers rompers continued their whimsy in the Acknowledgments section of their paper by thanking "H. Belloc for helpful consultations."

In the Rutgers study, the most haptophilic R group was —CH_2OH; it gave product consisting of 95% of the *cis* isomer **49**. At the other extreme, —$CONH_2$ was downright haptophobic; 90% of the product was *trans* (**48**). To simplify discussions of these effects, Thompson and Wong subsequently termed a process "proximofacial" if it takes place at the same molecular face as a group of potential influence (e.g., the R in **47**); a process at the remote face is "distofacial."[66] These terms are not meant to express any sense of transfer or conveyance from one site to another, as do suprafacial and antarafacial.

Thompson *et al.* also observed that CH_2OH could control stereoselectivity in *chemical* reduction of a double bond. For example, action of $LiAlH_4$ on **50** gave overwhelmingly the *cis* dihydro isomer **51**. They envisaged an initial aluminum

1) $LiAlH_4$
2) H_2O

50   **51**

alcoholate complex that swivels toward the olefin link for a *syn* delivery of hydride. In Thompson's interpretation, the CH_2OH serves as a reagent "hinge" that guides the reductant suprafacially to the site of action.[67]* And while we're on hinges let's swing over to a different brand for a moment.

Olefins are usually flat, but they can be forced to fold like a hinge. For example, in the constrained polycycle *anti*-sesquinorbornene (**52**)[69] the alkene unit is essentially planar, whereas in the *syn*-isomer (**53**)[70] the skeleton creases slightly along the π-bond, like a book beginning to close. The extent of "hinge bend" depends on several factors, including attached substituents.[71] Paul Bartlett, at Texas Christian University, pointed out that among sesquinorbornenes, the *syn*–oxa analog **54** exhibited one of the largest hinge bends. Appropriately enough, in their Texas laboratory this bronco was called "Big Bend."[72]

52   **53**   **54**

Now back to hapto. This prefix can also denote affinity between a metal atom and an unsaturated ligand. In 1968, F. Albert Cotton (then at MIT) proposed that **55** be named (1, 2, 3, 4-tetrahaptocyclooctatetraene)-tricarbonyl iron, because the iron atom is fas-

---

*A laboratory is not the only place where chemists fret about reductions. Editors of journals, hard pressed to keep down publication costs, frequently cajole authors to reduce the length of verbose manuscripts. And, vigilant referees sometimes provide the ammunition. Marshall Gates, professor at Rochester University and Editor of the *Journal of the American Chemical Society* from 1950–1969, recalled an instance when a referee complained that a paper was much too long. The critic maintained that if the manuscript could not be reduced considerably then it should be oxidized completely.[68]

55   **56**

57   **58**

tened to carbons 1–4 of the ring.[73] Compound **56** is *hexahapto* because the chromium associates with six of cyclooctatetraenes's eight carbons. When Roald Hoffmann and co-workers, in 1978, considered the *motion* of a group *across* a cyclopentadienyl ring, they referred to it as a "haptotropic shift."[74] (The interconversion of **57** and **58** provides an example.[75]) The suffix "tropic" (from the Greek *tropos*) means "changing or otherwise responding to a stimulus." We hope the term stimulates you to turn to the final chapter.

REFERENCES AND NOTES

1. Jencks, W.P. *Advan. Enzymology* **1975**, *43*, 219–410.
2. Solomons, T.W.G. *Organic Chemistry*, 3rd ed. (John Wiley & Sons, New York, 1984), pp. 865–867.
3. Abraham, E.P. *Scient. Amer.* **1981**, *244*, June, pp. 76–86.
4. Jencks, W.P.; Page, M.I. *Biochem. and Biophys. Res. Commun.* **1974**, *57*, 887–892.
5. Storm, D.R.; Koshland, D.E., Jr. *J. Am. Chem. Soc.* **1972**, *94*, 5805–5814.
6. Dafforn, A.; Koshland, D.E., Jr. *Proc. Nat. Acad. Sci. U.S.* **1971**, *68*, 2463–2467. For some newer angles on orbital steering see Menger, F.M. *Tetrahedron* **1983**, *39*, 1013–1040.
7. Jencks, W.P.; Page, M.I. *Proc. 8th FEBS Meeting, Amsterdam, The Netherlands,* **1972**, *29*, 45.
8. Nowak, T.; Mildvan, A.S. *Biochemistry* **1972**, *11*, 2813–2818.
9. Woodward, R.B. *Pure Appl. Chem.* **1968**, *17*, 519–547; see p. 545.
10. Jencks, W.P., private communication, March, 1976.
11. *The Encyclopedia of Philosophy* (The Macmillan Co. and The Free Press, New York, 1967), vol. 8, p. 307.
12. Allred, E.L.; Beck, B.R.; Mumford, N.A. *J. Am. Chem. Soc.* **1977**, *99*, 2694–2700, footnote 17.
13. Tanner, D.D., private communication, July, 1977.
14. Gassman, P.G.; Williams, F.J. *J. Am. Chem. Soc.* **1972**, *94*, 7733–7741.
15. Gassman, P.G., private communication, November, 1975.
16. Wulfman, D.S.; Poling, B. In *Reactive Intermediates*, Abramovitch, R.A., (Ed.), (Plenum Press, New York, 1980), p. 322.
17. Eliel, E.L.; Della, E.W.; Williams, T.H. *Tetrahedron Lett.* **1963**, 831–835.
18. Eliel, E.L.; Kandasamy, D. *J. Org. Chem.* **1976**, *41*, 3899–3904.
19. Winstein, S.; Holness, N.J. *J. Am. Chem. Soc.* **1955**, *77*, 5562–5578.
20. Corey, E.J.; Sneen, R.A. *J. Am. Chem. Soc.* **1956**, *78*, 6269–6278.
21. Klyne, W.; Prelog, V. *Experientia* **1960**, *16*, 521–523.
22. (a)Persoons, C.J.; Verwiel, P.E.J.; Ritter, F.J.; Talman, E.; Nooijen, P.J.F.; Nooijen, W.J. *Tetrahedron Lett.* **1976**, 2055–2058, (b) Talman, E.; Verwiel, P.E.J.; Ritter, F.J.; Persoons, C.J. *Isr. J. Chem.* **1978**, *17*, 227–235.
23. Grob, C.A. *Angew. Chem. Int. Ed. Engl.* **1969**, *8*, 535–546.
24. Fraser, R.R.; Champagne, P.J. *J. Am. Chem. Soc.* **1978**, *100*, 657–658.

25. Fraser, R.R.; Champagne, P.J. *Can. J. Chem.* **1980**, *58*, 72–78.

26. (a) Deslongchamps, P. *Stereoelectronic Effects in Organic Chemistry*, Organic Chemistry Series, Baldwin, J.E., (Ed.), (Pergamon Press, Oxford, 1983), vol. 1, (b) Deslongchamps, P. *Tetrahedron* **1975**, *31*, 2463–2490.

27. Kirby, A.J. *Acc. Chem. Res.* **1984**, *17*, 305–311.

28. Sauers, R.R.; Whittle, J.A. *J. Org. Chem.* **1969**, *34*, 3579–3582.

29. Nakazaki, M.; Naemura, K.; Yoshihara, H. *Bull. Chem. Soc. Jpn.* **1975**, *48*, 3278–3284.

30. (a) Nakazaki, M.; Naemura, K.; Hashimoto, M. *Bull. Chem. Soc. Jpn.* **1983**, *56*, 2543–2544, (b) For a review, see Nakazaki, M., in *Topics in Stereochemistry*, Eliel, E.L.; Wilen, S.H.; Allinger, N.L. (Eds.), (John Wiley & Sons, New York, 1984), vol. 15, pp. 199–251.

31. Martin, H.-D.; Pföhler, P. *Angew. Chem. Int. Ed. Engl.* **1978**, *17*, 847–848.

32. Martin, H.-D., private communication, November, 1985.

33. Maier, W. F., private communication, August, 1985.

34. Jeffrey, D.A.; Maier, W.F. *Tetrahedron Lett.* **1984**, *40*, 2799–2802.

35. Recounted in *Journal of Irreproducible Results* **1977**, *23*, 29. Dr. W.V. Metanomski kindly brought it to our attention in private correspondence, March, 1986.

36. Lord Kelvin, *The Baltimore Lectures on Molecular Dynamics and the Wave Theory of Light* (C.J. Clay and Sons, London, 1904), pp. 436, 619.

37. Cahn, R.S.; Ingold, C.K.; Prelog, V. *Angew. Chem. Int. Ed. Engl.* **1966**, *5*, 385–415.

38. Mislow, K.; Bickart, P. *Isr. J. Chem.* **1976/77**, *15*, 1–6.

39. Corey, E.J. *Pure Appl. Chem.* **1967**, *14*, 19–37.

40. Hanessian, S. *Total Synthesis of Natural Products: The "Chiron" Approach*, Organic Chemistry Series, Baldwin, J.E., (Ed.), (Pergamon Press, Oxford, 1983), vol. 3.

41. Hanessian, S., private communication, August, 1984. We thank Professor Hanessian for reprints and background information.

42. (a) Allen, R.H. *Star Names: Their Lore and Meaning* (Dover Publications, New York, first published 1963), (b) Goodwin, T.E. *Chem. Eng. News* **1985**, January 14, pp. 4–5.

43. Dreiding, A.S.; Wirth, K. *Match* **1980**, No. 8, 341–352.

44. Hoffmann, R.; Swaminathan, S.; Odell, B.G.; Gleiter, R. *J. Am. Chem. Soc.* **1970**, *92*, 7091–7097.

45. Bunnett, J.F., private communication, March, 1983.

46. Bunnett, J.F. *Surv. Prog. Chem.* **1969**, *5*, 53–93.

47. Bunnett, J.F. Sridharan, S. *J. Org. Chem.* **1979**, *44*, 1458–1463; Bunnett, J.F.; Sridharan, S.; Cavin, W.P. *J. Org. Chem.* **1979**, *44*, 1463–1471.

48. Dickerson, R.E.; Geis, I. *The Structure and Action of Proteins* (W.A. Benjamin, Inc., Menlo Park, CA, 1969), pp. 87–88.

49. Simmons, H.E.; Fukunaga, T. *J. Am. Chem. Soc.* **1967**, *89*, 5208–5215.

50. Hoffmann, R.; Imamura, A.; Zeiss, G.D. *J. Am. Chem. Soc.* **1967**, *89*, 5215–5220.

51. Goldstein, M.J.; Hoffmann, R. *J. Am. Chem. Soc.* **1971**, *93*, 6193–6204.

52. (a) Gleiter, R.; Haider, R. *Tetrahedron Lett.* **1983**, *24*, 1149–1152, (b) For a review on spiroconjugation and other transannular interactions see Martin, H.-D.; Mayer, B. *Angew. Chem. Int. Ed. Engl.* **1983**, *22*, 283–314.

53. (a) Grutzner, J.B.; Winstein, S. *J. Am. Chem. Soc.* **1972**, *94*, 2200–2208, (b) Goldstein, M.J.; Natowsky, S. *J. Am. Chem. Soc.* **1973**, *95*, 6451–6452, (c) Moncur, M.V.; Grutzner, J.B. *J. Am. Chem. Soc.* **1973**, *95*, 6449–6451, (d) Goldstein, M.J.; Tomoda, S.; Whittaker, G. *J. Am. Chem. Soc.* **1974**, *96*, 3676–3678, (e) Goldstein, M.J.; Dinnocenzo, J.P.; Ahlberg, P.; Engdahl, C.; Paquette, L.; Olah, G.A. *J. Org. Chem.* **1981**, *46*, 3751–3754, (f) Goldstein, M.J.; Wenzel, T.T. *Helv. Chim. Acta* **1984**, *67*, 2029–2036.

54. Cohen, B.J.; Kraus, M.A.; Patchornik, A. *J. Am. Chem. Soc.* **1977**, *99*, 4165–4167; *ibid.* **1981**, *103*, 7620–7629.

55. Patchornik, A., private communication, January, 1984.

56. Amit, B.; Ben-Efraim, D.A.; Patchornik, A. *J. Am. Chem. Soc.* **1976**, *98*, 843–844.

57. Pass, Sh.; Amit, B.; Patchornik, A. *J. Am. Chem. Soc.* **1981**, *103*, 7674–7675.

58. Fridkin, M.; Patchornik, A.; Katchalski, E. *J. Am. Chem. Soc.* **1966**, *88*, 3164–3165.

59. Shai, Y.; Patchornik, A. "Peptide Synthesis with Insoluble Polymeric Reagents: The use of a Soluble Acyl Carrier ('Shadchan')." Paper presented at the 50th Anniversary Meeting of the Israel Chemical Society, April, 1984 (unpublished).

60. Burwell, R.L., Jr. *Chem. Rev.* **1957**, *57*, 895–934.

61. (a) March, J. *Advanced Organic Chemistry*, 3rd ed. (John Wiley & Sons, New York, 1985), pp. 696–697, (b) Somorjai, G.A. *Adv. Catal.* **1977**, *26*, 1–68.

62. Thompson, H.W.; Naipawer, R.E. *J. Am. Chem. Soc.* **1973**, *95*, 6379–6386.
63. Ingold, C.K. *Structure and Mechanism in Organic Chemistry* (G. Bell and Sons, London, 1953), pp. 200–201.
64. Johnson, M.P.; Trippett, S. *J. Chem. Soc., Perkin Trans. 1* **1982**, 191–195.
65. Thompson, H.W., private communication, August, 1975.
66. Thompson, H.W.; Wong, J.K. *J. Org. Chem.* **1985**, *50*, 4270–4276.
67. (a) Thompson, H.W.; McPherson, E. *J. Org. Chem.* **1977**, *42*, 3350–3353, (b) Thompson, H.W.; Shah, N.V. *J. Org. Chem.* **1983**, *48*, 1325–1328.
68. Gates, M. private communication, July, 1977.
69. Bartlett, P.D.; Blakeney, A.J.; Kimura, M.; Watson, W.H. *J. Am. Chem. Soc.* **1980**, *102*, 1383–1390.
70. Paquette, L.A.; Carr, R.V.C.; Böhm, M.C.; Gleiter, R. *J. Am. Chem. Soc.* **1980**, *102*, 1186–1188.
71. (a) Watson, W.H.; Galloy, J.; Bartlett, P.D.; Roof, A.A.M. *J. Am. Chem. Soc.* **1981**, *103*, 2022–2031, (b) Hagenbuch, J.–P.; Vogel, P.; Pinkerton, A.A.; Schwarzenbach, D. *Helv. Chim. Acta* **1981**, *64*, 1818–1832, (c) Ermer, O.; Bödecker, C.–D. *Helv. Chim. Acta* **1983**, *66*, 943–959, (d) Spanget–Larsen, J.; Gleiter, R. *Tetrahedron* **1983**, *39*, 3345–3350, (e) Houk, K.N.; Rondan, N.G.; Brown, F.K. *Isr. J. Chem.* **1983**, *23*, 3–9, (f) Watson, W.H. *Stereochemistry and Reactivity of Systems Containing π Electrons* (Verlag Chemie International, Deerfield Beach, FL, 1983).
72. Bartlett, P.D., private communication, April, 1983.
73. Cotton, F.A. *J. Am. Chem. Soc.* **1968**, *90*, 6230–6232.
74. Anh, N.T.; Elian, M.; Hoffmann, R. *J. Am. Chem. Soc.* **1978**, *100*, 110–116.
75. Ustynyuk, Yu. A.; Kisin, A.V.; Pribytkova, I.M.; Zenkin, A.A.; Antonova, N.D. *J. Organometal. Chem.* **1972**, *42*, 47–63.

Chapter 20

LEFTOVERS

So far, we've tried to group our tidbits more or less into themes; yet the title of this chapter is our admission that some items either did not seem to fit anywhere or that we came upon them at a late stage. But last does not mean least, so let us sample these leftovers. Even the first one may convince you they can be nourishing.

Would you believe a very common vitamin was once named for the fact that its structure was unknown?[1] This nutrient abounds in citrus fruits, cabbage, tomatoes, strawberries, etc. Yes, it is vitamin C, obtained crystalline by biochemist Albert Szent Györgyi,* then at Cambridge University. Around 1925 he had become attracted to plant chemistry after perceiving a similarity between the darkened color of damaged fruit and that of humans afflicted with disorders of the adrenal glands.[2] Looking into why certain plants do not turn brown, Szent–Györgyi observed that the juice of nonbrowning species could slow the discoloration of susceptible ones. In 1927 the young doctor isolated and analyzed the substance that delayed browning.[3] It was carbohydrate-like and acidic, but its structure was still a mystery. So he termed it "ignosic acid," related to an unknown sugar he called "ignose" (from Latin *ignorare*, which means "to have no knowledge of," and "ose," a common suffix in nomenclature of sugars).

Dr. Szent–Györgyi first proposed "ignose" in a manuscript submitted to *Biochemical Journal*.[4] But the editor, Sir Arthur Harden, took him to task for joking about a scientific matter and offered instead the name "hexuronic acid" to reflect the substance's acidity (pKa 4.17[5]) and its powerful reducing ability. (The ending "-uronic acid" is a cognomen for aldehydo-carboxylic acids derived from hexoses and the like.) Szent–Györgyi was not happy with "hexuronic acid" because the name carried structural connotations that were premature. So he countered by appealing, as it were, to higher authority, saying that if "ignose" is inappropriate then the substance might well be called "Godnose."[4] However, Editor Harden refused to kneel, so Dr. Szent–Györgyi was obliged to use "hexuronic acid" in his publications for a few years.[6]

When chemists at the Universities of Birmingham[7] and Zurich[8] later proved that the substance was not in fact a uronic acid, Szent–Györgyi and Walter Haworth† christened it "ascorbic acid" because it proved to be the agent known earlier as vitamin C, whose antiscurvy action had been well established (*a scor*butus = without scurvy).[9]

---

*Nobel Prize in medicine, 1937.
†Nobel Prize in chemistry, 1937, shared by Walter N. Haworth and Paul Karrer.

Incidentally, the American Medical Association generally disfavors chemical names that cite the disease being treated and proposed, instead, "cevitamic acid" (from C-vitam(in) + ic). That name never caught on.[10] In 1933, Birmingham researchers established the structure of ascorbic acid (vitamin C) as **1**,[7] and it was synthesized that same year by Haworth and Hirst.[11] A British commemorative postage stamp (**2**) issued in 1977 stylizes a cross-section of a citrus fruit along with a formula for vitamin C.[12]

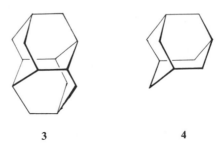

1   2

Much about this vitamin is still a mystery and controversial,[13] including the possibility that large doses can reduce the symptoms of the common cold[14] as well as the risk and severity of cancer.[15] The vitamin C story started with "ignose" and "Godnose." Where will it end? Who knows?

How about a name that celebrates a meeting of chemists? Sir Derek Barton (then at Imperial College London), a principal architect of chairs and boats (chapter 5), titled the esthetically pleasing molecule **3**, "congressane."[16] To appreciate this choice, let us turn back time a few years.

3   4

In 1933 at Prague, researchers S. Landa and V. Machácek discovered **4** as a component of petroleum.[17] They examined it crystallographically and called it "adamantane," because its C_{10} skeleton represented the repeating unit in that precious form of carbon known as diamond.* *Adamas* is Greek for "diamond"; and an "adamant" is an impenetratably hard substance (from Greek *a* ("not") and *damant* ("conquerable")). Four years later, Oskar Böttger of the University of Dresden synthesized a derivative with this skeleton; curiously, however, he did not cite the Czechoslovakian paper and suggested that the parent hydrocarbon **4** be called "diamantane."[19] Fortunately, this belated name for **4** never caught on. Landa and Machácek had clearly established priority; and since "di" often means "two," Böttger's

_____

*Another form of diamond was found in Canyon Diablo meteorites and has also been made artificially from graphite by high pressure/temperature techniques.[18] Its crystal structure is that of wurtzite (ZnS), so this form is called "hexagonal (or wurtzite) diamond." The repeating unit in hexagonal diamond is, in fact, the carbon skeleton of iceane (p. 153). Jewel thieves have long referred to stolen gems as "ice," so perhaps they known more about their trade than we think.

word diamantane might cause confusion. Incidentally, in 1966 Czechoslovakia issued a commemorative postage stamp depicting a ball-and-stick model of adamantane.[20]

In 1963, Swiss savant Vladimir Prelog wrote of his fascination with structure 3 (which had not yet been synthesized),[21] while independently Sir Derek Barton suggested it be adopted as the logo for the 19th International Congress of Pure and Applied Chemistry, to be held in London that year.[22] It did become the official emblem, and Sir Derek aptly christened it "congressane." The gauntlet was openly thrown down to chemists: Synthesize it!

Paul Schleyer, then at Princeton, was in excellent position to take up the challenge. After all, he had already prepared the $C_{10}H_{16}$ stabilomer adamantane (4) by isomerizing 5 with aluminum halides.[23] (Of structural isomers, the one with the lowest free energy of formation is called the "stabilomer." This word was coined by Schleyer *et al.* and stems from the German *stabil* for "stable."[24]) The $5 \rightarrow 4$ conversion was based on his reasoning that an alkane with the correct number of rings and carbons (in this

5   4

case three and ten, respectively) could give an adamantane skeleton (via a series of rearranging carbocations[25]), because such a scaffold is virtually free of torsional strain.[26] Applying this logic to reach the target congressane (3) he needed a starting molecule with 5 rings and 14 carbons. Before anyone could say "congressane," the bold doctor, Chris Cupas, and David Trecker photodimerized norbornene (6) to the pentacycle 7. With aluminum chloride they isomerized this dimer to congressane; the emblem had become a reality![27] (Granted, the yield was initially only 1% but was vastly improved in 1974.[28]) A husband–wife team at the U.S. Naval Research Laboratory, Jerome and Isabella Karle, settled the structure of 3 by X-ray crystallography.[29]

6   7   3

In 1966, Otto Vogl, Burton Anderson, and Donald Simons of the Du Pont Company (with Professor Schleyer's blessing) introduced a naming system for polyadamantanoid hydrocarbons based on the number of adamantane units.[30] Thus, congressane became "diamantane," an old proposed name for 4.[19] ("The more things change, the more they remain the same."[31]) Extending the principle of thermodynamic drive, Dr. Schleyer's enforcers have since synthesized triamantane (8),[32] and four (an appropriate number) chemists in Ireland (William Burns, M. Anthony McKervey, Thomas Mitchell, and John Rooney) have prepared 9, one of the three possible tetramantanes.[33] Sparkling achievements indeed.

Professor Schleyer and postdoctorate Eiji Ōsawa, from Japan, had tried to make a tetramantane back in 1968 by isomerizing an alkane having the proper number of rings (9) and carbons (22).[34] This time, however, that dramatic approach delivered an undesired isomer, 10, whose structure was pinned down by X-ray analysis. (For a

8   **9**   **10**

different perspective of **10** rotate the page clockwise to bring the dashed line horizontal.) In their publication, they gave this newborn the caconym "bastardane," because it conformed to a dictionary definition of "bastard," namely "of a kind similar to, but inferior to, or less typical than, the standard."[34] At the time Dr. Ōsawa carried out the experiments, he was not familiar with the English word bastard. When his labmates tried to explain its more common definition, he exclaimed: "Ah, an unwanted child!" That expressed perfectly the groups's feeling about **10**; so, to them the name was doubly appropriate.[35] By the way, you should not confuse bastardane with "bastadins," which are not unwanted at all but are metabolites obtained from a sponge, *Ianthella basta*. An Australian team isolated and christened several of them in 1980.[36]

"Eureka" is a much more desirable emotional response to the outcome of an experiment. It comes from the Greek *heurēka* ("I have found it") and supposedly was first exclaimed by Archimedes. David Ollis' smiths (at the University of Sheffield) and collaborators at May & Baker, Ltd., in England, felt this same exhilaration when they obtained acid **11** by degradation of a complex antibiotic, $C_{61}H_{88}Cl_2O_{33} \cdot H_2O$, called flambamycin. They designated **11** "eurekamic acid,"[37]

$$HO_2C \qquad \begin{array}{c} HOCHCH_3 \\ | \\ C-OH \\ | \\ COCH_3 \end{array}$$

11

And what about the name "flambamycin" for the compound that gave up life for the sake of **11**? It seems that Professor Ollis discussed his investigations on this antibiotic with Dr. Georges Jolles, research director at Sante Division, Rhône–Poulenc, France, while dining at the restaurant La Crêpe Flambée, in Paris. The meal must have been delicious, for they named the antibiotic after the restaurant.[37] That endorsement is worth more than three stars in the *Michelin Guide*. Furthermore, a hydrolysis product of flambamycin was dubbed "bamflalactone," simply because "bamfla" is an anagram of "flamba."[37,38] Evidently, scrambling the letters was easier that finding another recherché dining spot.

Eric Block (State University of New York, Albany) nicely blended gastronomy and medicinal folklore in his research on sulfur-containing molecules. Collaborating with teams at the University of Delaware and the center for Biophysics and Biochemistry in Caracas, Venezuela, Professor Block pursued the chemistry and physiology of garlic and onions. Botanically, these two members of the lily family are known as *Allium sativum* and *Allium cepa*, respectively. (Fittingly, Latin *allium* may come from a Celtic word meaning "pungent.")[39]

According to age-old beliefs, these foods possess therapeutic ingredients. For example, garlic supposedly protects against stroke, coronary thrombosis, and hardening of arteries; and it also displays antibacterial and antifungal properties.

In early chemical studies of distillates from *Allium sativum*, Theodor Wertheim obtained some unpleasant sulfur compounds, whose structures contained the CH_2=CH—CH_2 moiety. In 1844 he proposed calling this group "allyl," a word now well entrenched in chemistry.[40]

Research over the years demonstrated that a key precursor in garlic is (+)-S-allyl-L-cysteine sulfoxide (12). Stoll and Seebeck initially isolated this amino acid and named it "alliin." It was the first-discovered optically active natural product that is chiral at sulfur as well as at carbon. Alliin is colorless and odorless; in fact, an intact garlic bulb also has little or no odor. But, cutting or crushing the bulb brings alliin into contact with an enzyme (alliinase, also written as allinase) that breaks it down to 2-propenesulfenic acid (13). Two of these molecules in turn go on to produce allyl 2-propenethiosulfinate (14). Cavallito and co-workers were able to isolate 14 by mild

| 12 | 13 | 14 | 15 |

extraction of garlic and called it "allicin." It emits a potent odor and reputedly is one of the main culprits responsible for the smell of garlic.[39] But chemically, allicin is rather unstable, and its derived products include some having antithrombotic activity. Block and collaborators showed that one of these anticlotting agents has structure 15 (E and Z isomers).[41] Seeking a simple name for this odorless, sulfur-rich triene and anxious to avoid overuse of the prefix alli,[42] they christened 15 "ajoene." *Ajo* (pronouned "aho") is Spanish for "garlic." Perceptive mechanistic reasoning led the Block bunch to a successful synthesis of ajoene by simple hydrolysis of allicin.[41]

Inorganic chemists also know their gastronomy. In the 1960s when M. Fredrick Hawthorne and students tilled the fecund field of transition metal complexes at the University of California, Riverside, they sometimes broke for lunch at a Mexican restaurant, La Paloma. Its pleasing menu and atmosphere induced Professor Hawthorne to think Spanish when naming new carborane ligands bred in his laboratory.[43] Skeletal shape of the polyhedral ligand, a pinch of imagination, and linguistic help from La Paloma's cook proved the right recipe for simplifying nomenclature in that burgeoning discipline. For example, $B_{10}CH_{11}^{-3}$ and $B_9C_2H_{11}^{-2}$ became "carbollide" and "dicarbollide" ions, respectively. Both are related to a hypothetical $B_{11}H_{11}^{-4}$ parent, whose contour led Hawthorne *et al.* to baptize it the "ollide" ion (from Spanish *olla* for "kettle or jar").[44] The basketlike, icosahedral fragment $B_8C_2H_{10}^{-4}$ made its debut as a "dicarbacanastide" ion, sponsored by the Spanish noun *canasta* meaning "basket."[45] And they called $B_7C_2H_9^{-2}$ a "dicarbazapide" ion because its outline looks like something you could put your foot into. (*Zapato* is Spanish for "shoe.")[46] Examples of dicarbollyl, dicarbacanastyl, and dicarbazapyl ligands appear schematically in **16**, **17**, and **18**, respectively. We are glad the researchers at Riverside patronized La Paloma but can't help wonder what their ligands would be called had Professor Hawthorne opted for oriental cuisine.

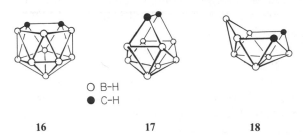

○ B-H
● C-H

16   **17**   **18**

When Switzerland's Vladimir Prelog in 1968 first referred to compounds of type **19** as "vespirenes,"[47] chemists may have thought he was praying for good results; vespers are evening prayers. However, a year later, his research team revealed that the "ve-" derives from *V*iererpunktsymmetri*e*gruppe D$_2$ and the "-spirene" from *spir*obi-fluor*ene*.[48] Thus, the appellation for this brood of compounds embodies symmetry as well as constituent parts. The two fluorene moieties (fluorene is **20**) are evident in **19** and are welded spiro-fashion. *Viererpunktsymmetriegruppe* (German for four-point symme-

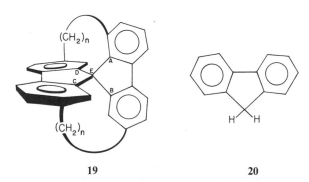

19   **20**

try group) means that **19** is a structure for which four symmetry operations are possible. One of these is the "identity" operation (often denoted I), in which we do nothing at all to the molecule; and, naturally, the outcome is equivalent to the original structure.[49] Also, **19** when rotated 180° around a horizontal axis (the bisector of angles DEC and AEB), gives a structure the same as the original. Thus, this axis has twofold (360°/180° = 2) symmetry, denoted C$_2$.[50] The molecule owns two other C$_2$ axes; one bisects angles AEC and DEB, the other bisects angles CEB and DEA. A figure with three equivalent C$_2$ axes belongs to symmetry group D$_2$.[47,49]

However, even with these symmetry elements the molecule is chiral. Professor Prelog's people demonstrated this stereochemical oddity dramatically by preparing both enantiomers of [6,6]vespirene (**19**, $n = 6$).[48] The specific optical rotations at 436 nm were + 1630° and − 1630°! That's chirality with a capital "C," so Dr. Prelog's vespers were indeed answered.

Earlier, we facetiously hoped a cyclone would disperse carbon monoxide evolved in a reaction (p. 219). A much more reliable method for removing this poisonous gas is to force it through a mixture of metal oxides that catalyze its oxidation to carbon dioxide at room temperature. Such mixtures are known as "hopcalites," because they were developed near the end of World War I by teams headed by Joseph Frazer from The

Johns *Hop*kins University and Gilbert N. Lewis and William C. Bray from the University of *Ca*lifornia, in collaboration with chemists at the Fixed Nitrogen Research Laboratory at American University.[51] The U.S. Government sponsored that research because of the pressing need for a gas mask to absorb carbon monoxide, which was claiming many victims of incomplete combustion in battleship turrets and in machine-gun pits.[52] One of these catalysts, "Hopcalite I," consists of MnO_2, CuO, Co_2O_3, and Ag_2O. Hats off to our inorganic colleagues for honoring their schools.

We learned earlier (chapter 6) that the names croconic acid and rhodizonic acid stem from the colors of their salts (yellow and rose, respectively). But what about salts that are themselves colorless and yet sport colorful names? Such is the case for 4, 4'-bipyridinium ions like **21** studied by Leonor Michaelis and colleagues at Rockefeller Institute in the 1930s.[53,54] They observed that reduction of **21** proceeds in stages to

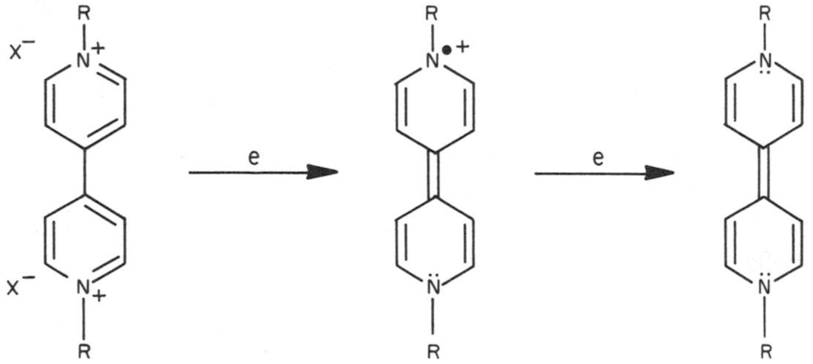

21	**22**	**23**

produce a highly colored, semiquinoid radical cation **22** and then a colorless reduction product **23**. Michaelis called precursor salts like **21** "viologens" (sometimes spelled "viologenes"[55]) because the colored intermediates generated from them were intensely violet. In time, the colored species themselves became mellifluently labeled "violenes."[55]

The π-systems of a pair of separate violenes can sometimes interact weakly with each other, intra- or intermolecularly, without forming a σ-bond;[55] and similar behavior had also been observed with π-delocalized neutral radicals.[56] Such reversible "dimers" were dubbed "π-mers" by Edward Kosower (who handed us bimane, p. 16) and fellow researchers;[56,57] and the process became "π-merization."[58] When Greek π is rendered "pi," the latter term amounted to "pimerization."[55] The phenomenon is quite general but depends strongly on solvent, temperature, concentration, etc. So, keep in mind that pimerization is not just epimerization with the first letter missing —and the two don't sound alike either.

Other colorfully titled molecules include the blue liquid hydrocarbon "azulene", **24**. Spanish *azul* and French *azur* mean "blue." Similarly, colors inspired the class name "fulvene," which Johannes Thiele of the Munich Academy of Sciences coined in 1900

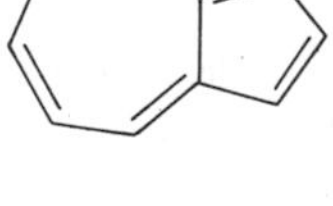

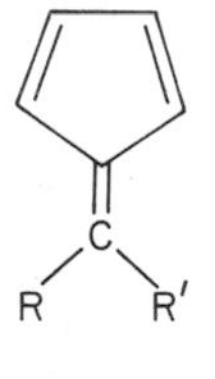

a   R = R' = H
b   R = R' = CH_3
c   R = CH_3 , R'= Ph
d   R = R' = Ph

24	**25**

for **25a** and its derivatives.[59] Fulvous means "dull brownish yellow" or "tawny." Fulvene itself (**25a**) is yellow;[60] and analogs **25b**, **c** and **d** are bright orange, orange, and red, respectively.[59]* In 1949, Ronald Brown of the University of Melbourne, Australia, built on Dr. Thiele's term by adopting "fulvalene" for **26**.[62] Prefixes may be hung onto either of the parent names to identify ring size; for example, **27** ($n = 1$) is triafulvene;[63] **27** ($n = 3$) is heptafulvene; and **28** ($m = n = 3$) is heptafulvalene. The unsymmetrical analog

        26   27   28

28 ($m = 1, n = 2$) is triapentafulvalene.[64] Does this last one seem familiar? It should; we met it before as calicene (p. 60). And you may remember hearing Beethoven's pentaundecafulvene (**28**, $m = 2, n = 5$) concerto; this is, after all, fidecene (p. 103). Jūro Ojima and colleagues in Japan have spanned our planet with a rainbow of really big fulvalenes and fulvenes.[65]

Earlier we noted that polarization in the exocyclic alkene unit of calicenes lends

            25b   29

aromatic character to both rings. As a result, calicenes have much higher dipole moments than do ordinary unsaturated hydrocarbons (p. 60). This trait applies to other fulvalenes and also to fulvenes. For example, polarization of the exocyclic double bond in ordinary (penta-) fulvenes dumps negative charge on the ring and gives it 6 π (aromatic) character. As a result of this high polarity, fulvenes combine with reagents that do not attack ordinary alkenes. For example, in 1934, Karl Ziegler and his collaborators at the University of Heidelberg added phenyllithium to **25b** and got **29**.[66]

Horst Prinzbach stretched the nomenclature even further by referring to **30** as "vinylogous fulvalenes" because of their extra exocyclic vinyl group.[67] For example, **30** ($m = 2, n = 3$) is a vinylogous sesquifulvalene (sesqui, because n is one and one-half times as large as in fulvalene itself [$n = 2$]). The Prinzbach probers showed that this 14 π-electron compound undergoes thermal conrotatory electrocyclization followed by two 1, 5-sigmatropic hydrogen hops to give **31**.[68] A concerted conrotation is actually

---

*In the 1950s, fulvene analogs were intensely investigated by theoreticians Alberte and Bernard Pullman at the Institut du Radium in Paris, in collaboration with experimentalist Ernst Bergmann, then at Weizmann Institute's Daniel Sieff laboratories, in Rehovot. Their joint efforts on nonbenzenoid aromatics produced an avalanche of publications. Professor Bergmann was much impressed at the Pullmans' ability to predict spectral properties of fulvenes by quantum-mechanical calculations. And, at one time he even proposed that fulvenes be rebaptized "pulvenes."[61]

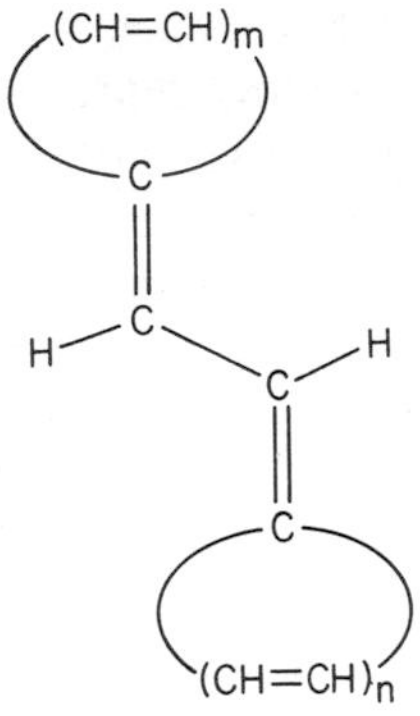

30

symmetry-forbidden; but the outcome is nonetheless conrotatory because, in this case, the allowed disrotatory route would result in severe steric strain.[68] Fortunately, vinylogous sesquifulvalene has understanding p-orbitals.

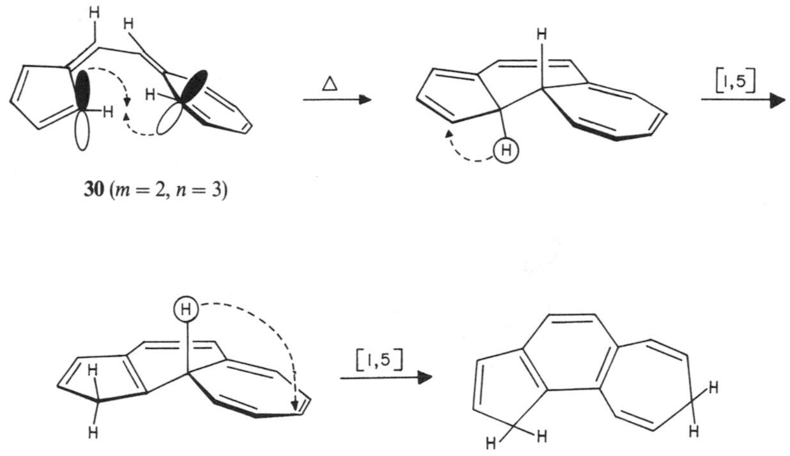

30 ($m = 2, n = 3$)

31

If molecules have colors and shapes, what about atoms? Why, of course! Just look at commercial molecular models to be convinced that elements possess color and form. A trend of sorts may have been set as early as April 7, 1865 by August Wilhelm Hofmann, a noted organic chemist and Director of the Royal College of Chemistry in London.* In a lecture titled "On the Combining Power of Atoms," delivered before the Royal Institution of Great Britain, Hofmann assembled balls and sticks to portray CH_4, $CHCl_3$, and other compounds of carbon. He said,

> I will on this occasion select my illustration from that most delightful of games, croquet. Let the croquet balls represent our atoms and let us distinguish the atoms of different elements by different colours. The white balls are hydrogen, the green ones chlorine atoms, the atoms of fiery oxygen are red, those of nitrogen blue; the carbon atoms are naturally represented by black balls.

---

*Hofmann spent the second half of his prolific research career at the University of Berlin. His name endures in organic chemistry in connection with thermal decomposition of quaternary ammonium hydroxides ("Hofmann elimination") and with conversion of primary amides to amines ("Hofmann degradation").

The record of Hofmann's lecture is one of the earliest references to molecular models in the chemical literature.[69] His atomic color code has since mutated, so these days we commonly see other hues and shades. For example, stylish sets from the Paris firm of SASM feature red hydrogens. light blue oxygens, dark blue nitrogens, yellow fluorines, green chlorines, brown bromines, and violet iodines.[70] And why not? Isn't Paris noted for its fashionable models?

By the way, biologists are color conscious too. Otherwise why would three classes of mutants of *E. coli* phage T4 be named "amber," ochre," and "opal?"[71] For this story Harris Bernstein at the University of Arizona took us back to 1960 at the California Institute of Technology, where he had just completed his Ph.D. in biochemical genetics.[72] One evening he ventured back to the laboratory to entice his friends, graduate student Charles Steinberg and postdoctorate Richard Epstein, to come along to a movie. But they were too deeply immersed in a mutant hunt involving painstaking transfer of phage from petri dishes to tester plates. And instead, they persuaded Harris to pass up the movie and pitch in to help pick plaques. By morning, their efforts led to discovery of a new category of mutants; so Charles and Richard rewarded Harris for his assistance by naming the mutant "amber" (*bernstein* is German for "amber," a yellow precious stone). Professor Bernstein recalls that through faulty laboratory technique his own plates proved a total loss, so he really made no positive contribution to the amber chase. But the name stuck. And so did the trend to go with colors, because related mutants discovered later became "ochre" and "opal."[71]

In a similar vein, the beautiful blue of a sapphire stone, and a personal predilection for that color, may have compelled Harvard's Robert Woodward to coin "sapphyrin" for pentapyrrolic systems having one direct link and four bridging methines (e.g., as in **32**). He reported such an analog in 1966 at a conference on aromaticity held at Sheffield,[73] but details were not published until 1983.[74] Meanwhile, in 1972, British chemists from the Universities of Nottingham and of Sussex synthesized **32** and dispelled any doubts about the genuineness of its blood line. Onomastically faithful, this closed conjugated, $22\,\pi$-electron macrocycle showed up as dark blue prisms.[73] They also constructed oxa- and thiasapphyrins having furan and thiophene rings in place of one or more pyrroles.[75]

The fraternity of chemical gemologists released yet more precious stones. For instance, in 1970, Professor Woodward and graduate student Michael King fashioned a glassy, green beauty, **33**, having one less methine bridge than a sapphyrin.[76] And in Britain, Alan W. Johnson's staff skillfully polished off some oxa-analogs of **33**.[73] The Britishers advertised their merchandise simply as "norsapphyrins"; but the Harvard duo, with a keen eye for the market, dubbed **33** a "smaragdyrin" (Latin *smaragdus* means "emerald"). Professor King was happy to go with green.[77] After all, he is part Irish, the emerald is his birthstone, and chemistry is a gem of a science.

$\bullet = CH_3$

$\bullet\bullet = CH_2CH_3$

32   33   34

In 1983, Albert Gossauer and Hans Rexhausen of Berlin's Technical University prepared a related 22 π-electron pentacycle with methine bridges between all the pyrrole units, typical of the situation in ordinary porphyrins. For the parent skeleton (**34**) in this new family they proposed the name "pentaphyrin." It deserves a separate title, because this structure sports its own shade of blue.[78]

Those who shop for greens and blues can always find satisfaction at pyrrole-based outlets around the world. For example, at Michigan State University, Eugene LeGoff and graduate student Bob Berger introduced a line of macrocycles (**35**) containing an odd number of olefinic carbons between the pyrrole rings of a porphyrin.[79] They synthesized the first prototype (a 22 π-electron loop) in 1978; and in 1986 Professor LeGoff and student Otho Weaver brought forth the 26π-conjugated analog **36**.[80]

35   36   37

Being of sound mind, Dr. LeGoff was not keen to promote this merchandise through systematic nomenclature. He preferred a convenient class name that spoke of the expanded flat π-network. A linguistic hunt into several different languages proved unrewarding, until Professor LeGoff turned to Greek and sought help from resident expert Gerasimos Karabatsos. Professor Karabatsos, who was then Department Chairman, pointed out that *platys* in Greek means "broad, flat." Perfect, thought LeGoff, who then blended in the suffix from porphyrin to creat "platyrin."[81] The number of methine carbons in each bridge are stuck up front. Thus, the bluish-violet beauty **36** is a [1, 5, 1, 5]platyrin; and if you prefer green, ask for model [1, 3, 1, 3]. The Michigan State merchants also informed customers that "homoporphyrins"[82] could be sold as [1, 1, 1, 2]platyrins.[79]

At Cologne University, Emanuel Vogel and associates Matthias Köcher, Hans Schmickler, and Johann Lex also developed a keen eye for color, but they did it by restructuring the core "porphin" skeleton to the isomer **37**, whose inner π-track is more rectangular than square. Professor Vogel's group viewed **37** as having structural features both of porphyrins and acenes, so this appealing macrocycle became "porphycene." NMR studies disclosed that its four pyrroles are effectively equivalent through fast NH tautomerism. Crystals of porphycene are violet; its solutions in organic solvents are blue and fluoresce beautifully in the red-violet.[83] Now, lest we become too dazzled, let's set color aside and get back into shape.

Stereoelectronic control has been invoked to explain hydrogen–deuterium exchange in a ketone with the "twistane" (**38**) skeleton (p. 268). The name "twistane" for tricyclo[4.4.0.3^{3,8}]decane goes back to the 1962 work of Howard Whitlock at the University of Wisconsin; he was the first to synthesize the parent hydrocarbon.[84,85] Having been indoctrinated to the virtue of descriptive names by his colleague Howard Zimmerman, who gave us the bicycle reaction (p. 72), barrelene (p. 58), and rod compounds (p. 156). Professor Whitlock set out to find one for **38**.[86] With molecular models we can construct **38** and thereby *reconstruct* Dr. Whitlock's deliberations.

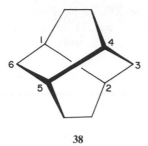

38

Start with cyclohexane and connect carbons 1 and 4 with a two-carbon bridge. This span holds the initial ring in a boat conformation. If we now join C2 and C5 by another two-carbon chain we must twist, or screw, the cyclohexane boat considerably. It ends up in a shape that Professor William Johnson, then at Wisconsin, appropriately dubbed "twist boat."[87] So Dr. Whitlock considered for **38** the possible titles "twistane" and "screwane." At the time, a dance named "the twist" was popular in the United States, so "twistane" won.[86]

The Hawaiian hula is another type of dance that calls for a good deal of pelvic twist. And according to Robert S.H. Liu, some polyenes gyrate that way when the music is right. Professor Liu and his students at the University of Hawaii, Manoa, were enmeshed in studies of highly conjugated alkenes, including photic and thermal isomerizations. They strived to understand why some double-bond sites are prone to undergo $E \rightleftharpoons Z$ more than others, because such geometry changes underlie the chemistry of the vision process.

The phenomenon of vision begins when rhodopsin in rod cells absorbs light and becomes bathorhodopsin. Rhodopsin (**39**) is a polyunsaturated aldehyde (Z-11-retinal) bound as a Schiff base to a membrane protein, opsin. The opsin and its ionic charges

39   **40**

inhibit isomerization of the π-link closest to the imine. And, near the other end of the polyene the E, 7-double bond also resists configurational change, probably for want of space around the trimethylcyclohexene. Also, possibly, a hydrophobic pocket in the protein immobilizes that nonpolar tail. But the regions close to carbons 10 and 11 are less encumbered, and it is these sites that act out the opening scenes when light strikes rhodopsin.

The visual process involves a complicated sequence with much not understood. But one aspect seemed clear: the cycle of events takes place too fast to permit much translational motion. Several research teams have pondered how a polyene can accommodate configurational as well as conformational changes given its restricted freedom in the protein. Dr. Liu hit upon an explanation, for which he credits his undergraduate class in organic chemistry—albeit indirectly.[88]

It happened one afternoon in February, 1984, while Professor Liu sat at his desk during scheduled office hours. He was waiting for undergraduates to come in with questions about his course lectures. No one showed up. So, to keep occupied, the solitary Professor reached for a set of ball and stick models, put together a five-carbon section, and began to flop it around. The learned chemist was adept at twirling individual bonds to get from a W to a U shape (see p. 178 for this terminology). And he also knew how to "bicycle-pedal" atoms, as recommended years earlier by Arieh Warshel (p. 73).[89] But after a while Dr. Liu found he could interconvert W and U forms with less agitation, namely by a modest, but simultaneous, torsion about the two middle bonds.[88] Suddenly, the light struck, and he realized the chemical implications of this digital exercise. Excited, he cornered Dr. Alfred Asato, a long-time co-worker, and before long the two were immersed in the nitty gritty of polyene contortions in rhodopsin.

We can follow the essence of the geometric maneuver with drawings **41–43**. Five sp^2 carbons on a flat surface start with W geometry. With centers 2 and 4 fixed, a concerted twist about the two inner bonds pivots the middle carbon up and over, while the terminal atoms slide sideways on the original surface. The molecular volume should change but little during the entire performance, and Dr. Liu thought it held the key to Z

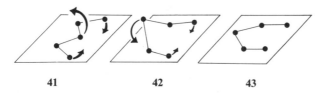

| 41 | 42 | 43 |

→ E and any accompanying conformational adjustments that light triggers upon rhodopsin. Liu and Asato sifted through the known facts about vision before convincing themselves their model fitted.

When it came time to go public with their mechanism, mentor Liu wanted to avoid a mundane description like "concerted rotation of two bonds." So he coined the fetching—but apt—"hula twist." Then he drafted a paper with hula twists sprinkled everywhere from the title to the summary. But Dr. Asato cautioned his boss that the name might not fly for the simple reason that scientists, by and large, display little sense of humor.* And Dr. Liu also recalled that an earlier try of his to dub some new visual pigment analogs "alohaopsin I, II,..." had been greeted with little enthusiasm by some workers in the field. Taking all this to heart, Professor Liu expurgated hula twist from the manuscript and replaced it with "CT-n (concerted twist at center n)." Still, with a twinkle in his eye—and perhaps a glimmer of hope—Dr. Liu inserted this footnote: "The research group at Hawaii prefers to use the term HT-n: hula twist at center n." The article (including footnote) was duly published, in 1985.[90] To their surprise, the scientific community sensed the island rhythm and picked up on the term hula twist, especially after *Chemical and Engineering News* featured it in a writeup.[91]

The Liu–Asato research article pointed out that whereas hula twist at the center of a five-atom segment (swivel of the navel, as it were) interconverts W and U, similar gyration on a sickle array produces a different sickle.[90] Moreover, hula twisting a carbon next to the end (i.e., a hip pivot) transforms a sickle to either a W or U, according

---

*Actually, chemists do have a good sense of humor but lose it when they serve as referees.

to which hip flips. For illustrations of typical hip actions, follow the solid circles in **44** and **45**.

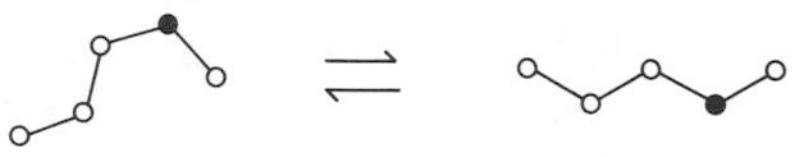

44

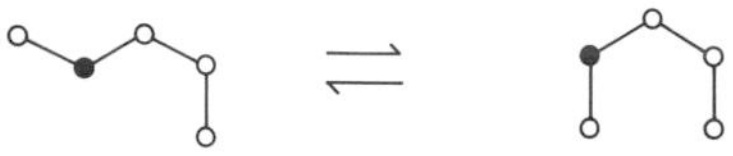

45

Reconciling these details with literature data led the Honolulu choreographers to suggest that the photic event in vision involves Z, 11-retinal Schiff base (see **39**) going to its E, 11 isomer having a 10, s-*cis* conformation. Accordingly, they proposed structure **40** for bathorhodopsin and also conjectured on the identity of subsequent transformation products.[90] They also extended the HT-*n* concept to bacteriorhodopsin and related areas of vision chemistry.[92]

One final touch to this episode from Hawaii. When Professor Liu described these ideas at a departmental seminar, the room was packed with undergraduates from his class. He had invited them, promising to reveal how they had played a role in his research. At the end of the talk Dr. Liu flashed a slide that showed his office, empty; and he thanked his undergraduates for not showing up that February afternoon. After all, they had left him alone and he learned the hula twist.[88]

Joseph Bunnett (patriarch of cine substitution, chapter 18) also turned to the world of dance for help. In 1971, his group (then at Brown University) treated 1,2,4-tribromobenzene (**46**) with potassium anilide and found **47–51** among the products.[93,94] This movement of bromines within a molecule and from one molecule to another reminded Professor Bunnett of dancers pirouetting on a stage, so he dubbed the reaction "base-catalyzed halogen dance."[94] He even transformed this vision into a reality. In 1971, Dr. Bunnett and Althea Short, a dance instructor at the University of California, Santa Cruz, trained an ensemble of students to demonstrate the mechanism of the reaction through an actual choreographic routine. A videotape of this performance was made and is available in the videotape library of the University of California, Santa Cruz.[95] In 1980, Dr. Bunnett's researchers found that chlorines also

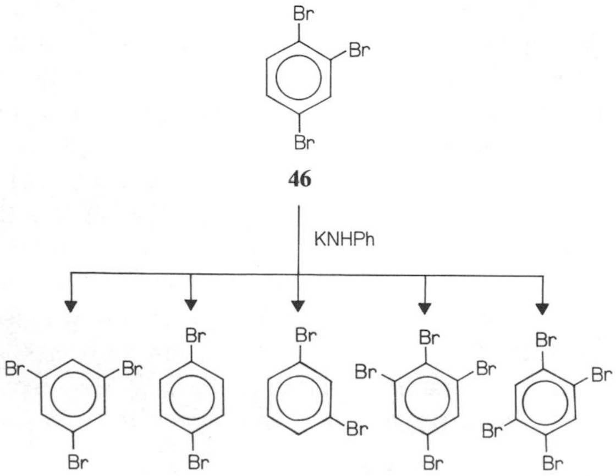

46

KNHPh

47   **48**   **49**   **50**   **51**

enjoy this type of recreation.[96] So in future musicals, dancing may even be done by chlorines instead of chorines.

Professor Bunnett and his troupe did not restrict their artistic talents to the dance. When he and Francis Kearley studied benzyne formation by action of KNH_2 on dihalobenzenes, they found that two different halogens will vie with each other for departure. It seems the heavier halogen needs the exercise, so it bows out while the lighter one prefers to stay put. To describe their findings, these chemists were moved to publish an article written entirely in verse![97] Hats off to troubadours Bunnett and Kearley for this novel break from the ordinary. And, a salute to *Journal of Organic Chemistry* editor Frederick Greene, who permitted publication in that style. In a footnote, however, Professor Greene cautioned would-be bards that further poetic manuscripts face an uncertain future in his editorial office.[97]

In 1968, Alex Nickon and postdoctorate Gopal Pandit synthesized tetracycle **52**.[98] The cyclopropane ring rests on *three ax*ial pillars fixed to a cyclohexane chair. Hence, at the suggestion of graduate student Douglas Covey, Dr. Nickon christened **52** "triaxane." If a second three-membered ring were to hang similarly from the bottom side of the chair, all axial bonds would be used up and the resulting heptacycle (**53**) could be termed "peraxane." This structure appeared in the Nickon–Pandit publication and in a subsequent one by Rangaswamy Srinivasan of IBM's Research Center.[99] A top view of **53** projects a Star of David inside a hexagon. So Eric Lord, a postdoctorate in Irving Borowitz's laboratory, suggested that **53** might be called "Mogen Davidane," as *Mogen David* is Hebrew for "Star of David." At that time, Dr.

52

53

Borowitz was (appropriately enough) at Yeshiva University in New York.[100] A stellar group headed by Ken-ichi Hirao at Hokkaido University, Sapporo, synthesized a dimethylperaxane in 1982.[101]

Peter Garratt and his student, John White, at University College London, saw a link between Nickon's triaxane (**52**) and Eaton's peristylane **54** (p. 51).[102] The relationship

54

becomes clear if we compare their top projections. In both cases, the carbons of an inner *n*-membered ring clasp alternate corners of an outer 2*n*-membered ring. Hence, Garratt and White regard both compounds as [*n*]peristylanes; **52** is [3]peristylane and **54** is [5]peristylane. For those who like to collect complete sets, Leo Paquette's company at Ohio State University has manufactured [4]peristylane.[103]

Earlier, we discussed the "bull" in bullvalene (chapter 10). What about the other end? Before pondering the suffix "valene," as used in 1964 by Heinz Viehe (then at Union Carbide European Research Associates, Brussels, more recently at Belgium's University of Louvain),[104,105] we should explain "valence isomerism." This last term was introduced formally in 1958 by Cyril Grob and Peter Schiess[106] and was defined by Emanuel Vogel of the University of Cologne in 1963.[107] Valence isomerism proceeds "without migration of atoms or groups of atoms." Furthermore, "... the structural change consists solely of a 'reorganization' of the σ- and π-electrons within the framework of the molecule, accompanied by corresponding changes in atomic distances and bond angles."[107] Thermal interconversion of **55** and **56**, discovered by Arthur Cope and co-workers at the Massachusetts Institute of Technology in 1952,[108] illustrates valence isomerism.

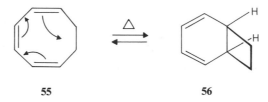

55   **56**

Then in 1964, Dr. Viehe defined "valenes" as molecules that can (at least on paper) undergo valence isomerization.[104,105] Among the simplest are the valence isomers of benzene, including "Dewar benzene" (**57**, proposed in 1867),[109] "Ladenburg benzene" (**58**, proposed in 1869 and now known as prismane[110]), and "benzvalene" (**59**, discussed

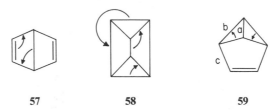

57   **58**   **59**

by Erich Hückel in 1937 as a planar, possible resonance contributor to benzene).[111] The curved arrows show formally how these valenes can revert to benzene but do *not* imply mechanisms. For example, to do it with **57** requires a disrotatory path (see **57** → **60**). But, as this electrocyclic reaction involves four electrons, a *concerted* change is symmetry forbidden. Therefore, the bond shuffling may not be synchronized at all. In

57   **60**

fact, diradical mechanisms have been suggested for some valence isomerizations.[112]

These three, historically significant valenes have been prepared in the laboratory. We have already encountered Thomas Katz's prismane (Ladenburg benzene) synthesis (chapter 6). Eugene van Tamelen and Socrates Pappas of Stanford University first put together Dewar benzene (**57**)[113]; and Kenneth Wilzbach, James Ritscher, and Louis Kaplan of Argonne National Laboratory gave us benzvalene (**59**).[114] The absence of symmetry-allowed pathways to benzene was a blessing for these three research teams; their valenes were stable enough to deal with despite severe angle strain.[115]

Some of Professor Viehe's other valenes are vinylogs (i.e., ethenologs) of benzvalene (**59**). By mentally inserting vinyl units into bonds a, c, and b, he generated the octavalenes **61**, **62**, and **63**.[104] Christl and Lang constructed **62**; and we met **63** before as calfene or semibullvalene (p. 14). Insert another vinyl unit into calfene and get bullvalene (**64**).

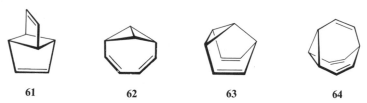

61	**62**	**63**	**64**

Do you recall (chapter 10) the fluxional nature of bullvalene predicted by William Doering[116] and first proven experimentally by Gerhard Schröder?[117] We now see that a fluxional molecule is a special case of rapid valence isomerization in which *all the structures are identical.*[116]

On the plains of Ohio and Iowa, chemists learned how to breed calfenes that don't become bullvalenes. Their genetic coup was to implant a butadiene into calfene

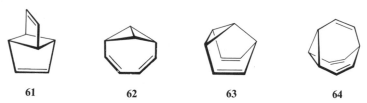

65	**66**	**67**

(semibullvalene) with the expectation that quasifluxion in semibullvalene would give way to a more stable valence bond offspring. Sure enough, when foremen Leo Paquette (Ohio State University) and Jon Clardy (Iowa State University) and their ranchhands generated the transient diaza progenitor **65**, molecular nitrogen obligingly dropped off. And, as hoped, it left behind not a pair of interconverting semibulls (**66a** ⇌ **66b**) but, instead, a lower-energy sibling, pentaene **67**.[118] An SOS from Dr. Paquette and graduate student Robert Wingard to the Ohio State Classics Department elicited the fact that Greek *elasson* means "less" or "smaller." So, they branded **67** "elassovalene" to remind everyone of its lower energy relative to **66**.[119] The elassovalene bloodline commands respect because of possible homoaromaticity in the 6pi cycloheptatriene stall—or even in the entire 10pi corral.[118,120]

Rowland Pettit's busy nest, adept at hatching prehistoric birds (p. 13), entered **68** in the fluxional sweepstakes.[121] (We met **68** in chapter 4, where Eiji Ōsawa tried to

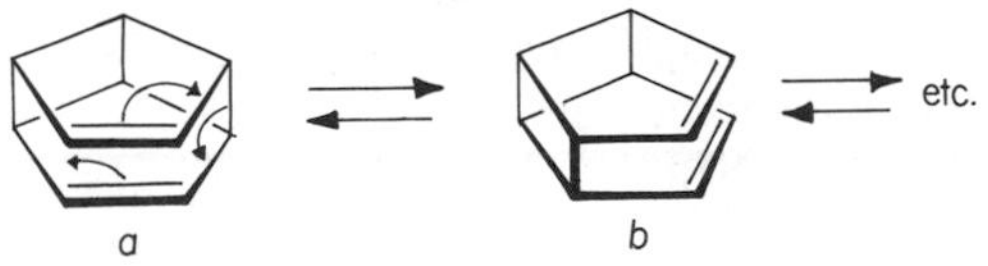

etc.

a   b

68

convert it to housane.[122]) The valence isomerizations **68a** → **68b** → etc. can continue *ad infinitum*, and all the structures are clones. In view of this possibility, the Pettit party dubbed **68** "hypostrophene"; *hypostrophe* is Greek for "a turning about or a reoccurrence."[121] They demonstrated the fluxional character of **68** ingeniously with the deuterated analog **69**.[121] This structure contains two olefinic H's and six hydrogens attached to saturated carbons, a 1:3 ratio. The ^{1}H NMR spectrum, however, showed this ratio to be 2:3, exactly what one would expect for unlabeled hypostrophene (four

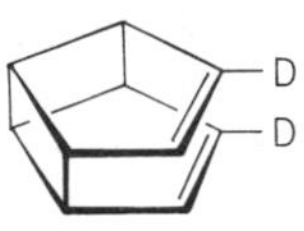

69

vinylic H's, six saturated ones). The deuterium had scattered randomly to all positions. A fine example of molecular roulette. "Round and round and round it goes, where it will stop, nobody knows!"

Churchane (p. 50), Buddha (p. 138), the Samson effect (p. 140), wolf and lamb reactions (p. 274), and Mogen Davidane (p. 295) show that chemistry and religion are compatible. For one more example, we turn to a landmark paper from Robert Woodward's laboratory.[123] It described the first total synthesis of a nonaromatic steroid (**70**) as well as of relay compounds that had already been elaborated to such important biochemicals as progesterone, testosterone, androsterone, cholesterol, and cortisone. Woodward's winners toiled arduously and tirelessly. Ketone **71** and its C10

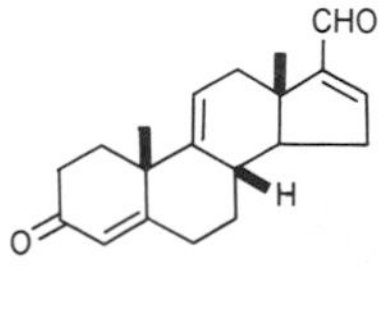

70 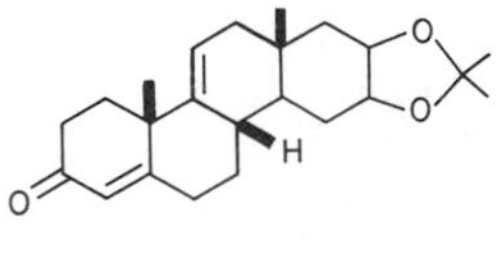   **71**

epimer, crucial intermediates in their scheme, were isolated for the first time on December 25, 1950. Franz Sondheimer, one member of the team, recalled they were so elated at this success that they referred to these epimers as α- and β-christmasterone. The name was not published but was used in lectures.[124] These chemists may have missed Christmas dinner but, clearly, Santa Claus brought them what they wanted.

For those who are always short of small change, chemistry can help out. It seems that a lepton was a small coin of ancient Greece (Greek *leptos* means "small"). From it, researchers at Imperial Chemical Industries Ltd. and the University of Sheffield, England, coined (literally) "isoleptic" to describe metal complexes, such as ($CH_2 =$ $CHCH_2)_4Sn$, in which all ligands are the same.[125] Michael Lappert and co-workers at the University of Sussex believed that "homoleptic" is more appropriate and allows the distinction "heteroleptic" when the bits are not all identical, as in $(CH_3)_3SiCl$.[126,127] The Sussex scientists used "isoleptic" to refer to the relationship between, say, $(CH_2=CHCH_2)_4Sn$ and $(CH_2=CHCH_2)_4Si$. The individual "coins" need not contain carbon and hydrogen; Richard Lagow's chemical numismatists at the University of Texas at Austin have minted homoleptic $Hg(SiF_3)_2$, $Bi(SiF_3)_3$, and $Te(SiF_3)_2$.[128] In these contexts, lepton is synonymous with ligand.[127] (Physicists know leptons as a family of subatomic particles that includes electrons—but that is another story.[129])

Throughout these chapters we have come to appreciate that a name given to anything can influence our expectations about it. But, as any student soon learns, a good many chemical names are downright deceptive—albeit not with malice aforethought. Professor Hugh Akers and undergraduates Meng Vang and Ali Yousepour at Lamar University in Texas drove this point home in a pithy article about misleading names.[130]

We have already met propyleine (p. 194) without a propyl group, mercaptans (p. 165) without mercury, and fluorene (20, p. 286) with no fluorine. This last hydrocarbon fluoresced (hence the name), but the effect was actually due to an impurity.[130] Now be aware that "theobromine" (72), from cacao beans, contains no bromine. The word derives from the genus name of the tree *Theobroma cacao*, which produces the bean.[130] Likewise, "iodinin" (73) is devoid of iodine and was named for its bacterium source

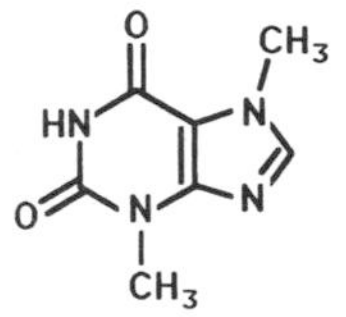

72 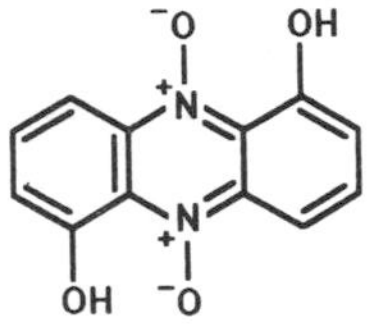   73 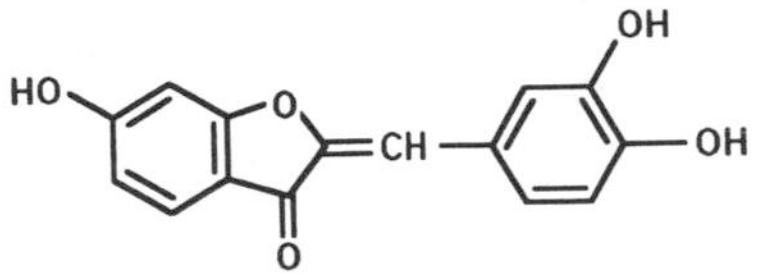   74

Chromobacterium iodinum.[131] There is nary a sulfur in "sulfuretin" (74; a yellow pigment that brightens the flower *Cosmos sulphureus*) or in "sulphurenic acid" (75; a steroid from *Polyporus sulphureus*).[130] Phosgene (Cl_2CO) contains no phosphorous. This deadly gas was made by irradiation of chlorine and carbon monoxide and the name comes from Greek *phos* ("light") and *genein* ("to produce").[132] There is no chlorine in chlorogenin (76), a spirostan from the lily *Chlorogalum pomeridianum*, and no gallium in gallic acid (77), obtained originally from oak galls.[130] (Incidentally, structures 72–77 were computer drawn with Professor Jih Ru Hwu's speedy "ChemPlate" method.[133]) The list goes on: no chromium in chromotropic acid or chromycin; no iodine in iodopsin; and no calcium in calciferol.[130]

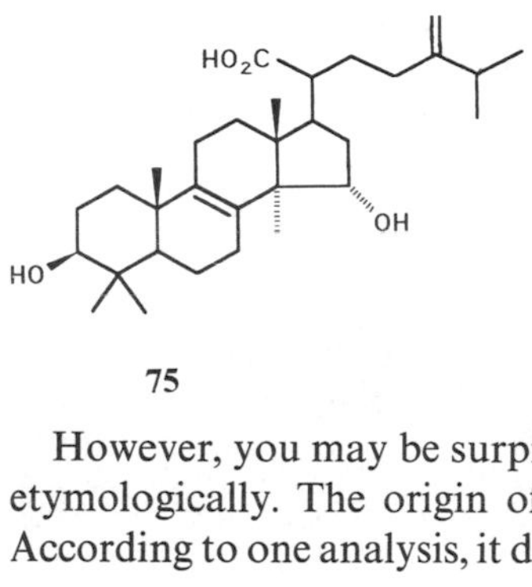

75 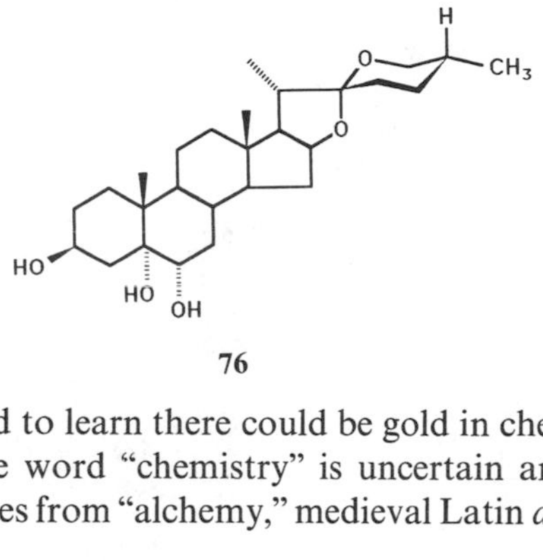   76 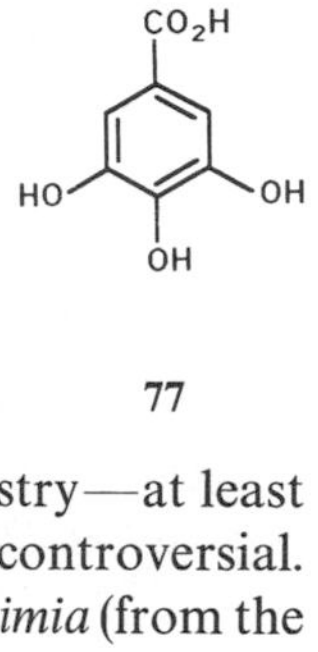   77

However, you may be surprised to learn there could be gold in chemistry—at least etymologically. The origin of the word "chemistry" is uncertain and controversial. According to one analysis, it derives from "alchemy," medieval Latin *alchimia* (from the

Arabic *al-*kīmīya). *Al* is simply the definite article in Arabic; but the root *kīm* (chem) is much of a mystery.[134]

One distinguished sinologist and science historian traced this root to China where, *ca.* 100–1000 AD, practitioners tried fervently to convert base metals to gold. Their obsession with this metal stemmed from a yearning to concoct a drug that would bestow longevity or even immortality upon its users. The Chinese believed that gold would likely be a component of any such elixir, because it is the one metal that does not corrode and that exists in the ground in an uncombined state.

The Chinese written character for gold when pronounced in many early Chinese dialects sound like "kim." This sound may have made its way over the Silk Road, the overland trade route from China to the Mediterranean, then elsewhere.[134] From "kim" to "chem" is no large jump; so gold may be at the heart of "chemistry." In any case, this metal destined our beloved science to immortality.

REFERENCES AND NOTES

1. Bader, A. Aldrich Chemical Co., private communication, January, 1976. Dr. Bader kindly drew our attention to this anecdote.
2. McGee, H. *On Food and Cooking* (Charles Scribner's Sons, New York, 1984), p. 152.
3. Szent-Györgyi, A.v. *Nature (London)*, **1927**, 782–783.
4. Szent-Györgyi, A., private communication, August, 1977.
5. *The Merck Index*, 9th ed. (Merck and Co., 1976), p. 111.
6. (a) Szent-Györgyi, A. *Biochem. J.* **1928**, *22*, 1387–1409, (b) Szent-Györgyi, A. *J. Biol. Chem.* **1931**, *90*, 385–393, (c) Svirbely, J.L.; Szent-Györgyi, A. *Nature (London)* **1932**, 576, 690, (d) Szent-Györgyi, A. *Nature (London)* **1932**, 943.
7. (a) Hirst, E.L. *J. Soc. Chem. Ind.* **1933**, *52*, 221, (b) Herbert, R.W.; Hirst, E.L.; Percival, E.G.V.; Reynolds, R.J.W.; Smith, F. *J. Chem. Soc.* **1933**, 1270–1290.
8. Karrer, P.; Salomon, H.; Schöpp, K.; Morf, R. *Helv. Chim. Acta* **1933**, *16*, 181–183.
9. (a) Szent-Györgyi, A.; Haworth, W.N. *Nature (London)* **1933**, 24; Szent-Györgyi, A. *Nature (London)* **1933**, 225–226.
10. (a) Asimov, I. *Words of Science and History Behind Them* (The New American Library, New York, 1959), (b) Flood, W.E. *The Dictionary of Chemical Names* (Philosophical Library, New York, 1963).
11. Ault, R.G.; Baird, D.K.; Carrington, H.C.; Haworth, W.N.; Herbert, R.; Hirst, E.L.; Percival, E.G.V.; Smith, F.; Stacey, M. *J. Chem. Soc.* **1933**, 1419–1423.
12. Gratton, R. *Philatelia Chimica* **1985**, 7, 30–33. Dr. Gratton (Rolland Inc., Mont-Rolland, Quebec, private communication, August, 1985) kindly furnished us this stamp for photography.
13. *Chem. Eng. News* **1980**, September 15, p. 37.
14. Pauling, L.C. *Vitamin C and the Common Cold* (W.H. Freeman and Co., San Francisco, CA, 1976).
15. Cameron, E.; Pauling, L. *Cancer and Vitamin C* (The Linus Pauling Institute of Science and Medicine, Menlo Park, CA, 1979).
16. Schleyer, P.v.R., private communications, October, 1975 and March, 1984.
17. Landa, S.; Macháček, V. *Collect. Czech. Chem. Commun.* **1933**, *5*, 1–5.
18. Bundy, F.P.; Kasper, J.S. *J. Chem. Phys.* **1967**, *46*, 3437–3446 and references cited there.
19. Böttger, O. *Chem. Ber.* **1937**, *70B*, 314–325.
20. Schreck, J.O. *J. Chem. Educ.* **1986**, *63*, 283–287.
21. Prelog, V. *Pure and Appl. Chem.* **1963**, *6*, 545–560.
22. Barton, D.H.R., personal discussion, July, 1982.
23. Schleyer, P.v.R. *J. Am. Chem. Soc.* **1957**, 79, 3292.
24. (a) Godleski, S.A.; Schleyer, P.v.R.; Ōsawa, E.; Inamoto, Y.; Fujikura, Y. *J. Org. Chem.* **1976**, *41*, 2596–2605, (b) Godleski, S.A.; Schleyer, P.v.R.; Ōsawa, E.; Wipke, W.T. *Prog. Phys. Org. Chem.* **1981**, *13*, 63–117.
25. (a) Schleyer, P.v.R.; Donaldson, M.M. *J. Am. Chem. Soc.* **1960**, *82*, 4645–4651, (b) Engler, E.M.; Farcasiu, M.; Sevin, A.; Cense, J.M.; Schleyer, P.v.R. *J. Am. Chem. Soc.* **1973**, *95*, 5769–5771.

26. Balaban, A.T.; Schleyer, P.v.R. *Tetrahedron* **1978**, *34*, 3599–3609.

27. Cupas, C.; Schleyer, P.v.R.; Trecker, D.J. *J. Am. Chem. Soc.* **1965**, *87*, 917–918.

28. Gund, T.M.; Ōsawa, E.; Williams, V.Z., Jr.; Schleyer, P.v.R. *J. Org. Chem.* **1974**, *39*, 2979–2987.

29. Karle, I.L.; Karle, J. *J. Am. Chem. Soc.* **1965**, *87*, 918–920.

30. Vogl, O.; Anderson, B.C.; Simons, D.M. *Tetrahedron Lett.* **1966**, 415–418.

31. Bartlett, J. *Familiar Quotations* (Little, Brown, and Co., Boston, MA, **1980**), p. 514.

32. Williams, V.Z., Jr.; Schleyer, P.v.R.; Gleicher, G.J.; Rodewald, L.B. *J. Am. Chem. Soc.* **1966**, *88*, 3862–3863.

33. Burns, W.; McKervey, M.A.; Mitchell, T.R.B.; Rooney, J. *J. Am. Chem. Soc.* **1978**, *100*, 906–911.

34. Schleyer, P.v.R.; Ōsawa, E.; Drew, M.G.B. *J. Am. Chem. Soc.* **1968**, *90*, 5034–5036.

35. Schleyer, P.v.R., private communication, October, 1975.

36. Kaslauskas, R.; Lidgard, R.O.; Murphy, P.T.; Wells, R.J. *Tetrahedron Lett.* **1980**, *21*, 2277–2280.

37. Ollis, W.D.; Smith, C.; Wright, D.E. *Tetrahedron* **1979**, *35*, 105–127.

38. Ollis, W.D., personal discussions, June, 1979, and July 1982.

39. Block, E. *Scient. Amer.* **1985**, *252* March, pp. 114–119.

40. Wertheim, T. *Ann. Chem. Pharm.* **1844**, *51*, 289–315.

41. Block, E.; Ahmad, S.; Jain, M.K.; Crecely, R.W.; Apitz–Castro, R.; Cruz, M.R. *J. Am. Chem. Soc.* **1984**, *106*, 8295–8296.

42. Block, E., personal discussion, May, 1985; private communication, June, 1985.

43. Hawthorne, M.F., private communication, June, 1984.

44. (a) Hawthorne, M.F.; Pilling, R.L. *J. Am. Chem. Soc.* **1965**, *87*, 3987–3988, (b) Hawthorne, M.F.; Young, D.C.; Andrews, T.D.; Howe, D.V.; Pilling, R.L.; Pitts, A.D.; Reintjes, M.; Warren, L.F. Jr.; Wegner, P.A. *J. Am. Chem. Soc.* **1968**, *90*, 879–896.

45. Francis, J.N.; Hawthorne, M.F. *J. Am. Chem. Soc.* **1968**, *90*, 1663–1664.

46. George, T.A.; Hawthorne, M.F. *J. Am. Chem. Soc.* **1969**, *91*, 5475–5483.

47. Prelog, V. *Chem. Brit.* **1968**, *4*, 382–387.

48. Haas, G.; Prelog, V. *Helv. Chim. Acta* **1969**, *52*, 1202–1218.

49. Jaffé, H.H.; Orchin, M. *Symmetry in Chemistry* (John Wiley & Sons, New York, 1965).

50. March, J. *Advanced Organic Chemistry*, 3rd ed. (John Wiley & Sons, New York, 1985), pp. 84–85.

51. Lamb, A.B.; Bray, W.C.; Frazer, J.C.W. *J. Ind. Eng. Chem.* **1920**, *12*, 213–221.

52. Taube, H. *Priestley Medal Address, Chem. Eng. News* **1985**, May 6, pp. 40–45.

53. Michaelis, L. *Biochem. Z.* **1932**, *250*, 564.

54. Michaelis, L.; Hill, E.S. *J. Gen. Physiol.* **1933**, *16*, 859.

55. Geuder, W.; Hünig, S.; Suchy, A. *Tetrahedron* **1986**, *42*, 1665–1677 and references cited there.

56. Kosower, E.M.; Hajdu, J. *J. Am. Chem. Soc.* **1971**, *93*, 2534–2535.

57. Kosower, E.M.; Teuerstein, A. *J. Am. Chem. Soc.* **1978**, *100*, 1182–1186.

58. Kosower, E.M.; Hajdu, J.; Nagy, J.B. *J. Am. Chem. Soc.* **1978**, *100*, 1186–1193.

59. Thiele, J. *Chem. Ber.* **1900**, *33*, 666–673.

60. Trost, B.M.; Cory, R.M. *J. Org. Chem.* **1972**, *37*, 1106–1110.

61. Pullman, B. *Int. J. Quantum Chem., Quantum Biol. Symp.* **1979**, *6*, 33–45.

62. Brown, R.D. *Trans. Faraday Soc.* **1949**, *45*, 296–300.

63. (a) Billups, W.E.; Lin, L.–J.; Casserly, E.W. *J. Am. Chem. Soc.* **1984**, *106*, 3698–3699, (b) Staley, S.W.; Norden, T.D. *J. Am. Chem. Soc.* **1984**, *106*, 3699–3700.

64. Bergmann, E.D. *Chem. Rev.* **1968**, *68*, 41–84.

65. (a) Ojima, J.; Itagawa, K.; Nakada, T. *Tetrahedron Lett.* **1983**, *24*, 5273–5276, (b) Kuroda, S.; Ojima, J.; Kitatani, K.; Kirita, M.; Nakada, T. *J. Chem. Soc., Perkin Trans. 1* **1983**, 2987–2995, (c) Ojima, J.; Itagawa, K.; Hamai, S.; Nakada, T.; Kuroda, S. *J. Chem. Soc., Perkin Trans. 1* **1983**, 2997–3004.

66. Ziegler, K.; Schäfer, W. *Liebigs Ann. Chem.* **1934**, *511*, 101–109.

67. Babsch, H.; Prinzbach, H. *Tetrahedron Lett.* **1978**, 645–648.

68. Prinzbach, H.; Babsch, H.; Hunkler, D. *Tetrahedron Lett.* **1978**, 649–652.

69. Described by Ollis, W.D. *Proc. R. Inst. Gr. Br.*, **1971**, *45*, 1–31. We thank Professor Ollis for a reprint of this paper.

70. SASM, 99 Rue Oberkampf, Paris.

71. Freifelder, D. *Molecular Biology and Biochemistry* (W.H. Freeman and Co., San Francisco, CA, 1978), p. 131.
72. Bernstein, H., private communication, April, 1982.
73. Broadhurst, M.J.; Grigg, R.; Johnson, A.W. *J. Chem. Soc., Perkin Trans. 1* **1972**, 2111–2116.
74. Bauer, V.J.; Clive, D.L.J.; Dolphin, D.; Paine, J.B., III; Harris, F.L.; King, M.M.; Loder, J.; Wang, S.–W.C.; Woodward, R.B. *J. Am. Chem. Soc.* **1983**, *105*, 6429–6436.
75. Broadhurst, M.J.; Grigg, R.; Johnson, A.W. *J. Chem. Soc., Perkin Trans. 1* **1972**, 1124–1136.
76. King, M.M., Ph.D. Dissertation, Harvard University, Cambridge, MA, 1970.
77. King, M.M., private communication, March, 1984.
78. Rexhausen, H.; Gossauer, A. *J. Chem. Soc., Chem. Commun.* **1983**, 275.
79. Berger, R.A.; LeGoff, E. *Tetrahedron Lett.* **1978**, 4225–4228.
80. LeGoff, E.; Weaver, O.G. *Abstracts of Papers, 192nd National Meeting American Chemical Society, Anaheim, CA* (American Chemical Society, Washington, DC, 1986), ORGN 304.
81. LeGoff, E., discussion by telephone, July, 1986.
82. Callot, H.J.; Schaeffer, E. *J. Org. Chem.* **1977**, *42*, 1567–1570.
83. (a) Vogel, E.; Köcher, M.; Schmickler, H.; Lex, J. *Angew. Chem. Int. Ed. Engl.* **1986**, *25*, 257–259, (b) O'Sullivan D. *Chem. Eng. News* **1986**, June 9, pp. 27–29.
84. Whitlock, H.W., Jr. *J. Am. Chem. Soc.* **1962**, *84*, 3412–3413. Full details appeared in Whitlock, H.W., Jr.; Siefken, M.W. *J. Am. Chem. Soc.* **1968**, *90*, 4929–4939.
85. For other syntheses and use of twistane as a pedagogic prototype for retrosynthetic analysis, see Hamon, D.P.G.; Young, R.N. *Aust. J. Chem.* **1976**, *29*, 145–161.
86. Whitlock, H.W., Jr., private communication, July, 1977.
87. Johnson, W.S.; Bauer, V.J.; Margrave, J.L.; Frisch, M.A.; Dreger, L.H.; Hubbard, W.N. *J. Am. Chem. Soc.* **1961**, *83*, 606–614.
88. Liu, R.S.H., private communications, July and August, 1985.
89. (a) Warshel, A. *Nature (London)* **1976**, *260*, 679–683, (b) Warshel, A. *Proc. Natl. Acad. Sci. U.S.A.* **1978**, *75*, 2558–2562, (c) Warshel, A.; Barboy, M. *J. Am. Chem. Soc.* **1982**, *104*, 1469–1476.
90. Liu, R.S.H.; Asato, A.E. *Proc. Natl. Acad. Sci. U.S.A.* **1985**, *82*, 259–263.
91. *Chem. Eng. News* **1985**, January 7, p. 40.
92. (a) Liu, R.S.H.; Mead, D.; Asato, A.E. *J. Am. Chem. Soc.* **1985**, *107*, 6609–6614, (b) Liu, R.S.H.; Browne, D.T. *Acc. Chem. Res.* **1986**, *19*, 42–48.
93. Bunnett, J.F.; Moyer, C.E., Jr. *J. Am. Chem.* **1971**, *93*, 1183–1190.
94. Bunnett, J.F. *Acc. Chem. Res.* **1972**, *5*, 139–147.
95. Bunnett, J.F., private communication, October, 1977.
96. Mach, M.H.; Bunnett, J.F. *J. Org. Chem.* **1980**, *45*, 4660–4666.
97. Bunnett, J.F.; Kearley, F.J., Jr. *J. Org. Chem.* **1971**, *36*, 184–186.
98. Nickon, A.; Pandit, D. *Tetrahedron Lett.* **1968**, 3663–3666.
99. Srinivasan, R. *Tetrahedron Lett.* **1972**, 4537–4540.
100. (a) Borowitz, I.J., private communication, January 1976, (b) Borowitz, I.J., private communication to Srinivasan, R., January, 1973. We thank Dr. Srinivasan for sharing this information.
101. Ken-ichi, H.; Ohuchi, Y.; Yonemitsu, O. *J. Chem. Soc., Chem. Commun.* **1982**, 99–100.
102. Garratt, P.J.; White, J.F. *J. Org. Chem.* **1977**, *42*, 1733–1736.
103. (a) Paquette, L.A.; Browne, A.R.; Doecke, C.W.; Williams, R.V. *J. Am. Chem. Soc.* **1983**, *105*, 4113–4115, (b) Paquette, L.A.; Fischer, J.W.; Brown, A.R.; Doecke, C.W. *J. Am. Chem. Soc.* **1985**, *107*, 686–691.
104. Viehe, H.G. *Angew. Chem. Int. Ed. Engl.* **1965**, *4*, 746–751.
105. Viehe, H.G., private communication, October, 1977.
106. Grob, C.A.; Schiess, P. *Angew. Chem.* **1958**, *70*, 502.
107. Vogel, E. *Angew. Chem. Int. Ed. Engl.* **1963**, *2*, 1–11.
108. Cope, A.C.; Haven, A.C., Jr. Ramp, F.L.; Trumbull, E.R. *J. Am. Chem. Soc.* **1952**, *74*, 4867–4871.
109. Dewar, J. *Proc. R. Soc. Edinburgh* **1866/1867**, *6*, 82–86.
110. (a) Viehe, H.G.; Merenyi, R.; Oth, J.F.M.; Senders, J.R.; Valange, P. *Angew. Chem. Int. Ed. Engl.* **1964**, *3*, 755–756, (b) *Chem. Eng. News* **1964**, December 7, pp. 38–39.
111. Hückel, E. *Z. Elektrochem.* **1937**, *45*, 752; *ibid.*, **1937**, *45*, 760.
112. Bryce–Smith, D. *Pure Appl. Chem.* **1968**, *16*, 47–63.
113. van Tamelen, E.E.; Pappas, S.P. *J. Am. Chem. Soc.* **1963**, *85*, 3297–3298.
114. Wilzbach, K.E.; Ritscher, J.S.; Kaplan, L. *J. Am. Chem. Soc.* **1967**, *89*, 1031–1032.

115. Kobayashi, Y.; Kumadaki, I. *Top. Curr. Chem.* **1984**, *123*, 103–150.
116. Doering, W.v.E.; Roth, W.R. *Tetrahedron* **1963**, *19*, 715–737.
117. Schröder, G. *Angew. Chem. Int. Ed. Engl.* **1963**, *2*, 481–482.
118. Paquette, L.A.; Liao, C.C.; Burson, R.L.; Wingard, R.E., Jr.; Shih, C.N.; Fayos, J.; Clardy, J. *J. Am. Chem. Soc.* **1977**, *99*, 6935–6945.
119. Paquette, L., private communication, September, 1980.
120. Askani, R.; Pelech, B. *Tetrahedron Lett.* **1980**, *21*, 1841–1844.
121. McKennis, J.S.; Brener, L., Ward, J.S.; Pettit, R. *J. Am. Chem. Soc.* **1971**, *93*, 4957–4958.
122. Ōsawa, E. *J. Org. Chem.* **1977**, *42*, 2621–2626.
123. Woodward, R.B.; Sondheimer, F.; Taub, D.; Heusler, K.; McLemore, W.M. *J. Am. Chem. Soc.* **1952**, *74*, 4223–4251.
124. Sondheimer, F., private communication, November, 1975.
125. O'Brien, S.; Fishwick, M.; McDermott, B.; Wallbridge, M.G.H.; Wright, G.A. *Inorg. Syn.* **1972**, *13*, 73–79.
126. Davidson, P.J.; Lappert, M.F.; Pearce, R. *Acc. Chem. Res.* **1974**, *7*, 209–217.
127. Lappert, M.F., private communication, July, 1979.
128. Bierschenk, T.R.; Juhlke, T.J.; Lagow, R.J. *J. Am. Chem. Soc.* **1982**, *104*, 7340–7341.
129. (a) Harari, H. *Scient. Amer.* **1983**, *248*, April, pp. 56–68, (b) Whitman, M. *J. Chem. Educ.* **1984**, *61*, 952–956, (c) Walker, D.C. *Acc. Chem. Res.* **1985**, *18*, 167–173.
130. Akers, H.A.; Vang, M.C.; Yousepour, A. *J. Chem. Educ.* **1986**, *63*, 255–256.
131. *The Merck Index*, 10th ed. (Merck and Co., Rahway, NJ, 1983), pp. 727–728.
132. Bailey, D.; Bailey, K.C. *An Etymological Dictionary of Chemistry and Mineralogy* (Edward Arnold and Co., London, 1929), p. 209.
133. Hwu, J.R.; Wetzel, J.M.; Robl, J.A. *J. Chem. Educ.* **1987**, *64*, 135–137.
134. (a) Butler, A.R.; Reid, R.A. *Chem. Brit.* **1986**, *22*, 311–312, (b) Benfey, O.T. *the pHilter* **1986**, *18*, No. 4, 1–4 (American Chemical Society Student Affiliates Newsletter).

Appendix A

BRIEF ETYMOLOGY OF SOME TRADITIONAL CHEMICAL NAMES*

acenaphthene from *ace*tylene and *naphth*ylene, from which it was synthesized by Berthelot in 1866. He named it in 1877.

acetal from *acet*aldehyde and *al*cohol, from which the acetal $CH_3CH(OC_2H_5)_2$ is made. The term dates back to 1869.

acetic acid from Latin *acetum* (vinegar), which in turn comes from Latin *acere* (to be sour).

acetone from Latin *acetum* (vinegar) and the Greek suffix "-one," which designates a female descendent, as in anemone (daughter of the wind; *anemos* is Greek for wind). Acetone was obtained from acetate salts by pyrolysis.

acridine from Latin *acer* (sharp). Acridine has a sharp odor and irritates the skin and mucous membranes. Graebe and H. Caro coined the name in 1877.

acrolein from Latin *acer* (sharp) and *olere* (smell). Acrolein has a sharp smell. Derived terms include acrylic, acrylate, etc.

adenine from Greek *adena* (gland). Adenine is present in pig pancreas and was named by A. Kossel in 1885.

adipic acid from Latin *adeps* (fat). Adipic acid forms during oxidation of various fats.

alcohol from Arabic *al koh'l*, which was finely divided antimony sulfide that was used as eye shadow. Subsequently, alcohol was also used to describe a liquid obtained by distillation and eventually (*ca.* 1850) as a class name.

aldehyde from *al*cohol and Latin "*dehyd*rogenatus" (dehydrogenated). Aldehydes can be made by removal of two hydrogen atoms from a primary alcohol.

aldol from *ald*ehyde and alcoh*ol*. Aldol ($CH_3CH(OH)CH_2CHO$) contains an aldehyde and an alcohol group. The term originated in 1874.

*The sources were primarily the following. (a) Ruske, R. *Einführung in die organische Chemie* (Verlag Chemie GmbH, Weinheim/Bergstr., Germany, 1970), (b) Bailey, D.; Bailey, K.C. *An Etymological Dictionary of Chemistry and Mineralogy* (Edward Arnold & Co., London, 1929), (c) Flood, W.E. *The Dictionary of Chemical Names* (Philosophical Library, New York, 1963). We are indebted to Dr. W. Val Metanomski (Chemical Abstracts Service) and Professor Martin Feldman (Howard University) for suggestions and for assistance on specific items.

aldomedon presumably from ace*tald*ehyde and di*medon*, from which it is prepared. This name for 2, 2′-ethylidenebis(5, 5-dimethyl-1-3-cyclohexanedione) was introduced by C. Neuberg and E. Reinfurth (*Biochem. Z.* **1920**, *106*, 281–291); apparently, it has not been used since 1929.

alicyclic from *ali*phatic and cyclic.

aliphatic from Greek *aleiphat* (fat). Fatty acids are open chained and characteristically noncyclic. The term was coined by Hofmann.

alkali from Arabic *al qalīy* (the roasted ashes). Substances we now recognize as inorganic bases originally came from calcined ashes of marine plants.

alkaloid from "alkali" and Greek *-o-eidēs* (like, in the form of). Alkaloids are basic compounds of plant origin. The term was introduced by a pharmacist, W. Meissner, in 1819.

alkane, alkene the -ane and -ene endings were devised by Hofmann in 1865 for hydrocarbons with formulas C_nH_{2n+2} and C_nH_{2n}, respectively.

alkyl from alk- and -yl. Alk- comes from "alkali" and stems from the similarity of the formula of an alkali metal halide (e.g., NaCl) and an organic halide (e.g., CH_3Cl). the suffix -yl derives from Greek *hylē* (the stuff from which a thing is made).

amide from *am*monia and the ending -ide, which is used in oxide, chloride, etc. Berzelius proposed the term, noting similarity among $NaNH_2$, Na_2O and NaCl.

amphoteric from Greek *amphoteros* (on both sides). Amphoteric compounds can react as both acids and bases.

amyl alcohol from Latin *amylum* (starch). It was discovered by C.W. Scheele in 1785 as a companion of ethanol in potato fermentation.

androgen from Greek *andros* (man). An androgen is a male sex hormone.

aniline from Portugese *anil* (indigo). Aniline was prepared by O. Underderben in 1826 by dry distillation of indigo.

anisole from Greek *anison* (anise or dill). The essential principle of anise is 1-(p-methoxyphenyl)propene; hence, compounds containing p-methoxyphenyl groups often have common names starting with anis-. Anisaldehyde and anisic acid are other examples.

anomer from Greek *ano* (upper) and *meros* (part). Cyclic structures for sugars like glucose were usually drawn with carbon chains vertical and with C1 at the top. Epimers at C1 (e.g., α- and β-glucose) were called anomers. (Riiber, C.N.; Sorensen, N.A. *Kgl. Norske Vidensk. Selsk. Skrifter*, **1933**, *No. 7*, 1–50.)

anthocyanin from Greek *anthos* (flower) and *kyanos* (dark blue). The name was introduced by L.C. Marquart in 1835 to designate the blue pigment of flowers. More generally, it can refer to plant pigments with a variety of colors.

anthracene from Greek *anthrax* (coal). Anthracene is a constituent of coal tar. The name is due to Laurent.

arabinose from *Gummi arabicum* (gum arabic), the resin of trees of the Acacia family. L-Arabinose (named in 1840) can be obtained from such resins.

arginine from Latin *argentum* (silver) and -ine (a common ending for nitrogen-containing compounds). Arginine forms a well-defined silver salt. The name was proposed by E. Schultze and E.Steiger.

aromatic from the fact that compounds isolated in the nineteenth century from essential oils (e.g., those of almonds and cinnamon) have a pleasant aroma. August Kekulé first suggested the term.

asparagine from Latin *asparagus*, which came from Greek *asparagos* (asparagus). The name dates back to 1813.

aspirin from *a*cetyl and *Spir*säure, an old name for salicylic acid. Aspirin is acetylsalicyclic acid. H. Dresser suggested the term in 1899.

azelaic acid from the fact it can be made by nitric acid ("*azote*" is French for nitrogen) oxidation of oleic acid ("*elai*on" is Greek for oil).

bakelite from Baekeland, the Belgian chemist who started industrial production of phenol formaldehyde resins in 1906.

bathochromic from Greek *bathos* (depth) and *chromatos* (color). In ultraviolet and visible absorption a bathochromic shift is a change to longer wavelength. A spectroscopic shift in the opposite direction (i.e., to shorter wavelength) is termed "hypsochromic," from Greek *hypsos* (height). "Hypochromic" and "hyperchromic" designate, respectively, a decrease and an increase in absorption intensity.

benzal from benzaldehyde; it refers to the benzylidene group, C_6H_5CH.

benzene from *Styrax benzoin*, a tree native to Sumatra and Java. The bark yields a resin, gum benzoin, from which "benzoic acid" was obtained. Péligot (1833) and E. Mitscherlich (1834) heated benzoic acid with lime to form C_6H_6; Mitscherlich named it "benzine." Liebig preferred "benzol," but Laurent (1835) proposed "benzene."

benzil from *benz*oin and -il. It is thus a "derivative of benzoin," from which it can be made by oxidation.

benzoin from gum benzoin, of which it is the major constituent.

benzophenone from benzo- (the C_6H_5CO—group appears in benzoic acid) and phenone (which means *phen*yl ket*one*).

biuret from Latin *bi*- (two) and "urea." Biuret ($NH_2CONHCONH_2$) can be formed from two molecules of urea by loss of NH_3.

borneol from Borneo. This alcohol is the essential ingredient of Borneo camphor, obtained from a tree that grows in Borneo and Sumatra.

buna rubber from "*bu*tadiene" and "Na" (sodium). It is made by sodium-initiated copolymerization of butadiene with styrene or acrylonitrile.

butyric acid from Latin *butyrum* (butter); it occurs in rancid butter. The term was devised by Chevreul in 1826.

cacodyl from Greek "*kakōdēs*" (ill-smelling). Compounds containing the $(CH_3)_2As$—, or cacodyl, group have unpleasant odors. The term was introduced by Berzelius in 1850.

cadaverin from Latin *cadaver* (corpse). Cadaverin has the odor of a decaying dead body.

caffeine from French *cafe* (coffee). The name dates back to 1830.

calciferol from "*calci*um," the Latin *fer* (to carry), and -ol (alcohol). Calciferol is essential for proper absorption of calcium and phosphorus, and thus for proper formation and growth of bones.

camphor from *kāfur*, an old Arabic trade name for camphor. The term was changed to *camphora* in medieval Latin.

capric, caproic, caprylic acids from Latin *caper* (goat). These acids occur as triglycerides in butter, especially that prepared from goat's milk.

carbazole from *carb*on, "az-" (nitrogen) and -ole (as in pyrrole). Named by C. Graebe and C. Glaser around 1872, it behaves like a hydrocarbon, contains nitrogen, and is related to pyrr*ole*.

carbinol from carbin (Kolbe's name for a methyl group) and -ol (alcohol). Kolbe coined this name for methyl alcohol around 1868 and suggested it be used as a basis for naming other alcohols.

carbolic acid from Latin *carb*o (coal), "*ol*eum" (oil), and the fact that it is acidic. Carbolic acid (phenol) is obtained in the oily fraction from distillation of coal tar.

carbonic acid from the fact that it contains carbon and gives an acidic aqueous solution. Lavoisier proposed the term in 1787.

carbonyl from *carbon*ic acid and -yl (from Greek *hylē*, meaning the stuff from which a thing is made). The carbonyl group appears in carbonic acid.

catechol from *Acacia catechu*, a shrub found in India and Burma.

cellobiose from *cell*ulose (the corresponding polymer), *bi* (two), and -*ose* (an ending that denotes a saccharide).

cellulose from Latin *cellula* (small room, cell). Cellulose is the major constituent of cell walls in plants. The name is probably due to Payen (*ca.* 1835).

cephalin from Greek *kephalé* (head). It is a phospholipid found in the brain.

chalcone (German chalkon) from Greek *chalkos* (copper or brass). A name for 1, 3-diphenyl-2-propen-1-one coined by Kostanecki, St. v.; Tambor, J. *Chem. Ber.* **1899**, *32*, 1921–1926. Ring-hydroxylated chalcones generally are reddish yellow in alkaline or in concentrated sulfuric acid solutions.

chelidonic acid from the juice of the plant *Chelidonium majus*. This Latin name is probably derived from Greek *chelidon* (swallow). According to legend, a swallow found the juice to restore sight to its young when blinded. The name dates to 1863.

chloramphenicol from *chloro* + *am*ide + *phen*yl + *ni*tro + gly*col*; these functional groups are present in the molecule.

chloranil from *chlor*ine + *anil*ine. It can be made from these chemicals.

chloroform from *chloro*- and *form*yl. The latter term was used before the nineteenth century to describe the CH radical.

chlorohydrins from *chlor*ine and *hydr*ic. Initially derived by replacement of one or more —OH groups of a polyhydric alcohol by chlorine.

chlorophyll from Greek *chlōro*s (green) and *phyll*on (leaf); it is the green substance in leaves. The name was introduced by Pelletier and Caventou in 1818.

chloropicrin from *chlor*ine and *picr*ic acid. It can be prepared by distillation of picric acid with bleaching powder (a source of chlorine).

chloroprene from chloro- and iso*prene*; it has a chlorine in place of the methyl group of isoprene.

cholesterol from Greek *chole* (gall, bile), *stere*os (solid), and -ol (alcohol). Cholesterol is the major constituent of gallstones.

choline from chole (see cholesterol). Choline is found in bile.

chrysene from Greek *chrys*os (gold) and -ene (as in benzene). Chrysene is colorless when pure, but it often contains contaminants that give it a yellow appearance.

cinnamic acid from Latin *cinnamomum*, an oil obtained from the bark of cinnamon trees. The main constituent of the oil, "cinnamaldehyde," can be oxidized to cinnamic acid.

citraconic acid from *citr*ic and it*aconic*; it can be derived from citric acid through loss of H_2O and CO_2 and is an isomer of itaconic acid.

citric acid from Latin *citr*us (citron, the genus that includes citrus fruits).

civetone from "*civet* cat," in which this ketone occurs.

collidine may come from the Greek *kolla* (glue). The name dates back to 1854.

creosote from Greek *kreo*- (combining form of kreas, flesh) and *sōzō* (to save). It is used in sheep dips, i.e., liquids into which sheep are dipped to destroy parasites.

crotonaldehyde from *Croton tiglium*. Oil from the seeds of this plant provides $CH_3CH=CHCOOH$, which was named "crotonic acid" in 1838. Hence, the aldehyde became crotonaldehyde.

cumic acid from the plant *Cuminum cyminum*, which yields an oil (cumin) that gives *p*-isopropylbenzoic acid upon oxidation. Hence, this acid was called cumic acid.

Distillation of cumic acid with lime gives isopropylbenzene, which was named "cumene" in 1863.

cyanogen from Greek *kyanos* (dark bule) and -gen (producer). This old name for the CN radical was suggested by Gay-Lussac early in the nineteenth century. He recognized it as an important constitutent of Prussian blue, $Fe_4[Fe(CN)_6]_3$, and of prussic acid (HCN).

cysteine, cystine from Greek *kystis* (pouch, bag, bladder). Cystine was found in a bladderstone; an "e" was arbitrarily inserted to give its precursor the name cysteine. The name dates back to 1843.

cytosine, cytidine from Greek *kytos* (hollow vessel; a cell, when used scientifically). Cytosine is one of the bases in nucleic acids, which occur in cells.

decalin from Greek *deka* (ten). It is decahydronaphthalene.

depsides from Greek *depso* (to tan). Depsides occur in decomposition products of tannins.

deutero- from Greek *deutero-* (second). The prefix appears in deuteroporphyrin (protoporphyrin with both vinyl groups replaced by hydrogen) but most commonly denotes the presence of deuterium.

dimedone (German dimedon) is 5, 5-dimethyl-1, 3-cyclohexanedione, the scientific name of the compound. The abbreviation was introduced by C. Neuberg and E. Reinfurth (*Biochem. Z.* **1920**, *106*, 281–291), seemingly from the words dimethyl diketone.

diosphenol from Greek *di*os (divine) and *osmé* (odor); it is a pleasant-smelling monoterpenoid ketone having an enolic (i.e., phenol-like) hydroxyl group.

durene from Latin *durus* (hard). It is a solid (i.e., hard) derivative of benzene.

dynamite from Greek *dynamis* (power or force).

dypnone (contraction of "dihypnone") stands for $PhCOCH{=}C(CH_3)Ph$, the aldol condensation product of acetophenone. Acetophenone was at one time used as a soporific under the name "hypnone," from Greek *hypnos* (sleep).

electron from electrine, a term used by G. Johnstone Stoney (*Phil. Mag.* (5), **1881**, *11*, 381–390) to refer to the smallest unit of electricity. Later it was also applied to the negative particle as we know it. Initially, the Greek word *ēlektron* denoted a metallic substance consisting of gold and silver; it also meant amber, and so formed the basis of the word electricity.

electrophile from Greek *philos* (loving). An electrophile seeks a negative region of a molecule (i.e., a site of high electron density). Similarly, a nucleophile is attracted to a positive region.

ellagic acid from ellag-, an anagram of the French *galle* (gallnut). Gallnuts are source of ellagic acid. The "-ic" is a common ending for acids. The name was coined in 1810.

enzyme from Greek *zyme* (yeast). Yeast is a source of several enzymes. The term dates back to 1881.

eosin from Greek *ēōs* (dawn). It is a rose-red dye, and thus resembles the sky at dawn. The term first dawned on someone in 1866.

epi- from Greek *epi-* (upon or close upon). Epimers are isomers that differ at only one carbon (e.g., glucose and mannose) and are thus closely related. The term also indicates an intramolecular connection ("bridge") as in epichlorohydrin, epoxide, epidioxide, etc.

equilin from Latin *equus* (horse). It is a female sex hormone first isolated from the urine of pregnant mares.

ergosterol from *ergot* (a diseased growth of seeds of rye and other grasses in which ergosterol is found). Greek *ster*eos (solid), and -ol (alcohol). Ergosterol is a solid alcohol. It was called "ergosterin" in 1889; the name was modified in 1906 when the compound was recognized to be a secondary alcohol.

erythrose from Greek *erythros* (red). Erythrin occurs in a lichen (*Roccella tinctoria*) that can be used to prepare red or purple archil dyes. Erythrin is the ester of orsellinic acid and erythritol, which can be oxidized to a tetrose sugar, erythrose.

ester from German *Essigäther* (acetic ether) an early name for ethyl acetate.

ether from Latin *aether*, after Greek *aithēr* (clear sky, upper space). An allusion to the clean smell and volatility of compounds derived by action of various acids (H_2SO_4, HCl, HOAc) on alcohol. The term is now generally restricted to Et_2O and as a class name.

ethyl from *eth*er and Greek h*ylē* (the material from which a thing is made). Ordinary ether consists of two ethyl groups attached to an oxygen atom. The term is due to Liebig.

flavone from Latin *flavus* (yellow). Many yellow plant dyes are derivatives of flavone.

fluo-, fluoro- from Latin *fluo-* (to flow). A number of minerals containing fluorine can be used as fluxes in smelting (i.e., they reduce the melting temperature and allow the ore to flow).

fluorescein from its brilliant yellow-green fluorescence.

folic acid from Latin *folium* (leaf). Folic acid is found in green leaves (e.g., spinach) and in grasses.

formic acid from Latin *formica* (ant). It is found in ants, bees, and stinging nettles.

Freons from "freeze"; this is the Du Pont Company's trade name for fluorine-containing refrigerants.

fructose from Latin *fruct*us (fruit) and -ose (common ending of sugars). It is found in many sweet fruits. The name was coined in 1864. This sugar was also called laevulose, because it was levorotatory.

fuchsine from the flower fuchsia, whose purple-red color it resembles.

Fuller's Earth a fine grained, earthy substance used in fulling (cleaning and thickening cloth). The name goes back to 1523.

fulminate from Latin *fulmen* (lightning). Mercuric fulminate, $Hg(CNO)_2$, explodes violently when heated or struck. The name originated in 1825.

fumaric acid from Latin *fumus* (smoke). It occurs in the plant *Fumaria officinalis*, which was burned by ancients who believed the fumes would ward off evil spirits.

furan (contraction of furfurane) obtained by decarbonylation of furfural.

furfural (furfuraldehyde) from Latin *furfur* (bran). It is produced by the action of acid on the polypentosides (pentosans) in corncobs, oat hulls, and rice hulls.

fusel oil from German *fusel* (bad brandy and other spirits). Fusel oil is a mixture (mostly C_5 alcohols) obtained by distillation of raw spirits.

galactose from Greek *galact-* (milk) and -ose (suffix denoting a sugar). Galactose is a sugar formed by hydrolysis of lactose, which is found in milk. The name was coined in 1869.

gallium usually attributed to *Gallia*, the Latin name for France where the element was discovered by Le Coq de Boisbaudran in 1875. Another (apocryphal?) account holds that he actually named the element after himself. His first name, Le Coq, is French for "cock" or "rooster," so he used the Latin word for cock (*gallus*) as a root. The story goes on that his choice of name was interpreted as nationalistic, not only because Latin *Gallia* means France but also because the cock is the emblem of that country. Perhaps in retaliation, German workers who discovered the next element named it germanium, after the Latin word (*Germanus*) for their own country. (Private communication from David W. Ball; see Appendix B.)

geminal from Latin *gemini* (twins). A geminal dihalide has two halogens on the same carbon, as in CH_3CHCl_2.

gestagen from Latin *gestatio* (pregnancy). A gestagen is a substance involved in prepregnancy events, such as ovulation.

glacial acetic acid from Latin *glacialis* (ice-like). Pure acetic acid solidifies in a cold room to ice-like crystals.

Glauber's salt ($Na_2SO_4 \cdot 10H_2O$) after the German chemist J.R. Glauber (1604–1670) who first made it artificially. The name was not introduced until 1736.

glucose from Greek *gleuckos* (sweet wine). Lowitz obtained this sugar from grapes in 1792. The suffix -ose, subsequently used in naming other sugars, stems from the common Greek ending -os. Peligot proposed the name glucose in 1838.

glucuronic acid from *gluc*ose, *ur*ine, and -onic (the usual suffix for a monoacid derived from a sugar). It is structurally related to glucose (having —COOH instead of —CH_2OH) and occurs in urine.

glutaconic acid from *glut*aric acid (which it gives upon hydrogenation) and citr*aconic acid* (with which it is isomeric).

glutamic acid from *glut*en (which provides glutamic acid upon hydrolysis), *amino* (a group present in glutamic acid), and -ic (a common ending for acids).

glutathione from *gluta*mic, one of the three amino acids (glutamic, cysteine, glycine) in the tripeptide glutathione, and *thio* (sulfur). Sulfur is present in cysteine.

glycerol from Greek *glykeros* (sweet). Glycerol has a sweet taste. The name dates back to 1884.

glycine from Greek *glykys* (sweet) and -ine (a common suffix for amine bases). Glycine tastes sweet.

glycogen from Greek *glykys* (sweet) and -gen (a suffix used to mean producer of). Glycogen produces the sugar glucose upon hydrolysis. The term was coined in 1860.

glycol from "glyc-," which has two-thirds as many letters as "glycer-." A glycol has two-thirds as many hydroxyl groups as glycerol.

glyoxal from *glyc*ol (which it gives upon reduction), *ox*alic acid (an oxidation product), and -al (aldehyde). The name dates back to 1858.

Grignard reagents after Victor Grignard (1871–1935), who over a period of years developed their use as synthetic reagents. Grignard was a student of Barbier, who actually first used them in synthesis.

guaiacol after *Guaiacum*, the genus of tropical trees and shrubs that produce a resin that affords guaiacol upon distillation. The name was concocted in 1864 from Spanish *guayaco*. The noted American chemist Ira Remsen (1846–1927) was a stickler for grammar and spelling and contended that no one could claim to be a true organic chemist who couldn't spell guaiacol.

guanine from its presence in guano (bird excrement). The name descended upon chemistry in 1850.

gulose from gul-, an anagram of glu; to show the close relationship between gulose and glucose.

gutta percha from Malay *getah percha* (gum of the Percha tree, which produces this rubbery substance). The term was introduced in 1845.

halogen from Greek *hals* (salt) and -gen (taken to mean producer of). The halogens can produce salts by reaction with metals.

hemoglobin from Greek *haima* (blood) and Latin *globus* (globe, ball). This heme-protein is found in blood, and the term globule was formally applied to a corpuscle of blood. The name dates back to 1869; its predecessor, haemato-globulin, is 24 years older.

hippuric acid from Greek *hippos* (horse) and -uric (from urine). The acid occurs in horse urine. The term was off and running in 1838.

histamine from Greek *histi*on (tissue) and amine. This amine is released in the body when certain tissues are irritated.

homologous from Greek *homos* (same) and *logos* (name). In a homologous series, compounds of the same family differ only by the number of $-CH_2-$ groups.

hormone from Greek *hormon* (to be a messenger). Hormones are chemical messengers that stimulate organs to carry out functions. The term dates back to 1906.

hydrindene from *hydr*ogenated *indene*; it is a synonym for indane.

-hydrol from *hydr*ogen and -ol (alcohol). It is used in the common names of certain secondary alcohols (e.g., benzhydrol, $(C_6H_5)_2CHOH$). Presumably, it refers to the H and OH on the same carbon.

hydrolysis from Greek *hydor* (water) and *lyein* (to dissolve).

hydroxamic acid from *hydrox*ylam*ine* and -ic (common ending for acids). They are N-hydroxyamides made from hydroxylamine and acyl compounds.

hypnone from Greek *hypnos* (sleep). Hypnone is a soporific and a hypnotic.

Iceland spar a form of calcium carbonate first imported from Iceland.

indene from *ind*ole, to which it is related structurally.

indigo from Greek *indikon* (Indian substance). It is a blue dye, once imported from India. The name goes back to 1555.

indole from indigo. While working on the constitution of indigo, Baeyer regarded C_8H_7N as a parent structure capable of forming various oxygen-containing derivatives. He named it indol (by analogy with benzol, the German name for the parent structure benzene). The spelling indole is preferred, so as not to connote an alcohol or phenol.

inositol from Greek "ino-" (muscle) and -ose (ending for sugars). It is a sweet substance, discovered by Scherer in 1850 in heart muscles of the ox. It was called inosite in 1857 but was later changed.

insulin from Latin *insula* (island). The hormone is produced in the islands of Langerhans, which are located in the pancreas.

isoprene was named in 1860 by its discoverer, C.G. Williams. He obtained it by the dry distillation of rubber but gave no explanation for the name.

itaconic acid from aconitic acid (from which it forms by decarboxylation) by transposition of the letters.

kaolin from Chinese *kao* (high) and *ling* (hill). This fine, white clay was first obtained from a mountain in North China named "kaoling." The name goes back to 1867.

keratin from Greek *keras* (horn); it is the chief protein of fingernails, hair, horn, and hoofs.

kerosene from Greek *keros* (wax); like wax, it is obtained from petroleum. It was named in 1854.

ketone from German *keton*, an adaptation of aceton (see acetone).

kieselguhr from German *kiesel* (gravel) and *guhr* (ferment).

lactam from *lact*one and *am*ide. Lactams are cyclic amides.

lactic acid from Latin *lac* (milk). Lactic acid is present in sour milk.

lactide from *lact*ic acid anhydr*ide*. It is formed by removal of two molecules of water from two molecules of lactic acid. The name was introduced in 1848.

lactol from *lact*one and alcoh*ol*; any structure containing both an alcohol and a lactone group (e.g., the intramolecular addition product of a COOH to a carbonyl). A lactol is not the reduction product of a lactone any more than a ketol is the reduction product of a ketone.

lactone (German lacton) from lactide, an early name used for a dimeric cyclic diester of lactic acid. Rudolph Fittig *et al.* (*Liebigs Ann. Chem.* **1880**, *200*, 21–96) proposed "lactone" for cyclic monoesters. Lactam, for cyclic amides, was a subsequent logical extension.

laevulose an original name for fructose, which rotates polarized light to the left (Latin *laevo* for left).

lanolin from Latin *lan*a (wool) and *ol*eum (oil). It is an oily material extracted from sheep's wool.

latex from Latin *latex* (liquid, fluid). It is the milky liquid exuded by certain scored plants. The substance coagulates into a rubber on exposure to air.

lecithin from Greek *lekithos* (yolk of an egg). It is a fatty substance found in egg yolks.

leucine from Greek *leuk*os (white) and -ine (suffix for an amine base). It was named by Braconnet in 1826, who obtained it in the form of white plates.

ligroin is the fraction of petroleum that distills from 90 to 120°C; the origin of the term is unknown.

linoleic acid from Latin *lin*um (flax) and *oleum* (oil). Linoleic acid is obtained by hydrolysis of linseed oil. Linseeds are the seeds that produce flax plants. The name linoleic was introduced in 1857.

lipid from Greek *lip*os (fat) and -oe*ides* (like, in the form of). Lipids comprise fats and related esters.

litmus from Dutch *lak* (lac, a red resin) and *moes* (pulp). It is red in acid and was originally sold as small cakes. The term goes back to 1502.

lutidine from toluidine, by dropping the "o" and rearranging the remaining letters. Both compounds have the molecular formula C_7H_9N.

lysine from Greek *lys*is (a loosening or dissolution) and -ine (suffix for an amine base). It was discovered by Dreschel among hydrolysis products of casein.

maleic and **malic acids** from Latin *malum* (apple). Malic acid is found in unripe apples and other fruits; it got its name in 1797. Maleic acid was named later; it can be made from malic acid by dehydration.

malonic acid can be made from malic acid by chromic acid oxidation. The name (as a variant of malic) was devised by Dessaignes in 1858.

mandelic acid from German *mandel* (almond). It is obtained by treatment of the glucoside amygdalin (present in bitter almonds) with hydrochloric acid.

mannitol from manna (the juice of the manna tree) and -ol (alcohol). Mannitol is the chief constituent of manna. It was named mannite in 1830.

mellitic acid from Latin *mel* (honey). The aluminum salt of mellitic acid (mellite) is found in peat as crystals the color of honey.

mesaconic acid from Greek *meso-* (middle); it was regarded as an intermediate in the isomerization of citraconic acid to itaconic acid. This term for methylfumaric acid goes back to 1854.

mesitylene from mesite, a liquid whose properties were once thought to hold a middle position (Greek *meso* is middle) between alcohol and ether. Mesite was later shown to be acetone, and mesitylene can be made by condensation of three molecules of acetone.

mesityl oxide from mesite (see mesitylene). It can be made by aldol condensation of two molecules of acetone; and it contains oxygen.

metabolism from Greek *metaballein* (to transform or change). Related terms are anabolism, from Greek *ana* (upward), and catabolism, from Greek *kata* (downward).

morphine from Morpheus, a god of dreams. Morphine can alleviate pain and cause sleep. The term was introduced in 1828.

morpholine from morphine. L. Knorr erroneously deduced that morphine contains an oxazine nucleus, so he named the simplest oxazine morpholine.

muscone from Latin *muscus* (musk). Muscone is the odorous principle of musk.

mustard gas received its spicy name from British troops during World War I because of its odor. It is poisonous and causes intense irritation and blistering.

mutarotation from Latin *mutare* (to change or modify).

naphtha was used in Latin and Greek for a flammable, volatile liquid derived from the distillation of coal tar.

naphthacene from *naphth*alene and anthr*acene*; this tetracycle may be regarded as anthracene with a naphthalene nucleus in place of one of the benzene rings.

naphthalene from naphtha and -ene (as in benzene). It crystallizes from heavy naphtha (i.e., the portion boiling at 170–240°C).

neoprene from Greek *neos* (new) and chloro*prene*. It was (at one time) a new polymer of chloroprene.

nicotine from *Nicotiana tabaccum*, the tobacco plant, which in turn was named for Jacques Nicot, who introduced tobacco into France in 1560. The term nicotine has been used since 1819.

ninhydrin perhaps from metathesis of some letters in hydrindene.

nitrile from modern Latin *nitrilis* (a nitrogen derivative). Nitriles contain the —CN group. The term dates to 1848.

novocaine from Latin *nov*us (new) and *cocaine*. It was, at one time, a new pain killer; cocaine had been used previously. The term has been in use since 1906.

nucleic acid from Latin *nucleus*. Nucleic acids occur in cell nuclei.

nucleophile see electrophile.

oestrone (often spelled estrone) from Greek *oistros* (gadfly). Metaphorically it refers to a sting or passion, and the oestrus is the period of sexual desire in females. As a result, the prefix oestr- is used in naming female sex hormones, such as oestrone.

olefin from Latin *oleum* (oil) and *facere* (to make). Olefins such as ethylene produced oily products when treated with halogens.

oleic acid from Latin *oleum* (oil). It is present, as triglycerides, in some vegetable oils.

ornithine from Greek *orni*s (bird) and -ine (ending for an amine base). Ornithine is an amino acid found in birds.

oxalic acid from Greek *oxys* (sharp, sour). The taste of oxalic acid can be described as sharp or sour.

oxime from *oxygen* and *imide*. Oximes, $R_2C=NOH$, have an oxygen atom inserted in an imide, $R_2C=NH$. The term dates back to 1891 and was shortened from oximide.

palmitic acid from Latin *palma* (palm). The acid's triglyceride, palmitin, occurs in may fats and oils, including palm oil.

pantothenic acid from Greek *pantothen* (from everywhere). R.J. Williams, in 1933, found this substance in many different biological tissues.

paraffin from Latin *parum affinis* (showing little affinity). Paraffins are chemically unreactive. The term has waxed poetic since 1872.

peptide from Greek *peptein* (cook, digest). Peptides form during digestion of proteins.

perylene from *peri*-dinaphth*ylene*; it consists of two naphthalene groups connected to each other through their peri positions (i.e., C1, C8).

petroleum from Latin *petra* (rock) and *oleum* (oil). Petroleum is an oily material often imbedded in rocks. The name has been around since 1526.

phenanthrene from (di)*phenyl* and *anthracene*. It can be looked upon as diphenyl with two-carbon bridge, and it is an isomer of anthracene.

phenyl from Greek *phaino-* (shining). The coal tar used to manufacture illuminating gas in the nineteenth century contained benzene, phenol, and other compounds with benzene rings. Laurent brought the French name *phényle* to light in 1841.

phenetole from *phenyl*, *eth*yl and -ole (from Latin *oleum*, oil). Phenetole, a liquid, contains phenyl and ethyl groups.

phenone from *phen*yl and -one (ketone). A phenone is a phenyl ketone; the general formula is C_6H_5COR.

phorone from cam*phor* and -one (ketone). The term originally described a ketone of the formula $C_9H_{14}O$ obtained from the distillation of calcium camphorate. It now is applied to an isomer, $(CH_3)_2C{=}CHCOCH{=}C(CH_3)_2$.

phosgene from Greek *phos* (light) and *genein* (to produce). Phosgene was originally produced by the action of the sun's rays on carbon monoxide and chlorine. The term first saw the light of day in 1812.

phthalic acid contraction of naphthalic acid, obtained by oxidation of naphthalene. The name dates back to 1857.

phytol from Greek *phyt*on (plant) and -ol (alcohol). It is an alcohol first isolated from chlorophyll by R. Willstätter.

picoline from Latin *pix* (pitch) and *oleum* (oil). Picolines occur in coal tar and bone oil. The name is on record since 1853.

picric acid from Greek *pikros* (bitter). Picric acid has an intensely bitter taste. The name entered the chemical scene in 1838.

pimelic acid from Greek *pimele* (fat). Pimelic acid forms upon oxidation of fats. The term was introduced in 1838.

pinacol from Greek *pinax* (tablet) and -ol (alcohol). Obtained by reduction of acetone, it is a diol, and its hydrate crystallizes in large tablets. An old name is pinacone (pinak + (acet)one).

pinacolone from pinacol and -one (ketone). It is a ketone made by treatment of pinacol with acid. Around 1866 it was called pinacoline from pinac + ol(eum) + ine.

piperidine from Latin *piper* (pepper) and -ine (suffix for an amine base). It is obtained by hydrolysis of piperine, an alkaloid found in peppers.

piperonal from Latin *piper* (pepper). This aromatic aldehyde can be made from a degradation product of piperine, an alkaloid in various kinds of pepper.

pivalic acid from *pi*nacolone (from which it was made by dichromate oxidation) and *valeric acid* (with which it is isomeric).

porphyrin from hematoporphyrin (Greek *porphyros*, meaning purple). Decomposition of a brownish pigment from blood called hematin provides hematoporphyrin, an iron-free red powder, so called because of its source and color.

pregnane from Latin *prae-* (before) and *gnascor, gnatus* (to be born). Pregnane is the alkane related to progesterone (the hormone that governs the processes of pregnancy).

prehnitic acid was named after Colonel von Prehn who first brought the mineral prehnite to Europe from the Cape of Good Hope. Prehnitic acid's crystals resemble those of prehnite.

progesterone from Latin *pro-* (for), *gestat*io (the carrying of young), *ster*ol and -one (ketone). It is a diketone with a steroid structure that prepares the uterus for child bearing.

proline from *pyrrol*idine and -ine (ending for an amine base). It is an amino acid containing a pyrrolidine ring.

propargyl alcohol from propionic and Latin *argentum* (silver). The terminal alkyne forms an insoluble salt with silver ion.

propionic acid from Greek *prōtos* (the first) and *piōn* (fat). It is the first carboxylic acid that shows a similarity to the fatty acids. Dumas, Malagute, and Leblanc introduced the French version of the name (acide propionique) in 1847; older names were metacetonic acid ("after-acetic" acid) and metacetic acid.

protein from Greek *prōteios* (primary). This name was introduced by G.J. Mulder, a Dutch chemist, at the suggestion of Berzelius. Mulder was the first to study systematically these substances, which he obtained from casein, fibrin, and egg albumin. He felt they were the most important constituents of plants and animals.

purine from Latin *purus* (pure) and *urina* (urine). Purine was named by E. Fischer in 1884 and can be prepared from uric acid, a constituent of urine.

putrescine from Latin *putrescere* (decay or rot), a vivid description of putrescine's odor.

pyracene from pyrene and acenaphthene. This name for a $C_{14}H_{12}$ hydrocarbon was coined by Karl Fleischer and Paul Wolff to emphasize its structural relationship to these two compounds. *Chem. Ber.* **1920**, *53*, 925–931.

pyracylene from pyracene and acenaphthylene. Ronald D. Brown suggested this name for a $C_{14}H_8$ hydrocarbon because of its structural similarity to these two unsaturated molecules. *J. Chem. Soc.* **1951**, 2391–2394.

pyrene from Greek *pyro* (fire) and -ene (as in benzene). It is an aromatic hydrocarbon found in coal tar. Its fiery name, introduced in 1839, refers to the fact that coal tar is made by destructive distillation.

pyridine from Greek *pyro* (fire). This amine was obtained by pyrolysis of bones and distillation of the derived oil.

pyrimidine from pyridine and amidine. The term was coined by A. Pinner in 1885.

pyrrole from Greek *pyrros* (fiery red) and Latin *oleum* (oil). Its vapors impart a red color to pine splinters moistened with hydrochloric acid. The name goes back to the year 1835.

pyruvic acid from Greek *pyro* (fire) and Latin *uva* (grape). It was formed by pyrolysis (distillation) of tartaric acid (from grapes).

quinoline from *quina* (an old name for Cinchona bark), Latin *oleum* (oil) and -ine (common suffix for amine bases). Distillation from KOH of the antimalarial drug quinine gives quinoline, an oily amine. The name, coined by Gerhardt, dates back to about 1845.

quinone from *quinic* and -one (ketone). This diketone was made by distillation of quinic acid with MnO_2 and H_2SO_4. Quinic acid, in turn, is found in Cinchona bark (*quina*). Berzelius suggested the name in 1840.

racemic acid from Latin *racemus* meaning bunch or cluster (as of grapes or berries); an early name for the isomer of tartaric acid (isolated from grapes) that was resolved into mirror-image forms. The name has been active since 1835.

raffinose from French *raffiner* (to refine) and -ose (suffix for sugars). It was obtained during the refining of sucrose by the barium process.

resorcinol from *res*in, orcin (3, 5-dihydroxytoluene; now called orcinol) and -ol (indicating a phenol). Resorcinol (m-dihydroxybenzene) forms when certain resins are fused with KOH.

rhamnose from Greek *rhamn*ose (buckthorn) and -ose (ending for sugars). It is obtained by hydrolysis of xantho-rhamnin, a glycoside in buckthorn berries.

riboflavin from *ribo*se and Latin *flav*us (yellow). It contains a side chain structurally similar to ribose, and it is yellow in color.

ribose from *a*rab*inose*, a chemical relative, by rearrangement of some of the letters.

saccharic acid from Latin *saccharum* (sugar). It is produced by the oxidation of certain hexose sugars. The name dates back to 1800.

saccharin from Latin *saccharum* (sugar). It is a very sweet compound discovered serendipitously in Ira Remsen's laboratory in 1879. It found extensive use as an artificial sweetener.

salicylic acid from Latin *salix* (willow). The acid and its esters are found in many plants.

saponification from Latin *sapo* (soap). Alkaline hydrolysis of fats produced soaps.

Schweizer's reagent ammoniacal solution of copper (II) hydroxide; named after M.E. Schweizer (1818–1860), Swiss chemist who discovered this solution dissolves cellulose. Kaufman, G.B. *J. Chem. Educ.* **1984**, *61*, 1095–1097.

sebacic acid from Latin *sebum* (tallow); it can be made from castor oil. The term was first used in 1790.

semicarbazide from Latin *semi*- (half), *carb*amide and hydr*azide*. Semicarbazide, $NH_2CONHNH_2$, can be thought of as part carbamide (urea, NH_2CONH_2) and part hydrazide ($CONHNH_2$).

skatole from Greek *skor* (manure), of which skatole is a constituent.

sorbitol from Latin *sorb*ium (service berry) and -ol (alcohol). It was originally obtained from the juice of service berries, and has six —OH groups in its structure.

spectrum is Latin for image, appearance.

squalene from Latin *squal*us (shark) and -ene. It is obtained from shark liver and is polyunsaturated.

starch from Middle English *sterche* (to stiffen). It is used to stiffen fabrics.

stearic acid from Greek *stear* (stiff fat, tallow). It is present, as the glyceride, in most animal and vegetable facts and oils.

steroid from Greek *stereos* (stiff, solid) and *-oeidēs* (like, in the form of). The term was first used (1827) as a part of the name cholesterol, a solid originally found in gall stones. It has since been used as a class name.

stilbene from Greek *stilbō* (to glitter) and -ene (as in benzene). It crystallizes in lustrous plates and prisms. Laurent coined the French term stilbene in 1843.

styphnic acid from Greek *styphnos* (contracting). The acid has an astringent effect. The name dates back to 1850.

suberic acid from Latin *suber* (cork). It can be obtained by oxidation of cork. The term floated into use in 1799.

succinic acid from Latin *succinum* (amber). The acid was originally observed in the distillate from amber. The named was applied in 1790.

sucrose from French *sucre* (sugar) and -ose (ending for sugars).

sulphone from *sulph*ur-ket*one*. Its structure, RSO_2R', was regarded as being similar to that of ketones, $RCOR'$. The term dates to 1872.

sultone internal ester of a hydroxysulfonic acid (i.e., the sulfur analog of a lactone). H. Erdmann introduced the term in 1888.

tar from English *teru* (derived from an Old Teutonic root meaning "tree"). Tar is formed by destructive distillation of wood, coal, and other organic materials.

tartaric acid from tartar (the potassium hydrogen tartrate that is present in grapes and forms a hard crust on wine casks during fermentation). The origin of the word "tartar" is obscure. Tartaric acid is made from tartar by addition of a strong acid.

taurine from Latin *taur*us (bull) and -ine (amine group). Taurine (aminoethylsulfonic acid) can be isolated from ox bile. The term came into use in 1842.

tautomerism from Greek *tauto* (the same) and *meros* (part). Tautomerism involves conversion to another structure with the same parts (i.e., atoms). Formerly also called merotropy and tropomerism.

teflon from *tetra*fluoroethylene and -on (a general suffix). Teflon is made by polymerization of tetrafluoroethylene.

terephthalic acid from *tere*binth (the tree that yields turpentine) and phthalic acid. It can be made by oxidation of turpentine and is an isomer of phthalic acid.

terpene from terpentin (old spelling of turpentine). Many plants (including the tree that yields turpentine) contain oily substances that give them their fragrance. These "essential oils" can be isolated by steam distillation. Terpenes are compounds structurally related to $C_{10}H_{16}$ present in such oils.

terylene from *tere*phthalic acid and eth*ylene* glycol. The trade name of a fabric made by polymerization of these two compounds was introduced in 1954.

testosterone from *test*is, *ster*oid, and horm*one*. It is the chief male sex hormone, is made in the testis, and is a steroid.

tetralin from its German name, *tetra*hydronaphtha*lin*. The English name is tetrahydronaphthalene.

thallium from Latin *thallus* (green bud or shoot). The element was first detected spectrally from its green line.

thebaine from Latin *theba*icus (of or pertaining to Thebes, an ancient city in Egypt) and -ine (suffix for an amine base). The alkaloid thebaine is found in opium, and in the nineteenth century much of the world's opium came from Egypt.

theobromine from the tree *Theobroma cacao*, whose name, in turn comes from Greek *theos* (god) and *brōma* (food).

theophylline from *Thea* (the tea plant), Greek *phyllon* (leaf), and -ine (amine ending). It is an alkaloid in tea leaves, and its name originated around 1894.

thiamin from Greek *thei*on (sulfur) and either *amine* or vit*amin*. It is a vitamin (B-1) and contains a sulfur and an amine group.

thionyl from Greek *theion* (sulfur) and *hylē* (the stuff from which a thing is made). The term refers to the SO group.

thiophene from Greek *theion* (sulfur) and -ene (as in benzene). It has sulfur and shows aromatic properties similar to those of benzene.

threose is an artificial epimer of erythrose. The name is a variant of erythrose.

tiglic acid from *Croton tiglium*, the plant from which it can be obtained. The name dates back to 1875.

tocopherols from Greek *tokos* (childbirth), Greek *pherō* (to carry), and -ol (alcohol). They are alcohols found in sources of vitamin E, which is necessary for fertility in males and the birth process in females.

tolane (German tolan) from toluene (German toluol). A name for diphenylacetylene coined by H. Limpricht and H. Schwanert (*Liebigs Ann. Chem.* **1868**, *145*, 330–350) based on its relationship to toluol and toluylen (their name for stilbene).

toluene (German toluol) from tolu and Latin *oleum* (oil). It was obtained by distillation of balsam from the Tolu tree.

trilene from *tri*chloroethy*lene*, for which it is a trade name.

trimellitic acid from *tri*basic (it is benzene-1, 2, 4-tricarboxylic acid) and mellitic acid (from which it can be made). Baeyer named it in 1870.

trimesic acid from *tri*basic (it is benzene-1, 3, 5-tricarboxylic acid) and *mesitylenic acid* (from which it can be made). In 1867, R. Fitting named it.

trioxane from tri- (three), *ox*ygen, and -ane. A cyclic crystalline polymer of formaldehyde; it has three oxygens and no unsaturation.

tropone from *trop*ine and -one (ketone). It contains a seven-membered carbocyclic ring like that in the amine tropine, itself obtained on hydrolysis of atropine. This last alkaloid was named after *Atropa belladonna* (Deadly Nightshade), in which it is found.

trypsin from Greek *tryō* (to rub) and pe*psin*. Trypsin and pepsin are both enzymes that catalyze the breakdown of proteins. The "rub" part of the name may stem from the fact that trypsin was obtained from the pancreas by rubbing it down with glycerin.

turpentine (spelled terbentyne or terebentyne in the fourteenth and fifteenth centuries) from Latin *turbentina* (the resin of the *Terebenth* tree). This resin yields oil of turpentine, but nowadays this mixture of terpenes comes from a variety of trees.

tyrosine from Greek *tyros* (cheese) and -ine (amine group). It is an amino acid found (among other places) in old cheese. The name was first used in 1857.

uracil from *ur*ea and *ac*id (or *ac*rylic). It can be made from urea and acrylic acid.

urethane from *ur*ea, *eth*er and -ane (a general suffix). It can be prepared from urea nitrate and ethanol. It was discovered by Dumas in 1833. Thinking it was a new ether, he proposed that name in 1838.

valeric acid from *Valeriana officinalis*, the plant in whose roots the acid is found. An early name was valerianic acid.

veratric acid from *Veratrum*. The acid occurs in various plants of the genus *Veratrum*.

veronal from the city Verona. According to legend, Emil Fischer (who codiscovered the drug with J. von Moring in 1903) was discussing a possible name with colleagues when he had to end the discussion to catch a train to Verona.

vicinal from Latin *vicinus* (neighboring, adjacent). A vicinal dihalide has halogens on neighboring carbons.

vinyl from Latin *vinum* (wine). Vinyl chloride, bromide, and so forth, may be considered derivatives of vinyl alcohol, $CH_2{=}CHOH$, which is an unstable tautomer of acetaldehyde. The term vinyl appeared in 1863.

vitamin from Latin *vita* (life) and *amine*. The name (initially vitamine) was coined by Kasimir Funk in 1912; the substances are needed for a healthy life, and Dr. Funk erroneously thought they were all amines.

xanthein from Greek *xanthos* (yellow). It is a yellow dye sometimes found in cell sap.

xanthene from Greek *xanthos* (yellow) and -ene (as in benzene). It is formed by reduction of xanthone, the parent of xanthene dyes, many of which are yellow. Xanthene contains two benzene rings.

xanthic acid from Greek *xanthose* (yellow). It gives a yellow precipitate with copper salts. The term dates back to 1831.

xanthine from Greek *xanthos* (yellow). This dihydroxypurine produces a yellow residue when a mixture with nitric acid is evaporated. The name originated in 1857.

xylene from Greek *xylon* (wood) and -ene (as in benzene). The xylenes (dimethylbenzenes) were originally obtained from wood tar.

xylose from Greek *xylon* (wood). This sugar occurs in many types of wood.

zeolite from Greek *zeō* (to seethe or boil) and *lithos* (stone). Zeolites are hard silicate materials that swell up and melt when heated.

Appendix B

ORIGINS OF ELEMENT NAMES*

TABLE B.1. Elements with Names of Obscure Origin

Element	Origin[a]
Gold	Sanskrit, *Jval*; Ang.–Sax., *gold*; ME, *guld*
Iron	Ang.–Sax., *iron*; ME, *iren*
Lead	Ang.–Sax., *lead*; ME, *leed*
Silver	Ang.–Sax., *seolfor*, *sylfer*
Sulfur	Sanskrit., *sulvere*; ME, *sulphre*
Tin	Ang.–Sax., *tin*; ME, *tin*
Zinc	Ang.–Sax., *zinc*

[a]Key to all tables: Ang.–Sax. = Anglo–Saxon, Eng. = English, Ger. = German, Gr. = Greek, L. = Latin, ME = Middle English, Sp. = Spanish, Swed. = Swedish.

TABLE B.2. Elements Named for Colors

Element	Origin
Bismuth	Ger., *weisse Masse*, white mass
Cesium	L., *caesius*, sky blue
Chlorine	Gr., *chloros*, greenish yellow
Chromium	Gr., *chroma*, color
Indium	the color indigo
Iodine	Gr., *iodes*, violet
Iridium	L., *iris*, rainbow
Praesodymium	Gr., *prasios* + *didymos*, green twin
Rubidium	L., *rubidos*, deepest red
Zirconium	Arabic, *zargun*, gold color

*The tables, reprinted here by permission, are from an article written by David W. Ball while he was a graduate student at Rice University (*J. Chem. Educ.* **1985**, *62*, 787–788).

TABLE B.3. Elements Named after People (Real or Mythical)

Element	Origin
Curium	Pierre and Marie Curie, discoverers of radioactivity
Einsteinium	Albert Einstein, originator of the theories of relativity
Fermium	Enrico Fermi, discoverer of nuclear reactions
Gadolinium	Johann Gadolin, a Finnish chemist who discovered yttrium
Lawrencium	Ernest O. Lawrence, a developer of the cyclotron
Mendelevium	Dmitri Mendeleev, developer of the periodic chart
Niobium	Niobe, an evil and blasphemous daughter of Tantalos (see below)
Nobelium	Alfred Nobel, founder of the Nobel Prizes and inventor of dynamite
Promethium	Prometheus, the Greek god who gave mankind fire
Tantalum	Tantalos, the Greek mythical figure banished to a tantalizing fate in Hades
Thorium	Thor, the Norse god of thunder
Titanium	the Titans, Greek gods
Vanadium	Vanadis, "wise woman" in Scandinavian mythology

TABLE B.4. Elements Named after Places

Element	Origin
Americium	the Americas
Berkelium	Berkeley, California
Californium	California
Copper	L., *cuprum*, from the island of Cyprus
Erbium	Ytterby, a town in Sweden
Europium	Europe
Francium	France
Gallium	L., *Gallia*, France
Germanium	Germany
Hafnium	L., *Hafnia*, Copenhagen
Holmium	L., *Holmia*, Stockholm
Lutetium	Lutetia, an ancient name for Paris
Magnesium	Magnesia, a district in Thessaly
Polonium	Poland
Rhenium	L., *Rhenus*, the Rhine
Ruthenium	L., *Ruthenia*, Russia
Scandium	L., *Scandis*, Scandinavia
Strontium	Strontian, a town in Scotland
Terbium	Ytterby, a town in Sweden
Thulium	Thule, an early name for Scandinavia
Ytterbium	Ytterby, a town in Sweden
Yttrium	Ytterby, a town in Sweden

TABLE B.5. Elements Named after Heavenly Bodies

Element	Origin
Cerium	the asteroid Ceres
Helium	Gr., *helios*, the sun
Mercury	the planet Mercury
Neptunium	the planet Neptune
Palladium	the asteroid Pallas
Phosphorus	Gr., *phosphoros*, light-bearing; a name applied to the planet Venus when appearing as a morning star
Plutonium	the planet Pluto
Selenium	Gr., *Selene*, moon
Tellurium	L., *tellus*, the earth
Uranium	the planet Uranus

TABLE B.6. Elements Having Names of Miscellaneous Origin

Element	Origin
Actinium	Gr., *aktinos*, beam or ray
Aluminum	L., *alumen*, alum (an astringent)
Antimony	Gr., *anti* + *monos*, not alone, not one
Argon	Gr., *argos*, inactive
Arsenic	Gr., *arsenikos*, male or masculine
Astatine	Gr., *astatos*, unstable
Barium	Gr., *barys*, heavy
Beryllium	Gr., *beryl*, beryl (a gem)
Bromine	Gr., *bromos*, stench
Cadmium	L., *cadmia*, calamine (a zinc ore)
Calcium	L., *calx*, lime (calcium oxide)
Carbon	L., *carbo*, coal or charcoal
Cobalt	Ger., *kobald*, evil spirit or goblin
Dysprosium	Gr., *dysprositos*, hard to get at
Fluorine	L., *fluere*, to flow
Hydrogen	Gr., *hydros* + *genes*, water-forming
Krypton	Gr., *kryptos*, hidden
Lanthanum	Gr., *lanthanein*, to lie hidden
Lithium	Gr., *lithos*, stone
Manganese	L., *magnes*, magnet
Molybdenum	Gr., *molybdos*, lead
Neodymium	Gr., *neo* + *didymos*, new twin
Neon	Gr., *neos*, new
Nickel	Ger., *kupfernickel*, niccolate (a mineral)
Nitrogen	L., *nitrium*; Gr., *nitron*, native soda
Osmium	Gr., *osme*, smell or odor
Oxygen	Gr., *oxys* + *genes*, acid-forming
Platinum	Sp., *platina*, silver
Potassium	Eng., *potash* (a potassium salt)
Protactinium	Gr., *protos*, first, + actinium (see above)
Radium	L., *radius*, ray
Radon	from radium (see above)
Rhodium	Gr., *rhodios*, roselike
Samarium	Eng., *samarskite* (a mineral)
Silicon	L., *silex*, flint
Sodium	Eng., *soda*
Technetium	Gr., *technitos*, artificial
Tungsten	Swed., *tung* + *sten*, heavy stone
Xenon	Gr., *xenon*, stranger

Appendix C

LATIN AND GREEK TERMS FOR NUMBERS

Number	Latin[a]	Greek[b]
1	*unus*	*eis, enas, mia, en, ena*
2	*duo, bini*	*duo*
3	*tres*	*treis, tria*
4	*quattuor*	*tessares, tessara*
5	*quinque*	*pente*
6	*sex*	*ex*
7	*septem*	*epta*
8	*octo*	*octo*
9	*novem*	*ennea*
10	*decem*	*deka, deca*
11	*undecim*	*endeka*
12	*duodecim*	*dodeka*
13	*tredecim*	*dekatreis*
14	*quattuordecim*	*dekatessares*
15	*quindecim*	*dekapente*
16	*sedecim*	*dekaex*
17	*septedecim*	*dekaepta*
18	*duedeviginti*	*dekaocto*
19	*undeviginti*	*dekaennea*
20	*viginti*	*eikosi*
21	*viginti et unus*	*eikosi-en(a)*
22	*viginti et duo*	*eikosi-do*
23	*viginti et tres*	*eikosi-treis*
24	*viginti et quattuor*	*eikosi-tessares*
25	*viginti et quinque*	*eikosi-pente*
100	*centum*	*ekaton*
1000	*millie (milia* or *millia* for more than one)	*xillioi*
1,000,000	*decies centena mil(1)ia*	*en ekatommyrion*

[a]Simpson, D.P. *Cassell's New Latin Dictionary* (Funk and Wagnalls Co., New York, 1960).
[b]Divry, G.C., Ed. *Divry's Modern English–Greek and Greek–English Desk Dictionary* (D.C. Divry Inc., New York, 1966), p. 762.

Appendix D

NUMERICAL PREFIXES IN CHEMICAL WORDS

1/2	semi-, hemi-, demi-	20	eicosa-
1	uni-, mono-, holo-	21	heneicosa-
$1\frac{1}{2}$	sesqui-	22	docosa-
2	bi-, di-, diplo-	23	tricosa-
21/2	hemipenta-, sester-	24	tetracosa-
3	ter-, tri-	25	pentacosa-
4	tetra-, quadri-	26	hexacosa-
5	penta-, quinto-	27	heptacosa-
6	hexa-, sexa-	28	octacosa-
7	hepta-, septa-	29	nonacosa-
8	octa-, okta-	30	triaconta-
9	nona-, ennea-	40	tetraconta-
10	deca-	50	pentaconta-
11	undeca-, hendeca-	60	hexaconta-
12	dodeca-	70	heptaconta-
13	trideca-, triskaideca-	80	octaconta-
14	tetradeca-	90	nonaconta-
15	pentadeca-	100	hecta-
16	hexadeca-	101	henhecta-
17	heptadeca-	102	dohecta-
18	octadeca-	110	decahecta-
19	nonadeca-	120	eicosahecta-
		200	dicta-

Appendix E

GREEK ALPHABET

Name of Letter	Capital	Lower Case	Transliteration[a]
alpha	A	α	a
beta	B	β	b
gamma	Γ	γ	g
delta	Δ	δ	d
epsilon	E	ε	e short
zeta	Z	ζ	z
eta	H	η	e long
theta	Θ	θ	th
iota	I	ι	i
kappa	K	κ	k, c
lambda	Λ	λ	l
mu	M	μ	m
nu	N	ν	n
xi	Ξ	ξ	x
omicron	O	o	o short
pi	Π	π	p
rho	P	ρ	r
sigma	Σ	σ	s
tau	T	τ	t
upsilon	Υ	υ	y
phi	Φ	ϕ	f
chi	X	χ	ch (as in German echt)
psi	Ψ	ψ	ps
omega	Ω	ω	o long

[a]Transliteration according to *The Merck Index*, 9th ed. (Merck and Co., Rahway, NJ, 1976), p. MISC-48

Appendix F

NOBEL PRIZES IN THE SCIENCES

Alfred Bernhard Nobel was born October 21, 1833, in Stockholm. As a child he was sickly and was not allowed to participate in sports. He had but two terms of formal schooling. In 1842, his family moved to St. Petersburg (now Leningrad) where Alfred's father, a technician and inventor, manufactured submarine mines and torpedoes for the Russian government. In St. Petersburg, Alfred had a private tutor for several years, but he was largely self-taught and never attended a university. About 1850–52 he traveled abroad and visited the United States, where he gained some training under a noted engineer, John Ericsson. In 1859, the family left Russia to resettle in Sweden. There Nobel worked on explosives in his father's research laboratory and subsequently prospered from numerous inventions. Alfred became multilingual (five languages) and always answered his own mail longhand in the language of the recipient. He never married; but he supported a mistress and was obliged to do so until her death because she threatened to publish a hoard of love letters from him.

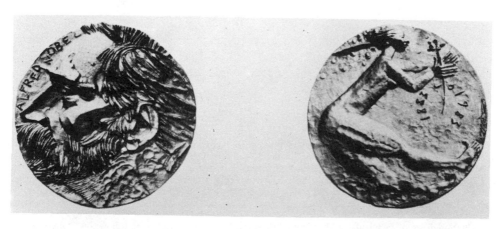

The left photograph shows a profile of Alfred Nobel on the obverse of a commemorative centennial medal struck in 1983 by the Milano Mint of Italy. The stylized figure on the reverse side symbolizes the theme: Science for Humanity 1883–1983. Both photographs were kindly provided by the medal's United States distributor, Medacoin International, Las Vegas, Nevada.

325

Alfred Nobel was an inventive genius and held about 355 patents in different countries. He not only invented dynamite, blasting caps, blasting gelatin, and ballistite (one of the earliest smokeless powders) but also owned patents dealing with synthetic rubber, leather, and artificial silk. Nobel was fond of literature and wrote several novels and plays, but with little success. He amassed a fortune from the manufacture of dynamite and other explosives and from oil produced at fields in Baku, Russia. His later years brought increasing illness, nervousness, and anguish from guilt feelings. He abhorred war and hated the thought that his explosives could bring about much injury and death even though he had developed them for peaceful use. Alfred Nobel died December 10, 1896, in San Remo, Italy, where he operated a laboratory. He willed the bulk of his estate to endow five annual prizes to persons who conferred the greatest benefit on mankind in the fields of physics, chemistry, physiology or medicine, literature, and peace. Nobel stipulated that the physics and chemistry prizes be awarded by the Swedish Academy of Science, physiology or medicine by the Karolinska Institute of Stockholm, literature by the Swedish Academy, and peace by a committee elected by the Norwegian Parliament. (At that time, Norway was in a united kingdom with Sweden.) It took several years to draw up the detailed rules and the official wording of statutes that govern the selection of Nobel laureates, so the awards did not begin until 1901. The peace prize may go to organizations as well as to individuals. In 1968, Riksbank, the central bank of Sweden, created a sixth award, Alfred Nobel Memorial Prize in Economic Science, to commemorate the bank's 300th anniversary. The first economics winner was announced in 1969.

Nobel prizes are not bestowed posthumously. Each award may be shared but only up to a maximum of three recipients so as to maintain a high level for the honor and the honorarium. An individual may win more than once, either in the same category or in different ones. Those who decline an award are still considered Nobel laureates. Winners receive a gold medal, a diploma, and a substantial sum of money at an annual ceremony held in Stockholm (in Oslo, for the peace prize) on December 10, the anniversary date of Alfred Nobel's death. The actinide element Nobelium (atomic number 102) was named in his honor.* Table F. 1 provides a listing of all chemistry, physics, and physiology/medicine prize recipients and their fields of research.‡

---

*Biographical data on Alfred Nobel and the prize winners were compiled from several standard encyclopedias and also from the following sources: Vanderbilt, B.M. *Abstracts of Papers*, 170th National Meeting of the American Chemical Society, Chicago, IL. (American Chemical Society, Washington, DC, 1975), HIST 6. Zuckerman, H. *Scientific Elite* (The Free Press, A Division of Macmillan Publishing Co., New York, 1977), *List of the Nobel Prize Winners*, distributed by the Swedish Royal Academy of Sciences, the Nobel Committees, Stockholm, 1984.

‡Nationalities of recipients at the time of the awards are abbreviated as follows: A, American; Ar, Argentine; Au, Austrian; Aul, Australian; B, British; Be, Belgian; C, Canadian; Ch, Chinese; Cz, Czech; D, Dutch; Da, Danish; F, French; Fi, Finnish; G, German; H, Hungarian; I, Italian; In, Indian; Ir, Irish; J, Japanese; N, Norwegian; P, Portugese; Pa, Pakistani; R, Russian; Sc, Scottish; Sp, Spanish; Swe, Swedish; Swi, Swiss; SA, South African.

TABLE F.1. Nobel Prizes Awarded in Chemistry, Physiology and Medicine, and Physics

Year	Chemistry	Physiology or Medicine	Physics
1901	Jacobus Henricus Van't Hoff (D) for laws of chemical dynamics and osmotic pressure	Emil von Behring (G) for discovering diphtheria antitoxin	Wilhelm K. Röntgen (G) for discovering X-rays
1902	Emil Fischer (G) for synthesizing sugars, purine derivatives and peptides	Sir Ronald Ross (B) for work on malaria	Hendrik Antoon Lorentz and Pieter Zeeman (D) for discovering the Zeeman effect of magnetism on light
1903	Svante August Arrhenius (Swe) for work in ionization of electrolytes	Niels Ryberg Finsen (Da) for treating diseases with light	Henri Antoine Becquerel, Marie Curie, and Pierre Curie (F) for discovering radioactivity and studying uranium
1904	Sir William Ramsay (B) for discovering helium, neon, xenon, and krypton, and placing them in the periodic table	Ivan Petrovich Pavlov (R) for work on the physiology of digestion	Baron Rayleigh (B) for discovering argon and studying gas densities
1905	Adolph von Baeyer (G) for research on dyes and other organic compounds	Robert Koch (G) for work on tuberculosis, including discoveries of the tubercule bacillus and tuberculin	Phillip Lenard (G) for studying properties of cathode rays
1906	Henri Moissan (F) for preparing fluorine and developing the electric furnace	Camillo Golgi (I) and Santiago Ramon y Cajal (Sp) for studying nerve tissue	Sir Joseph John Thomson (B) for studying electrical discharge through gases
1907	Edward Buchner (G) for biochemical research and discovering cell-less fermentation	Charles Louis Alphonse Laveran (F) for studying diseases caused by protozoans	Albert A. Michaelson (A) for inventing optical instruments and measuring the speed of light
1908	Ernest Rutherford (B) for discovering that alpha rays break down atoms and for other work in radioactivity	Paul Ehrlich (G) and Elie Metchnikov (R) for work on immunity	Gabriel Lippman (F) for work in color photography
1909	Wilhelm Ostwald (G) for work on catalysis, equilibrium, and reaction rates	Emil Theodor Kocher (Swi) for his work on the thyroid gland	Guglielmo Marconi (I) and Karl Ferdinand Braun (G) for developing the wireless telegraph
1910	Otto Wallach (G) for research on alicyclic substances	Albrecht Kossel (G) for studying cell chemistry, proteins and nucleic substances	Johannes D. van der Waals (D) for studying relationships of liquids and gases
1911	Marie Curie (F) for discovering radium and polonium	Allvar Gullstrand (Swe) for studying the refraction of light through the eye	Wilhelm Wien (G) for studying blackbody radiation

TABLE F.1. (*Continued*)

Year	Chemistry	Physiology or Medicine	Physics
1912	Francois Auguste Victor Grignard (F) for inventing Grignard reagents; and Paul Sabatier (F) for discovery of catalytic hydrogenation	Alexis Carrel (F) for suturing blood vessels and grafting vessels and organs	Nils Dalén (Swe) for inventing automatic gas regulators for lighthouses
1913	Alfred Werner (Swi) for his coordination theory	Charles Robert Richet (F) for studying allergies caused by foreign substances	Heike Kamerlingh Onnes (D) for liquefying helium and for other cryogenic studies
1914	Theodore W. Richards (A) for determining atomic weights of many elements	Robert Bárány (Au) for work on the equilibrium organs of the inner ear	Max T.F. von Laue (G) for using crystals to measure X-rays
1915	Richard Willstätter (G) for work on chlorophyll and other coloring matter in plants	No award	Sir William Henry Bragg and Sir William L. Bragg (G) for X-ray crystallography
1916	No award	No award	No award
1917	No award	No award	Charles Barkla (B) for studying diffusion of light and radiation of X-rays from elements
1918	Fritz Haber (G) for the process of synthesizing ammonia from nitrogen and hydrogen	No award	Max Planck (G) for stating the quantum theory of light
1919	No award	Jules Bordet (Be) for discoveries on immunity	Johannes Stark (G) for the Stark effect of electric fields on spectra
1920	Walther Nernst (G) for studies of heat changes in chemical reactions	August Krogh (Da) for discovering the system of action of blood capillaries	Charles E. Guillaume (F) for discovering several alloys
1921	Frederick Soddy (B) for studying radioactive substances and isotopes	No award	Albert Einstein (G) for explaining the photoelectric effect and for contributions to mathematical physics
1922	Francis W. Aston (B) for discovering many isotopes using a mass spectrograph	Archibald V. Hill (B) for discovering heat production by muscles; and Otto Meyerhof (G) for his theory on the production of lactic acid in muscles	Niels Bohr (Da) for studying the structure of atoms and their radiations
1923	Fritz Pregl (Au) for microanalysis of organic substances	Sir Frederick Grant Banting (C) and John J.R. Macleod (Sc) for discovering insulin	Robert A. Millikan (A) for measuring the charge on an electron and for work on the photoelectric effect
1924	No award	William Einthoven (D) for inventing the electrocardiograph	Karl M.G. Siegbahn (Sw) for work with the X-ray spectroscope

Year			
1925	Richard Zsigmondy (G) for studies on colloids	No award	James Franck and Gustav Hertz (G) for laws on the collision between an electron and an atom
1926	Theodor Svedberg (Swe) for work on dispersions, including colloids	Johannes Fibiger (Da) for discovering a cancer-causing parasite	Jean Baptiste Perrin (F) for studying the discontinuous structure of matter and determining atomic radii
1927	Heinrich O. Wieland (G) for studies of gall acids and related substances	Julius Wagner von Jauregg (Au) for discovering the fever treatment for paralysis	Arthur H. Compton (A) for the Compton effect on X-rays reflected from atoms; and Charles T.R. Wilson (B) for the tracing of paths of ions
1928	Adolf Windaus (G) for studying sterols and their connection with vitamins	Charles Nicolle (F) for work on typhus	Owen W. Richardson (B) for the thermionic effect (i.e., electron emission by hot metals)
1929	Sir Arthur Harden (B) and Hans August Simon von Euler-Chelpin (G) for work on fermentation and enzymes	Christiaan Eijkman (D) for discovering vitamins that prevent beriberi; and Sir Frederick G. Hopkins (B) for discovering vitamins that help growth	Louis Victor de Broglie (F) for his theory of wave-particle duality
1930	Hans Fischer (G) for studies on coloring matter of blood and leaves	Karl Landsteiner (A) for discovering blood types	Sir Chandrasekhara Venkata Raman (In) for discovering a new effect in radiation from elements
1931	Carl Bosch and Friedrich Bergius (G) for inventing methods for manufacturing ammonia and liquefying coal	Otto H. Warburg (G) for discovering that enzymes help respiration in tissues	No award
1932	Irving Langmuir (A) for studies of monomolecular films on surfaces	Edgar D. Andrian and Sir Charles S. Sherrington (B) for work on the function of neurons	Werner Heisenberg (G) for founding quantum mechanics
1933	No award	Thomas H. Morgan (A) for studies on the function of chromosomes in heredity	Paul Dirac (B) and Ervin Schrödinger (Au) for uncovering new forms of atomic theory
1934	Harold Clayton Urey (A) for discovering deuterium	George Minot, William P. Murphy and George H. Whipple (A) for discoveries on liver treatment for anemia	No award
1935	Frederic Joliot and Irene Joliot–Curie (F) for synthesizing new radioactive elements	Hans Spemann (G) for discovering the organizer effect in growth of embryos	Sir James Chadwick (B) for discovering the neutron
1936	Peter J.W. Debye (D) for studies on dipole moments and diffraction of electrons and X-rays in gases	Sir Henry H. Dale (B) and Otto Loewi (Au) for discoveries on chemical transmission of nerve impulses	Carl David Anderson (A) for discovering the positron; and Viktor F. Hess (Au) for discovering cosmic rays

TABLE F.1. (*Continued*)

Year	Chemistry	Physiology or Medicine	Physics
1937	Sir Walter N. Haworth (B) for studies of carbohydrates and vitamin C; and Paul Karrer (Swi) for studies of carotenoids, flavins, and vitamins A and B₂	Albert Szent-Györgyi (H) for work on oxidation in tissues, vitamin C, and fumaric acid	Clinton Davisson (A) and George Thomson (B) for discovering diffraction of electrons by crystals
1938	Richard Kuhn (G) for work on carotenoids and vitamins (declined)	Corneille Heymans (Be) for studies of regulation of respiration	Enrico Fermi (I) for discovery of transuranium elements
1939	Adolph Butenandt (G) for studies of sex hormones (declined); and Leopold Ruzicka (Swi) for studies of polymethylenes	Gerhard Domagk (G) for discovering prontosil, the first sulfa drug (declined)	Ernest O. Lawrence (A) for inventing the cyclotron and working on artificial radioactivity
1940–1942	No award	No award	No award
1943	Georg von Hevesy (H) for using isotopes as tracers	Henrik Dam (Da) for discovering vitamin K; and Edward Doisy (A) for synthesizing it	Otto Stern (A) for studying atoms with molecular beams
1944	Otto Hahn (G) for discoveries in atomic fission	Joseph Erlanger and Herbert Gasser (A) for work on single nerve fibers	Isidor Isaac Rabi (A) for studies of magnetic properties of atomic nuclei
1945	Arturi Virtanen (Fi) for inventing new methods in agricultural biochemistry	Sir Alexander Fleming, Howard W. Florey, and Ernst B. Chain (B) for discovery and isolation of penicillin	Wolfgang Pauli (Au) for the Pauli exclusion principle
1946	James B. Sumner (A) for finding that enzymes can be crystallized; and Wendell M. Stanley and John H. Northrop (A) for preparing enzymes and virus proteins in pure form	Hermann Joseph Muller (A) for discovering that X-rays can produce mutations	Percy Williams Bridgman (A) for work in the field of very high pressures
1947	Sir Robert Robinson (B) for research on plant substances of biological importance	Carl F. Cori and Gerti Cori (A) for work on glycogen conversion; and Bernardo Houssay (Ar) for studying the pancreas and pituitary gland	Sir Edward V. Appleton (B) for exploring the ionosphere
1948	Arne Tiselius (Swe) for discoveries on the nature of serum proteins	Paul Mueller (Swi) for discovering that DDT is an insecticide	Patrick M.S. Blackett (B) for discoveries in cosmic radiation
1949	William Frances Giauque (A) for studying reactions at very low temperatures	William R. Hess (Swi) for discovering how organs are controlled by certain parts of the brain; and Antonio E. Moniz (P) for originating prefrontal lobotomy	Hideki Yukawa (J) for discovering the meson

Year			
1950	Otto Diels and Kurt Alder (G) for the Diels–Alder reaction	Philip S. Hench, Edward C. Kendall (A), and Taddeus Reichstein (Swi) for discoveries on cortisone and ACTH	Cecil Frank Powell (B) for a photographic method for studying atomic nuclei and for studies of mesons
1951	Edwin M. McMillan and Glenn T. Seaborg (A) for discovering plutonium and other transuranium elements	Max Theiler (SA working in U.S.) for developing a yellow fever vaccine	Sir John D. Cockroft (B) and Ernest T.S. Walton (Ir) for transmutations of atomic nuclei by accelerated atomic particles
1952	Archer J.P. Martin and Richard Synge (B) for developing partition chromatography	Selman A. Waksman (A) for discovering streptomycin	Felix Bloch and Edward Mills Purcell (A) for studying magnetic properties of atomic nuclei
1953	Hermann Staudinger (G) for discoveries in polymer chemistry	Fritz Albert Lipmann (A) and Hans Adolf Krebs (B) for discoveries in biosynthesis and metabolism	Frits Zernike (D) for inventing the phase contrast microscope
1954	Linus Pauling (A) for work on the chemical bond	John F. Enders, Thomas H. Weller, and Frederick C. Robbins (A) for inventing a simple method to grow polio virus in test tubes	Max Born (G) for research in quantum mechanics; and Walther Bothe (G) for discoveries made with his coincidence method
1955	Vincent du Vigneaud (A) for the synthesis of hormones	Hugo Theorell (Swe) for discoveries on the nature and action of oxidation enzymes	Willis E. Lamb, Jr. (A) for studies of the hydrogen spectrum; and Poly–Karp Kusch (A) for determining the magnetic moment of the electron
1956	Sir Cyril Hinshelwood (B) and Nikolei N. Semenov (R) for work on chemical chain reactions	Andre F. Cournand, Dickinson W. Richards, Jr. (A), and Werner Forssmann (G) for using a catheter to chart the heart's interior	John Bardeen, Walter H. Brattain, and William Shockley (A) for inventing the transistor
1957	Lord (Alexander) Todd (B) for work on the protein composition of cells	Daniel Bovet (I) for discovering antihistamines	Tsung Dao Lee and Chen Ning Yang (Ch, worked in U.S.) for disproving the law of conservation of parity
1958	Frederick Sanger (B) for deducing the structure of insulin	George Wells Beadle and Edward Lawrie Tatum (A) for work in biochemical genetics; and Joshua Lederberg (A) for studies of genetics in bacteria	Pavel A. Cherenkov, Ilya M. Frank, and Igor Y. Tamm (R) for discovering and interpreting the Cherenkov effect pertaining to high-energy particles
1959	Jaroslav Heyrovský (Cz) for developing polarography	Severo Ochoa and Arthur Kornberg (A) for synthesis of nucleic acids	Emilio Segré and Owen Chamberlain (A) for demonstrating the existence of the antiproton
1960	Willard F. Libby (A) for a method of radioactive dating	Sir Macfarlane Burnett (Aul) and Peter B. Medawar (B) for work on immunity reactions	Donald A. Glaser (A) for inventing the bubble chamber to study subatomic particles
1961	Melvin Calvin (A) for research in photosynthesis	Georg von Békésy (A) for showing how the ear distinguishes among sounds	Robert Hofstadter (A) for studies of nucleons; and Rudulf L. Mössbauer (G) for research on gamma rays

TABLE F.1. (Continued)

Year	Chemistry	Physiology or Medicine	Physics
1962	John Cowdery Kendrew and Max Ferdinand Perutz (B) for studies on globular proteins	James D. Watson (A), Francis H. Crick, and Maurice H.F. Wilkins (B) for the structure of DNA	Lev Davidovitch Landau (R) for research on liquid helium
1963	Guilio Natta (I) and Karl Ziegler (G) for developing organometallic Ziegler–Natta polymerization catalysts	Sir John Carew Eccles (Aul) for work on transmission of nerve impulses; and Andrew Fielding Huxley and Lloyd Hodgkin (B) for their description of the behavior of nerve impulses	Eugene Paul Wigner (A) for work on atomic nuclei and elementary particles; and Maria Geppert Mayer (A) and J. Hans Jensen (G) for work on the structure of atomic nuclei
1964	Dorothy C. Hodgkin (B) for X-ray studies of compounds such as vitamin B_{12} and penicillin	Konrad E. Bloch (A) and Feodor Lynen (G) for work on cholesterol and fatty acid metabolism	Charles H. Townes (A), Nicolai G. Basov (R), and Alexander M. Prokhorov (R) for developing lasers and masers
1965	Robert Burns Woodward (A) for contributions to organic synthesis	Francois Jacob, Andre Lwoff, and Jacques Monod (F) for work on genetic control of enzyme and virus synthesis	Sin–Itiro Tomonaga (J), Julian S. Schwinger, and Richard P. Feynman (A) for basic work in quantum electrodynamics
1966	Robert S. Mulliken (A) for developing the molecular orbital theory	Francis Peyton Rous (A) for discovering a carcinogenic virus; and Charles B. Huggins (A) for using hormones to treat cancer	Alfred Kastler (F) for work on the energy levels of atoms
1967	Manfred Eigen (G), Ronald G.W. Norrish, and George Porter (B) for techniques that measure rates of rapid reactions	Ragnar Granit (Swe), H. Keffer Hartline, and George Wald (A) for work on the chemical and physiological aspects of vision	Hans Albrecht Bethe (A) for contributions to the theory of nuclear reactions, especially those producing energy in stars
1968	Lars Onsager (A) for the theory of reciprocal relations of various kinds of thermodynamic activity	Robert W. Holley, H. Gobind Khorana, and Marshall W. Nirenberg (A) for explaining how genes determine cell function	Luis W. Alvarez (A) for work on subatomic particles
1969	Derek H.R. Barton (B) and Odd Hassel (N) for work relating reactions to molecular geometry	Max Delbrück, Alfred Hershey, and Salvador Luria (A) for work with bacteriophages	Murray Gell–Mann (A) for work on classification and interactions of nuclear particles
1970	Luis Federico Leloir (Ar) for discovering compounds that effect energy storage in living things	Julius Axelrod (A), Bernard Katz (B), and Ulf Svante von Euler (Swe) for discovering the role played by certain chemicals in transmitting nerve impulses	Hannes Olof Gosta Alfven (Swe) for studying electrical and magnetic effects in fluids that conduct electricity; and Felix Neel (F) for discoveries of magnetic properties that apply to computer memories
1971	Gerhard Herzberg (C) for work on the structure of certain molecules and free radicals	Earl W. Sutherland, Jr. (A) for discovering how hormones act, including the discovery of cyclic AMP, which influences hormone action	Dennis Gabor (B) for developing holography, a type of three-dimensional photography

Year			
1972	Christian B. Anfinsen, Stanford Moore, and William H. Stein (A) for contributions to the chemistry of enzymes	Gerald M. Edelman (A) and Rodney R. Porter (B) for discovery of the chemical structure of antibodies	John Bardeen, Leon N. Cooper, and John Robert Schrieffer (A) for work on superconductivity
1973	Geoffrey Wilkinson (B) and Ernst Fischer (G) for work on organometallic compounds	Nikolaas Tinbergen (D), Konrad Z. Lorenz (D), and Karl von Frisch (Au) for studies on animal behavior	Ivar Giaever (A), Leo Esaki (J), and Brian Josephson (B) for work on electron tunneling through semiconductors and superconductors
1974	Paul John Flory (A) for work in polymer chemistry	Christian de Dawe (Be), Albert Claude, and George E. Palade (A) for pioneer work in cell biology	Antony Hewish (B) for discovering pulsars; and Sir Martin Pyle (B) for using small radio telescopes to see into space
1975	John Warcup Cornforth (Aul, worked in Britain) for stereochemical studies of enzyme-catalyzed reactions; and Vladimir Prelog (Swi) for studies in stereochemistry	David Baltimore (A), Renato Dulbecco (I, worked in U.S.), and Howard H. Temin (A) for discoveries on the interaction of tumor viruses and the cell's genetic material	Aage N. Bohr and Ben R. Mottelson (Da), and James Rainwater (A) for research on nonspherical atomic nuclei
1976	William N. Lipscomb (A) for studies of boron hydrides	Baruch S. Blumberg (A) for investigations on hepatitis B virus; and D. Carlton Gajdusek (A) for research on slow-acting viruses	Burton Richter and Samuel C.R. Ting (A) for independently discovering the J or pi particle, a unique form of subatomic matter
1977	Ilya Prigogine (Be) for contributions to nonequilibrium thermodynamics	Roger Guillemin, Andrew Schally, and Rosalyn Yallow (A) for work on the role of hormones in body chemistry	Philip W. Anderson, John H. van Vleck (A), and Sir Nevill F. Mott (B) for contributions to the development of semiconductor devices
1978	Peter Mitchell (B) for studies of cellular energy transfer	Werner Arber (Swi), Daniel Nathans (A), and Hamilton O. Smith (A) for discoveries in molecular genetics	Pyotr Kapitsa (R) for research in low-temperature physics; and Arno Penzias and Robert Wilson (A) for discovering cosmic microwave background radiation
1979	Herbert C. Brown (A) and Georg Wittig (G) for hydroboration and the Wittig reaction, respectively	Allan MacLeod McCormack (A) and George Newbold Hormsfield (B) for contributions to the computerized axial tomographic (CAT) scanner	Sheldon L. Glashow (A), Steven Weinberg (A), and Abdus Salam (Pa) for unifying the weak nuclear force and the force of electromagnetism
1980	Paul Berg (A), Walter Gilbert (A), and Frederick Sanger (B) for studies on nucleic acid structure	Beruj Benacerraf (A), George D. Snell (A), and Jean Dausset (F) for discoveries on genetic regulation of the immune system	James W. Cronin and Val L. Fitch (A) for research on subatomic particles
1981	Kenichi Fukui (J) and Roald Hoffmann (A) for a theory that permits prediction of the structures of products of concerted reactions and conditions needed to carry them out	Roger W. Sperry (A), Torsten N. Wiesel (Swe), and David H. Hubel (C, naturalized A) for elucidating certain aspects of brain function	Kai Siegbahn (Swe) for contributions to high-resolution electron spectroscopy; and Nicolaas Bloembergen and Arthur Shawlaw (A) for contributions to laser spectroscopy

TABLE F.1. (*Continued*)

Year	Chemistry	Physiology or Medicine	Physics
1982	Aaron Klug (B) for developing the crystallographic electron microscope and studying nucleic acid–protein complexes	Sune K. Bergström (Swe), Bengt I. Samuelsson (Swe), and John R. Vane (B) for research on prostaglandins	Kenneth G. Wilson (A) for applying renormalization group theory to phenomena such as phase changes
1983	Henry Taube (A) for mechanistic studies of oxidation–reduction of metal complexes	Barbara McClintock (A) for discovery of mobile genetic elements in her research on Indian corn	Subrahmanyan Chandrasekhar (A) for theoretical studies of physical processes pertinent to the structure and evolution of stars; and William A. Fowler (A) for theoretical and experimental studies of nuclear reactions important in formation of chemical elements in the universe
1984	R. Bruce Merrifield (A) for developing a method of solid-phase synthesis of peptides	Niels K. Jerne (Da) for theories elucidating specificity, development, and regulation of the immune system; and Georges J.F. Köhler (West G) and Cesar Milstein (Ar) for developing a method to produce monoclonal antibodies with predetermined specificity using hybridoma cells	Carlo Rubbia (I) and Simon van der Meer (D) for decisive contributions to the discovery of the field particles W and Z, communicators of weak interaction
1985	Jerome Karle and Herbert A. Hauptman (A) for developing direct mathematical methods used to obtain X-ray crystal structures of organic and inorganic molecules	Michael S. Brown and Joseph L. Goldstein (A) for discoveries related to cholesterol metabolism and cholesterol-related diseases	Klaus von Klitzing (West G) for discoveries of the effect of magnetism on electrons and their movement
1986	Dudley R. Herschbach and Yuan T. Lee (A) for developing a method for studying chemical reactions through use of directed molecular beams; and John C. Polanyi (C) for using infrared chemiluminescence to study detailed energy disposal during chemical reactions	Stanley Cohen (A) and Rita Levi-Montalcini (dual I, A) for the discovery of growth factors, a finding that has increased the understanding of many diseases including tumors, developmental malformations, generative changes in senile dementia, and delayed wound healing	Ernst Ruska (G) for basic work in electron optics and the design of the first electron microscope; and Gerd Binnig (G) and Heinrich Rohrer (Swi) for designing the scanning tunneling microscope

INDEX